Structural Health Monitoring and Performance Evaluation of Bridges and Structural Elements

Structural Health Monitoring and Performance Evaluation of Bridges and Structural Elements

Armin Mehrabi
Seyed Saman Khedmatgozar Dolati

MDPI

Basel • Beijing • Wuhan • Barcelona • Belgrade • Novi Sad • Cluj • Manchester

Editors

Armin Mehrabi
Civil and Environmental Engineering
Florida International University
Miami
United States

Seyed Saman Khedmatgozar Dolati
Civil and Environmental Engineering
Florida International University
Miami
United States

Editorial Office
MDPI AG
Grosspeteranlage 5
4052 Basel, Switzerland

This is a reprint of articles from the Special Issue published online in the open access journal *Infrastructures* (ISSN 2412-3811) (available at: www.mdpi.com/journal/infrastructures/special_issues/0LRF202A18).

For citation purposes, cite each article independently as indicated on the article page online and as indicated below:

Lastname, A.A.; Lastname, B.B. Article Title. *Journal Name* **Year**, *Volume Number*, Page Range.

ISBN 978-3-7258-2266-9 (Hbk)
ISBN 978-3-7258-2265-2 (PDF)
doi.org/10.3390/books978-3-7258-2265-2

Cover image courtesy of Armin Mehrabi
The Sunshine Skyway Bridge in St. Petersburg, Florida, USA.

Contents

About the Editors

Armin Mehrabi

Dr. Armin B. Mehrabi, an ASCE Fellow, is an accomplished civil engineer with extensive expertise in bridge evaluation, inspection, and rehabilitation, particularly in cable-supported bridges and accelerated bridge construction, as well as advanced materials such as FRP and UHPC. He earned his Ph.D. in Civil Engineering from the University of Colorado, Boulder; MBA from Niagara University; and MS from the University of Tehran. His career spans various academic and professional roles, including positions at FIU, University of Nebraska-Lincoln, and as President of Bridge Engineering Solutions. Dr. Mehrabi has been recognized for his innovative work, including developing non-destructive evaluation techniques and securing patents related to bridge health monitoring and FRP applications. At FIU, Dr. Mehrabi serves as an Associate Professor and has held key leadership roles, such as Graduate Program Director and Research Director of the ABC-UTC, where he has led numerous research projects funded by federal and state agencies. His research focuses on bridge health monitoring, damage detection, and innovative materials and methods for structural assessment. He has secured significant research funding and has published extensively, with a book, over 51 journal papers, and dozens of articles in conference proceedings and magazines. Dr. Mehrabi has six patents related to structural systems and damage detection and prevention measures. Dr. Mehrabi's work is highly regarded, evidenced by his numerous awards, including being named one of the Top 25 Newsmakers of the Year 1997 by ENR. His thesis on Seismic Behavior of Masonry Infilled RC Frames won the Most Outstanding Thesis Award by The Masonry Society in 1997. His contributions extend to editorial roles in prestigious journals and involvement in the development of industry standards and guidelines. His expertise in bridge evaluation and structural health monitoring continues to significantly impact the field of civil engineering.

Seyed Saman Khedmatgozar Dolati

Seyed Saman Khedmatgozar Dolati holds a Ph.D. in Civil Engineering from Florida International University (FIU), and specializes in bridge and structural engineering. His expertise includes concrete and steel bridges, prestressed and precast concrete members, fiber-reinforced polymers (FRPs), advanced engineering materials, seismic design and rehabilitation, structural health monitoring (SHM), and non-destructive testing (NDT). He currently works as a Bridge Engineer at HDR and serves as a Postdoctoral Associate at FIU. Dr. Dolati has contributed significantly to federally funded research projects, focusing on innovative materials and construction methods for bridge engineering. He has worked on projects with the U.S. Department of Transportation, Federal Highway Administration (FHWA), and the Florida Department of Transportation to improve the durability of infrastructures. Dr. Dolati holds several U.S. patents for innovations in connection methods for structural elements. His research has been widely published in top academic journals and cited by numerous researchers and engineers. He has presented at several international conferences and has been recognized with awards like the Presidential Award, Outstanding Doctoral Degree Award, SEI Futures Fund Scholarship, Outstanding Scholar Award, and Best Journal Paper of the Year. Beyond research, Dr. Dolati is actively involved in mentoring students and leading professional organizations, including the American Society of Civil Engineers (ASCE) and the Natural Hazards Engineering Research Infrastructure (NHERI). His contributions to the field of civil engineering are further highlighted by his extensive teaching experience and leadership roles in industry conferences.

Preface

Bridges and other structures that provide critical services are vital components of surface infrastructure. However, much of this infrastructure, constructed during economic development, is now aging and in need of repair or replacement due to structural and functional deficiencies. To maintain their safety, it is essential to use reliable methods for health monitoring and performance evaluation to prepare preventive maintenance before damage becomes severe. Over the past few decades, several bridge failures have shown that conventional monitoring techniques are inadequate. As a result, more effective methods for damage detection and structural health monitoring (SHM) are needed. Accurate condition assessment and performance evaluation also enable decision-makers to prioritize maintenance and allocate resources to reinforce inadequate structures, enhancing overall resiliency. Recent research has focused on two main areas: improving methods for assessing current infrastructure conditions through SHM and non-destructive evaluation (NDE), and developing low-maintenance, smart, and durable alternatives for new structures.

Two major approaches have emerged for condition assessment: embedding sensors in structures during or after construction for continuous or periodic data collection and using NDE techniques to gather information on existing structures. Researchers aim to develop effective and efficient sensing systems that combine the advantages of sensor networks and NDE methods while keeping costs affordable and making data interpretation easier. Despite recent advancements, the demand for more advanced health monitoring and maintenance approaches remains high. Furthermore, current processes for sensor installation, operation, and data collection remain cumbersome, slow, and expensive.

This Special Issue compiles articles on a broad range of SHM, NDE, and damage detection methods for bridges and structures. It covers various techniques, such as hands-on non-destructive testing (NDT), non-contact sensors, load testing, and vibration analysis. This issue also highlights new algorithms for data interpretation and maintenance strategies, focusing on using SHM and NDE results for preventive measures. These demonstrate a growing interest in applying machine learning (ML) and deep learning (DL) algorithms for structural damage detection. This Special Issue hints at future developments in SHM, NDT, damage detection, and performance evaluation, with a follow-up Special Issue planned to explore these areas further. The planned Special Issue will go further to include remedial and rehabilitation measures to enhance resilience in both existing and future infrastructure, building on lessons learned from current practices.

We extend our heartfelt gratitude to the authors who contributed to this Special Issue and to the readers for their interest.

Armin Mehrabi and Seyed Saman Khedmatgozar Dolati
Editors

infrastructures

MDPI

Article

Damage Detection in External Tendons of Post-Tensioned Bridges

Dariya Tabiatnejad [1,*], Bardia Tabiatnejad [2], Seyed Saman Khedmatgozar Dolati [1] and Armin Mehrabi [1]

[1] Department of Civil and Environmental Engineering, Florida International University, Miami, FL 33174, USA; skhed004@fiu.edu (S.S.K.D.); amehrabi@fiu.edu (A.M.)

[2] Haley & Aldrich, Inc., San Jose, CA 95131, USA; bardia.tabiatnejad@alumni.ubc.ca

* Correspondence: dtabi003@fiu.edu

Abstract: This study investigates damage detection in the tendons of post-tensioned segmental box girder bridges, focusing on the vibration-based technique and its application in conjunction with the Precursor Transformation Matrix (PTM). Due to the critical role of bridge tendons in structural integrity, efficient and timely detection methods are essential. The methodology combines theoretical modeling with Finite Element Method (FEM) simulations and empirical data collection to evaluate the PTM's effectiveness in identifying tendon damage. Key results indicate that the PTM, when paired with vibration analysis, enhances the identification and localization of damage, proving to be a robust method in structural health monitoring. This approach not only speeds up damage detection but also potentially lowers maintenance costs by pinpointing specific damage sites, thereby preventing widespread structural failures. The findings offer a promising tool for ensuring the longevity and safety of post-tensioned bridges.

Keywords: post-tensioned box girder bridges; damage detection; finite element method; non-destructive test; precursor transformation

check for updates

Citation: Tabiatnejad, D.; Tabiatnejad, B.; Khedmatgozar Dolati, S.S.; Mehrabi, A. Damage Detection in External Tendons of Post-Tensioned Bridges. *Infrastructures* **2024**, *9*, 103. https://doi.org/10.3390/infrastructures9070103

Academic Editor: Joan Ramon Casas Rius

Received: 18 May 2024
Revised: 21 June 2024
Accepted: 25 June 2024
Published: 30 June 2024

1. Introduction

1.1. Tendon Damage Detection in Bridges and Its Importance

Damage in bridge tendons is a significant concern in the field of civil engineering, specifically regarding the maintenance and safety of prestressed concrete bridges. Bridge tendons, consisting mainly of post-tensioned cables made from high-strength steel, are susceptible to various forms of degradation that can critically affect the structural integrity and longevity of bridges.

Corrosion is one of the foremost mechanisms of tendon damage. The vulnerability of tendons to corrosion intensifies when they are exposed to aggressive environments, such as those with chlorides from deicing salts or marine atmospheres. This environmental exposure leads to the initiation and propagation of corrosion in the steel strands of the tendons, consequently reducing their cross-sectional area and load-carrying capacity. The progression of corrosion often remains undetected until significant damage has occurred due to the tendons being encapsulated in grout, which complicates regular inspection routines [1,2].

Stress corrosion cracking (SCC) is another critical issue, particularly in environments where the steel is subjected to tensile stresses in the presence of corrosive agents. The combination of mechanical stress and corrosive environments can induce cracking, which propagates through the tendons, leading to sudden and catastrophic failures. This issue is highlighted in the literature, where the need for improved monitoring techniques that can detect early signs of SCC in post-tensioned structures is emphasized [3,4].

Hydrogen embrittlement also poses a significant risk to the integrity of bridge tendons. This phenomenon occurs when hydrogen atoms penetrate the metal lattice of steel tendons, facilitating crack initiation and propagation under tensile stress. The susceptibility of tendons to hydrogen embrittlement is closely linked to the material's manufacturing processes and quality, highlighting the importance of stringent quality control measures during the production and construction phases [5].

In addition to material and environmental factors, mechanical damage due to overloading or accidental impacts during construction or while in service can also compromise tendon integrity. Such impacts can cause localized fractures or severe abrasions, diminishing the tendon's effectiveness in bearing loads and maintaining overall bridge stability.

Given the inherent challenges, timely inspections and appropriate maintenance are crucial for the durability and safety of bridge structures. This is especially true for the crucial external tendons, which carry significant tensile loads. Regular inspection techniques may struggle to identify damage in these tendons, as they are typically encased in plastic sheathing. For instance, Figure 1 illustrates a compromised section of a tendon in a post-tensioned segmental box girder bridge, highlighting the difficulty of detecting such damage. Thus, there is a pressing need for a swift and economical method to identify tendon damage. This research has assessed the efficiency of combining vibration-based damage detection with the Precursor Transformation Matrix (PTM) for inspecting external tendons. This innovative approach is gaining traction because it provides a quick and cost-effective solution for conducting essential assessments and finding damaged tendons while reducing disruption and expense. The suspect tendons and locations identified using this technique can then be subject to in-depth costlier non-destructive (NDT) and semi-destructive testing methods to better characterize the damages. Integrating cutting-edge technology with established engineering methods is a forward-thinking advancement toward more resilient and sustainable bridge maintenance. The focus of this paper is on external post-tensioning tendons.

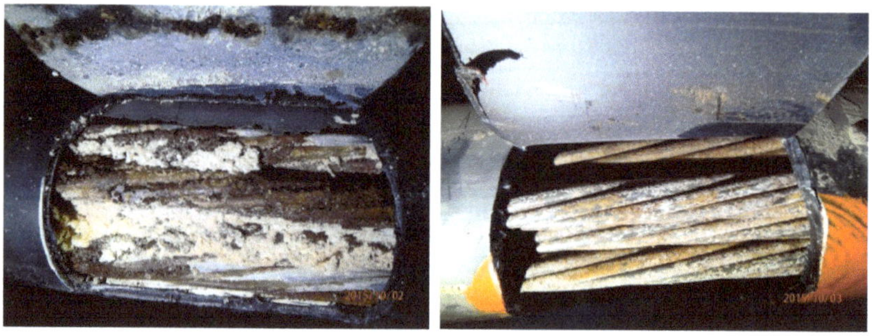

Figure 1. Damage in tendons in a post-tensioned segmental box girder bridge due to corrosion.

1.2. Precursor Transformation Matrix (PTM) Method for Damage Detection in Bridges

The Precursor Transformation Matrix (PTM) method is an analytical procedure proposed by Mehrabi et al. (1998) [6] for detecting and localizing damage in structures. PTM relates measured changes in a structure's condition under constant loading to possible damage sources through a transformation matrix. This matrix is generated by modeling precursor events like temperature changes or displacements on an undamaged analytical model of the structure. The resulting patterns of response change are compared to measured changes from the actual structure to identify likely damage locations and significance. PTM offers potential advantages over conventional vibration-based damage detection methods, including greater sensitivity and efficiency under static loading conditions. This approach also stands out for its cost-efficiency in structural damage detection. It was initially conceptualized for cable-stayed bridges but may apply to other structural types. Mehrabi et al. (1998) [6] presented theoretical derivations of PTM and demonstrated its feasibility through simulated case studies on trusses and a cable-stayed bridge model. Key considerations for implementation include proper analytical modeling, accounting for measurement errors, and having adequate sensor instrumentation. In summary, PTM provides a promising approach to structural health monitoring and damage detection by relating analytical models to measured structural response changes.

There is a significant knowledge gap concerning the interaction of tendon forces in structural systems, especially in understanding how these forces influence each other and how damage in one tendon can impact the forces in other tendons. This interaction is critical for maintaining the structural integrity of bridges and similar constructions. The research addresses this gap by integrating the Precursor Transformation Matrix (PTM) method with vibration-based damage detection techniques. This combination is anticipated to enhance the efficiency of detecting damaged tendons quicker, allowing for potentially reducing repair costs. By employing PTMs and vibration-based methods, the study seeks to develop a more effective and economical approach to structural health monitoring, particularly focused on improving the accuracy and speed of identifying tendon damage within complex structures.

2. Literature Review

2.1. Post-Tensioned Box Girder Bridges

Post-tensioned box girders have become a popular structural system for bridge construction [7], especially for medium to long spans [8]. In this bridge type, the superstructure normally consists of hollow box-shaped segments, typically made of reinforced concrete, that are post-tensioned using high-strength tendons after the concrete has hardened [9]. Post-tensioning places the concrete into compression, offsetting tensile stresses and enhancing serviceability, strength, and durability [10]. Compared to alternative systems, the advantages of post-tensioned box girders include high torsional stiffness, adaptability to skewed supports, and optimized material usage through tailored prestressing [11]. Other post-tensioning configurations have also been implemented, including non-segmental superstructure and application of PT in substructure bridge elements. PT tendons are used internally, where the tendon ducts are embedded in the concrete elements and grouted after application of post-tensioning, as well as externally, where the tendon ducts are exposed. However, this study focuses on PT bridges that use external tendons alone or in conjunction with internal tendons.

Segmental box girder bridges represent a distinctive methodology in bridge engineering, characterized by the assembly of the bridge from shorter segments to form a continuous box-shaped structure. A longitudinal section of a single span within a segmental box girder bridge system is depicted in Figures 2 and 3. In this modular construction approach, short, prefabricated segments are joined together to form the superstructure. These segments are commonly made from precast reinforced concrete and are interconnected through the application of post-tensioned high-strength steel tendons, giving the structure integrity and longevity.

The bridge segments are classified into various types, including typical segments, deviators, and end diaphragm (pier segments), as shown in Figures 2 and 3.

2.2. Damage Detection Techniques for Bridge Tendons

The integrity and safety of bridge structures are paramount for civil infrastructure, necessitating regular and precise monitoring to detect potential damage, particularly in tendons, which are critical for bridges' stability and load-bearing capabilities. Non-destructive Evaluation (NDE) techniques have emerged as pivotal in identifying such damage early, without compromising the structure's functionality or longevity.

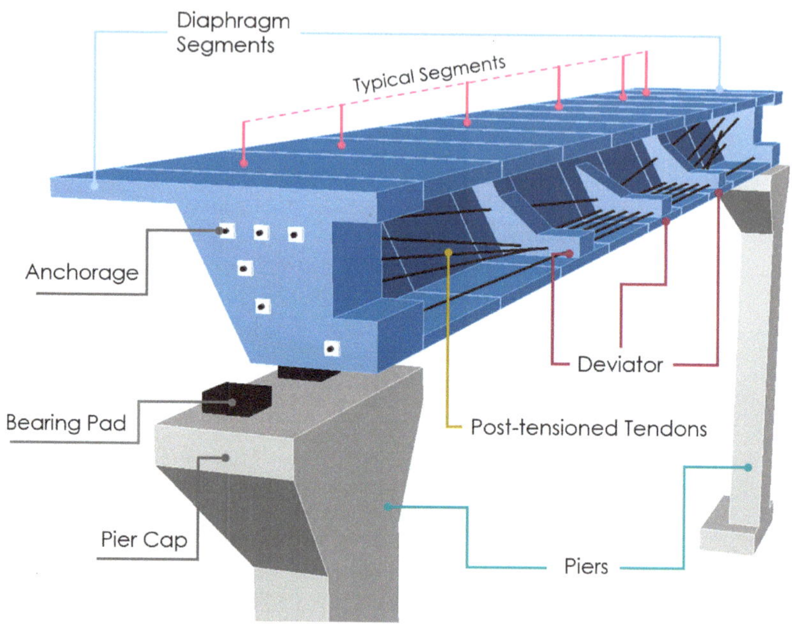

Figure 2. 3D view of a longitudinal section of a segmental box girder bridge.

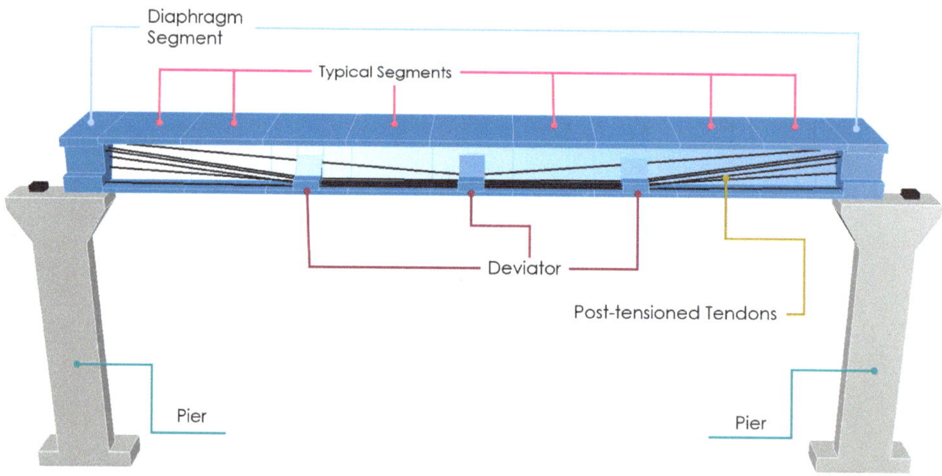

Figure 3. Side view of a longitudinal section of a segmental box girder bridge.

Bridge tendons, comprising steel strands or wires, are susceptible to various forms of damage, including corrosion, fatigue, and stress-related deterioration. Traditional inspection methods often entail invasive procedures that can further compromise the tendon's integrity. Consequently, NDE techniques have become increasingly favored for their ability to offer detailed insights into the condition of tendons without inflicting additional harm [12].

NDE techniques suitable for assessing external tendons can be organized into nine distinct groups, reflecting the technological and physical principles underlying their respective Non-Destructive Testing (NDT) instruments. These groups encompass visual inspection methods, mechanical wave and vibration-based techniques [13], infrared thermography, electrochemical analysis, electromagnetic approaches, ground-penetrating radar, radiographic methods, various uncategorized techniques, and sensor-based assessments [14]. References [15,16] provide a detailed overview, while Figure 4 illustrates the NDE techniques applicable to external post-tensioning tendons. The last column of the chart in this figure is related to the sensor-based methods. These methods form the basis for measuring the forces in tension elements that are the focus of this study. Among the sensors commonly used are fiber optic sensors, accelerometers, laser vibrometers, and strain gauges. Fiber optic sensors are lightweight, unobtrusive, and insensitive to electromagnetic interference. Accelerometers and laser vibrometers are employed to measure the frequency of vibration, define modal shapes, and determine the force within post-tensioning elements over time. The use of a strain gauge allows monitoring of the variation of stresses and, therefore, the forces of the tendons. Changes in structural behavior over time can be considered an indicator of deterioration. Modal analysis methods, such as the modal curvature evaluation method, have been used in general for detecting damage through the structure based on data obtained from structural health monitoring systems [17]. However, this method would not be effective in detecting damage to tendons unless each tendon was instrumented with a sensor. In this case, the method will match the basics of the PTM methodology. Nevertheless, additional sensors and monitoring would impose additional costs and efforts, making this method less practical.

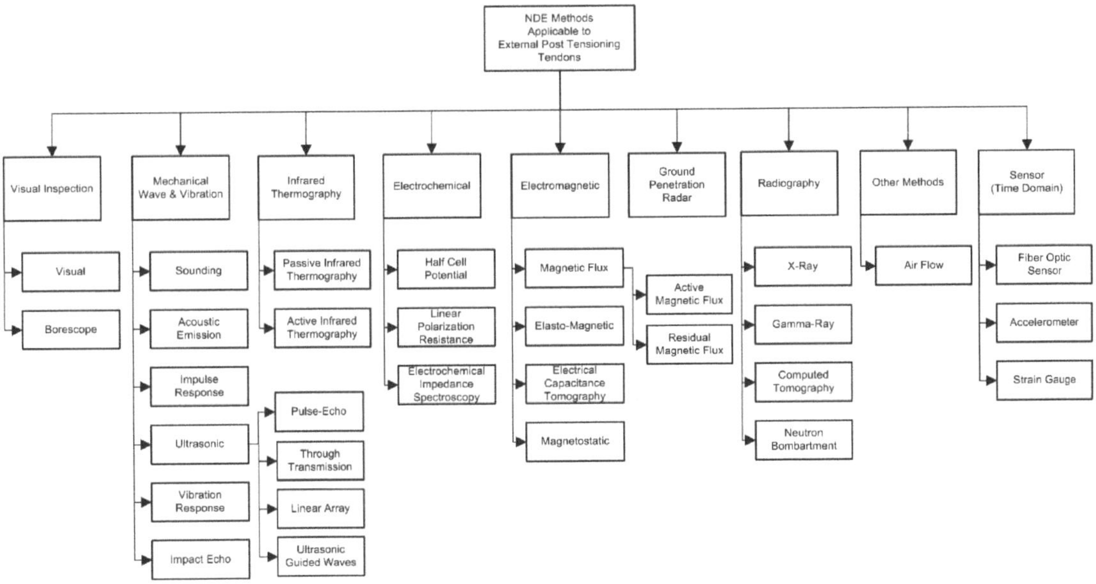

Figure 4. Categorization of Non-Destructive Evaluation Techniques suitable for External Post-Tensioning Systems [18].

Despite its many advantages, NDT has several limitations. Some NDT processes are costly and time-consuming. Moreover, the accuracy of NDT methods can be influenced by various factors, such as the skill and experience of the inspector, the condition of the structure, and environmental conditions during testing. For instance, visual inspection is highly dependent on the inspector's ability to detect and interpret defects, leading to results variability. Ultrasonic testing, while effective for detecting internal flaws, requires a skilled operator to interpret the results accurately and may not detect defects that are oriented parallel to the sound wave. Radiographic testing involves the use of ionizing radiation, which poses health risks and requires strict safety precautions. Additionally, some NDT methods, such as infrared thermography, are sensitive to environmental conditions and may produce less accurate results under unfavorable conditions. These limitations highlight the need for careful selection of NDT methods based on the specific requirements and conditions of the inspection task [18].

In post-tensioned concrete elements, such as segmental box girder bridges, it is crucial to detect damage in the tendons early to avoid costly repairs and prevent catastrophic incidents. Therefore, it is preferred to employ a technique that can detect damage globally in post-tensioned tendons, allowing for the identification of damaged tendons while they are still concealed in a plastic cover, as shown in Figure 1. This study demonstrates that combining the Post-Tensioning Method (PTM) and vibration method enables the detection of damaged tendons at an early stage and at a reasonable cost.

2.3. *The Role of FEM in Damage Detection*

The Finite Element Method (FEM) has emerged as a critical tool in the domain of structural engineering. The application of FEM in damage detection for structures and bridges is multifaceted, incorporating both direct and indirect approaches. Direct methods involve altering the finite element model based on directly observed and measured physical changes in the structure, such as cracks or deformations. For example, in a study by Sanda et al. (2018), an efficient finite element model updating technique for bridge damage detection was discussed. The authors emphasized integrating measurement data to refine the finite element model, thus enhancing its predictive accuracy in identifying damage locations and severities directly [19].

Indirect methods, on the other hand, do not rely on visible damage indicators but instead detect damage through changes in the structural response. These methods often involve monitoring changes in structural parameters such as frequencies, mode shapes, internal forces, and damping ratios. The premise is that damage will alter the stiffness, mass distribution, or energy dissipation characteristics of a structure, thereby affecting its dynamic properties, which can indicate the presence of damage indirectly.

A notable example of an indirect method is the work by Figueiredo et al. (2019), which presented a finite element-based machine-learning approach to detect damage in bridges under operational and environmental variations. This approach leverages changes in modal parameters and other indirect indicators, processed through machine learning algorithms, to identify potential damage without direct physical evidence [20].

Both direct and indirect methods offer valuable approaches to damage detection in structural engineering. Direct methods provide a straightforward way to incorporate visible damages into the FEM analysis. In contrast, indirect methods offer a non-invasive alternative that leverages changes in structural responses to infer the presence and severity of damages. The choice between direct and indirect methods depends on various factors, including the nature of the structure, the type of damage expected, and the available monitoring technology.

3. PTM Methodology

The Precursor Transformation Method (PTM) presents an approach for damage detection and the long-term health monitoring of structures. This method is grounded in identifying precursors, or the underlying causes, of observable changes in the structural state under consistent loading conditions, such as the static loads encountered in bridges. PTM addresses the limitations of existing structural monitoring methods, which struggle to interpret vast quantities of sensor data in a timely and cost-effective manner. By focusing on the causes of structural changes rather than the effects, PTM offers enhanced sensitivity and cost efficiency compared to traditional vibration-based or parameter estimation methods [6].

An analogy to the PTM method would be the method of pattern recognition. Each source develops a certain pattern as a response in such a method. When more than one source causes a single response, it is difficult to differentiate which sources have created the response. By recognizing the pattern of each source and then composing the response from multiple sources, one can detect the individual sources contributing to the measured response. In fact, the PTM method first compiles the pattern of damage sources and, by applying it to the response change measured using NDT, makes it possible to recognize the patterns within the response, thereby identifying the sources.

PTM was originally developed within the context of a research project focused on the condition assessment of bridge stay cables, sponsored by the Federal Highway Administration [6]. However, the principles of PTM are applicable to a broad range of structures beyond cable-stayed bridges. The method involves measuring selected structural response parameters at two different times to identify damage incurred during that period, allowing for real-time or near-real-time structural health monitoring.

The method's ability to detect damage without precise knowledge of load magnitudes and configurations, as long as the loading remains constant, represents a significant advantage. Despite its advantages, PTM is not without challenges. The construction of an accurate finite element model of the structure is crucial, and all significant potential damage sources must be considered to avoid misleading results. Moreover, the method must account for potential errors in modeling and measurement, which could introduce fictitious damage precursors. Optimization and machine learning techniques and considering more response parameters or repeated tests can help mitigate the impact of such errors.

In summary, the Precursor Transformation Method offers a promising avenue for the advanced monitoring and maintenance of structural health, with potential applications extending across various civil engineering domains. Its development and refinement could lead to more reliable and efficient methods for ensuring the safety and longevity of critical infrastructure.

The foundational principle of PTM is to relate measured alterations in the structural state—such as displacements, strains, or internal forces—to specific precursors through a transformation matrix. This matrix is constructed using a finite element model of the undamaged structure to predict how various damage scenarios, represented by externally imposed strains or displacements, would affect the structure. Despite introducing some approximations, this method simplifies the calculation process significantly and maintains accuracy within the bounds of measurement and computational precision.

The formulation of the PTM, as described by Mehrabi et al. (1998) [6], can be presented in general form:

$$\Delta T_{PREDICTED} = c_{FEM}^{-1} \Delta P_{MEASURED} \tag{1}$$

where C_{FEM} is the precursor transformation matrix generated by applying unit changes of temperature, strain, or displacement to each element, $\Delta P_{MEASURED}$ is the vector of changes in structural response parameters measured in the structure, such as change in forces of tendons, and $\Delta P_{PREDICTED}$ is the vector of predicted precursors in the form of changes of temperature, strain, or displacement to each element with which the transformation matrix is constructed. $\Delta P_{PREDICTED}$ is directly related to the changes in each element, resulting in the measured changes in structural response and indicating damage to each element.

How Does the Method Work

The core principle of PTM is that changes in the state of a structure (displacements, strains, internal forces) can be related to "precursors" through a transformation matrix. These precursors or causes are externally imposed events or damage sources, which are independent of the structure and the subsequent changes in the state of the structure due to those events. Examples of damage sources that can be modeled as precursor events include loss of material or stiffness, joint slippage, support settlements, and loosening of connection bolts.

The PTM approach steps are as follows:

1. Measurement of Structural Changes: The initial step involves measuring changes in the structural responses, such as displacements, strains, or internal forces, under a specific static loading configuration, such as self-weight load. Various methods

for measuring these responses are outlined in Section 2.2. For instance, vibration techniques can be employed to gauge the internal forces within tendons. To accurately assess the alterations in structural responses, it is essential to compare the initial, undamaged states with the responses observed post-damage. Consequently, it is advisable to conduct initial measurements of the structural behaviors soon after construction to establish a reference point for the structure's undamaged state. Moreover, regular monitoring of these responses is vital for the early detection of potential damage within structural elements.

2. Development of a Transformation Matrix: A transformation matrix is created by relating the measured changes in the structure's responses to external factors or precursors, such as temperature variations or imposed displacements that could represent potential damage scenarios.
3. Use of a Finite Element Model: An undamaged finite element model of the structure is employed to calculate patterns of change in the structure's responses due to the identified precursors. This model simplifies the calculation process and helps approximate the results.
4. Analysis of Structural Response to Precursors: The method involves analyzing how the structure's response parameters, such as displacements, strains, and internal forces, change due to imposed precursors, which are simulated as temperature changes or initial strains in the finite element model.
5. Identification of Damage Sources: By comparing the measured changes in the structure's state with the expected changes from the transformation matrix, the PTM approach can identify the locations and relative significance of potential damage sources.

In the case study that follows, the formulation and construction of the precursor transformation matrix are explained.

4. Case Study

4.1. Description of a Case Study of Bridge Model Dimensions and Tendon Layout

The main bridge structure of this study is a bridge span with a post-tensioned segmental box girder structural system that crosses a wide river connecting two large cities on two sides of the river. Approaching the main structure are viaducts connected to approach ramps and access roads on embankments on either side. Each span in the viaduct has roughly 115 ft. length and accommodates one traffic lane and two shoulders. For the application of PTM, one of the viaduct spans, as shown in Figure 5, is used as the bridge in the case study.

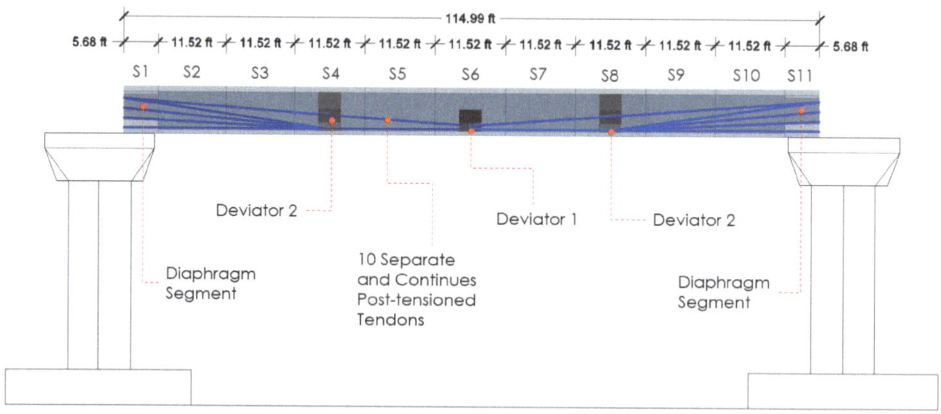

Figure 5. Longitudinal dimensions of case study segments.

Figure 5 illustrates a span composed of 11 distinct reinforced concrete segments, with each segment labeled by the letter 'S' followed by its segment number. End diaphragm segments (S1 and S11) with 5.68 ft. length have been located at two ends of the span. Figure 6 illustrates the cross-section details and dimensions for end diaphragm segments. The sequence of numbers after the letter "T" represents the positioning of the tendons within that specific section. Additionally, the span includes three deviators, each 11.52 feet long, categorized into two distinct types: Segment S6 represents deviator Type 1 (Figure 7), whereas segments S4 and S8 belong to deviator Type 2 (Figure 8). The rest of the segments, specifically S2, S3, S5, S7, S9, and S10, are designated as typical segments, each also extending 11.52 feet in length. The cross-section details of typical segments are shown in Figure 9.

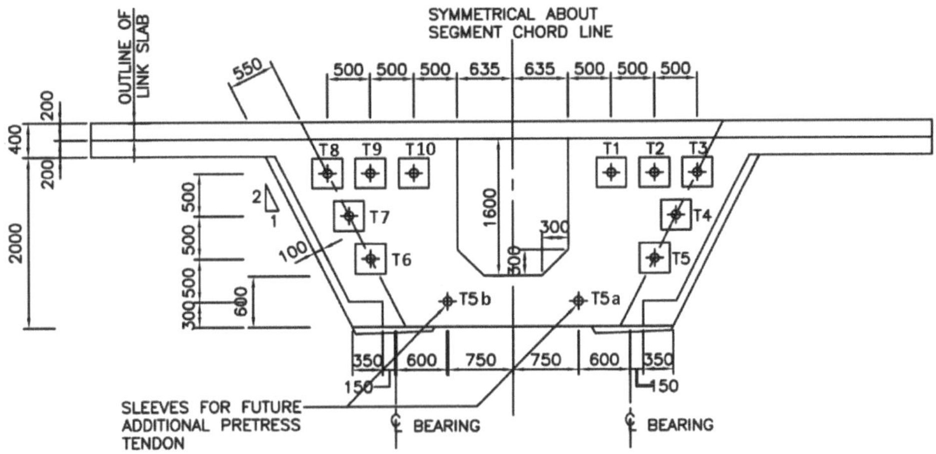

Figure 6. End diaphragm cross-section details for Segments S1 and S11.

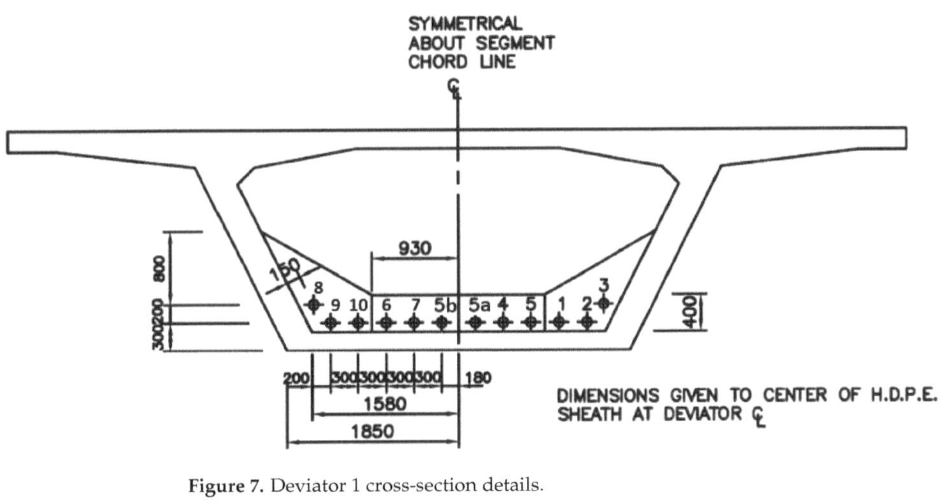

Figure 7. Deviator 1 cross-section details.

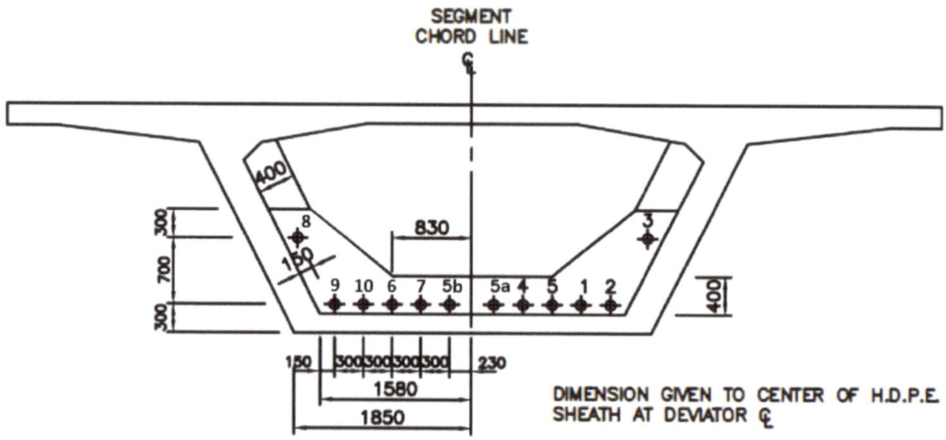

Figure 8. Deviator 2 cross-section details.

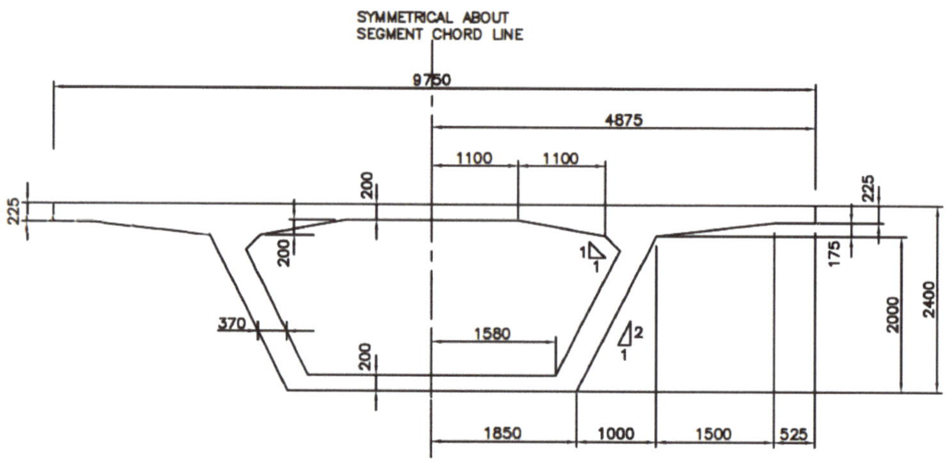

Figure 9. Typical segment cross-section details.

Also, the bridge of the case study has 10 separate and continuous external post-tensioned tendons, illustrated in Figure 5 by blue lines. These tendons, composed of high-strength steel strands, clasp the segments firmly together following the post-tensioning process. These tendons are anchored in end diaphragm segments and pass through the hollow areas of the typical segments, altering their directions as they pass through the holes in the deviator segments. Details about these tendons can be found in Table 1.

Table 1. Tendon properties.

Tendon	1	2	3	4	5	5a	5b	6	7	8	9	10
Area (in.^{2})	5.37	5.37	5.37	4.52	3.39	0	0	3.39	4.52	5.37	5.37	5.37

4.2. Description of the Bridge Finite Element Method (FEM) Model

Employing the Precursor Transition Matrix (PTM) method requires altering a single structural parameter, often presumed to be affected by damage. This entails conducting an analysis based on this altered parameter and comparing the outcomes with those of the structure's initial, undamaged state. Given the high sensitivity of this method to analytical outcomes, achieving accuracy necessitates the simulation of an exceptionally precise model. Consequently, this study opts for Finite Element Method (FEM) modeling techniques, utilizing the ABAQUS v2021 [21] software for detailed analysis.

Figure 10 presents a three-dimensional representation of the FEM model built using the ABAQUS v2021 [21] with the concrete segments rendered transparent to reveal the tendons, which are depicted in red. The geometry of the FEM model is based on the dimensions and details outlined in Section 4.1. The support system is modeled based on the assumption of simple support at one end and roller support at the opposite end, as depicted in Figure 11. Moreover, the concrete components are modeled with three-dimensional 8-node linear brick solid elements (C3D8 8 nodes), while truss and rebar components are modeled with 3D 2-node truss elements (T3D2), Ideal for structures composed of elements loaded only in tension or compression along their length.

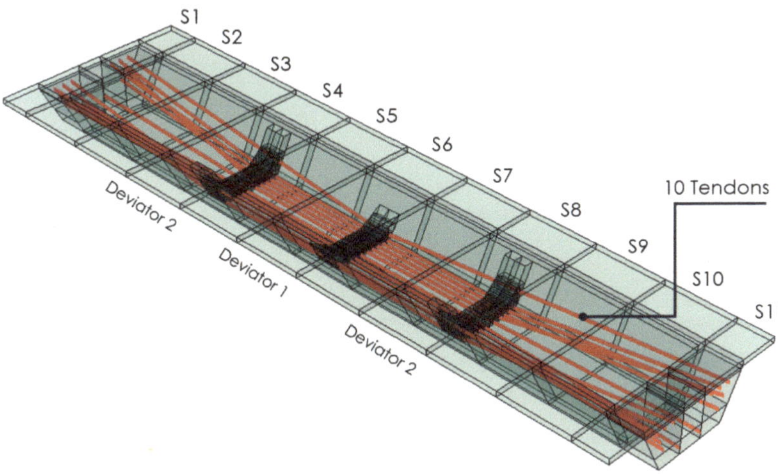

Figure 10. The FEM model (3D view).

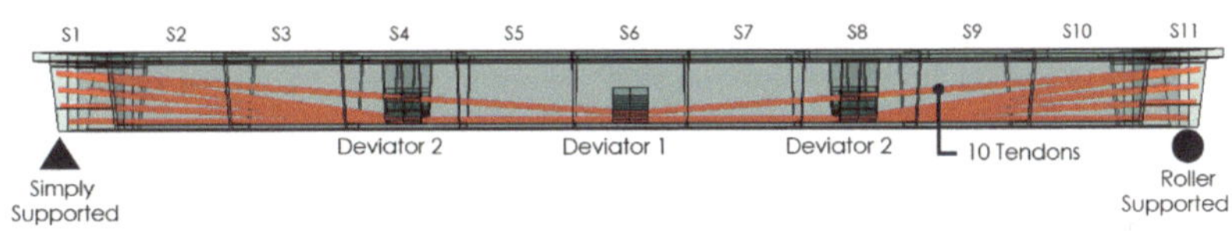

Figure 11. Supports in the FEM model (side view).

Tendons, as noted earlier, pass through the hollow areas of standard segments without attaching to them directly. Instead, they are anchored at the end diaphragm segments and pass through sleeves cast in the deviators.

To model the anchorage of the tendons within the end segments, both ends of all tendons have been embedded within the concrete in the end diaphragm segments. This technique could capture the behavior of the anchored tendon precisely. Also, to model the interaction between the tendons and the deviators so that the tendon can slide in the sleeves of the deviators, a translational connection to the deviators for the tendon with slot type has been defined. All translational degrees of freedom for this connection type have been constrained except the degree of freedom that is in the longitudinal direction of the bridge span. Figure 12 shows the tendons and segments that have such interaction.

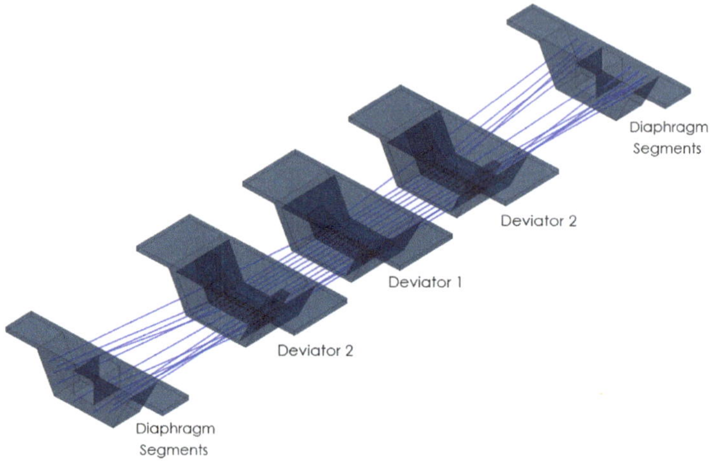

Figure 12. Concrete segments interacting with external tendons.

As mentioned earlier, all the concrete segments are reinforced concrete. Therefore, to capture this feature in FEM models, the steel reinforcement has been embedded in all the concrete segments as provided in the bridge drawings. Figure 13 shows the typical embedded steel rebars within a typical concrete segment.

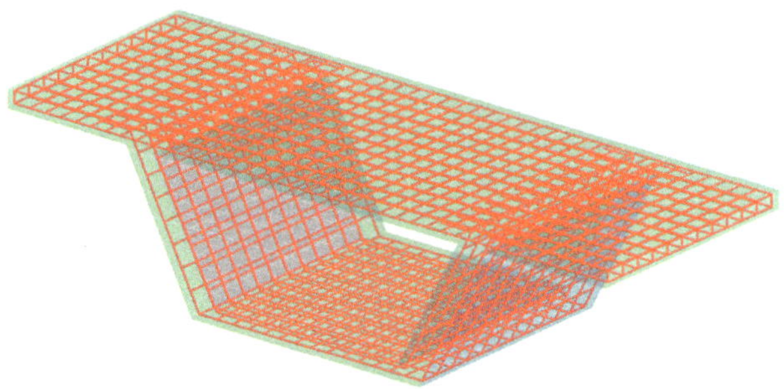

Figure 13. Steel reinforcement embedded in a typical segment.

Another connection that has been defined in the FEM model is the connection between the concrete box segments. Since no sliding or separation is expected between segments for the type of analysis performed for PTM, a tie connection is used between the concrete segments, as shown in Figure 14 with pink color.

4.3. *Verification of the FEM Model*

The verification process for the model is essential in the Finite Element Method (FEM) to show that the simulations are both accurate and reliable. In this study, the verification of the FE model was conducted by comparing the displacement at the mid-span along the bridge calculated using theoretical formulas with the displacement obtained from FE analysis using the constructed FE model.

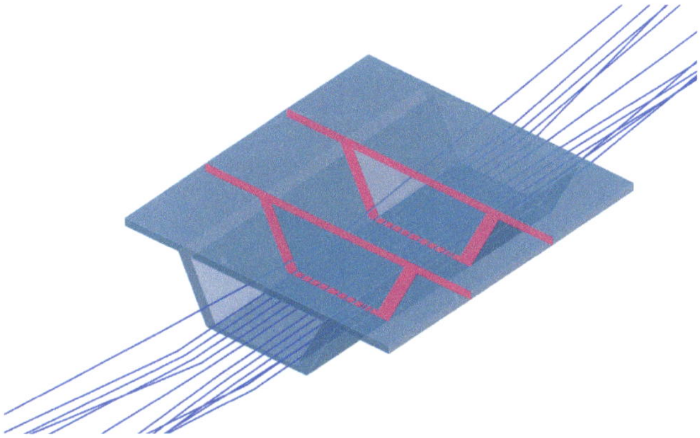

Figure 14. Rigid tie connection between concrete segments in FEM modeling.

Two types of loading have been assigned to the bridge for verification study, namely dead loads and post-tensioning, to simulate the original and undamaged situation of the bridge. The dead loads are applied to the deck surface of the bridge span considering the concrete weights and surface dead loads coming from asphalts and other components of the bridge (Figure 15). Post-tensioning is applied to the tendons by reducing the temperature of the truss elements representing the tendons. By reducing the temperature, the length of the tendons reduces and since they are embedded in the end diaphragm segments, internal axial force is generated in the tendons (Figure 16).

As the post-tensioning process begins, the concrete segments compress against one another, causing the entire bridge span to curve into an arch shape, which results in the upward movement of the span's midpoint. Subsequently, when the distributed concrete self-weight load is applied, this midpoint shifts downward. Therefore, to determine the final displacement of the midpoint, one must subtract these two calculated values from each other. However, for more accuracy, the FEM displacement results from both post-tensioning and concrete self-weight loads are checked against theoretical calculations.

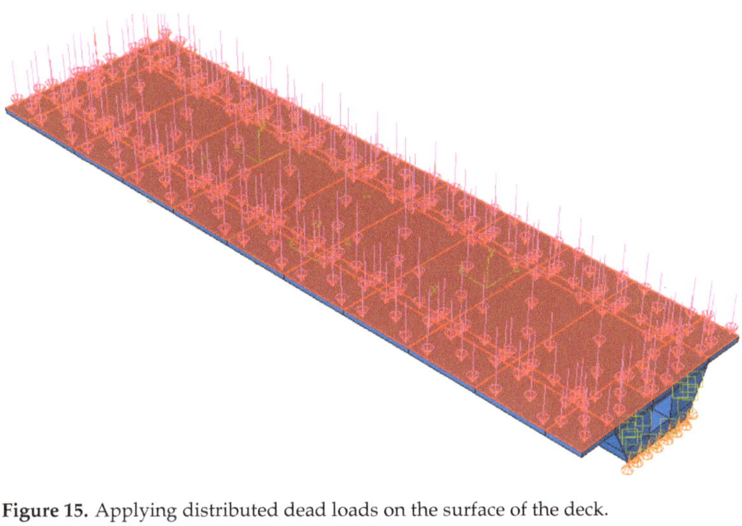

Figure 15. Applying distributed dead loads on the surface of the deck.

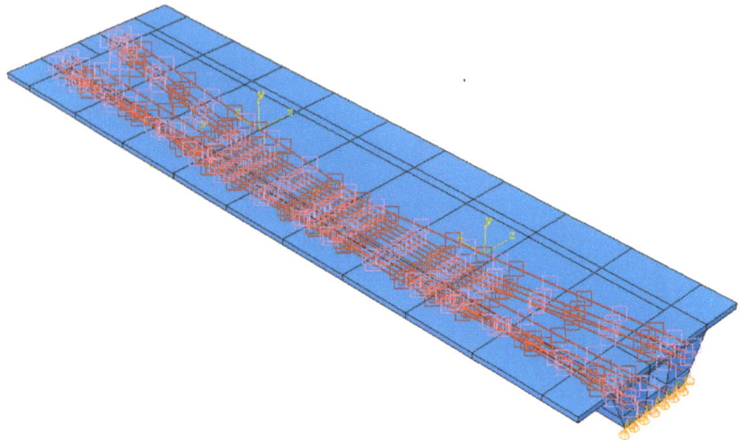

Figure 16. Applying post-tensioning force in tendons by decreasing temperature.

In the analysis, two distinct equations are used to calculate the deflections of simply supported post-tensioned concrete beams, with the differentiation based on the number of harping points, as shown in Figures 17 and 18. Harping or draping refers to the process of altering the position of all or some prestressing tendons along the beam's length. This adjustment in tendon positioning modifies their eccentricity within the section. Reducing the tendons' eccentricity at a member's ends diminishes the moment and tensile stresses induced by prestressing at those extremities [22]. Within these equations, P stands for the axial force in kips, which differs from one tendon to another; *L* denotes the bridge's span length, fixed at 1379.92 inches; E is the concrete's modulus of elasticity, valued at 3600 kips/in.^{2}; and I represents the moment of inertia for the concrete section, calculated to be 8,872,480 in.^{4}. Furthermore, e_e and e_c represent the eccentricities at the beam's supports and its center, respectively. In Table 2, the computations for each tendon were conducted individually. By summing up the deflections resulting from the axial forces of each tendon, the overall deflection attributable to post-tensioning can be determined equal to 3.43 inches. Meanwhile, as can be seen in Figure 19, the short-term deflection determined through the FEM ABAQUS model stands at 3.56 inches, which is merely four percent greater than the theoretical value, indicating a good agreement. This demonstrates the reliability and accuracy of the FEM model.

Double harping point

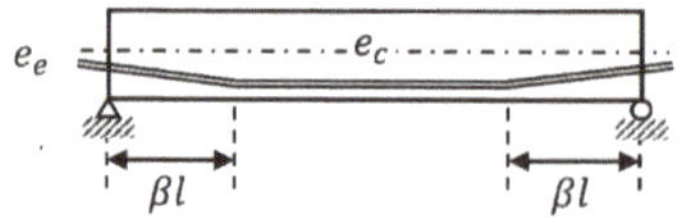

$$\Delta = \left(\frac{e_c}{8} - \frac{\beta^2}{6}(e_c - e_e)\right)\frac{Pl^2}{EI}$$

Figure 17. Theoretical deflection for a post-tensioned concrete beam with a double harping point.

Single harping point (at midspan)

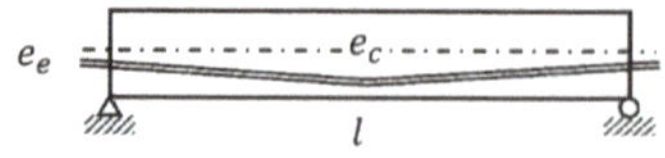

$$\Delta = \left(\frac{2e_c + e_e}{24}\right)\frac{Pl^2}{EI}$$

Figure 18. Theoretical deflection for a post-tensioned concrete beam with a single harping point.

Table 2. Deflection calculation in the case study bridge due to tendon post-tensioning.

Tendon Number	Area in.^{2}	Stress kips/in.^{2}	P = Area × Stress kips	e_c in.	e_e in.	Δ in.	Used Formula
1	5.364	242.6	1301.28	42.11	−9.07	0.243	(a)
2	5.364	236.4	1268.03	49.98	−9.07	0.405	(b)
3	5.364	236.1	1266.63	49.98	−9.07	0.405	(b)
4	3.383	235.1	795.52	49.98	30.31	0.282	(b)
5	4.522	236.1	1067.69	49.98	10.61	0.360	(b)
5a	0.200	234.2	46.85	49.98	0	0.017	(b)
5b	0.200	234.2	46.85	49.98	0	0.017	(b)
6	4.522	236.1	1067.69	49.98	10.61	0.360	(b)
7	3.383	235.1	795.52	49.98	30.31	0.282	(b)
8	5.364	236.1	1266.63	49.98	−9.07	0.405	(b)
9	5.364	236.4	1268.03	49.98	−9.07	0.405	(b)
10	5.364	242.6	1301.28	42.11	−9.07	0.243	(a)
					∑Δ =	3.426	in.

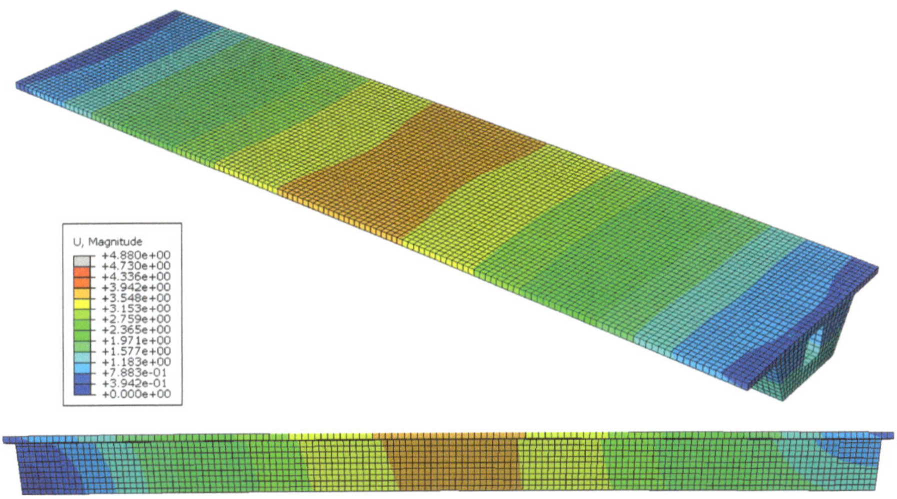

Figure 19. FEM results for the deflection in the case study bridge due to post-tensioning.

The subsequent step involves applying the concrete self-weight load. Given that this load is distributed, it is feasible to compute the short-term deflection using the mechanics of material formulation for the deflection of a simply supported beam under a distributed load. The deflection due to the concrete's self-weight is calculated in this way to be 1.11 inches. Comparatively, the FEM model's deflection result for this load is 1.12 inches, demonstrating a close agreement. The comparison of results for both types of loading verifies the FEM model to be reliable for this study.

Furthermore, Figure 20 illustrates the bridge deflection from the combined effects of post-tensioning of the tendons and the concrete's self-weight. The FEM model's mid-span deflection result for this final stage is 2.44 inches, while the calculated deflection is 2.31 inches, again demonstrating a close agreement.

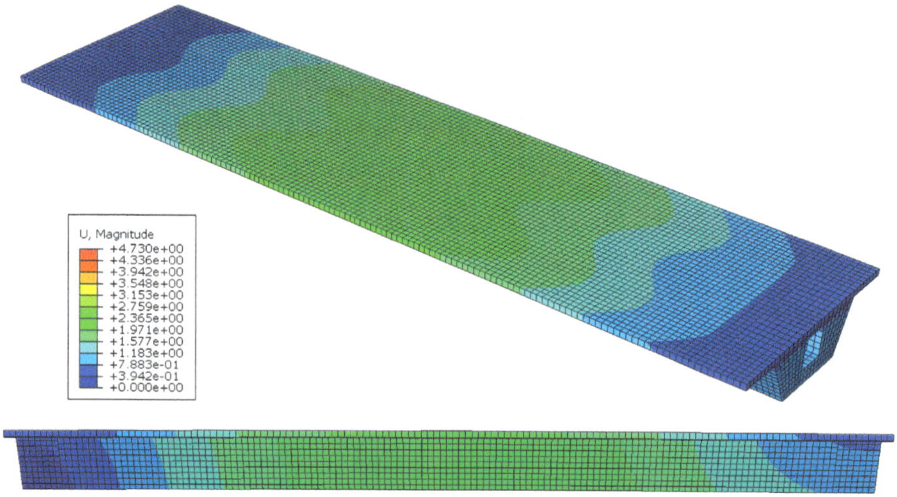

Figure 20. FEM deflection results in the case study bridge due to post-tensioning and self-weight.

4.4. Constructing the Precursor Transformation Matrix (PTM) for the Bridge of Case Study

To generate the precursor transformation matrix (C_{FEM}), it is necessary to follow the steps mentioned in Section How Does the Method Work. The initial step involves measuring changes in the structural responses, such as displacements, strains, or internal forces, under a specific static loading configuration such as self-weight load. Various methods for measuring these responses are outlined in Section 2.2. For instance, vibration (or other) techniques can be employed to gauge the internal forces within tendons. To accurately assess the alterations in structural responses, it is essential to compare the initial, undamaged states with the responses observed post-damage. Consequently, it is advisable to conduct initial measurements of the structural behaviors soon after construction to establish a reference point for the structure's undamaged state. Moreover, regular monitoring of these responses is vital for the early detection of potential damage within structural elements.

The first step involves collecting the axial forces from all tendons using the FEM model to represent the undamaged state of the case study bridge, which is subjected to both post-tensioning loads and static dead load. To achieve this, the axial stresses in the tendons are computed and then multiplied by their respective cross-sectional areas. Figure 21 displays the axial stresses in half of the bridge's tendons as an example. Table 3 lists the axial force values shown as vector FI, where FI_i represents the initial undamaged force in Tendon i.

In the next step, the precursor transformation matrix is constructed following a scenario where a specific tendon (Tendon i of n, n being the total number of tendons in the cross-section of the bridge) undergoes a 10% reduction in the cross-section (damage unit induced in Tendon i) due to hypothetical damage in its cross-sectional area. It is understood that damage to tendons rarely translates to a uniform cross-section reduction along its entire length, and it would most likely be localized in the limited length of the tendon. However, presenting a 10% reduction of the cross-section here intends to generate changes in the structural response parameter (tendon force) for the construction of the precursor transformation matrix and should not necessarily be viewed as such damage being possible. It just represents any type of damage that will reduce the stiffness of the tendon and relax the post-tensioning force in that tendon.

Figure 21. FEM axial stress results for half of the case study bridge tendons.

Table 3. Axial forces in tendons for the undamaged condition under dead load.

Tendon Number	1	2	3	4	5	6	7	8	9	10
Axial Force (kips)	321.4	313.5	313.2	250.1	287.3	287.4	250.2	313.2	313.5	321.4

In the FEM bridge model, this change in the stiffness of the tendon is simulated by a hypothetical reduction of the area of the *i*-th tendon by 10%. After analyzing the FEM model, the axial forces in all tendons are obtained and placed in the *i*-th row in an n-by-n matrix called the force matrix, or FM. Each row in this matrix then represents a separate analysis, with diagonal elements reflecting the tensile forces in the tendon subjected to the 10% cross-sectional area reduction. Specifically, the element at the intersection of the *i*-th row and *j*-th column, FM_{ij}, indicates the internal force in the *j*-th tendon when only the *i*-th tendon has experienced a 10% reduction. This process is replicated for all involved tendons, leading to a 10 × 10 force matrix due to the involvement of 10 operational tendons. The completed force matrix is presented in Table 4.

Table 4. Axial tendon force (kips) after experiencing a 10% reduction in cross-section each time.

	1	2	3	4	5	6	7	8	9	10
1	289.9	314.3	313.9	250.7	287.9	287.9	250.6	313.7	314.0	321.7
2	322.2	283.0	314.1	250.8	288.1	288.1	250.7	313.9	314.2	321.9
3	322.1	314.4	282.8	250.8	288.1	288.1	250.8	313.9	314.2	321.9
4	321.8	314.1	313.7	225.6	287.8	287.9	250.6	313.7	314.0	321.7
5	321.9	314.1	313.8	250.7	259.2	287.9	250.6	313.7	314.0	321.7
6	321.7	314.0	313.7	250.5	287.8	259.3	250.7	313.9	314.2	321.9
7	321.7	313.9	313.6	250.5	287.7	288.0	225.6	313.8	314.1	321.8
8	321.9	314.2	313.9	250.7	288.0	288.2	250.9	282.8	314.5	322.1
9	321.8	314.1	313.8	250.7	287.9	288.3	250.9	314.2	283.1	322.2
10	321.7	313.9	313.6	250.5	287.7	288.1	250.7	314.0	314.3	289.9

After the force matrix is established, the Precursor Transformation Matrix (C_{FEM}) is generated by subtracting the force values in each row of the force matrix from those of the corresponding undamaged state. The C_{FEM} is presented in Table 5. In the C_{FEM} therefore, the value in row *i* and column j (C_{FEMij}) represents the change of force in the *j*-th tendon due to a reduction in the cross-section of the *i*-th tendon.

Table 5. Precursor Transformation Matrix (C_{FEM}) for axial tendon forces (kips).

	1	2	3	4	5	6	7	8	9	10
1	−31.5	0.8	0.7	0.6	0.7	0.5	0.4	0.5	0.4	0.3
2	0.8	−30.4	1.0	0.7	0.8	0.6	0.6	0.7	0.7	0.4
3	0.7	1.0	−30.4	0.7	0.8	0.7	0.6	0.7	0.7	0.5
4	0.4	0.6	0.6	−24.5	0.6	0.4	0.4	0.5	0.5	0.3
5	0.5	0.7	0.6	0.6	−28.1	0.5	0.4	0.5	0.5	0.3
6	0.3	0.5	0.5	0.4	0.5	−28.1	0.6	0.6	0.7	0.5
7	0.3	0.5	0.5	0.4	0.4	0.6	−24.5	0.6	0.6	0.4
8	0.5	0.7	0.7	0.6	0.7	0.8	0.7	−30.4	1.0	0.7
9	0.4	0.7	0.7	0.6	0.7	0.8	0.7	1.0	−30.4	0.8
10	0.3	0.4	0.5	0.4	0.5	0.6	0.6	0.7	0.8	−31.5

4.5. Results and Discussion

Table 5 displays the C_{FEM} for internal tendon force changes in kips. Normalization is advisable to enhance the interpretability of this matrix by converting it to a unitless format. This can be achieved by calculating the percentage change in the axial forces of the tendons. This is performed by dividing the change in the axial force of each tendon in its current state (C_{FEMij}), by the axial force in its initial undamaged state (FI_j), then multiplying the result by 100 to express it as a percentage. Table 6 presents C_{FEM} normalized in this way. This matrix reveals that when a tendon undergoes hypothetical damage equivalent to 10% cross-sectional loss, the axial internal force within that tendon decreases by approximately 10%. On the other hand, the internal axial forces in the other tendons remain stable, with only slight increases up to a maximum of 0.3%. This level of change is basically within the tolerance of any measurement method, i.e., it is not detectable at all. From these observations, it can be concluded that in the damage detection process for tendons of a segmental box girder bridge, a detected reduction in the internal force of a specific tendon can be attributed directly to damage within that tendon, rather than changes resulting from damage in others. In other words, there is no interrelation between tendons, and each tendon acts independently in this respect with damage to one tendon not noticeably affecting the internal forces of other tendons. This insight is crucial for engineers, enabling them to identify and locate damage quickly and accurately. This phenomenon can be attributed to the fact that the tendons are post-tensioned, and to the significant axial stiffness of the bridge superstructure. Accordingly, while PTM is a powerful tool for damage detection in certain types of structures and can also be employed for tendons of PT bridges, because of the simplicity demonstrated with this study, the use of PTM is not necessary for damage detection in PT tendons and a straightforward approach can be used. That is, if force change is detected in a tendon utilizing vibration or other methods, that will be an indication of damage in that same tendon. Accordingly, a simple process is proposed for damage detection in external tendons of post-tensioned concrete bridges as shown in Figure 22.

Table 6. Percentage of change in forces over the initial forces.

	1	2	3	4	5	6	7	8	9	10
1	−9.8	0.2	0.2	0.2	0.2	0.2	0.2	0.2	0.1	0.1
2	0.2	−9.7	0.3	0.3	0.3	0.2	0.2	0.2	0.2	0.1
3	0.2	0.3	−9.7	0.3	0.3	0.2	0.2	0.2	0.2	0.1
4	0.1	0.2	0.2	−9.8	0.2	0.2	0.2	0.1	0.1	0.1
5	0.2	0.2	0.2	0.2	−9.8	0.2	0.2	0.2	0.2	0.1
6	0.1	0.2	0.2	0.2	0.2	−9.8	0.2	0.2	0.2	0.2
7	0.1	0.1	0.1	0.2	0.2	0.2	−9.8	0.2	0.2	0.1
8	0.1	0.2	0.2	0.2	0.2	0.3	0.3	−9.7	0.3	0.2
9	0.1	0.2	0.2	0.2	0.2	0.3	0.3	0.3	−9.7	0.2
10	0.1	0.1	0.2	0.2	0.2	0.2	0.2	0.2	0.2	−9.8

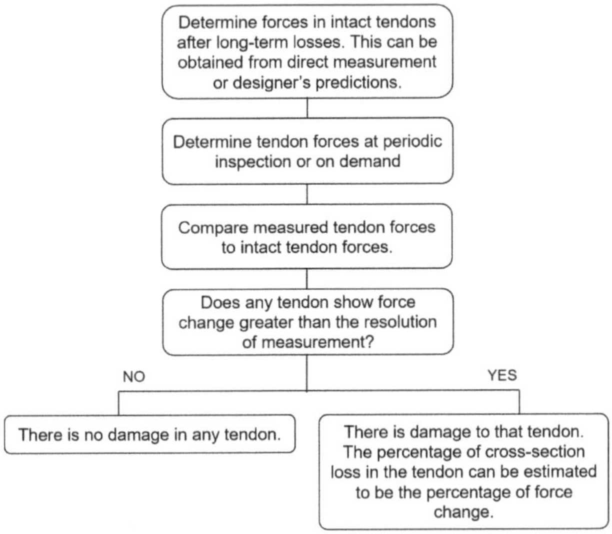

Figure 22. The process of tendon damage detection proposed in this study.

5. Conclusions

This research has explored damage detection in the tendons of post-tensioned segmental box girder bridges, utilizing a range of techniques including the Precursor Transformation Matrix (PTM), Finite Element Method (FEM) analysis, and a vibration-based method. The following conclusions can be made from the paper:

- There is a need for a global damage detection method to accurately determine the location of damage, thereby avoiding the implementation of costly and time-consuming NDT methods on a larger scale. One such method is the Post-Tensioning Method (PTM).
- The findings of this study indicate that sophisticated methodologies, such as the development of complex Finite Element Method (FEM) models and the application of the PTM for tendon inspection, are not necessary for the effective monitoring of bridge tendon integrity.
- The research reveals that damage in one tendon does not impact the internal force in other tendons. This crucial discovery implies that any observed changes in tendon forces can be directly attributed to the affected tendon alone.
- The isolation of effects simplifies and streamlines the damage detection process, making it both more efficient and less time-consuming.
- Monitoring changes in internal forces using the vibration-based method is sufficient for damage detection. This method can be applied immediately after construction and periodically during regular inspections.
- By periodically monitoring internal forces over time, any significant changes indicating potential damage can be quickly identified, ensuring timely maintenance.
- To facilitate the application of the damage detection methods described in this paper, continuous monitoring of tension forces in the tendons using vibration sensors or other types of sensors can be implemented.
- The outcome of this study enhances the ability to rapidly and accurately pinpoint damage, significantly reducing the potential risks associated with delayed maintenance.

The case study presented in this paper uses a practical PT bridge span with a number of tendons that is common in design. Although the results may vary slightly for other configurations, the PTM and conclusion of this paper are valid until the level of damage is significant enough to result in the separation of concrete segments. The latter case is not the focus of this study since it is detectable by regular inspections and shall be avoided by using methods such as PTM to prevent irreversible damage.

Author Contributions: Conceptualization, D.T., S.S.K.D., A.M. and B.T.; methodology, D.T., S.S.K.D., A.M. and B.T.; software, D.T., S.S.K.D., A.M. and B.T.; validation, D.T., S.S.K.D., A.M. and B.T.; formal analysis, D.T., S.S.K.D., A.M. and B.T.; investigation, D.T., S.S.K.D., A.M. and B.T.; resources, D.T., S.S.K.D., A.M. and B.T.; data curation, D.T.; writing—original draft preparation, D.T., S.S.K.D., A.M. and B.T.; writing—review and editing, D.T., S.S.K.D., A.M. and B.T.; visualization, D.T., S.S.K.D., A.M. and B.T.; supervision, A.M.; project administration, A.M.; funding acquisition, A.M. All authors have read and agreed to the published version of the manuscript.

Funding: This research received no external funding.

Data Availability Statement: The data presented in this study are available on request from the corresponding author.

Conflicts of Interest: Author Bardia Tabiatnejad was employed by the company Haley & Aldrich, Inc. The remaining authors declare that the research was conducted in the absence of any commercial or financial relationships that could be construed as a potential conflict of interest.

References

1. Broomfield, J.P. *Corrosion of Steel in Concrete: Understanding, Investigation and Repair*; CRC Press: Boca Raton, FL, USA, 2023.
2. Lau, K.; Permeh, S.; Ghahfarokhi, M.S.; Faridmarandi, S.; Azizinamini, A. *Establishing Inspection Periods and Preliminary Recommendations for Preventive Measures for Mid-Rise Buildings near Coastal and Inland South Florida Environments*; Florida International University (FIU): Miami, FL, USA, 2023.
3. Hauser, J.A.I.I.; Crooker, T.W. *An Investigation of Stress-Corrosion Cracking Susceptibility in Candidate Steels for Tension Leg Platform Tendons*; U.S. Naval Research Laboratory: Washington, DC, USA, 1986.
4. Lau, K.; Permeh, S.; Faridmarandi, S.; Azizinamini, A. Review of Mitigation of Carbonation-Induced Corrosion with UHPC. In Proceedings of the AMPP Annual Conference, Denver, CO, USA, 19–23 March 2023; AMPP: Houston, TX, USA, 2023.
5. Scannell, W.T. *Applicability of Cathodic Protection in Prevention of Corrosion Damage to Steel Tendons in Prestressed Concrete (High Strength Steel, Hydrogen Embrittlement, Brittle Cracking)*; Florida Atlantic University: Boca Raton, FL, USA, 1986.
6. Mehrabi, A.B.; Tabatabai, H.; Lotfi, H.R. Damage detection in structures using precursor transformation method. *J. Intell. Mater. Syst. Struct.* **1998**, *9*, 808–817. [CrossRef]
7. Ibarhim, A.; Salim, H.; Rahman, N.A. Progressive collapse of post-tensioned box girder bridges under blast loads using applied element method. In Proceedings of the Structures Congress, Chicago, IL, USA, 29–31 March 2012.
8. Harvey, D.I. Precast Post-Tensioned Segmental Box Girder Bridges in Vancouver. In Developments in Short and Medium Span Bridge Engineering. In Proceedings of the Third International Conference on Short and Medium Span Bridges, Canadian Society for Civil Engineering, American Concrete Institute, Canadian Institute of Steel Construction, Toronto, ON, Canada, 7–10 August 1990.
9. Kong, F.K.; Evans, R.H. *Reinforced and Prestressed Concrete*; CRC Press: Boca Raton, FL, USA, 2017.
10. Harajli, M.H. Strengthening of concrete beams by external prestressing. *PCI J.* **1993**, *38*, 76–88. [CrossRef]
11. Menn, C. Analysis and Design of Bridge Superstructures. In *Prestressed Concrete Bridges*; Birkhäuser: Basel, Switzerland, 1990; pp. 211–275.
12. Doebling, S.W.; Farrar, C.R.; Prime, M.B.; Shevitz, D.W. *Damage Identification and Health Monitoring of Structural and Mechanical Systems from Changes in Their Vibration Characteristics: A Literature Review*; Los Alamos National Lab. (LANL): Los Alamos, NM, USA, 1996. [CrossRef]
13. Poorghasem, S.; Bao, Y. Review of robot-based automated measurement of vibration for civil engineering structures. *Measurement* **2023**, *207*, 112382. [CrossRef]
14. Tan, X.; Poorghasem, S.; Huang, Y.; Feng, X.; Bao, Y. Monitoring of pipelines subjected to interactive bending and dent using distributed fiber optic sensors. *Autom. Constr.* **2024**, *160*, 105306. [CrossRef]
15. Hurlebaus, S.; Hueste, M.; Karthik, M.; Terzioglu, T. *Condition Assessment of Bridge Post-Tensioning and Stay Cable Systems Using NDE Methods*; Transportation Research Board of the National Academies, Texas A&M Transportation Institute: College Station, TX, USA, 2016.
16. Azizinamini, A.; Gull, J. *FDOT Protocol for Condition Assessment Of Steel Strands in Post-Tensioned Segmental Concrete Bridges: Final Report Volume II. Florida*; Department of Transportation: Tallahassee, FL, USA, 2012.
17. Ditommaso, R.; Iacovino, C.; Auletta, G.; Parolai, S.; Ponzo, F.C. Damage Detection and Localization on Real Structures Subjected to Strong Motion Earthquakes Using the Curvature Evolution Method: The Navelli (Italy) Case Study. *Appl. Sci.* **2021**, *11*, 6496. [CrossRef]
18. Taeby, M.; Mehrabi, A.B. Risk-based selection of inspection method for external post-tensioning system of bridges. *Appl. Sci.* **2022**, *12*, 7103. [CrossRef]

19. Saidou Sanda, M.; Gauron, O.; Turcotte, N.; Lamarche, C.-P.; Paultre, P.; Talbot, M.; Laflamme, J.-F. Efficient finite elements model updating for damage detection in bridges. In *Experimental Vibration Analysis for Civil Structures: Testing, Sensing, Monitoring, and Control 7*; Springer: Berlin/Heidelberg, Germany, 2018.
20. Figueiredo, E.; Moldovan, I.; Santos, A.; Campos, P.; Costa, J.C. Finite element–based machine-learning approach to detect damage in bridges under operational and environmental variations. *J. Bridge Eng.* **2019**, *24*, 04019061. [CrossRef]
21. Dassault Systemes Simulia Corp. *ABAQUS, v. 2021*; Dassault Systèmes: Vélizy-Villacoublay, France, 2021.
22. Wilden, H. *PCI Design Handbook: Precast and Prestressed Concrete*; Precast/Prestressed Concrete Institute: Chicago, IL, USA, 2010.

infrastructures

Review

Operational Modal Analysis on Bridges: A Comprehensive Review

Hamed Hasani and **Francesco Freddi** *

Department of Engineering and Architecture, University of Parma, 43121 Parma, Italy; hamed.hasani@unipr.it
* Correspondence: francesco.freddi@unipr.it

Abstract: Structural health monitoring systems have been employed throughout history to assess the structural responses of bridges to both natural and man-made hazards. Continuous monitoring of the integrity and analysis of the dynamic characteristics of bridges offers a solution to the limitations of visual inspection approaches and is of paramount importance for ensuring long-term safety. This review article provides a thorough, straightforward examination of the complete process for performing operational modal analysis on bridges, covering everything from data collection and preprocessing to the application of numerous modal identification techniques in both the time and frequency domains. It also incorporates advanced methods to address and overcome challenges encountered in previous approaches. The paper is distinguished by its thorough examination of various methodologies, highlighting their specific advantages and disadvantages, and providing concrete illustrations of their implementation in practical settings.

Keywords: operational modal analysis (OMA); bridge monitoring; structural health monitoring (SHM); damage detection; dynamic analysis; modal identification techniques; artificial intelligence (AI); machine learning

check for updates

Citation: Hasani, H.; Freddi, F. Operational Modal Analysis on Bridges: A Comprehensive Review. *Infrastructures* **2023**, *8*, 172. https://doi.org/10.3390/infrastructures8120172

Academic Editors: Armin Mehrabi and Seyed Saman Khedmatgozar Dolati

Received: 2 November 2023
Revised: 27 November 2023
Accepted: 2 December 2023
Published: 4 December 2023

1. Introduction

Infrastructures have undergone continuous improvement over time due to their critical functions and societal significance. Building, maintaining, monitoring, and reconstructing the structures have posed challenges to technicians [1]. Throughout history, a fundamental concern has been the pursuit of more effective safety measurements. It became increasingly crucial to identify, evaluate, and address potential risks, degradation, or collapse as the structures age and deteriorate and to thoroughly assess the financial and logistical aspects of rectifying structural deficiencies [2,3].

Continuous monitoring of bridge health, particularly in areas prone to natural disasters such as Japan [4], has led to the development of bridge management systems (BMSs), the integration of bridge monitoring within a centralized framework [5–9]. Increasing safety, reducing redundant costs, and enhancing management efficiency can be achieved through the implementation of BMSs.

Traditional bridge monitoring methods are based on physical inspections, which inherently possess limitations. They are time-consuming, laborious, and potentially unreliable [10]. Moreover, any delayed action or neglect of maintenance might lead to considerable future costs, specifically for structures with special importance [11]. To overcome these shortcomings, bridge structural health monitoring (BSHM) techniques have been employed, using both static and dynamic approaches. Dynamic monitoring has gained significant attention thanks to advancements in data acquisition technology, the increased accessibility of sensors, cost efficiency, and ease of installation. Additionally, concerns regarding data storage and sensor longevity have been mitigated by technological progress. Improvements in signal processing and data manipulation methods have also contributed. Through the use of sensing technologies, data acquisition systems, and analysis techniques, structural responses can be monitored, anomalies can be detected, and health statuses can be assessed through dynamic analysis.

Moreover, recently, due to the emergence and integration of advanced technologies, specifically machine learning (ML) methods [12], BSHM has undergone an evolution. Data collection, analysis, and visualization have been made easier, leading to a significant improvement in monitoring capabilities, accuracy, and efficiency. Bridge monitoring is expected to continue evolving, as these technologies are advanced, enabling more proactive and sustainable approaches to bridge maintenance and management.

Damage is regarded as a systematic alteration from an initial state that adversely affects the system's behavior. Bridge damage is often categorized in a vertical hierarchy that must be correctly followed to reach a precise estimation [12]. The most commonly used strategy was proposed by Rytter [13].

As illustrated in Figure 1, structural damage can be classified into five stages, each associated with a question. It begins with estimating the presence, advances to determining the location, type, and severity of the damage, and ultimately predicting the susceptible zone in the structure that is likely to be affected. These steps exhibit a significant interdependence; for instance, an erroneous procedure in determining the presence or localization of structural damage may result in imprecise quantification of damage severity, consequently leading to inaccurate prognostication [13].

To follow the hierarchy represented in Figure 1, the bridge's structural behavior should be first assessed through static or dynamic approaches. From a static point of view, the focus is on assessing the bridge's structural stability and load capacity. In this regard, to determine any potential weaknesses or anomalies, parameters such as strain, displacement, and stress distribution are measured. However, incorporating such hierarchical approaches within the context of dynamic analysis allows us not only to evaluate the structural dynamic behavior under varying loads and conditions, but also to identify and assess any potential damage that may occur during the bridge's operational lifespan.

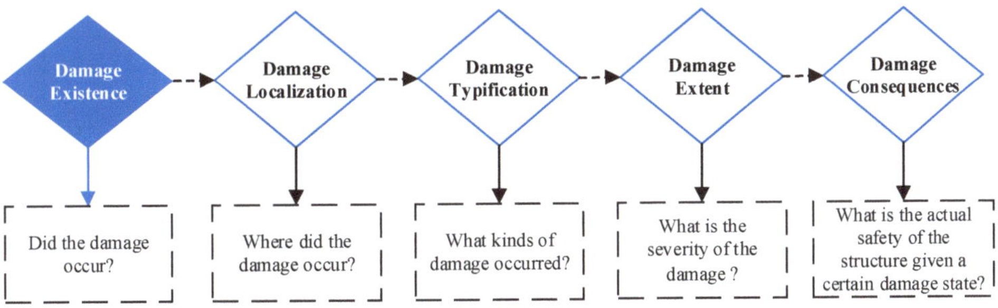

Figure 1. Damage hierarchy.

Similar to its static counterpart, dynamic analysis of structures is categorized into short- and long-term evaluations. Short-term assessment involves analyzing structural behavior during or after specific events, such as earthquakes, load testing of bridges, ambient responses to live loads, etc. On the other hand, long-term evaluation is an ongoing, periodic process that entails continuously observing and assessing the structural behavior of a bridge. This aims to identify any deviations or deterioration that may necessitate maintenance or repair, typically involving a combination of data collection, analysis, and reporting [14].

Figure 2 depicts an overview of the procedural stages required for dynamic BSHM. The initial phase involves determining the appropriate number and types of sensors for capturing relevant data. Decisions are made regarding the strategic placement of the sensors to ensure reliable and precise results. Once the sensors are in place, data collection commences, involving the continuous recording of information related to the structural behavior, including vibration, strain, temperature, and other. Subsequently, the collected data undergoes processing and transformation into a usable format. In the following step, analytical methods such as signal processing and statistical techniques are applied to comprehend the structure's dynamic behavior by determining dynamic parameters. Information management encompasses the storage, organization, and clear presentation of the analyzed data to facilitate decision-making. In the structural health examination step, the computed dynamic parameters are employed to determine the potential damage type, pinpoint its location, and assess severity. This involves comparing the current dynamic behavior with the initial state of the structure to identify any deviations. In the decision-making stage, informed decisions are made regarding the structural health, necessitating assessments of whether maintenance, repair, or further investigation is necessary. If structural deficiencies are identified, examination and repair procedures are implemented to address any detected damage, ensuring structural safety and longevity [12].

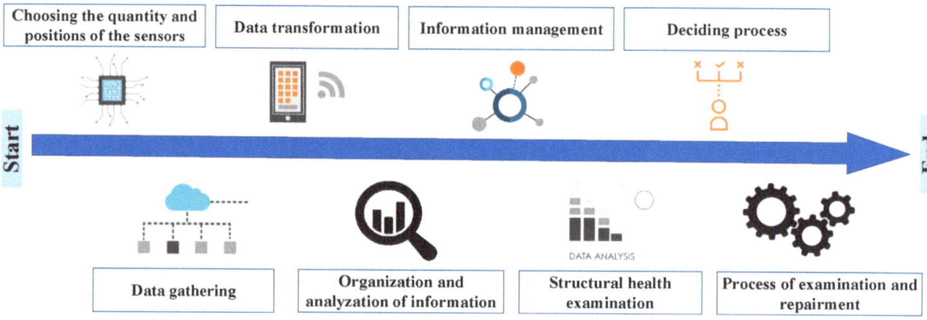

Figure 2. Overview of the structural dynamic monitoring steps [12].

Figure 3 is the schematic representation illustrating the constituent parts of the present review paper. It provides a comprehensive overview of operational modal analysis (OMA), encompassing the entire process from acquiring structural data to their utilization in BSHM applications. The paper starts by discussing the instrumentation and data acquisition system (Section 2), followed by a data preprocessing step, which involves the elimination of undesired frequencies and artifacts, as well as discussing synchronization (Section 3). Section 4 focuses on modal identification techniques in time and frequency domains, while Section 5 describes postprocessing methods, using identified modal parameters to detect any potential damage over time. Finally, Sections 6–8 explore advanced techniques aimed at improving and compensating for the limitations of previous methods, thereby advancing capabilities in bridge health assessment.

This review paper provides a comprehensive overview of the entire process involved in performing OMA on bridges. It encompasses various steps, including data acquisition and preprocessing, the application of diverse modal identification techniques in both time and frequency domains, and postprocessing. Additionally, advanced methodologies are integrated to address and overcome limitations observed in earlier approaches. The paper provides information about each methodology, emphasizing their specific merits and limitations, and offers concrete examples of their implementation in real-world scenarios. What distinguishes this paper is its extensive coverage of research in the field, making it one of the most thorough literature reviews available.

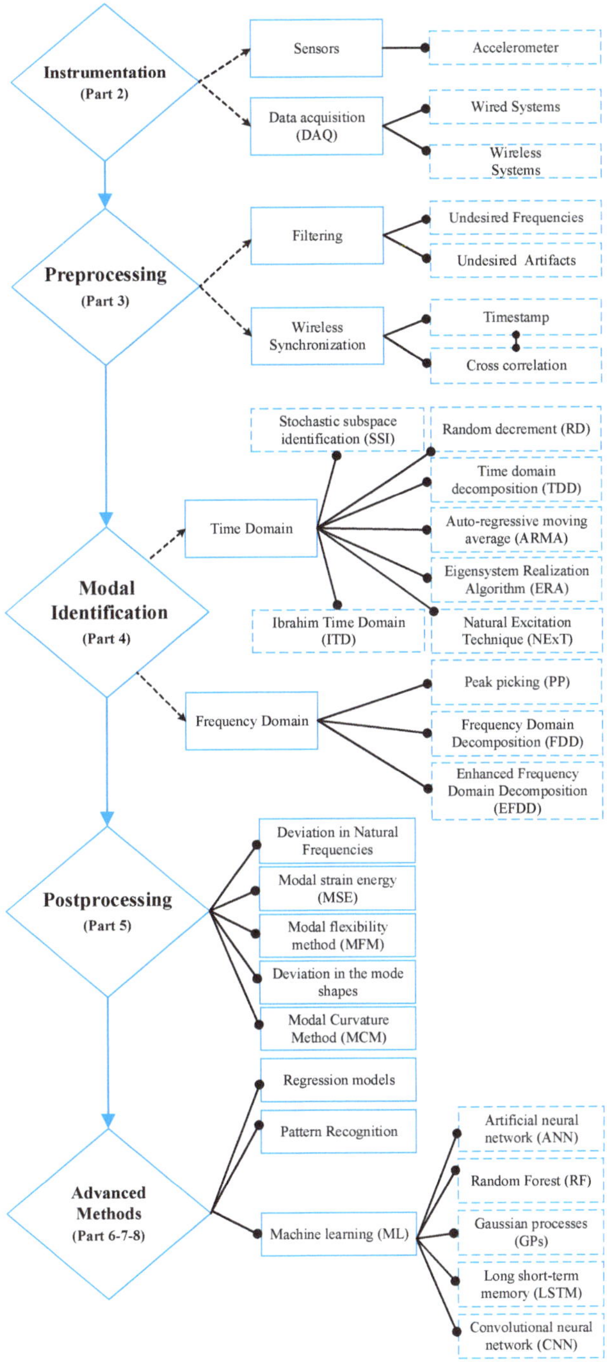

Figure 3. Overview of the sections in the paper.

Figure 4 illustrates the contributing papers that were used in the compilation of this review article based on their publication years, 1990 to 2023. The depicted growth in the publications over the recent years can be attributed to several factors, such as the advancements in data acquisition technology, which enhance the applicability and accuracy of OMA on bridges. The increased accessibility of sensors, coupled with improvements in cost efficiency and ease of installation, has further fueled this. Moreover, technological progress has addressed concerns related to data storage and sensor longevity, contributing to a more robust and sustainable OMA framework. Additionally, the evolution of signal processing methods has significantly correlated to the positive trends observed in the graph.

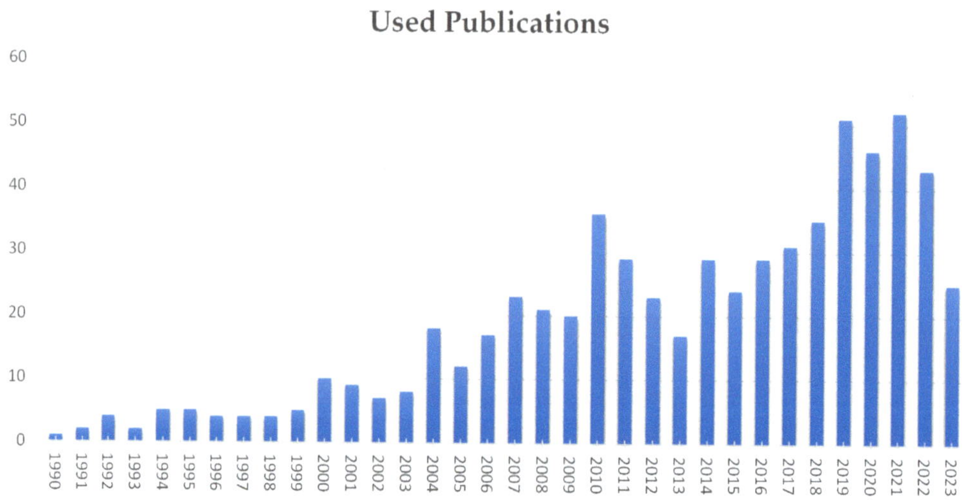

Figure 4. Contribution of the papers in this study.

2. Instrumentation

To evaluate the dynamic behavior of structures, instrumentation plays a vital role in gathering data. It comprises a network of sensors that capture data through a data acquisition component [15]. They can be collected by strategically positioning sensors along the bridge, including strain gauges (SGs) [16], displacement [17], velocity transducers [18], and accelerometers [19]. On the other hand, understanding the impact of environmental changes on bridge behavior necessitates the adoption of environmental sensors such as temperature [20], humidity [21], etc.

The primary step in planning dynamic measurements involves selecting the precise location and direction of the sensors and determining the total number of measurements. This choice can rely on experience or be based on computational simulations, utilizing finite element models (FEMs). Alternatively, it may be based on predictions of the dynamic response derived from simple beam, plate, or shell theories. Regardless of the measurement type, each transducer should possess sufficient sensitivity to detect the expected operating signals. In cases where transducers must be placed in noisy areas with poorly understood sources of vibration, or when the test is expensive or impractical to repeat, it may be necessary to use redundant adjacent transducers with different sensitivities. This precaution ensures the achievement of the desired test data even if some transducers fail. Additionally, an important consideration is to maintain a sufficiently large channel count to provide a suitable background for robust identification processes [14].

Sensing systems are divided into two categories: wired and wireless. In terms of data transition, wired sensors employ physical cables [22], while wireless sensors rely on radio frequencies or Bluetooth [23]. Although wireless sensing systems offer numerous advantages over their wired counterparts, including easy installation, flexibility for relocation, remote monitoring, and cost efficiency, they also encounter a distinct set of challenges. Wireless sensors utilize batteries as their power source. The limited energy supply from batteries poses a constraint for wireless transmissions. Moreover, wireless data transmission exhibits lower reliability compared to wired systems, resulting in relatively lower data transfer rates for wireless systems. Moreover, in wireless systems, synchronizing the data recorded by various channels is more intricate compared to cable-based systems that employ a central system clock housed at the data server [24]. To overcome transmission delays and propagation [25], various techniques, such as time-stamp estimation and correlation functions, are employed, which will be discussed later in this paper.

2.1. Accelerometers

An accelerometer is used to measure linear acceleration, from monitoring seismic activity to detecting system freefall. In OMA, the dynamic response is mainly recorded via accelerometers, which are mainly classified into these types: piezoelectric [26], piezoresistive [27], and microelectromechanical system (MEMS) accelerometers [28,29]. As a result of technological advancements in MEMS accelerometers, their application has proliferated. MEMSs are an appropriate alternative to other accelerometers due to their low power consumption, high sensitivity, affordability, and sufficient sampling frequency.

2.2. Data Acquisition (DAQ)

Before dynamic analysis techniques can be used to evaluate recorded data, the analogue signals generated by sensors must be converted into digital form using analogue-to-digital converters (ADCs). The performance of the ADC is generally determined by the number of bits available in its internal processor [14]. The market offers a variety of data acquisition systems, each distinguished by their own attributes, configurations, and pricing levels. Selecting appropriate DAQ systems becomes complex when factors beyond budget constraints come into play.

3. Preprocessing of the Data

To improve data quality, raw data must undergo preprocessing to eliminate noise and artifacts before extracting dynamic characteristics. Filtering techniques, such as low-pass, high-pass, bandpass, or decimation, are employed to eliminate undesired frequencies, thereby increasing the signal-to-noise ratio by enhancing the frequency range of interest. Additionally, a detrending filter is used to remove or reduce the impact of low-frequency trends or drifts in the data. Furthermore, external interference and errors in measurement may cause artifacts in data acquisition. In this regard, techniques such as signal interpolation, outlier detection, and data smoothing are employed [24].

On the other hand, while dealing with multiple sensors, to achieve accurate dynamic characteristics, specifically mode shapes [30], precise data alignment and synchronization are necessary. When dealing with wireless and wired sensing systems, several factors should be considered. The recorded data from wired sensors can be directly managed and synchronized due to their physically cabled connection to a centralized DAQ system, providing precise timing. In contrast, achieving synchronization for data recorded by multiple sensors is intricate in wireless connections, due to their inherent nature. Consequently, time-stamp estimation and cross-correlation functions are utilized to evaluate propagation and adjust the transmission delays [31,32]. Nevertheless, it is important to highlight that when utilizing the timing synchronization function (TSF), which depends on a dedicated timer managed by the network access point (i.e., the router) for time-stamp estimation, a slight timing error of one millisecond may arise during the data alignment process. This minor misalignment is unlikely to result in significant complications in the modal identification results.

4. Dynamic Analysis of the Bridge

To evaluate the long-term structural behavior of the bridge, dynamic characteristics, including natural frequencies, mode shapes, and damping ratios, are monitored over time. This procedure involves the application of modal identification methods, as explained in Section 4.1.

4.1. Modal Identification Methods

Experimental modal analysis (EMA) and operational modal analysis (OMA) are two widely employed approaches for modal identification. The application of EMA [33–35] includes applying controlled input excitations with predetermined force magnitudes at specified frequencies, which yields the advantage of mitigating undesired sources of noise in the recorded structural responses. Moreover, applied loading can be adjusted to align with the specific testing criteria. With regard to bridges, vibration excitation can be achieved through eccentric rotating mass vibrators or an impulsive shaker [35]. On the other hand, OMA [36–44], also known as output-only modal analysis, ambient modal identification, or in-operation modal analysis, relies on vibration data collected while the structure is under its operational state. OMA becomes practical when the structure's size prohibits artificial excitation or when the system cannot undergo a complete interruption in operation. OMA relies only on response data since the excitation sources are not known. One of the advantages of OMA over prior methods is its independence from input excitation. This makes it valuable when the structure is subjected to random excitation from the ambient environment, such as occurs with bridges. Furthermore, OMA can be used for real-time monitoring of a structure while it is in use. In addition, bridges, towers, and high-rise buildings that pose accessibility or cost challenges can be monitored with OMA [14].

There are other non-destructive tests (NDTs) in addition to visual inspection that have been used for bridge inspection, such as ultrasonic testing (UT), radiographic testing (RT), magnetic particle testing (MPT), eddy current testing (ECT), etc. [45,46]. OMA offers several advantages over other NDT methods, including its ability to provide a comprehensive assessment by identifying both the overall dynamic characteristics of the entire structure and specific localized issues. Its non-intrusive and non-destructive nature is crucial for preventing damage or disturbance to the bridge's natural vibration modes. Moreover, its real-time data acquisition capabilities enable the detection of structural damage at an early stage, before it becomes visible or causes significant problems. OMA is also cost-effective and enhances safety in the inspection process.

In modal identification, a significant challenge lies in the demand for computational power and efficiency. This is more obvious when processing hundreds of output data points at the same time. OMA emerged as an important topic in the late 1980s, especially along with the evolution of computational technologies [47]. Considerable progress in OMA has been made since the early 1990s [24].

Two main techniques are used for modal analysis in OMA: the frequency domain and the time domain. In the time domain, the emphasis is on observing the structural reaction over a period of time, revealing short-term behaviors and temporal phenomena. On the other hand, in the frequency domain, time-domain data is translated into frequency components using techniques such as Fourier transform, which allows natural frequencies, mode shapes, and damping ratios to be identified [24].

4.2. Frequency Domain Analysis

The frequency domain offers a wider range of applications compared to time-domain techniques, primarily due to the inherent stability of structural characteristics in this domain [14]. While in the time domain, the focus is on free responses that manifest throughout the entire considered time duration; in the frequency domain, each mode exhibits prominence within a limited frequency band, leading to the presence of a "natural modal decomposition". This decomposition emerges from the consideration of distinct frequency bands where individual modes exert dominance [24].

There are three main modal identification methods in the frequency domain for BSHM: peak picking (PP), frequency domain decomposition (FDD), and enhanced frequency domain decomposition (EFDD).

4.2.1. Peak Picking (PP)

PP stands as the first OMA technique that enabled the evaluation of structural behavior using frequency spectra [48–53]. In this method, the identification process of natural frequencies is based on detecting the peaks within the power spectrum plot. This method is practical when structural modes are well separated and damping is minimal. While the employment of the method is straightforward, it may yield misleading results when closely spaced modes occur [54]. However, during the modal analysis considering real and complex structures such as a bridge, closely spaced modes are invariably present [55]. To address this limitation, a frequency domain technique known as frequency domain decomposition (FDD) was introduced, which will be described here later. The PP method was adopted as a modal identification technique for assessing the dynamic characteristics of the Z24 Bridge, a reinforced concrete bridge in Switzerland, by Peeter et al. [55]. They evaluated the effectiveness of PP in comparison with other modal analysis techniques used by participating laboratories. Guo et al. [56] used this method to analyze the dynamic responses of the Xihoumen Bridge in China under various wind conditions, such as typhoons and regular winds, and they also compared the PP results with time-domain methods. Furthermore, Zong et al. [49] employed PP in the dynamic analysis of concrete-filled steel tubular (CFST) half-through arch bridges to determine the natural frequencies, while Feng et al. [57] utilized PP for the modal extraction of two instrumented highway bridges. The application of the PP technique is emphasized in the aforementioned research [58–68].

4.2.2. Frequency Domain Decomposition (FDD)

FDD was introduced as an extension of PP based on singular value decomposition (SVD) to make the power spectral density (PSD) matrix out of the response spectra matrix. This method was enhanced by Brinker et al. [69] in 2000, which led to the current FDD method. FDD is a widely accepted method capable of handling closely spaced modes, nonlinear systems, and nonstationary signals [14]. It is considered nonparametric since it does not require previously known mathematical models to identify the modal parameters. In this method, while natural frequencies are the singular values obtained from the PSD matrix, mode shapes are estimated using singular vectors [24]. The main drawbacks of using the FDD method are the accuracy of the natural frequencies, which are limited by the resolution of the FFT, and its inability to accurately calculate the damping ratio of the structure. Malekjafarian et al. [70] and Elhattab et al. [71] used FDD to identify the mode shapes of a bridge from vehicle-induced dynamic response signals. To investigate the truck effects on the natural frequencies and mode shapes of a post-tensioned concrete box girder bridge, Akköse et al. [72] employed FDD. Brincker et al. [73] applied FDD as modal identification, utilizing a spectral density matrix transformed into single-degree-of-freedom (SDOF) systems, validated by the Great Belt Bridge, Denmark. Another study by Wu et al. combined FDD with strain mode identification to address damage assessment challenges in reinforced concrete beams [74]. Research by Alamdari et al. [75] employed FDD for detecting progressive damage by reducing the dimensionality space. OBrien [76] adapted short-time FDD for damage detection from slow-moving vehicles and compared simulated structures with known parameters. In a comparative study, FDD analysis of the ambient vibration data of the Vasco da Gama Bridge in Portugal was conducted and compared with the results of PP and SSI by Cunha et al. [77]. Choe et al. [78] assessed a wireless sensor network-based SHM system's accuracy through FDD. Weng et al. [79] employed FDD and SSI for modal frequency extraction from cable–deck interactions. Using an automated FDD algorithm, another paper developed an OMA tool for easily integrating default estimates of modal parameters into commercial software packages [80]. FDD was also compared with a ground-based interferometric radar approach for bridge assessment through natural frequency estimation from displacement measurements by Michel et al. [81]. Moreover, these papers discussed the implementation of the FDD method [82–86].

4.2.3. Enhanced Frequency Domain Decomposition (EFDD)

As mentioned, through the application of the FDD method, natural frequencies and mode shapes can be accurately identified, even in cases where modes are closely spaced. However, its significant drawback is the inability to provide estimations for damping ratios. To overcome this limitation, an improved version of FDD, named enhanced frequency domain decomposition (EFDD) [87], was introduced. EFDD proved more precise capabilities in the estimation of mode shapes, damping ratios, and natural frequencies compared to conventional FDD. EFDD identifies the PSD function close to the resonance peaks [88,89] and converts it to the time domain using inverse discrete Fourier transform (IDFT). This involves estimating the natural frequencies by counting zero crossings over time and determining damping ratios through logarithmic decrements from the normalized autocorrelation function [89]. Kasimzade et al. [41] explored the modal parameters of a steel model bridge using the EFDD method, and the effectiveness of the method was proven in determining these parameters accurately for BSHM. Jacobsen et al. [89] also employed EFDD to handle harmonic effects during modal extraction, resulting in more robust estimates of structural characteristics. Furthermore, in another study, EFDD was used for stone masonry arch bridge modal identification in Turkey's northeastern region by Gonen et al. [90], with a comparative analysis of the strengths and limitations of different techniques. Moreover, AltuniŞik et al. [91] compared EFDD and SSI, noting close agreement between their results. The referenced papers also mentioned EFDD as the modal identification method used [92–96].

4.3. *Time Domain Analysis*

Due to the limitations in the frequency resolution of spectral estimates and leakage errors in the frequency domain identification approaches, time-domain methods were subsequently developed. In the time domain, the identification problem is considered full rank. The fact that all modes are present at any given time can also be considered a drawback, as it can make the estimation process more complex. The advantage of time domain identification is that it is easier to acquire bias-free data. This is because the free decays used are not as sensitive to noise as they are in the frequency domain. Moreover, time domain methods provide better results when a large frequency range or multiple number of modes exist in the data [14,97].

The most widely used time domain methods in OMA are random decrement (RD), Ibrahim time domain (ITD), the eigensystem realization algorithm (ERA), the autoregressive moving average (ARMA), time domain decomposition (TDD), stochastic subspace identification (SSI), and the natural excitation technique (NExT).

4.3.1. Random Decrement (RD)

RD is an output-only method in the time domain introduced by Cole in 1968 [98], which focuses on the averaging method [99,100]. In this method, real-time interpretation is achieved by translating the data time history into a sum of autocorrelation functions through generalized harmonic analysis of a single-degree-of-freedom system. RD's methodology is demonstrated by a high-speed digital computer to analyze the structural dynamic response. Within the domain of RD, considerations are given to the effects of finite time, the filtering of several degrees of freedom, the approach to stability boundaries, nonlinear systems, and structural failure. RD provides an accurate result for damping ratios, natural frequencies, mean square values, and half-power points of the input [98]. The application of RD for concrete bridge girder damage location identification using fiber Bragg grating (FBG) sensors is illustrated in [101]. Furthermore, RD was employed by Kordestani et al. [102] for BSHM, particularly under the influence of a moving load, which was validated by experimental examples. Using the Nanjing Yangtze River Bridge in China for method verification, He et al. [103] used the RD-based empirical mode decomposition (EMD) technique for modal identification in vibrational data. Wu et al. [104] introduced a mode separation technique, followed by the development of a multiple random decrement method to isolate free vibration responses while excluding excitation effects. The proposed methodology was applied to ambient velocity measurements from the cables of the Chi-Lu Bridge in Taiwan. In another study, Kaloop et al. [105] utilized RD to estimate the impulse response of the tower's displacement of a bridge using data from a global positioning system (GPS). These papers also contributed to the utilization of the RD method [106–109].

4.3.2. Ibrahim Time Domain (ITD)

As one of the first methods for modal identification in multiple-output systems, the ITD method was developed by Ibrahim in 1973 [110–112] for implementation in OMA, where free decays are obtained from random responses. ITD estimates the structural modal parameters such as the natural frequencies, damping ratios, and mode shapes from the time function data via the construction and solving of an eigenvalue problem. The time functions can be either free decays from the structure, impulse response functions, or pseudo free decays [113]. To investigate highly coupled systems with severe modal interference, ITD can identify modes with relatively small contributions to the response [114]. Siringoringo et al. [115] compared ITD with other time domain methods for analyzing a suspension bridge's ambient vibration. Huang [114] utilized ITD to identify vibration frequencies, mode shapes, and damping ratios of a concrete box girder bridge, noting the strong correlation between free and ambient vibration tests. The ITD method formed the basis for extracting dynamic characteristics from a laboratory cable-stayed bridge model, enhancing its feasibility and practicality, in a study by Liu et al. [116]. Asmussen et al. [117] and Fujino et al. [118] combined ITD with the RD technique for modal parameter extraction on the Queensborough Bridge in Canada and the Hakucho Bridge in Japan, respectively, effectively estimating eigenfrequencies, damping ratios, and mode shapes from ambient data. Cable-stayed bridge identification was further explored by Wu et al. [119], who introduced a mode separation technique utilizing ITD for parameter identification. The application of ITD can also be followed in these papers [120–122].

4.3.3. Eigensystem Realization Algorithm (ERA)

ERA was initially proposed by NASA's Langley Research Center in 1984, and it generates a system realization using the time domain response, (multi) input and (multi) output data. It uses a minimal realization [123], which means that it has the smallest possible number of states while still representing the input–output relationship of the actual system. This makes the model more efficient and easier to analyze [115]. It was first implemented with impulse excitation and then later with ambient vibration data using NExT. ERA is considered one of the most accurate identification methods for output-only measurement and has been applied to several types of civil engineering structures under operational conditions [123,124]. In the study by Qin et al. [125], ERA was applied as the modal identification of the Tsing Ma Bridge in Hong Kong, using ambient testing data. To enhance accuracy, techniques such as Chebyshev digital filtering and RD functions were also employed to reduce noise and transform ambient responses into free vibrations. The improved ERA incorporated cross-correlation functions, eigenvalue decomposition, and a Hankel matrix for modal identification, successfully identifying 79 modes with complex modes due to uneven damping. Furthermore, ERA combined with the modal similarity index (MSI) and mode energy level (MEL) criteria ensured reliable mode identification in a paper by Zhang et al. [126]. In addition, the introduction of the modal response contribution index (MRCI) for ERA by Qu et al. [127] addressed challenges posed by noise and spurious modes, enhancing accuracy. Moreover, ERA was utilized for damage identification in a full-scale steel stringer bridge, compared with complex mode indicator function (CMIF) results [128]. Other research also employed ERA for modal identification [105,129,130].

4.3.4. Autoregressive Moving Average (ARMA)

ARMA is another modal identification technique in the time domain that has the capability to predict future values in a time series by utilizing both past values and prediction errors. According to [131], mathematician Walker introduced the ARMA model in 1931. It is a developed model of a linear time-invariant system under white noise excitation, which has an assumption of a stationary measured response. In this model, the coefficient matrix is derived from a multivariate time series. If the objective is to determine the dynamic characteristics of the system, this matrix is transformed into a state transition matrix within a stochastic state space model. This transformation serves as the foundation for extracting modal parameters. Additionally, in scenarios with multiple input excitations, the vector ARMA or ARMAV model is implemented [132]. The paper by Chen et al. [133] used ARMA in modelling the spatial correlation of traffic excitation in bridge structures. The model was able to capture the frequency-dependent nature of the excitation spectrum density matrix, which resulted in more accurate estimates of the structural properties. Moreover, Erdoğan et al. [134] employed ARMA to model the stochastic components of the time series of the bridge's movements, capturing the random fluctuations in the data and identifying the different frequencies that were present. In another case study, to estimate the vibration modes of the Golden Gate Bridge in the US, ARMA was utilized by Pakzad et al. [135] to fit a time series to a linear combination of past values and a white noise excitation. Fang et al. [136] used this identification method to estimate and predict deflection data of the typical monitoring site of the Masangxi Bridge in China. ARMA was applied to a concrete-filled steel tube (CFST) arch bridge located in China by Lu et al. [137]. They investigated the effects of variations in the strength and diameter of the strands on the reliability assessment of the suspenders. These papers also contributed to the application of the ARMA method [138–141].

4.3.5. Time Domain Decomposition (TDD)

TDD is based on an approach with a single degree of freedom in the time domain. This method initially extracts the mode shapes and then identifies the corresponding natural frequencies. The approach assumes that the solution to the governing equations of motion can be separated into functions of time only and space only, which leads to a more efficient and simpler technique. Although TDD needs frequency information for natural frequency extraction and filter design, the most computationally expensive part of the method is processing time domain data [142]. Kim et al. [143] proposed a new technique for extracting modal parameters of a railway bridge using TDD. The technique was tested on a two-span steel composite girder bridge, which is a typical Korean high-speed railway bridge system. In another paper, TDD was employed as part of the methodology by Daniotti et al. [144] for the damping estimation of traffic-induced vibrations on a long-span suspension bridge. Furthermore, Salokhe et al. [145] applied TDD to the dynamic response data of a girder bridge to extract modal frequencies. This information was then utilized, along with visualization and noise reduction techniques, for damage detection, particularly with the assistance of SVD. Park et al. [146] used TDD for mode shape extraction from ambient vibration measurements of a cable-stayed bridge. Moreover, Huang et al. [147] used a simulated finite element model of the Lavic Road Overcrossing Bridge in California, US, demonstrating how TDD and the methodology can effectively update the structural properties while accounting for measurement uncertainties.

4.3.6. Stochastic Subspace Identification (SSI)

SSI is another time-domain modal identification method of OMA, developed by Van Overschee and De Moor in 1991 [148], which enables derivation of a state-space model for complex dynamic systems under stochastic excitation directly from measured data [149,150]. In contrast to other techniques such as ARMAV, it has less computational complexity. It is categorized into two algorithms: data-driven SSI (SSI-DATA) and covariance-driven SSI (SSI-COV). Both are capable of estimating system modal parameters; however, SSI-COV excels in providing more accurate estimations of damping ratios compared to SSI-DATA [151]. Nan Jin et al. [152] used the short-time SSI (ST-SSI) framework for estimating bridge frequencies from passing vehicles' dynamic responses. Yang et al. [153] also utilized SSI to analyze bridge frequencies with rough road surfaces and multiple vehicles, modifying observability matrices to isolate the bridge characteristics. Boonyapinyo et al. [154] highlighted the superior SSI-DATA method for flutter derivative estimation, particularly in certain bridge configurations. In [155], Brownjohn et al. analyzed the ambient vibration data from the Humber Bridge in the United Kingdom using SSI, comparing results with other methods and a previous test from 1985. Furthermore, Li et al. [156] simulated bridge excitation and validated modal parameters through numerical and experimental means using SSI. Zhou et al. [157] employed SSI in wind tunnel testing, showing improved modal parameter identification for the Oujiang Bridge in China. Loh et al. [158] used SSI-COV to analyze a real-world arch bridge, employing a stabilization diagram for accurate modal parameters. Wu et al. [159] addressed mode identifiability complexities in cable-stayed bridges under traffic and wind excitation, setting amplitude thresholds. Moreover, Tran et al. [160] applied Combined Deterministic SSI alongside other methods for large-scale bridge identification. The SSI technique finds widespread use in assessing bridge performance through modal analysis [79,161–172].

4.3.7. Natural Excitation Technique (NExT)

The NExT method is one of the OMA methods that was initially employed by James et al. [173]. This technique was first used for EMA and then for OMA. NExT uses cross-spectra of ambient vibration responses to generate impulse response functions (IRFs). Similarly to other OMA methods, this also requires the structural response caused by environmental excitation at multiple locations of the structure. Subsequently, using the data time histories, the correlation function is calculated, which is an important tool for studying systems subjected to ambient excitation [115]. The NExT method can be considered an OMA technique when combined with a time-domain multi-input multioutput (MIMO) algorithm, such as the ERA, the extended Ibrahim time-domain (EITD) method, and the polyreference complex exponential (PRCE) techniques [47]. Farrar and James provided an in-depth derivation of the NExT concept in [174]. Yang et al. addressed the challenge of multiple-channel asynchronous responses in BSHM [175] by enhancing NExT by minimizing phase slopes and maximizing linear dependencies of modal components, effectively aligning multiple-channel responses and achieving precise mode shapes. Another study by Kim et al. [176] employed NExT to identify bridge damping ratios from nonstationary ambient vibration data. By stationarizing the process, the mean and standard deviation of this dynamic parameter for the first vertical mode decrease, proposing a technique that reduces uncertainties in damping identification. Additionally, the NExT method has been utilized in other papers [177–180].

5. Damage Detection Using Modal Parameters

The objective of this phase is to determine whether damage is actually present in a structure. The algorithms used in this step are of two types: a model-driven approach and a data-driven approach. While the first includes an iterative process of numerical model updating, the other method directly compares the derived dynamic characteristics. To quantify and detect damage, damage-sensitive features (DSFs) are quantifiable parameters derived from sensor data. DSFs depend on alterations in dynamic responses such as shifting natural frequencies, mode shapes, strain levels, damping properties, and modifying energy dissipation [181,182]. They are based on dynamic features such as natural frequencies, mode shapes, modal beam curvatures, modal flexibilities, and modal strain energy.

5.1. Changes to the Natural Frequencies

As one of the common methods, analyzing the temporal evolution of natural frequencies in structures, such as bridges, is a tool for damage detection, which results from changes in stiffness and mass. The extent of these alterations depends on the location, severity, and types of damage, as well as the sensitivity of the instrumentation used for data collection. Early work by Adams et al. [183] in the late 1970s introduced the concept of identifying damage through shifts in natural frequencies, later expanded by Crawly & Adams and Salawu in 1979 [184,185]. In another study, the investigation of alterations in natural frequencies was undertaken to anticipate prestressing losses in both concrete post-tensioned girder bridges and steel post-tensioned bridges. In particular, employing the fundamental frequency is not suggested for concrete girder bridges experiencing minor cracking and for steel girder bridges with tendon deviators spanning their lengths. Conversely, the use of the fundamental frequency is endorsed for evaluating their flexural stiffness [186,187]. Messina et al. proposed the damage location assurance criterion (DLAC) and the multiple damage location assurance criterion (MDLAC) while considering the alteration in the natural frequencies [188], which were used to locate predefined damage locations using 10 to 12 modes. Nevertheless, in this case, spurious locations could also arise. As another approach, the single damage indicator (SDI), developed by Kim et al. [189], detects cracks and quantifies their severity by assessing frequency changes. Complex structures pose challenges in localizing damage using modal frequencies [190,191], although regular geometries can employ this method effectively, requiring damaged and undamaged states and an appropriate number of frequencies [192,193]. From another perspective, minor damage has a limited effect on larger structures' natural frequencies, making this method less efficient [191]. Additionally, natural frequencies can change in the same way if damage occurs at two symmetrical locations in the same structure [193]. Comparable alterations can emerge from mass changes or environmental factors. To address environmental effects on natural frequencies, techniques such as principal component analysis (PCA) [194], Kalman filters [195], and the Mahalanobis squared distance [196] are employed to mitigate the influence of environmental factors on measured frequencies.

5.2. Changes to the Mode Shapes

Compared to natural frequencies, mode shapes are less influenced by environmental factors, and provide spatial insights for damage localization purposes [13]. Various techniques have emerged over time, including the modal assurance criterion (MAC), developed by Allemang and Brown [197] in 1982, which utilizes eigenvector orthogonality to identify alterations in mode shapes [198,199]. Later, Kim et al. [200] developed the coordinate modal assurance criterion (COMAC), using modal node displacement for damage detection and localization. Salawu and Williams [201] evaluated the application of MAC and COMAC, and reported their effectiveness, although spurious damage indications were also observed. However, COMAC often results in errors in the deterioration detection of beam structures, which was observed by Salgado et al. [202], with scaling and polarity errors. Putting aside this limitation, the application of COMAC remains diverse in engineering fields, which leads to its integration with other approaches in addition to civil engineering fields [203,204]. The efficiency of COMAC is based on the correct identification of relevant modes contributing to correlation. With regard to this concern, Hunt et al. [205], in 1992, proposed the enhanced coordinate modal assurance criterion (ECOMAC). ECOMAC improves upon COMAC with the computation of the mean deviations of modal amplitudes of each node to enhance the accuracy and reliability of damage detection [182,206].

5.3. Modal Curvature Method (MCM)

In 1991, the MCM was proposed by Pandey et al. [207] with the aim of exploring the second derivative of modal curvatures for deeper mode shape monitoring using the curvature–flexural stiffness correlation. Through the increase in the values of modal curvature, this method detects the reduction in stiffness by comparing damaged and undamaged states. Ho and Ewins [208] developed the MCM by improving abnormal curvature change differentiation. Despite the advantages of using this method, it faces some limitations. Numerous sensors and modes are demanded for a precise higher mode definition [209], and curvature estimation from vibration data will contain errors by a central difference approximation, which will also be exacerbated via high-frequency noise [210]. Although increasing the number of samples results in noise mitigation, truncation errors may also occur [211]. Furthermore, the comparison of the MCM with strain measurements reveals inherent errors [212], discouraging sole reliance on the MCM for damage identification. The integration of the MCM with other DSF parameters such as frequencies can enhance the accuracy of damage detection and localization, as proved by Capecchi et al. [213], who combined MCM with sparse sensor layouts for improved damage sensitivity in an arch bridge. Later, mode shape curvature squared (MSCS) was suggested by Wahab and De Roeck [214], exploring practical applications of modal curvature and the curvature damage factor (CDF), leveraging the Z24 bridge analysis involving simulated and real data [182,206].

5.4. Modal Strain Energy (MSE)

MSE, integral to various modal-based DSFs, is exemplified by the damage index method (DIM) [215]. The energy stored while a structure undergoes deformation is called the modal strain energy. The DIM approach is based on Bernoulli–Euler beam theory to detect the damage by monitoring the reductions in modal strain, which represent the diminished energy storage capacity. Moreover, MSE facilitated a quantitative assessment of the damage extent, i.e., enabling a crack size estimation. An investigation by Kim et al. [216] containing an FE beam model showed the superiority of MSE over frequency-based damage indicators. Additionally, the research by Yam et al. [217] proved similarities in displacement modes and strain modes. One of the shortcomings of MSE is that while strain modes are highly sensitive to local-scale structural changes, higher modes are less effective at detecting damage. Additionally, similarly to the challenges of implementing MCM, the DIM's reliance on modal curvatures presents the same limitations, including the need for central difference approximation. Notably, the DIM involves the acquisition of continuous strain values through curve fitting between sensor locations, a procedure that may inadvertently obscure local damage [182,209].

5.5. Modal Flexibility Method (MFM)

The MFM was first proposed in 1994 by Pandey and Biswas [218] and subsequently applied to bridge structures by Toksoy and Aktan [219]. It is based on the inverse relationship between stiffness and flexibility matrices, offering a brief representation of structural behavior with fewer modes in comparison to the stiffness matrices. This feature increases the damage sensitivity of the method, especially in easily extractable lower modes, as demonstrated by Wang et al. [220]. By integrating the modal curvature method with the MFM, Zhang and Aktan [221] extended the method's damage sensitivity. Furthermore, using an experimental test, Lu et al. [222] evaluated the implementation of modal flexibility with the flexibility curvature of a beam. Although it increased the sensitivity to local damage, challenges in pinpointing flexibility peaks under multiple damage scenarios arise [182].

5.6. DSFs Application

Several studies have focused on assessing a bridge's structural health to estimate the severity and location of damage using different DSFs. Talebinejad et al. [223] applied the ECOMAC, DIM, MCM, and MFM on a cable-stayed bridge, subjected to varying noise and excitation levels. It was demonstrated that the contamination of noises hindered damage sensitivity, allowing only significant damage events to be detected. Ndambi et al. [224] evaluated reinforced concrete beams, testing frequencies, using the MFM, MAC, COMAC, and DIM in a laboratory setting. Both the COMAC and DIM methods excelled in localizing the damage but struggled to accurately quantify incremental damage progression. In another study, to estimate the mode shapes, using the MCM and MFM, Cruz and Salgado [225] used a composite bridge simulation model with vibration data. All applied DSFs were successful in detecting and localizing the damage under no noise contamination. However, under the contamination of noise, the performance of the mentioned DSFs decreased noticeably. With regard to the real data, in higher modes of the structure, clear changes were observed, but not for the lower modes, even in the presence of severe damage. According to Cruz and Salgado, among all utilized DSFs, the MCM and MFM showed the highest levels of detection and localization capabilities. Moreover, Fan and Qiao [226] demonstrated the prior findings in their paper by assessing the frequencies, mode shapes, MCM, and DIM. They confirmed that in higher modes, the applied DSFs, particularly modal curvature-based parameters, are more sensitive to damage than lower modes. They also found that the performance of all damage features decreased when noise was added. It can be concluded from the mentioned investigations that the application of DSFs presents varying performance and is susceptible to noise, posing limitations in real-world applications due to assumptions of a linear stationary structural vibration response, a concern for nonstationary vehicle-induced excitations on damaged bridges. Putting aside these limitations, DSFs can still be useful for the assessment of structures with regular geometry, especially when used in conjunction with other methods [182].

5.7. Shortcomings of the Preceding Methods and the Solution

The challenge in assessing the structural health of bridges is to distinguish between normal and abnormal changes in their dynamic properties over time. Abnormal changes result from deviations in material properties, such as loss of stiffness, which is indicative of damage. In contrast, normal variations in the data are primarily caused by environmental and operational factors. It is important to recognize that these factors can affect the accuracy of the results when using the DSFs mentioned and should not cause false alarms in the monitoring system. External conditions such as changes in temperature and humidity or traffic loads can determine variations in stiffness, which in turn affect the dynamic characteristics of the bridge. Temperature fluctuations cause daily frequency shifts of approximately 5% and seasonal shifts of 10% [227–229]. As mentioned, the frequency alterations derived from fluctuations in temperature pose a challenge in distinguishing them from structural damage [230,231]. The study by Farrar et al. [232] revealed that this can mask the actual damage in gradually deteriorated bridge girders. Although a decrease in the natural frequencies, caused by the reduction in the stiffness, is expected while damage occurs, the observed results were different. Instead, in the initial damage scenarios, caused by the laboratory temperature, the frequencies initially increased before decreasing. Moisture impact can even make it worse, having revealed nonlinear modal frequency distributions for the Z-24 Bridge during freezing temperatures due to frozen moisture in the structure, according to studies by Peeters and De Roeck [227]. With operational conditions contributing over 5% of fluctuations [233], operational variability, more than environmental factors, impacts daily frequency fluctuations [234]. In 2001, Kim et al. [235] found that the natural frequency variations induced by traffic are negligible for medium- to long-span bridges, while they significantly affect shorter-span bridges, as reported in 2012 [236]. As an additional factor that affects the welded connections during their operation, fatigue-induced stress cycles can be considered. Orthotropic steel decks, with complex geometry and subjected to stochastic traffic loads, exhibit complex stress fields. Alcover et al. [237] determined a linear relationship between stress cycles for orthotropic steel deck joints and traffic volume, specifically rib-to-deck welded joints. Song and Ding [238] also worked on the correlations between stress amplitudes and ambient temperatures due to material property changes caused by temperature variations.

In conclusion, the effects of environmental and operational factors on the application of DSFs for structural damage detection are challenging and inevitable. To increase the accuracy and reliability of the previously mentioned damage detection process, the dominant effects of these variables should be mitigated as much as possible. In this regard, Sohn [239] and Xia et al. [240] explored the application of data normalization techniques to address these limitations. It should be noted that these techniques consider scenarios with both environmental or operational data (implementing regression models) and without (employing pattern recognition methods). The goal is to mitigate the effect of external factors, normalize the data, and improve the reliability and precision of damage estimation and detection processes [182].

6. Regression Models

In real-world applications, developing complex models to reduce the impact of external factors on the data is often impractical. Therefore, instead of delving deep into the physics of the problem, it is advisable to rely on it as a black-box model. These models have their parameters fine-tuned using extensive datasets, allowing the establishment of relationships between potential influencing factors and dynamic characteristics, which can be accomplished with the implementation of regression analysis. This statistical technique is employed to establish a connection between dependent and independent (predictor) variables to gain insight into how each predictor (model inputs) influences the dependent variable (model output). For instance, when creating regression models to understand the influence of temperature on natural frequencies, it is crucial to construct one for each frequency. This comprehensive approach involves considering a wide range of variations, including data from both summer and winter periods [241]. This was demonstrated by Peeters et al. [242], who employed linear regression analysis, particularly the ARX model (autoregressive (AR) model with exogenous input) to remove the temperature effect of the identified vibration frequencies from a bridge in Leuven, Belgium. Due to their capabilities in modelling the shift changes in frequency due to below-zero temperatures, multilinear regression models can be employed as a suitable option. A regression error value of approximately 5% is generally used to reduce the effect of erroneous data [243]. In another study conducted by Sohn et al. [244], the recorded data from the Alamosa Canyon Bridge in the US was used to construct a model which characterized the alterations in eigenfrequencies caused by fluctuating temperatures. The model was then utilized to establish confidence intervals for the frequencies corresponding to a new temperature profile. However, in complex data scenarios, advanced regression techniques, including polynomial, ridge, lasso, elastic net, Bayesian, support vector regression (SVR), decision tree, random forest (RF), and gradient boosting regression, are used, which enable the analysis of the complex environmental and operational factors, exemplified by the polynomial regression method used by Ding and Li [245] when mapping the modal frequency variability in long-span suspension bridges. Moreover, Hassan et al. [246] employed multiple regression for bridge health prediction, while Mangalathu et al. [247] employed lasso regression to identify seismic demand models for bridges. A Bayesian vibration-based approach was proposed by Kim et al. [248] for long-term BSHM, and a study by Laory et al. [249] developed regression trees, neural networks, and support vector regression models for analyzing the Tamar Suspension Bridge in the UK.

7. Pattern Recognition (PR)

Farrar et al. [232], in 1994, defined PR as a method to highlight the changes in the frequency response functions measured using cracked (damaged) and uncracked (undamaged) bridges. PR in BSHM involves applying methods such as statistical (SPR) [250], machine learning, time–frequency analysis, and data mining [251,252]. Differentiating between patterned or structural changes in the intact and damaged state under operational and environmental variations is the primary objective of the SPR paradigm. Its process can be outlined as consisting of four stages: assessing operational performance, collecting data, extracting and generating features, and using statistical models for feature classification. Establishing rigid boundaries between these stages can be challenging. It is crucial to recognize that several important processes come into play within the data acquisition, feature extraction, and statistical modelling aspects of the SPR, including data normalization, data cleansing, data fusion, and data compression [253]. The first one involves separating sensor reading variations induced by real damage from those caused by varying operational and environmental conditions, effectively addressing data contamination with external factors. Data cleansing is the practice of selectively choosing data for inclusion or exclusion from the feature selection process. Data fusion, however, combines information from multiple sensors to increase damage detection accuracy. Finally, the last one focuses on reducing the data dimensions or the features extracted from the data. The goal is to facilitate the efficiency of the storage, and improve the statistical quantification of parameters [251]. Figueiredo et al. [251] emphasized implementing SPR and ML to enhance the performance and safety of bridges. Furthermore, while using statistical pattern recognition, Hu et al. [254] conducted vibration-based SHM on a prestressed concrete box girder bridge. In another paper, Haritos et al. [255] investigated the application of statistical pattern recognition and system identification for BSHM, suggesting a combined approach for comprehensive damage assessment. Datteo et al. [256] focused on employing statistical pattern recognition and PCA to reduce the complexity in the analysis of vibration data, exemplified through the G. Meazza stadium in Milan, Italy. Cheung et al. [257] also used pattern recognition in field tests while noting localization challenges and the importance of signal conditioning.

8. Machine Learning (ML)

BSHM systems employ ML to compare two states. It works with constructing a model to extract specific characteristics from the desired dataset with the training and testing stages. As an ideal condition, the used database should be the representation of the complete array of structural excitations. Furthermore, employing signal processing methods such as noise filtering and data normalization can directly contribute to the generation of a highly calibrated dataset. ML finds it application for BSHM in damage detection of sensor-detected data [258].

Figure 5 illustrates the workflow needed to develop an ML model for SHM. The process begins with an excitation step, where the structure undergoes any type of stimuli. Subsequently, data acquisition captures relevant information from the structure's response. The acquired data then undergoes normalization to standardize its scale and facilitate consistent analysis. Following normalization, data cleaning eliminates any noise or outliers. The process then involves data compression through dimensionality reduction techniques to streamline information while retaining its crucial features. Feature extraction identifies and highlights essential characteristics in the data, contributing to the subsequent stage of data fusion, where multiple data sources are integrated for a comprehensive understanding. The final stage, pattern recognition, employs ML algorithms to detect and interpret patterns in the fused data, ultimately enabling effective structural health monitoring [259].

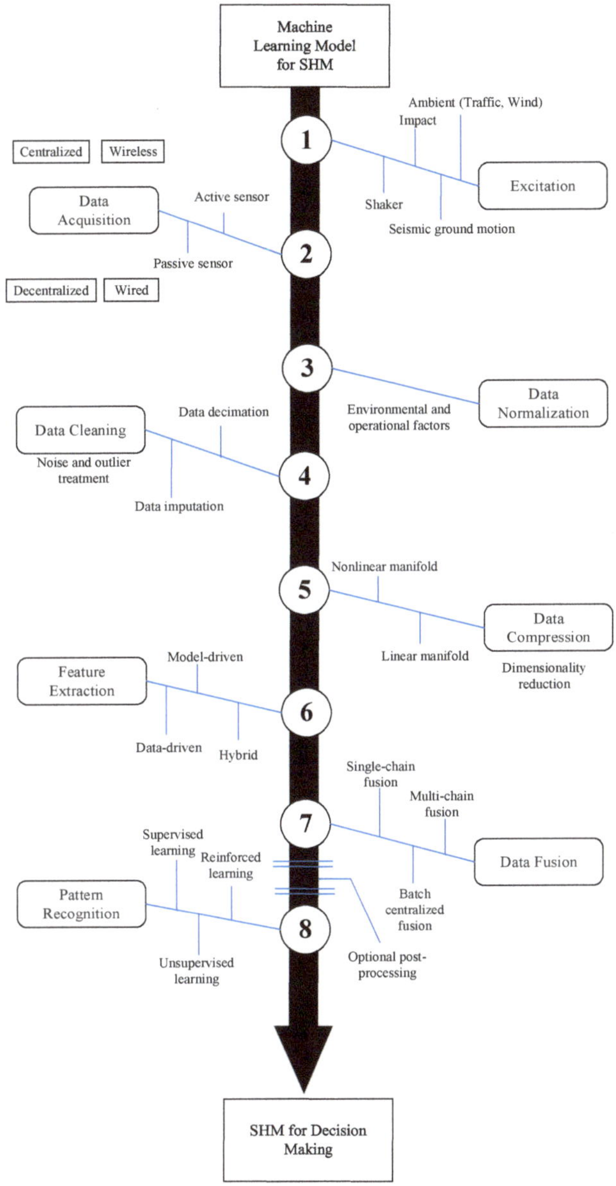

Figure 5. Overview of essential and optional steps in data-driven SHM systems with ML enhancements [259,260].

Through the implementation for the case studies, ML approaches have demonstrated their capability in addressing the following issues with regard to BSHM [261]:

- They can effectively describe physically complex correlations;
- Automatic detection and compensation of sensor faults are achievable;
- Trained models can be transferred to other structures/problems with similar boundary conditions;
- The separation of factors affecting specific structural behaviors is facilitated;

- Future structural behavior can be estimated based on previously predicted values.

The most commonly used ML techniques used in BSHM are artificial neural networks (ANNs), support vector machines (SVMs), random forests (RFs), Gaussian processes (GPs), convolutional neural networks (CNNs), and long short-term memory (LSTM) neural networks.

8.1. Artificial Neural Network (ANN)

An ANN is a computational network inspired by biological systems that uses basic concepts of the human brain. These networks can be trained for pattern identification and information classification that are similar to human cognitive processes [12,262,263]. In ANNs, information is transmitted between the neurons, which play the communication roles between layers, determined by their structure and synaptic weights. Input layers are user-defined, considering elements such as geometries, mechanical/structural properties, and physical parameters. On the other hand, hidden layers emerge as a consequence of computations based on the input layers, finally culminating in the generation of output layers as the final results [264–266]. For BSHM for prediction and anomaly detection, Kwon et al. [267] combined ANNs with the building information modelling (BIM) method, which effectively identified element anomalies for maintenance when applied to a field cable-stayed bridge in Korea. In another study, Huang et al. [268] analyzed effective factors, investigating ANNs' effectiveness in predicting bridge deck deterioration. Furthermore, to tackle the challenges of incomplete and imprecise data, Mehrjoo et al. [269] discussed the numerical analyses of an ANN-based method for the intensity estimation of damage in truss bridge joints. Favarelli et al. [270], and Weinsteinet al. [271], Nguyen et al. [272], and Jayasundara [273], employed ANN-based anomaly detection with bridge vibrational data, and Xu et al. [274] developed an ANN-based two-step algorithm for vibration-based damage identification (VBDI) applied on the Crowchild Bridge located in Alberta, Canada. Eftekhar et al. [275] integrated ANNs with proper orthogonal decomposition (POD) and subspace-based damage indicators, respectively. Nguyen et al. [276] leveraged ANNs with residual FRFs and PCA for SHM of a concrete arch beam replica. Tran [277] also provided a thesis focused on the damage detection accuracy of ANN-based methods.

8.2. Support Vector Machines (SVMs)

SVMs, first developed by Cortes and Vapnik [278] in 1995, are a supervised ML method that is used in regression and classification, operating within a high-dimensional space to divide the data into two distinct groups. Using iterative training, this method is used to enhance the performance by fine-tuning the separation between the data points and the hyperplane. SVMs excel in dealing with challenges caused by nonlinear and high-dimensional datasets, even with the limited sampling size. BSHM employs SVMs to distinguish between normal and abnormal conditions based on annotated data. Through this capability, structural irregularities can be identified early [12]. Pan et al. [279] and Alamdari et al. [280] adopted SVMs to distinguish between damaged and undamaged cases and detect anomalies in dynamic response data. Additionally, Pan et al. [281] used SVMs in combination with enhanced feature extraction techniques such as the wavelet transform, the Hilbert–Huang transform (HHT), and the Teager–Huang transform (THT) for detecting damage in cable-stayed bridges. Navamuel et al. [282] used SVMs in an SHM alert system to differentiate between undamaged and damaged scenarios based on modal properties and temperature variations. Kim et al. [283] utilized SVMs for automated detection, recovery, and isolation of faulty data in OMA using wireless sensor networks (WSNs) for cable-stayed bridge damping estimation. Gui et al. [284] proposed optimization algorithm-based SVMs for large-scale structural damage detection, resulting in an improved damage detection method compared to traditional methods, particularly the one optimized using a genetic algorithm. These papers also contributed to the utilization of SVMs in their research [285–288].

8.3. Random Forest (RF)

Breiman, in 1996 [289], proposed a nonparametric and tree-based ensemble method as a random forest using multiple decision tree models. To generate each ensemble member, the RF uses the bagging technique, a method for data collection from multiple training datasets. Bagging randomly chooses samples from the decision tree space in a consistent manner. This method is vastly used for both classification and regression purposes and is among the most frequently utilized ML algorithms [12,290–292]. As a result of its inherent ability to handle enormous datasets while delivering resilient predictions, this algorithm gained considerable attention within the domain of BSHM. In their study, Arnold et al. [293] investigated the application of the RF as an approach for detecting events and classifying time series in bridge monitoring. Lei et al. [294] used RF algorithms to evaluate vibration-based seismic damage for regional bridges. The model demonstrated the prediction performance with over 90% accuracy, identifying critical parameters for seismic design and disaster prevention. Furthermore, these papers also contributed to the utilization of the RF in bridge monitoring [295–297].

8.4. Gaussian Processes (GPs)

As explained by Rasmussen and Williams [298], GPs represent a generalized Gaussian probability distribution. The application of this method aims at data regression and prediction. It has a stochastic process and is classified into a mean and a covariance function, often applied for the definition of a priori distribution in Bayesian inference [299]. Indeed, a stochastic process can be defined as the probability distribution of a set of random variables that are in sync with the input data provided as a group of random variables [300]. Chalouhi et al. [301] used a GP-based ML framework for damage detection by modelling deck accelerations. Moreover, a simplified treed Gaussian process (TGP) was applied by Zhang et al. [302] to address the nonstationary behavior in BSHM using motorcycle accident and Z24 bridge SHM data. Da Silva et al. [303] employed Gaussian process regression (GPR) applied on a bridge under varying temperatures. Furthermore, O'Connor et al. [304] developed a cyber-enabled wireless SHM system using GPR for highway bridges. Moreover, GPs were utilized by Moravej et al. [305] for the calibration of the Bayesian model and updating.

8.5. Convolutional Neural Network (CNN)

The CNN is a widely used deep learning method known for its advantages, including shared weights, local connections, and subsampling, and is a promising approach for data anomaly detection that enables the extraction of valid characteristics from data automatically. Its structure consists of an input layer, a convolutional layer, a pooling layer, and a fully connected layer. The second layer's function is the feature extraction from the input database via a user-defined filter matrix. On the other hand, the pooling layer reduces the spatial data dimensions, and the fully connected layers perform classification tasks. The CNN's performance depends on the balanced training samples. While facing the anomaly patterns in a complex system, the tuning process might be time-consuming and laborious [12,306,307]. Nguyen et al. [308] used a CNN to detect and localize structural damage based on changes in modal curvature, exemplified through the Bo Nghi bridge in Vietnam, as a case study. In another study by Zhang et al. in [306], a CNN was applied to analyze the acceleration data recorded from a bridge, in combination with the feature extraction capabilities of statistical features for anomaly detection and classification. Shajihan et al. [309] employed CNNs to classify SHM data faults from sensors using three-channel input data, achieving accuracy and recall on an unseen dataset. Moreover, Duan et al. [310] applied CNNs for damage identification in hanger cables of a tied-arch bridge. Zou et al. [311] proposed a CNN model by incorporating temporal features from the gated recurrent unit (GRU) model, demonstrating significant improvement in structural damage identification compared to other models. Yessoufou et al. [312] used a one-class convolutional neural network (OC-CNN) model capable of detecting bridge damage across various vehicle weights and speeds. Chamangard et al. [313] utilized compact one-dimensional (1D) CNNs with transfer learning to detect damage accurately, even with limited training data, achieving high accuracy when sufficient data were available. Li et al. [314] combined a CNN with short- and long-term memory neural networks to detect bolt-nut losses in steel bridges. Khodabandehlou et al. [315] used a CNN to predict the predefined damage states (including extent and location) with accuracy using vibration response data from a reinforced concrete highway bridge model. Pamuncak et al. [316] applied a CNN to estimate the structural response in real-world data from the Suramadu bridge monitoring system in Indonesia. The study by Lee et al. [317] also employed a CNN to achieve 87.3% accuracy in real-time damage localization for bridges. Furthermore, Teng et al. [318] proposed a method where a diverse population of bridge structures is created, and a CNN is used to extract damage features.

8.6. Long Short-Term Memory (LSTM) Networks

LSTM networks introduced a new feature over recurrent neural networks (RNNs), named "gated cells", by Hochreiter and Schmidhuberin in 1997 [319]. Three distinct gate mechanisms are incorporated within this method: the input, the output, and the forget gates, which collectively represent an additional control system for signal flow regulation into and out of the model to collect crucial features over time. Consequently, it has the ability to manage information flow via gated cells and is effective in learning complex and nonlinear relationships between factors such as environmental conditions [320–322]. By Shin et al. [323], a C-LSTM network integrated with a CNN and LSTM was applied to identify driving segments on bridges using vibration data, focusing on vibration peaks at bridge joints for damage detection. Moreover, Yue et al. [324] employed an enhanced Stack-LSTM-CNN mode in identifying the abnormal temperature-induced deflections in cable-stayed bridges. Hou et al. [325] proposed a warning framework using LSTM based on the BIM platform for the early detection of hazardous components with bridge monitoring data. Furthermore, Yang et al. [326] employed LSTM networks to model multisensory mapping relationships in SHM, addressing challenges in handling long-term and multidimensional series data. With respect to analyzing the bridge deflection, Guo et al. [327] used LSTM for predicting and comparing this parameter, and Duan et al. [328] proposed a novel approach to reconstruct bridge dynamic displacements using the strain and acceleration data source, overcoming the limitations of direct measurement techniques. Yue et al. [329] used LSTM networks to model a digital regression for temperature-induced deflection in cable-stayed bridge main girders, outperforming linear regression models. Zhao et al. [330] employed LSTM regression networks for in-service bridge monitoring, mapping temperature-induced strains, dynamic displacements, and vehicle-induced strains. In another paper, Sharma et al. [331] assessed the impact of ambient temperature on structures for anomaly detection while using LSTM networks. In the same way, Wang et al. [332] analyzed deflection and temperature data from the Chongqing Egongyan Rail Transit Suspension Bridge, China, using LSTM to detect potential damage signs. To recover the missing structural temperature data from the Nanjing Dashengguan Yangtze River Bridge in China, Liu et al. [333] applied LSTM. Zhao et al. [334] employed LSTM networks for the early detection of cracks in prestressed concrete box girder bridges using live-load strain data. Moreover, while using bidirectional LSTM (BiLSTM), Lu et al. [335] and Yang et al. [336] contributed to enhancing the BSHM.

In conclusion, our exploration of various ML models in the context of bridge structural health monitoring demonstrates a commendable level of precision. The overall low errors, including mean absolute error (MAE), root mean square error (RMSE), mean squared error (MSE), etc., across the reviewed literature, attest to the efficacy of ML approaches in enhancing the accuracy of structural health monitoring. The recommendation for the application of machine learning models in the dynamic monitoring of bridges is further underscored by the general trend of favorable outcomes observed in the surveyed papers. To enhance the reliability and usability of ML models employed in various aspects of BSHM, it is crucial to incorporate a considerable volume of data during the model construction process. Additionally, when developing an ML model for bridges, it is imperative to ensure its applicability across diverse scenarios. This is essential, as instances have been observed where the generated model is exclusively suitable for specific real structure scenarios with particular geometric or physical attributes.

It is also important to note that recent technological advancements and enhancements in various approaches, such as laser scanning (LiDAR) [337], terrestrial laser scanning (TLS) [338–346], unmanned aerial vehicles (UAVs/drones) [347–349], photogrammetry [350], and ground-penetrating radar (GPR) [341], have significantly facilitated the data capturing and establishment of a continuous data monitoring system for bridges. These developments have encouraged researchers and stakeholders to leverage 3D models of bridges and create their digital twins (DTs). These innovative approaches empower us to efficiently assess the structural health of bridges, identify potential issues, and plan maintenance or repairs with minimal disruption. LiDAR, employing laser pulses, generates a highly precise 3D point cloud that represents the bridge's geometry, making it suitable for regional-level assessments. TLS captures detailed 3D images using stationary laser scanners, creating a point cloud through laser beams emitted in horizontal and vertical planes. UAVs, equipped with LiDAR sensors or cameras, provide a flexible and cost-effective solution for monitoring bridges, accessing challenging areas for regular inspections without disrupting traffic. GPR utilizes electromagnetic waves to image and investigate structures such as bridges, and photogrammetry involves capturing bridge images from various angles to create a 3D point cloud, subsequently transformed into a mesh model for comprehensive bridge monitoring and the development of digital twins.

9. Conclusions

In recent years, OMA has increasingly supplanted EMA due to its advantages and fewer implementation restrictions, which include the independence from the input excitation (no external source required), applicability during bridge operation, suitability for monitoring complex structures, cost-effectiveness, real-time monitoring capabilities, and lack of impact on the structural integrity. This article provides a comprehensive review of operational modal analysis applied for bridge health monitoring. It covers various aspects such as instrumentation and data acquisition systems, preprocessing steps including filtering techniques, and modal identification techniques explored in both the time and frequency domains, such as SSI, RD, TDD, ARMA, ERA, NExT, ITD, PP, FDD, EFDD, postprocessing methods, including deviation in the natural frequencies and mode shapes, MSE, the MFM, and the MCM. Additionally, advanced techniques like regression models, pattern recognition, and machine learning methods such as ANN, RF, GP, LSTM, and CNN are discussed to overcome previous limitations. Each approach is showcased, outlining its specific advantages and disadvantages, and practical implementation examples are provided.

From the authors' perspective, it is crucial to recognize and follow certain essential principles to attain accurate and reliable OMA outcomes for BSHM:

- Instrumentation plays a foundational role in OMA. The selection of the sensing system must align precisely with the specific requirements, and the placement and installation of the sensors on the bridge should be performed with attention to the expected results. Achieving accurate modal identification outcomes requires the careful handling of instrumentation.
- When considering wireless networking as the data acquisition system while employing multiple sensors on the bridge, it is crucial to ensure the precise synchronization of data recorded. Even a minor discrepancy in data synchronization can lead to errors in the identified modal parameters, particularly mode shapes. In such cases, the accuracy of bridge damage detection can be compromised.
- In terms of achieving optimal reliability, efficiency, precision, and applicability in various modal identification techniques within the time or frequency domain, the SSI

method stands out as the preferred approach among researchers in the time domain, whereas EFDD excels in the frequency domain.
- The effects of environmental and operational factors on the application of DSFs for bridge damage detection are challenging and inevitable. To increase the accuracy and reliability of the damage detection process, the dominant effects of these variables should be mitigated as much as possible.
- To ensure the precision of the results, two key considerations emerge. First, it is advisable to implement multiple distinct DSF approaches simultaneously, enabling the localization and assessment of damage severity. Second, employing regression and pattern recognition techniques can mitigate the influence of environmental and operational factors on the data.
- Utilizing advanced methods such as machine learning proves precision in achieving OMA results for BSHM. However, when constructing the mathematical models for these methods, it is crucial to incorporate a diverse dataset. This ensures that the resulting model remains applicable across various bridges, accommodating variations in their geometric and material characteristics.

Looking ahead to the future, thanks to the technological advancements and increased sensor accessibility, bridge health monitoring systems will continue to progress. From an instrumentation perspective, wireless monitoring systems will gradually replace their wired counterparts due to advantages such as ease of installation, flexibility of relocation, remote monitoring, and cost-effectiveness. In addition, significant advances in signal processing techniques will improve the quality of input data and optimize data analysis for modal identification. The integration of state-of-the-art methods, especially those involving AI, has the potential to further enhance the performance of modal-based DSFs. In addition, the integration of AI with technologies such as drones, robots, etc., opens up new possibilities for engineers to bridge SHM systems. Despite significant advances in AI, there remains a scientific gap in understanding how environmental and operational variations affect the effectiveness of structural health monitoring using these innovative technologies. While previous studies have provided valuable insights, further research efforts are recommended to investigate this issue more comprehensively in the future.

In addition to the previously discussed points, the authors emphasize two crucial considerations in their perspective. First, while innovative approaches in analytical and scientific methods are proposed and validated through academic endeavors, it is imperative to ensure that the application of these techniques remains practical and falls within the operational capacity of technicians in daily practices. It is not sufficient for these advancements to merely exist in theory; they must be accessible and implementable in real-world scenarios. Secondly, there exists an ongoing necessity to train technicians comprehensively. Their understanding and proficiency in utilizing these novel methods for dynamic identification techniques in civil engineering should be cultivated. This training is essential to bridge the gap between theoretical knowledge and practical application, ensuring that technicians are not only aware of these advancements but are also adept at integrating them into their professional workflows. Consequently, enhancing a well-trained workforce becomes a critical component in the successful adoption and implementation of these cutting-edge techniques within the realm of civil engineering.

Funding: This research received no external funding.

Conflicts of Interest: The authors declare no conflict of interest.

References

1. Pizarro, A.; Manfreda, S.; Tubaldi, E. The Science behind Scour at Bridge Foundations: A Review. *Water* **2020**, *12*, 374. [CrossRef]
2. Malekjafarian, A.; McGetrick, P.J.; OBrien, E.J. A Review of Indirect Bridge Monitoring Using Passing Vehicles. *Shock. Vib.* **2015**, *2015*, 286139. [CrossRef]
3. Sarmadi, H.; Entezami, A.; Yuen, K.-V.; Behkamal, B. Review on Smartphone Sensing Technology for Structural Health Monitoring. *Measurement* **2023**, *223*, 113716. [CrossRef]

4. Fujino, Y.; Siringoringo, D.M. Bridge Monitoring in Japan: The Needs and Strategies. *Struct. Infrastruct. Eng.* **2011**, *7*, 597–611. [CrossRef]
5. Akgul, F. A BMS Development Project with an Integrated Inspection Program. In Proceedings of the 6th International Conference on Bridge Maintenance, Safety and Management (IABMAS), Stresa, Italy, 8–12 July 2012.
6. Miyamoto, A.; Kawamura, K.; Nakamura, H. Bridge Management System and Maintenance Optimization for Existing Bridges. *Comput.-Aided Civ. Infrastruct. Eng.* **2000**, *15*, 45–55. [CrossRef]
7. Miyamoto, A.; Kawamura, K.; Nakamura, H. Development of a Bridge Management System for Existing Bridges. *Adv. Eng. Softw.* **2001**, *32*, 821–833. [CrossRef]
8. Lee, J.; Sanmugarasa, K.; Blumenstein, M.; Loo, Y.-C. Improving the Reliability of a Bridge Management System (BMS) Using an ANN-Based Backward Prediction Model (BPM). *Autom. Constr.* **2008**, *17*, 758–772. [CrossRef]
9. Gholami, M.; Sam, A.R.B.M.; Yatim, J.M. Assessment of Bridge Management System in Iran. *Procedia Eng.* **2013**, *54*, 573–583. [CrossRef]
10. Chupanit, P.; Phromsorn, C. The Importance of Bridge Health Monitoring. *Proc. World Acad. Sci.* **2012**, *6*, 135–138.
11. Rashidi, M.; Mohammadi, M.; Sadeghlou Kivi, S.; Abdolvand, M.M.; Truong-Hong, L.; Samali, B. A Decade of Modern Bridge Monitoring Using Terrestrial Laser Scanning: Review and Future Directions. *Remote Sens.* **2020**, *12*, 3796. [CrossRef]
12. Zinno, R.; Haghshenas, S.S.; Guido, G.; VItale, A. Artificial Intelligence and Structural Health Monitoring of Bridges: A Review of the State-of-the-Art. *IEEE Access* **2022**, *10*, 88058–88078. [CrossRef]
13. Rytter, A.A. Vibration Based Inspection of Civil Engineering Structures. Ph.D. Thesis, University of Aalborg, Aalborg, Denmark, 1993.
14. Brincker, R.; Carlos, E. *Ventura Introduction to Operational Modal Analysis*; John Wiley & Sons: Hoboken, NJ, USA, 2015.
15. Wenzel, H. *Health Monitoring of Bridges*; John Wiley & Sons: Vienna, Austria, 2009; ISBN 9780080542539.
16. Teixeira de Freitas, S.; Kolstein, H.; Bijlaard, F. Structural Monitoring of a Strengthened Orthotropic Steel Bridge Deck Using Strain Data. *Struct. Health Monit.* **2012**, *11*, 558–576. [CrossRef]
17. Vicente, M.; Gonzalez, D.; Minguez, J.; Schumacher, T. A Novel Laser and Video-Based Displacement Transducer to Monitor Bridge Deflections. *Sensors* **2018**, *18*, 970. [CrossRef]
18. Huber, M.S.; Bay, J.A.; Halling, M.W.; Womack, K.C. Use of Velocity Transducers in Low-Frequency Modal Testing of Structures. *Nondestruct. Eval. Highw. Util. Pipelines IV* **2000**, *3995*, 414–424. [CrossRef]
19. Ragam, P.; Sahebraoji, N.D. Application of MEMS-Based Accelerometer Wireless Sensor Systems for Monitoring of Blast-Induced Ground Vibration and Structural Health: A Review. *IET Wirel. Sens. Syst.* **2019**, *9*, 103–109. [CrossRef]
20. Xu, Y.L.; Chen, B.; Ng, C.L.; Wong, K.Y.; Chan, W.Y. Monitoring Temperature Effect on a Long Suspension Bridge. *Struct. Control Health Monit.* **2010**, *17*, 632–653. [CrossRef]
21. Rad, H.N.; Sani, S.S.; Rad, M.N. A New Inexpensive System for SHM of Bridge Decks Using Wireless Sensor Networks Based on Measurements of Temperature and Humidity. In Proceedings of the 2015 2nd International Conference on Knowledge-Based Engineering and Innovation (KBEI), Tehran, Iran, 5–6 November 2015; IEEE: Piscataway, NJ, USA, 2015; pp. 998–1009.
22. Bayraktar, A.; Altunişik, A.C.; Türker, T. Structural Condition Assessment of Birecik Highway Bridge Using Operational Modal Analysis. *Int. J. Civ. Eng.* **2016**, *14*, 35–46. [CrossRef]
23. Zhang, G.; Moutinho, C.; Magalhães, F. Continuous Dynamic Monitoring of a Large-span Arch Bridge with Wireless Nodes Based on MEMS Accelerometers. *Struct. Control Health Monit.* **2022**, *29*, e2963. [CrossRef]
24. Rainieri, C.; Fabbrocino, G. *Operational Modal Analysis of Civil Engineering Structures: An Introduction and Guide for Applications*; Springer: New York, NY, USA, 2014; ISBN 1493907670.
25. Römer, K.; Blum, P.; Meier, L. Time Synchronization and Calibration in Wireless Sensor Networks. In *Handbook of Sensor Networks: Algorithms and Architectures*; Wiley: New York, NY, USA, 2005; pp. 199–237. [CrossRef]
26. Chen, Y.; Xue, X. Advances in the Structural Health Monitoring of Bridges Using Piezoelectric Transducers. *Sensors* **2018**, *18*, 4312. [CrossRef]
27. Ghemari, Z. Study and Analysis of the Piezoresistive Accelerometer Stability and Improvement of Their Performances. *Int. J. Syst. Assur. Eng. Manag.* **2017**, *8*, 1520–1526. [CrossRef]
28. Bedon, C.; Bergamo, E.; Izzi, M.; Noè, S. Prototyping and Validation of MEMS Accelerometers for Structural Health Monitoring—The Case Study of the Pietratagliata Cable-Stayed Bridge. *J. Sens. Actuator Netw.* **2018**, *7*, 30. [CrossRef]
29. Sabato, A. Pedestrian Bridge Vibration Monitoring Using a Wireless MEMS Accelerometer Board. In Proceedings of the 2015 IEEE 19th International Conference on Computer Supported Cooperative Work in Design (CSCWD), Calabria, Italy, 6–8 May 2015; IEEE: Piscataway, NJ, USA, 2015; pp. 437–442.
30. Chen, Y.; Zheng, X.; Luo, Y.; Shen, Y.; Xue, Y.; Fu, W. An Approach for Time Synchronization of Wireless Accelerometer Sensors Using Frequency-Squeezing-Based Operational Modal Analysis. *Sensors* **2022**, *22*, 4784. [CrossRef]
31. Shi, F.; Tuo, X.; Yang, S.; Li, H.; Shi, R. Multiple Two-Way Time Message Exchange (TTME) Time Synchronization for Bridge Monitoring Wireless Sensor Networks. *Sensors* **2017**, *17*, 1027. [CrossRef]
32. Fu, Y.; Mechitov, K.; Hoang, T.; Kim, J.R.; Memon, S.A.; Spencer, B.F. Efficient and High-precision Time Synchronization for Wireless Monitoring of Civil Infrastructure Subjected to Sudden Events. *Struct. Control Health Monit.* **2021**, *28*, e2643. [CrossRef]
33. Maguire, J.; Severn, R. Assessing the Dynamic Properties of Prototype Structures By Hammer Testing. *Proc. Inst. Civ. Eng.* **1987**, *83*, 769–784. [CrossRef]

34. Şahin, A.; Bayraktar, A. Forced-Vibration Testing and Experimental Modal Analysis of a Steel Footbridge for Structural Identification. *J. Test. Eval.* **2014**, *42*, 20130166. [CrossRef]
35. Zwolski, J.; Bien, J. Modal Analysis of Bridge Structures by Means of Forced Vibration Tests. *J. Civ. Eng. Manag.* **2011**, *17*, 590–599. [CrossRef]
36. Chen, G.-W.; Omenzetter, P.; Beskhyroun, S. Operational Modal Analysis of an Eleven-Span Concrete Bridge Subjected to Weak Ambient Excitations. *Eng. Struct.* **2017**, *151*, 839–860. [CrossRef]
37. Kvåle, K.A.; Øiseth, O.; Rønnquist, A. Operational Modal Analysis of an End-Supported Pontoon Bridge. *Eng. Struct.* **2017**, *148*, 410–423. [CrossRef]
38. Brownjohn, J.M.W.; Au, S.-K.; Zhu, Y.; Sun, Z.; Li, B.; Bassitt, J.; Hudson, E.; Sun, H. Bayesian Operational Modal Analysis of Jiangyin Yangtze River Bridge. *Mech. Syst. Signal Process.* **2018**, *110*, 210–230. [CrossRef]
39. Rainieri, C.; Notarangelo, M.A.; Fabbrocino, G. Experiences of Dynamic Identification and Monitoring of Bridges in Serviceability Conditions and after Hazardous Events. *Infrastructures* **2020**, *5*, 86. [CrossRef]
40. Benedettini, F.; Gentile, C. Operational Modal Testing and FE Model Tuning of a Cable-Stayed Bridge. *Eng. Struct.* **2011**, *33*, 2063–2073. [CrossRef]
41. Kasimzade, A.; Tuhta, S.; Aydin, H.; Gunday, F. Determination of Modal Parameters on Steel Model Bridge Using Operational Modal Analysis. In Proceedings of the 8th International Steel Structures Symposium, Konya, Turkey, 24 October 2019.
42. Kudu, F.N.; Bayraktar, A.; Bakir, P.G.; Turker, T.; Altunisik, A.C. Ambient Vibration Testing of Berta Highway Bridge with Post-Tension Tendons. *Steel Compos. Struct.* **2014**, *16*, 21–44. [CrossRef]
43. Ali, A.; Sandhu, T.Y.; Usman, M. Ambient Vibration Testing of a Pedestrian Bridge Using Low-Cost Accelerometers for Shm Applications. *Smart Cities* **2019**, *2*, 20–30. [CrossRef]
44. Salehi, M.; Demirlioglu, K.; Erduran, E. Evaluation of the Effect of Operational Modal Analysis Algorithms on Identified Modal Parameters of Railway Bridges. In Proceedings of the IABSE Congress, Ghent 2021: Structural Engineering for Future Societal Needs, Ghent, Belgium, 22–24 September 2021; pp. 441–449. [CrossRef]
45. Lee, S.; Kalos, N.; Shin, D.H. Non-Destructive Testing Methods in the U.S. for Bridge Inspection and Maintenance. *KSCE J. Civ. Eng.* **2014**, *18*, 1322–1331. [CrossRef]
46. Farhangdoust, S.; Mehrabi, A. Health Monitoring of Closure Joints in Accelerated Bridge Construction: A Review of Non-Destructive Testing Application. *J. Adv. Concr. Technol.* **2019**, *17*, 381–404. [CrossRef]
47. Zahid, F.B.; Ong, Z.C.; Khoo, S.Y. A Review of Operational Modal Analysis Techniques for In-Service Modal Identification. *J. Braz. Soc. Mech. Sci. Eng.* **2020**, *42*, 398. [CrossRef]
48. Sanchez, H.G.; Nova, F.R.; Gonzalez-Estrada, O.A. Implementation of the Operational Modal Analysis Technique in a Power Transmission Shaft. *J. Phys. Conf. Ser.* **2019**, *1247*, 012032. [CrossRef]
49. Zong, Z.H.; Jaishi, B.; Ge, J.P.; Ren, W.X. Dynamic Analysis of a Half-through Concrete-Filled Steel Tubular Arch Bridge. *Eng. Struct.* **2005**, *27*, 3–15. [CrossRef]
50. Bayraktar, A.; Altunişik, A.C.; Birinci, F.; Sevim, B.; Türker, T. Finite-Element Analysis and Vibration Testing of a Two-Span Masonry Arch Bridge. *J. Perform. Constr. Facil.* **2010**, *24*, 46–52. [CrossRef]
51. Gentile, C. Modal and Structural Identification of a R.C. Arch Bridge. *Struct. Eng. Mech.* **2006**, *22*, 53–70. [CrossRef]
52. Chopra, A.K. *Dynamics of Structures: Theory and Applications to Earthquake Engineering*; Earthquake Spectra; Prentice Hall: Hoboken, NJ, USA, 2012; Volume 635–636.
53. Heylen, W.; Lammens, S.; Sas, P. *Modal Analysis Theory and Testing*; Katholieke Universiteit Leuven: Leuven, Belgium, 1995.
54. Zhang, L.; Wang, T.; Tamura, Y. A Frequency–Spatial Domain Decomposition (FSDD) Method for Operational Modal Analysis. *Mech. Syst. Signal Process.* **2010**, *24*, 1227–1239. [CrossRef]
55. Peeters, B.; Ventura, C.E. Comparative Study of Modal Analysis Techniques for Bridge Dynamic Characteristics. *Mech. Syst. Signal Process.* **2003**, *17*, 965–988. [CrossRef]
56. Guo, J.; Jie, C.; Min, H.; Zhu, J.; Qing, Y. Monitoring—Based Evaluation of Dynamic Characteristics of a Long Span Suspension Bridge under Typhoons. *J. Civ. Struct. Health Monit.* **2021**, *11*, 397–410. [CrossRef]
57. Feng, M.Q.; Kim, D.K.; Yi, J.-H.; Chen, Y. Baseline Models for Bridge Performance Monitoring. *J. Eng. Mech.* **2004**, *130*, 562–569. [CrossRef]
58. Bayraktar, A.; Altunişik, A.C.; Sevim, B.; Türker, T.; Domaniç, A.; Taş, Y. Vibration Characteristics of Kömürhan Highway Bridge Constructed with Balanced Cantilever Method. *J. Perform. Constr. Facil.* **2009**, *23*, 90–99. [CrossRef]
59. Jin, S.-S.; Jeong, S.; Sim, S.-H.; Seo, D.-W.; Park, Y.-S. Fully Automated Peak-Picking Method for an Autonomous Stay-Cable Monitoring System in Cable-Stayed Bridges. *Autom. Constr.* **2021**, *126*, 103628. [CrossRef]
60. Zimmerman, A.T.; Swartz, R.A.; Lynch, J.P. Automated Identification of Modal Properties in a Steel Bridge Instrumented with a Dense Wireless Sensor Network. In *Bridge Maintenance, Safety, Management, Health Monitoring and Informatics, Proceedings of the 4th International Conference on Bridge Maintenance, Safety and Management, Seoul, Republic of Korea, 13–17 July 2008*; Taylor & Francis Group: London, UK, 2008; p. 345. [CrossRef]
61. Gentile, C.; Gallino, N. Ambient Vibration Testing and Structural Evaluation of an Historic Suspension Footbridge. *Adv. Eng. Softw.* **2008**, *39*, 356–366. [CrossRef]
62. Ren, W.-X.; Peng, X.-L. Baseline Finite Element Modeling of a Large Span Cable-Stayed Bridge through Field Ambient Vibration Tests. *Comput. Struct.* **2005**, *83*, 536–550. [CrossRef]

63. Ren, W.-X.; Zatar, W.; Harik, I.E. Ambient Vibration-Based Seismic Evaluation of a Continuous Girder Bridge. *Eng. Struct.* **2004**, *26*, 631–640. [CrossRef]
64. Galvín, P.; Domínguez, J. Dynamic Analysis of a Cable-Stayed Deck Steel Arch Bridge. *J. Constr. Steel Res.* **2007**, *63*, 1024–1035. [CrossRef]
65. Chen, G.; Beskhyroun, S. Dynamic Properties of an Eleven-Span Motorway Bridge at Different Levels of Excitation. In Proceedings of the New Zealand Society for Earthquake Engineering Annual Conference, Auckland, New Zealand, 21–23 March 2014; pp. 1–9.
66. Bayraktar, A.; Can Altunişik, A.; Sevim, B.; Türker, T. Ambient Vibration Tests of a Steel Footbridge. *J. Nondestr. Eval.* **2010**, *29*, 14–24. [CrossRef]
67. Wang, L.; Li, B.; Hou, J.; Ou, J. Modal Analysis of Dongying Yellow River Bridge for Long Term Monitoring. In *Sensors and Smart Structures Technologies for Civil, Mechanical, and Aerospace Systems*; Tomizuka, M., Ed.; SPIE: Bellingham, DC, USA, 2011; p. 79811Z.
68. Lei, Y.; Shen, W.A.; Song, Y.; Wang, Y.; Lynch, J.P. Intelligent Sensors with Application to the Identification of Structural Modal Parameters and Steel Cable Forces. In Proceedings of the Earth & Space 2008, Long Beach, CA, USA, 4 September 2008; American Society of Civil Engineers: Reston, VA, USA, 2008; pp. 1–8.
69. Brincker, R.; Zhang, L.; Andersen, P. Modal Identification from Ambient Responses Using Frequency Domain Decomposition. In Proceedings of the 18th International Modal Analysis Conference (IMAC), San Antonio, TX, USA, 7–10 February 2000.
70. Malekjafarian, A.; OBrien, E.J. Identification of Bridge Mode Shapes Using Short Time Frequency Domain Decomposition of the Responses Measured in a Passing Vehicle. *Eng. Struct.* **2014**, *81*, 386–397. [CrossRef]
71. Ahmed, E.; Uddin, N.; OBrien, E. Identifying Localized Bridge Damage Using Frequency Domain Decomposition. In Proceedings of the 26th ASNT Research Symposium, Jacksonville, FL, USA, 13–16 March 2017.
72. Akköse, M.; Gomez, H.C.; Feng, M.Q. Modal Response Identification of a Highway Bridge under Traffic Loads Using Frequency Domain Decomposition (FDD). *Chall. J. Struct. Mech.* **2017**, *3*, 63. [CrossRef]
73. Brincker, R.; Ventura, C.E.; Andersen, P. Damping Estimation by Frequency Domain Decomposition. In Proceedings of the IMAC 19: A Conference on Structural Dynamics, Kissimmee, FL, USA, 5–8 February 2001.
74. Wu, J.; Li, H.; Ye, F.; Ma, K. Damage Identification of Bridge Structure Based on Frequency Domain Decomposition and Strain Mode. *J. Vibroeng.* **2019**, *21*, 2096–2105. [CrossRef]
75. Makki Alamdari, M.; Anaissi, A.; Khoa, N.L.D.; Mustapha, S. Frequency Domain Decomposition-Based Multisensor Data Fusion for Assessment of Progressive Damage in Structures. *Struct. Control Health Monit.* **2019**, *26*, e2299. [CrossRef]
76. OBrien, E.J.; Malekjafarian, A. A Mode Shape-Based Damage Detection Approach Using Laser Measurement from a Vehicle Crossing a Simply Supported Bridge. *Struct. Control Health Monit.* **2016**, *23*, 1273–1286. [CrossRef]
77. Cunha, A.; Caetano, E.; Brincker, R.; Andersen, P. Identification from the Natural Response of Vasco Da Gama Bridge. In Proceedings of the IMAC 22, International Modal Analysis Conference, Dearborn, MI, USA, 26–29 January 2004.
78. Cho, S.; Park, J.; Jung, H.J.; Yun, C.B.; Jang, S.; Jo, H.; Spencer, B.F.; Nagayama, T.; Seo, J.W. Structural Health Monitoring of a Cable-Stayed Bridge Using Acceleration Data via Wireless Smart Sensor Network. In *Bridge Maintenance, Safety, Management and Life-Cycle Optimization, Proceedings of the 5th International Conference on Bridge Maintenance, Safety and Management, Philadelphia, PA, USA, 11–15 July 2010*; CRC Press: Boca Raton, FL, USA, 2010; Volume 6, pp. 158–164.
79. Weng, J.-H.; Loh, C.-H.; Lynch, J.P.; Lu, K.-C.; Lin, P.-Y.; Wang, Y. Output-Only Modal Identification of a Cable-Stayed Bridge Using Wireless Monitoring Systems. *Eng. Struct.* **2008**, *30*, 1820–1830. [CrossRef]
80. Brincker, R.; Andersen, P.; Jacobsen, N.J. Automated Frequency Domain Decomposition for Operational Modal Analysis. In *Conference Proceedings of the Society for Experimental Mechanics Series*; Springer: Cham, Switzerland, 2007.
81. Michel, C.; Keller, S. Determining and Investigating the Variability of Bridges' Natural Frequencies with Ground-Based Radar. *Appl. Sci.* **2022**, *12*, 5354. [CrossRef]
82. Torbol, M. Real-Time Frequency-Domain Decomposition for Structural Health Monitoring Using General-Purpose Graphic Processing Unit. *Comput.-Aided Civ. Infrastruct. Eng.* **2014**, *29*, 689–702. [CrossRef]
83. Brincker, R.; Andersen, P.; Zhang, L. Modal Identification and Damage Detection on a Concrete Highway Bridge by Frequency Domain Decomposition. In Proceedings of the Structural Engineers World Congress, Yokohama, Japan, 9–12 October 2002; pp. 1–8.
84. Yan, B.F.; Miyamoto, B. Wavelet-Aided Frequency Domain Decomposition Method for Dynamic Monitoring of Bridge Structure. *J. Civ. Struct. Health Monit.* **2005**.
85. Xing, S.; Halling, M.W.; Barr, P.J.; Meng, Q. Ambient Vibration Analysis of a Highway Bridge with Subspace Method and Frequency Domain Decomposition. In Proceedings of the 14th World Conference on Earthquake Engineering, Beijing, China, 12–17 October 2008.
86. Elahi, A.R.; Cardoni, A.; Domaneschi, M.; Cimellaro, G.P. Automating the Frequency Domain Decomposition Technique Using the Modal Assurance Criterion. In Proceedings of the 17th World Conference on Seismic Isolation (17WCSI), Turin, Italy, 11–15 September 2022; Springer: Cham, Switzerland, 2023; pp. 1063–1073.
87. Brincker, R.; Zhang, L.; Andersen, P. Modal Identification of Output-Only Systems Using Frequency Domain Decomposition. *Smart Mater. Struct.* **2001**, *10*, 441–445. [CrossRef]
88. Sevim, B.; Bayraktar, A.; Altunişik, A.C.; Adanur, S.; Akköse, M. Modal Parameter Identification of a Prototype Arch Dam Using Enhanced Frequency Domain Decomposition and Stochastic Subspace Identification Techniques. *J. Test. Eval.* **2010**, *38*, 1. [CrossRef]

89. Jacobsen, N.J.; Andersen, P.; Brincker, R. Using Enhanced Frequency Domain Decomposition as a Robust Technique to Harmonic Excitation in Operational Modal Analysis. In Proceedings of the ISMA2006: International Conference on Noise and Vibration Engineering, Lueven, Belgium, 18 September 2006; Volume 6, pp. 3129–3140.
90. Gonen, S.; Soyoz, S. Dynamic Identification of Masonry Arch Bridges Using Multiple Methodologies. In *Conference Proceedings of the Society for Experimental Mechanics Series*; Springer: Cham, Switzerland, 2021; pp. 37–47.
91. AltuniŞik, A.C.; Bayraktar, A.; Sevim, B. Analytical and Experimental Modal Analyses of a Highway Bridge Model. *Comput. Concr.* **2013**, *12*, 803–818. [CrossRef]
92. Ghalishooyan, M.; Shooshtari, A.; Abdelghani, M. Output-Only Modal Identification by in-Operation Modal Appropriation for Use with Enhanced Frequency Domain Decomposition Method. *J. Mech. Sci. Technol.* **2019**, *33*, 3055–3067. [CrossRef]
93. Mohammadi, N.; Nasirshoaibi, M. Modal Parameter Identification of a Three-Storey Structure Using Frequency Domain Techniques FDD and EFDD and Time Domain Technique SSI: Experimental Studies and Simulations. *J. Vibroeng.* **2017**, *19*, 2759–2776. [CrossRef]
94. Altunişik, A.C.; Bayraktar, A.; Sevim, B. *Operational Modal Analysis of a Scaled Bridge Model Using EFDD and SSI Methods*; NISCAIR-CSIR: New Delhi, India, 2012; Volume 19.
95. Kasımzade, A.; Tuhta, S.; Aydın, H.; Günday, F. Investigation of Modal Parameters on Steel Model Bridge Using EFDD Method. In Proceedings of the 2nd International Conference on Technology and Science, Thimphu, Bhutan, 14–16 November 2019.
96. Chu, I.; Amin, M.N.; Kim, J.-K. Prediction Model for the Hydration Properties of Concrete. *Comput. Concr.* **2013**, *12*, 377–392. [CrossRef]
97. Cunha, Á.; Caetano, E.; Magalhães, F.; Moutinho, C. Dynamic Identification and Continuous Dynamic Monitoring of Bridges: Different Applications along Bridges Life Cycle. *Struct. Infrastruct. Eng.* **2018**, *14*, 445–467. [CrossRef]
98. Cole, J.R.H. On-the-Line Analysis of Random Vibrations. In Proceedings of the 9th Structural Dynamics and Materials Conference, Palm Springs, CA, USA, 1–3 April 1968. [CrossRef]
99. Liu, C.; Teng, J.; Liu, J. Improvement of the Decentralized Random Decrement Technique in Wireless Sensor Networks. In Proceedings of the 2014 Word Congress on Advance in Civil, Environmental, and Materials, Busan, Republic of Korea, 24–28 August 2014; pp. 24–28.
100. Vandiver, J.K.; Dunwoody, A.B.; Campbell, R.B.; Cook, M.F. A Mathematical Basis for the Random Decrement Vibration Signature Analysis Technique. *J. Mech. Des.* **1982**, *104*, 307–313. [CrossRef]
101. Morsy, R.; Marzouk, H.; Haddara, M.; Gu, X. Multi-Channel Random Decrement Smart Sensing System for Concrete Bridge Girders Damage Location Identification. *Eng. Struct.* **2017**, *143*, 469–476. [CrossRef]
102. Kordestani, H.; Xiang, Y.-Q.; Ye, X.-W.; Jia, Y.-K. Application of the Random Decrement Technique in Damage Detection under Moving Load. *Appl. Sci.* **2018**, *8*, 753. [CrossRef]
103. He, X.H.; Hua, X.G.; Chen, Z.Q.; Huang, F.L. EMD-Based Random Decrement Technique for Modal Parameter Identification of an Existing Railway Bridge. *Eng. Struct.* **2011**, *33*, 1348–1356. [CrossRef]
104. Wu, W.-H.; Chen, C.-C.; Liau, J.-A. A Multiple Random Decrement Method for Modal Parameter Identification of Stay Cables Based on Ambient Vibration Signals. *Adv. Struct. Eng.* **2012**, *15*, 969–982. [CrossRef]
105. Kaloop, M.R.; Hu, J.W. Dynamic Performance Analysis of the Towers of a Long-Span Bridge Based on GPS Monitoring Technique. *J. Sens.* **2016**, *2016*, 7494817. [CrossRef]
106. Huang, Z.; Li, Y.; Hua, X.; Chen, Z.; Wen, Q. Automatic Identification of Bridge Vortex-Induced Vibration Using Random Decrement Method. *Appl. Sci.* **2019**, *9*, 2049. [CrossRef]
107. Wen, Q.; Hua, X.G.; Chen, Z.Q.; Niu, H.W.; Wang, X.Y. AMD-Based Random Decrement Technique for Modal Identification of Structures with Close Modes. *J. Aerosp. Eng.* **2018**, *31*, 04018057. [CrossRef]
108. Morsy, R.; Marzouk, H.; Gu, X. The Use of Multichannel Random Decrement for Monitoring of Concrete Bridge Girders. In Proceedings of the SMAR 2015: Third Conference on Smart Monitoring, Assessment & Rehabilitation of Civil Structures, Antalya, Turkey, 7–9 September 2015.
109. Morsy, R.; Marzouk, H.; Gu, X.; Elshafey, A. Use of the Random Decrement Technique for Nondestructive Detection of Damage to Beams. *Mater. Struct.* **2016**, *49*, 4719–4727. [CrossRef]
110. Brincker, R.; Olsen, P.; Amador, S.; Juul, M.; Malekjafarian, A.; Ashory, M. Modal Participation in Multiple Input Ibrahim Time Domain Identification. *Math. Mech. Solids* **2019**, *24*, 168–180. [CrossRef]
111. Ibrahim, S.R.; EC, M. A Method for Direct Identification of Vibration Parameters from the Free Response. In *The Shock and Vibration Bulletin*; NASA: Washington, DC, USA, 1977.
112. Ibrahim, S.R. A Time Domain Vibration Test Technique. Ph.D. Thesis, University of Calgary, Calgary, AB, Canada, 1973.
113. Malekjafarian, A.; Brincker, R.; Ashory, M.R.; Khatibi, M.M. Modified Ibrahim Time Domain Method for Identification of Closely Spaced Modes: Experimental Results. In *Conference Proceedings of the Society for Experimental Mechanics Series*; Springer: New York, NY, USA, 2012; pp. 443–449.
114. Huang, C.S.; Yang, Y.B.; Lu, L.Y.; Chen, C.H. Dynamic Testing and System Identification of a Multi-Span Highway Bridge. *Earthq. Eng. Struct. Dyn.* **1999**, *28*, 857–878. [CrossRef]
115. Siringoringo, D.M.; Fujino, Y. System Identification of Suspension Bridge from Ambient Vibration Response. *Eng. Struct.* **2008**, *30*, 462–477. [CrossRef]

116. Liu, C.; Teng, J. Modal Analysis of a Cable-Stayed Bridge Model Using a Modified Ibrahim Time Domain Algorithm. *J. Vibroeng.* **2014**, *16*, 3033–3044.
117. Asmussen, J.C.; Ibrahim, S.R.; Brincker, R. *Random Decrement and Regression Analysis of Traffic Responses of Bridges*; Aalborg University: Aalborg, Denmark, 1995.
118. Fujino, Y.; Abe, M.; Shibuya, H.; Yanagihara, M.; Sato, M.; Nakamura, S.-I.; Sakamoto, Y. Forced and Ambient Vibration Tests and Vibration Monitoring of Hakucho Suspension Bridge. *Transp. Res. Rec. J. Transp. Res. Board* **2000**, *1696*, 57–63. [CrossRef]
119. Wu, W.H.; Liao, C.A.; Chen, C.C. Parameter Identification for Multiple Modes of Cable-Stayed Bridge Cables Using Ambient Vibration Measurements from a Single Station. *WIT Trans. Model. Simul.* **2007**, *46*, 691–700. [CrossRef]
120. Zulkifli, E.; Widarda, D.R. Modal Parameters Identification of Light Rail Bridges. *IOP Conf. Ser. Mater. Sci. Eng.* **2019**, *669*, 012027. [CrossRef]
121. Jiang, H.; Meng, G.W.; Xiong, B.W. Modal Parameters Identification of Bridge Structure Based on Excitation of Wind Load. *Appl. Mech. Mater.* **2013**, *405–408*, 1660–1663. [CrossRef]
122. Wang, Y.-C.; Chen, C.-H. Dynamic Characteristics of a Cable-Stayed Bridge Measured from Traffic-Induced Vibrations. In Proceedings of the International Conference of Numerical Analysis and Applied Mathematics, Kos, Greece, 19–25 September 2012; pp. 2278–2281.
123. Juang, J.-N.; Pappa, R.S. An Eigensystem Realization Algorithm for Modal Parameter Identification and Model Reduction. *J. Guid. Control. Dyn.* **1985**, *8*, 620–627. [CrossRef]
124. Chiang, D.-Y.; Lin, C.-S. Identification of Modal Parameters from Ambient Vibration Data Using Eigensystem Realization Algorithm with Correlation Technique. *J. Mech. Sci. Technol.* **2010**, *24*, 2377–2382. [CrossRef]
125. Qin, Q.; Li, H.B.; Qian, L.Z.; Lau, C.-K. Modal Identification of Tsing Ma Bridge by Using Improved Eigensystem Realization Algorithm. *J. Sound Vib.* **2001**, *247*, 325–341. [CrossRef]
126. Zhang, G.; Ma, J.; Chen, Z.; Wang, R. Automated Eigensystem Realisation Algorithm for Operational Modal Analysis. *J. Sound Vib.* **2014**, *333*, 3550–3563. [CrossRef]
127. Qu, C.-X.; Mei, D.-P.; Yi, T.-H.; Li, H.-N. Spurious Mode Distinguish by Modal Response Contribution Index in Eigensystem Realization Algorithm. *Struct. Des. Tall Spec. Build.* **2018**, *27*, e1491. [CrossRef]
128. Catbas, F.N.; Lenett, M.; Brown, D.L.; Doebling, S.W.; Farrar, C.R.; Turer, A. Modal Analysis of Multi-Reference Impact Test Data for Steel Stringer Bridges. In Proceedings of the International Modal Analysis Conference—IMAC, Orlando, FL, USA, 3–6 February 1997; Volume 1, pp. 381–391.
129. Yang, X.-M.; Yi, T.-H.; Qu, C.-X.; Li, H.-N.; Liu, H. Automated Eigensystem Realization Algorithm for Operational Modal Identification of Bridge Structures. *J. Aerosp. Eng.* **2019**, *32*, 04018148. [CrossRef]
130. Szafrański, M. A Dynamic Vehicle-Bridge Model Based on the Modal Identification Results of an Existing EN57 Train and Bridge Spans with Non-Ballasted Tracks. *Mech. Syst. Signal Process.* **2021**, *146*, 107039. [CrossRef]
131. Lu, W.; Dong, J.; Pan, Y.; Li, G.; Guo, J. Damage Identification of Bridge Structure Model Based on Empirical Mode Decomposition Algorithm and Autoregressive Integrated Moving Average Procedure. *Arch. Civ. Eng.* **2022**, *68*, 653–667. [CrossRef]
132. Rainieri, C.; Fabbrocino, G. Operational Modal Analysis: Overview and Applications. *Strateg. Reduct. Seism. Risk* **2008**, 29–44.
133. Chen, Y.; Feng, M.Q.; Tan, C.-A. Modeling of Traffic Excitation for System Identification of Bridge Structures. *Comput.-Aided Civ. Infrastruct. Eng.* **2006**, *21*, 57–66. [CrossRef]
134. Erdoğan, H.; Gülal, E. The Application of Time Series Analysis to Describe the Dynamic Movements of Suspension Bridges. *Nonlinear Anal. World Appl.* **2009**, *10*, 910–927. [CrossRef]
135. Pakzad, S.N.; Fenves, G.L. Statistical Analysis of Vibration Modes of a Suspension Bridge Using Spatially Dense Wireless Sensor Network. *J. Struct. Eng.* **2009**, *135*, 863–872. [CrossRef]
136. Fang, R.; Jianxi, Y. Application of the ARMA Model to Bridge Structural Health Monitoring. *Intell. Autom. Soft Comput.* **2010**, *16*, 755–762. [CrossRef]
137. Lu, D.G.; Zhang, W.H.; Zhao, Z. Health Monitoring Data Modeling and Reliability Prediction of an Actual Bridge Based on ARMA Model. In *Maintenance, Safety, Risk, Management and Life-Cycle Performance of Bridges, Proceedings of the 9th International Conference on Bridge Maintenance, Safety and Management, IABMAS 2018, Melbourne, Australia, 9–13 July 2018*; CRC Press: Boca Raton, FL, USA, 2018; pp. 2089–2096. [CrossRef]
138. Boonyapinyo, V.; Miyata, T.; Yamada, H. Advanced Aerodynamic Analysis of Suspension Bridges by State-Space Approach. *J. Struct. Eng.* **1999**, *125*, 1357–1366. [CrossRef]
139. Wang, W.; Zhang, Y.; Ouyang, H. Influence of Random Multi-Point Seismic Excitations on the Safety Performance of a Train Running on a Long-Span Bridge. *Int. J. Struct. Stab. Dyn.* **2020**, *20*, 2050054. [CrossRef]
140. Xiong, C.; Lu, H.; Zhu, J. Operational Modal Analysis of Bridge Structures with Data from GNSS/Accelerometer Measurements. *Sensors* **2017**, *17*, 436. [CrossRef]
141. Vu, V.H.; Thomas, M.; Lakis, A.A.; Marcouiller, L. Identification of Modal Parameters by Operational Modal Analysis for the Assessment of Bridge Rehabilitation. In Proceedings of the 2nd International Operational Modal Analysis Conference, IOMAC 2007, Copenhagen, Denmark, 30 April–2 May 2007.
142. Kim, B.H.; Stubbs, N.; Park, T. A New Method to Extract Modal Parameters Using Output-Only Responses. *J. Sound Vib.* **2005**, *282*, 215–230. [CrossRef]

143. Kim, B.H.; Lee, J.; Lee, D.H. Extracting Modal Parameters of High-Speed Railway Bridge Using the TDD Technique. *Mech. Syst. Signal Process.* **2010**, *24*, 707–720. [CrossRef]
144. Daniotti, N.; Cheynet, E.; Jakobsen, J.B.; Snæbjörnsson, J. Damping Estimation from Full-Scale Traffic-Induced Vibrations of a Suspension Bridge. In Proceedings of the ASCE International Conference on Computing in Civil Engineering, Atlanta, GA, USA, 17–19 June 2019.
145. Salokhe, N.; Thakre, P.; Awale, R.N.; Kambale, S. Vibration Based Damage Detection Using Overall Frequency Response and Time Domain Decomposition. In Proceedings of the 2016 IEEE International Conference on Engineering and Technology (ICETECH), Coimbatore, India, 17–18 March 2016; IEEE: Piscataway, NJ, USA, 2016; pp. 991–994.
146. Park, J.; Park, C.; Kang, H.; Jo, B. Dynamic Characteristics of Seohae Cable-Stayed Bridge Based on Results from Long-Term Structural Health Monitoring. In Proceedings of the IABSE Symposium: Improving Infrastructure Worldwide, Weimar, Germany, 19–21 September 2007; pp. 212–215.
147. Huang, Q.; Gardoni, P.; Hurlebaus, S. Updating Structural Properties Using Modal Parameters Considering Measurement Errors. In Proceedings of the Structures Congress 2009, Austin, TX, USA, 29 April 2009; American Society of Civil Engineers: Reston, VA, USA, 2009; pp. 1–9.
148. De Moor, B.; Van Overschee, P.; Suykens, J. Subspace Algorithms for System Identification and Stochastic Realization. In Proceedings of the MTNS, Kobe, Japan, 17 June 1991.
149. Van Overschee, P.; De Moor, B. *Subspace Identification for Linear Systems: Theory, Implementation and Applications*; Kluwer Academic Publishers: Alphen aan den Rijn, The Netherlands, 1996.
150. Peeters, B. System Identification and Damage Detection in Civil Engineering. Ph.D. Thesis, Katholieke Universiteit Leuven, Leuven, Belgium, 2000.
151. Maddipour Farrokhifard, M.; Hatami, M.; Venkatasubramanian, V.M. Performance of Stochastic Subspace Identification Methods in Presence of Forced Oscillations. In Proceedings of the 2019 International Conference on Smart Grid Synchronized Measurements and Analytics (SGSMA), College Station, TX, USA, 21–23 May 2019; IEEE: Piscataway, NJ, USA, 2019; pp. 1–8.
152. Jin, N.; Yang, Y.B.; Dimitrakopoulos, E.G.; Paraskeva, T.S.; Katafygiotis, L.S. Application of Short-Time Stochastic Subspace Identification to Estimate Bridge Frequencies from a Traversing Vehicle. *Eng. Struct.* **2021**, *230*, 111688. [CrossRef]
153. Yang, Y.B.; Chen, W.-F. Extraction of Bridge Frequencies from a Moving Test Vehicle by Stochastic Subspace Identification. *J. Bridge Eng.* **2016**, *21*, 04015053. [CrossRef]
154. Boonyapinyo, V.; Janesupasaeree, T. Data-Driven Stochastic Subspace Identification of Flutter Derivatives of Bridge Decks. *J. Wind. Eng. Ind. Aerodyn.* **2010**, *98*, 784–799. [CrossRef]
155. Brownjohn, J.M.W.; Magalhaes, F.; Caetano, E.; Cunha, A. Ambient Vibration Re-Testing and Operational Modal Analysis of the Humber Bridge. *Eng. Struct.* **2010**, *32*, 2003–2018. [CrossRef]
156. Li, J.; Zhu, X.; Law, S.; Samali, B. Indirect Bridge Modal Parameters Identification with One Stationary and One Moving Sensors and Stochastic Subspace Identification. *J. Sound Vib.* **2019**, *446*, 1–21. [CrossRef]
157. Zhou, Y.; Jiang, X.; Zhang, M.; Zhang, J.; Sun, H.; Li, X. Modal Parameters Identification of Bridge by Improved Stochastic Subspace Identification Method with Grubbs Criterion. *Meas. Control (U. K.)* **2021**, *54*, 457–464. [CrossRef]
158. Loh, C.-H.; Chen, M.-C.; Chao, S.-H. Stochastic Subspace Identification for Operational Modal Analysis of an Arch Bridge. In *Sensors and Smart Structures Technologies for Civil, Mechanical, and Aerospace Systems 2012*; Tomizuka, M., Yun, C.-B., Lynch, J.P., Eds.; SPIE: Bellingham, DC, USA, 2012; p. 834504.
159. Wu, W.H.; Wang, S.W.; Chen, C.C.; Lai, G. Mode Identifiability of a Cable-Stayed Bridge under Different Excitation Conditions Assessed with an Improved Algorithm Based on Stochastic Subspace Identification. *Smart Struct. Syst.* **2016**, *17*, 363–389. [CrossRef]
160. Tran, T.T.X.; Ozer, E. Synergistic Bridge Modal Analysis Using Frequency Domain Decomposition, Observer Kalman Filter Identification, Stochastic Subspace Identification, System Realization Using Information Matrix, and Autoregressive Exogenous Model. *Mech. Syst. Signal Process.* **2021**, *160*, 107818. [CrossRef]
161. Kvåle, K.A.; Øiseth, O.; Rönnquist, A. Covariance-Driven Stochastic Subspace Identification of an End-Supported Pontoon Bridge Under Varying Environmental Conditions. In *Conference Proceedings of the Society for Experimental Mechanics Series*; Springer: Cham, Switzerland, 2017; pp. 107–115.
162. Wu, W.-H.; Wang, S.-W.; Chen, C.-C.; Lai, G. Application of Stochastic Subspace Identification for Stay Cables with an Alternative Stabilization Diagram and Hierarchical Sifting Process. *Struct. Control Health Monit.* **2016**, *23*, 1194–1213. [CrossRef]
163. Zhou, X.; Cao, L.; Khan, I.; Li, Q. Data Inspecting and Denoising Method for Data-Driven Stochastic Subspace Identification. *Shock. Vib.* **2018**, *2018*, 3926817. [CrossRef]
164. Bayraktar, A.; Türker, T.; Altunişik, A.C. Experimental Frequencies and Damping Ratios for Historical Masonry Arch Bridges. *Constr. Build. Mater.* **2015**, *75*, 234–241. [CrossRef]
165. Fan, J.; Zhang, Z.; Hua, H. Data Processing in Subspace Identification and Modal Parameter Identification of an Arch Bridge. *Mech. Syst. Signal Process.* **2007**, *21*, 1674–1689. [CrossRef]
166. Ren, W.; Zhao, T.; Harik, I.E.; Asce, M. Experimental and Analytical Modal Analysis of Steel Arch Bridge. *J. Struct. Eng.* **2004**, *130*, 9445. [CrossRef]
167. Loh, C.H.; Chen, M.C. Modeling of Environmental Effects for Vibration-Based SHM Using Recursive Stochastic Subspace Identification Analysis. *Key Eng. Mater.* **2013**, *558*, 52–64. [CrossRef]

168. Hsu, W.-T.; Loh, C.-H.; Chao, S.-H. Uncertainty Calculation for Modal Parameters Used with Stochastic Subspace Identification: An Application to a Bridge Structure. In *Sensors and Smart Structures Technologies for Civil, Mechanical, and Aerospace Systems*; Lynch, J.P., Ed.; SPIE: Bellingham, DC, USA, 2015; p. 94350Q.
169. Altunişik, A.C.; Bayraktar, A.; Sevim, B.; Özdemir, H. Experimental and Analytical System Identification of Eynel Arch Type Steel Highway Bridge. *J. Constr. Steel Res.* **2011**, *67*, 1912–1921. [CrossRef]
170. Döhler, M.; Lam, X.-B.; Mevel, L. Uncertainty Quantification for Modal Parameters from Stochastic Subspace Identification on Multi-Setup Measurements. *Mech. Syst. Signal Process.* **2013**, *36*, 562–581. [CrossRef]
171. Yang, X.-M.; Yi, T.-H.; Qu, C.-X.; Li, H.-N.; Liu, H. Performance Assessment of a High-Speed Railway Bridge through Operational Modal Analysis. *J. Perform. Constr. Facil.* **2021**, *35*, 04021091. [CrossRef]
172. Can, A.; Bayraktar, A.; Sevim, B. Output-Only System Identification of Posttensioned Segmental Concrete Highway Bridges. *J. Bridge Eng.* **2011**, 259–266. [CrossRef]
173. James, G.H., III; Carne, T.G.; Lauffer, J.P. *The Natural Excitation Technique (NExT) for Modal Parameter Extraction from Operating Wind Turbines*; Technical Report; Sandia National Labs.: Albuquerque, NM, USA, 1993.
174. Farrar, C.R.; James III, G.H. System Identification from Ambient Vibration Measurements on a Bridge. *J. Sound Vib.* **1997**, *205*, 1–18. [CrossRef]
175. Yang, X.-M.; Yi, T.-H.; Qu, C.-X.; Li, H.-N. Modal Identification of Bridges Using Asynchronous Responses through an Enhanced Natural Excitation Technique. *J. Eng. Mech.* **2021**, *147*, 04021106. [CrossRef]
176. Kim, S.; Kim, H.-K. Damping Identification of Bridges Under Nonstationary Ambient Vibration. *Engineering* **2017**, *3*, 839–844. [CrossRef]
177. Pan, C.; Ye, X.; Mei, L. Improved Automatic Operational Modal Analysis Method and Application to Large-Scale Bridges. *J. Bridge Eng.* **2021**, *26*, 04021051. [CrossRef]
178. Kim, S.; Park, J.; Kim, H.-K. Damping Identification and Serviceability Assessment of a Cable-Stayed Bridge Based on Operational Monitoring Data. *J. Bridge Eng.* **2017**, *22*, 04016123. [CrossRef]
179. Kim, S.; Kim, H.K.; You, Y.C. Enhanced Damping Estimation for Cable-Stayed Bridges Based on Operational Monitoring Data. *Struct. Eng. Int.* **2018**, *28*, 308–317. [CrossRef]
180. Heo, G.H.; Lee, G.; Lee, D.G.; Lee, C.O.; Kim, M.G. An Analysis of the Characteristics of SPG Bridges Using NExT and ERA for Real-Time Monitoring. *Key Eng. Mater.* **2004**, *270–273*, 2012–2017. [CrossRef]
181. Catbas, F.N.; Gul, M.; Burkett, J.L. Conceptual Damage-Sensitive Features for Structural Health Monitoring: Laboratory and Field Demonstrations. *Mech. Syst. Signal Process.* **2008**, *22*, 1650–1669. [CrossRef]
182. Moughty, J.J.; Casas, J.R. A State of the Art Review of Modal-Based Damage Detection in Bridges: Development, Challenges, and Solutions. *Appl. Sci.* **2017**, *7*, 510. [CrossRef]
183. Adams, R.D.; Walton, D.; Flitcroft, J.E.; Short, D. Vibration Testing as a Nondestructive Test Tool for Composite Materials. In *Composite Reliability, Philadelphia: American Society for Testing and Materials*; ASTM International: West Conshohocken, PA, USA, 1975.
184. Cawley, P.; Adams, R.D. The Location of Defects in Structures from Measurements of Natural Frequencies. *J. Strain Anal. Eng. Des.* **1979**, *14*, 49–57. [CrossRef]
185. Salawu, O.S. Detection of Structural Damage through Changes in Frequency: A Review. *Eng. Struct.* **1997**, *19*, 718–723. [CrossRef]
186. Limongelli, M.P.; Siegert, D.; Merliot, E.; Waeytens, J.; Bourquin, F.; Vidal, R.; Le Corvec, V.; Gueguen, I.; Cottineau, L.M. Damage Detection in a Post Tensioned Concrete Beam—Experimental Investigation. *Eng. Struct.* **2016**, *128*, 15–25. [CrossRef]
187. Bonopera, M.; Chang, K.; Tullini, N. Vibration of Prestressed Beams: Experimental and Finite-Element Analysis of Post–Tensioned Thin-Walled Box-Girders. *J. Constr. Steel Res.* **2023**, *205*, 107854. [CrossRef]
188. Messina, A.; Williams, E.J.; Contursi, T. Structural Damage Detection by a Sensitivity and Statistical-Based Method. *J. Sound Vib.* **1998**, *216*, 791–808. [CrossRef]
189. KIM, J.-T.; STUBBS, N. Crack Detection in Beam-Type Structures Using Frequency Data. *J. Sound Vib.* **2003**, *259*, 145–160. [CrossRef]
190. Lee, Y.-S.; Chung, M.-J. A Study on Crack Detection Using Eigenfrequency Test Data. *Comput. Struct.* **2000**, *77*, 327–342. [CrossRef]
191. Casas, J.R.; Aparicio, A.C. Structural Damage Identification from Dynamic-Test Data. *J. Struct. Eng.* **1994**, *120*, 2437–2450. [CrossRef]
192. Pau, A.; Greco, A.; Vestroni, F. Numerical and Experimental Detection of Concentrated Damage in a Parabolic Arch by Measured Frequency Variations. *J. Vib. Control* **2011**, *17*, 605–614. [CrossRef]
193. Montazer, M.; Seyedpoor, S.M. A New Flexibility Based Damage Index for Damage Detection of Truss Structures. *Shock. Vib.* **2014**, *2014*, 460692. [CrossRef]
194. Deraemaeker, A.; Reynders, E.; De Roeck, G.; Kullaa, J. Vibration-Based Structural Health Monitoring Using Output-Only Measurements under Changing Environment. *Mech. Syst. Signal Process.* **2008**, *22*, 34–56. [CrossRef]
195. Erazo, K.; Sen, D.; Nagarajaiah, S.; Sun, L. Vibration-Based Structural Health Monitoring under Changing Environmental Conditions Using Kalman Filtering. *Mech. Syst. Signal Process.* **2019**, *117*, 1–15. [CrossRef]
196. Deraemaeker, A.; Worden, K. A Comparison of Linear Approaches to Filter out Environmental Effects in Structural Health Monitoring. *Mech. Syst. Signal Process.* **2018**, *105*, 1–15. [CrossRef]

197. Allemang, R.J.; Brown, D.L. A Correlation Coefficient for Modal Vector Analysis. In Proceedings of the 1st SEM International Modal Analysis Conference, Orlando, FL, USA, 8–10 November 1982.
198. Allemang, R.J. The Modal Assurance Criterion—Twenty Years of Use and Abuse. *Sound Vib.* **2003**, *37*, 14–21.
199. Greś, S.; Döhler, M.; Mevel, L. Uncertainty Quantification of the Modal Assurance Criterion in Operational Modal Analysis. *Mech. Syst. Signal Process.* **2021**, *152*, 107457. [CrossRef]
200. Kim, J.H.; Jeon, H.S.; Lee, C.W. Application of the Modal Assurance Criteria for Detecting and Locating Structural Faults. In Proceedings of the 10th International Modal Analysis Conference, San Diego, CA, USA, 3–7 February 1992.
201. Salawu, O.S.; Williams, C. Bridge Assessment Using Forced-Vibration Testing. *J. Struct. Eng.* **1995**, *121*, 161–173. [CrossRef]
202. Salgado, R.; Cruz, P.J.S.; Ramos, L.F.; Lourenco, P.B. *Comparison between Damage Detection Methods Applied to Beam Structures*; IABMAS'06; Taylor and Francis: Abingdon, UK, 2006.
203. Balsamo, L.; Mukhopadhyay, S.; Betti, R.; Lus, H. Damage Detection Using Flexibility Proportional Coordinate Modal Assurance Criterion. In *Topics in Modal Analysis, Volume 7, Proceedings of the 31st IMAC, A Conference on Structural Dynamics*; Springer: New York, NY, USA, 2014; pp. 1–8.
204. Zhao, J.; Zhang, L. Structural Damage Identification Based on the Modal Data Change. *Int. J. Eng. Manuf.* **2012**, *2*, 59–66. [CrossRef]
205. Hunt, D.L. Application of an Enhanced Coordinate Modal Assurance Criterion. In Proceedings of the 10th International Modal Analysis Conference, San Diego, CA, USA, 3–7 February 1992.
206. Ho, V.L. Hybrid Computational Intelligence Approach for Defect Identification in Vietnamese Structures. Ph.D. Thesis, Ghent University, Ghent, Belgium, 2022.
207. Pandey, A.K.; Biswas, M.; Samman, M.M. Damage Detection from Changes in Curvature Mode Shapes. *J. Sound Vib.* **1991**, *145*, 321–332. [CrossRef]
208. Ho, Y.K.; Ewins, D.J. Structural Damage Identification with Mode Shapes. In Proceedings of the European COST F3 Conference on System Identification & Structural Health Monitoring, Madrid, Spain, 6–9 June 2000.
209. Farrar, C.R.; Worden, K. *Structural Health Monitoring—A Machine Learning Perspective*; John Wiley & Sons: Hoboken, NJ, USA, 2013.
210. Hamming, R. *Digital Filters*; Courier Corporation: North Chelmsford, MA, USA, 1989.
211. Sazonov, E.; Klinkhachorn, P. Optimal Spatial Sampling Interval for Damage Detection by Curvature or Strain Energy Mode Shapes. *J. Sound Vib.* **2005**, *285*, 783–801. [CrossRef]
212. Chance, J.; Tomlinson, G.R.; Worden, K. A Simplified Approach to the Numerical and Experimental Modelling of the Dynamics of a Cracked Beam. In Proceedings of the 12th International Modal Conference, Honolulu, HI, USA, 31 January–3 February 1994.
213. Capecchi, D.; Ciambella, J.; Pau, A.; Vestroni, F. Damage Identification in a Parabolic Arch by Means of Natural Frequencies, Modal Shapes and Curvatures. *Meccanica* **2016**, *51*, 2847–2859. [CrossRef]
214. Wahab, M.A.; De Roeck, G. Damage Detection in Bridges Using Modal Curvatures: Application to a Real Damage Scenario. *J. Sound Vib.* **1999**, *226*, 217–235. [CrossRef]
215. Stubbs, N.; Kim, J.T.; Topole, K. An Efficient and Robust Algorithm for Damage Localization in Offshore Platforms. In Proceedings of the ASCE 10th Structures Congress, San Antonio, TX, USA, 13–15 April 1992.
216. Kim, J.-T.; Ryu, Y.-S.; Cho, H.-M.; Stubbs, N. Damage Identification in Beam-Type Structures: Frequency-Based Method vs Mode-Shape-Based Method. *Eng. Struct.* **2003**, *25*, 57–67. [CrossRef]
217. Yam, L.Y.; Leung, T.P.; Li, D.B.; Xue, K.Z. Theoretical and Experimental Study of Modal Strain Analysis. *J. Sound Vib.* **1996**, *191*, 251–260. [CrossRef]
218. Pandey, A.K.; Biswas, M. Damage Detection in Structures Using Changes in Flexibility. *J. Sound Vib.* **1994**, *169*, 3–17. [CrossRef]
219. Toksoy, T.; Aktan, A.E. Bridge-Condition Assessment by Modal Flexibility. *Exp. Mech.* **1994**, *34*, 271–278. [CrossRef]
220. Wang, J.Y.; Ko, J.M.; Ni, Y.Q. Modal Sensitivity Analysis of Tsing Ma Bridge for Structural Damage Detection. In Proceedings of the SPIE's 5th Annual International Symposium on Nondestructive Evaluation and Health Monitoring of Aging Infrastructure, International Society for Optics and Photonics, Newport Beach, CA, USA, 6–8 March 2000.
221. Zhang, Z.; Aktan, A.E. The Damage Indices for Constructed Facilities. In Proceedings of the International Modal Analysis Conference, Nashville, Tennessee, 13–16 February 1995.
222. Lu, Q.; Ren, G.; Zhao, Y. Multiple Damage Location with Flexibility Curvature and Relative Frequency Change for Beam Structures. *J. Sound Vib.* **2002**, *253*, 1101–1114. [CrossRef]
223. Talebinejad, I.; Fischer, C.; Ansari, F. Numerical Evaluation of Vibration-Based Methods for Damage Assessment of Cable-Stayed Bridges. *Comput.-Aided Civ. Infrastruct. Eng.* **2011**, *26*, 239–251. [CrossRef]
224. Ndambi, J.-M.; Vantomme, J.; Harri, K. Damage Assessment in Reinforced Concrete Beams Using Eigenfrequencies and Mode Shape Derivatives. *Eng. Struct.* **2002**, *24*, 501–515. [CrossRef]
225. Cruz, P.J.S.; Salgado, R. Performance of Vibration-Based Damage Detection Methods in Bridges. *Comput.-Aided Civ. Infrastruct. Eng.* **2009**, *24*, 62–79. [CrossRef]
226. Wei, F.; Qiao, P. Vibration-Based Damage Identification Methods: A Review and Comparative Study. *Struct. Health Monit.* **2011**, *10*, 83–111. [CrossRef]
227. Peeters, B.; De Roeck, G. One-Year Monitoring of the Z24-Bridge: Environmental Effects versus Damage Events. *Earthq. Eng. Struct. Dyn.* **2001**, *30*, 149–171. [CrossRef]

228. Askegaard, V.; Mossing, P. *Long Term Observation of Rc-Bridge Using Changes in Natural Frequency*; Nordic Concrete Federation: Helsinki, Finland, 1988.
229. Cornwell, P.; Farrar, C.R.; Doebling, S.W.; Sohn, H. Environmental Variability of Modal Properties. *Exp. Tech.* **1999**, *23*, 45–48. [CrossRef]
230. Giraldo, D.F.; Dyke, S.J.; Caicedo, J.M. Damage Detection Accommodating Varying Environmental Conditions. *Struct. Health Monit.* **2006**, *5*, 155–172. [CrossRef]
231. Xia, Y.; Hao, H.; Zanardo, G.; Deeks, A. Long Term Vibration Monitoring of an RC Slab: Temperature and Humidity Effect. *Eng. Struct.* **2006**, *28*, 441–452. [CrossRef]
232. Farrar, C.R.; Baker, W.E.; Bell, T.M.; Cone, K.M.; Darling, T.W.; Duffey, T.A.; Eklund, A.; Migliori, A. *Dynamic Characterization and Damage Detection in the I-40 Bridge over the Rio Grande*; Los Alamos National Lab: Los Alamos, NM, USA, 1994.
233. Soyoz, S.; Feng, M.Q. Long-Term Monitoring and Identification of Bridge Structural Parameters. *Comput.-Aided Civ. Infrastruct. Eng.* **2009**, *24*, 82–92. [CrossRef]
234. Cross, E.J.; Koo, K.Y.; Brownjohn, J.M.W.; Worden, K. Long-Term Monitoring and Data Analysis of the Tamar Bridge. *Mech. Syst. Signal Process.* **2013**, *35*, 16–34. [CrossRef]
235. Kim, C.Y.; Jung, D.S.; Kim, N.S.; Yoon, J.G. Effect of Vehicle Mass on the Measured Dynamic Characteristics of Bridges from Traffic-Induced Test. In Proceedings of the IMAC XIX—19th International Modal Analysis, Kissimmee, FL, USA, 5–8 February 2001.
236. Kim, C.W.; Kawatani, M.; Hao, J. Modal Parameter Identification of Short Span Bridges under a Moving Vehicle by Means of Multivariate AR Model. *Struct. Infrastruct. Eng.* **2012**, *8*, 459–472. [CrossRef]
237. Farreras-Alcover, I.; Chryssanthopoulos, M.K.; Andersen, J.E. Regression Models for Structural Health Monitoring of Welded Bridge Joints Based on Temperature, Traffic and Strain Measurements. *Struct. Health Monit.* **2015**, *14*, 648–662. [CrossRef]
238. Song, Y.; Ding, Y. Fatigue Monitoring and Analysis of Orthotropic Steel Deck Considering Traffic Volume and Ambient Temperature. *Sci. China Technol. Sci.* **2013**, *56*, 1758–1766. [CrossRef]
239. Sohn, H. Effects of Environmental and Operational Variability on Structural Health Monitoring. *Philos. Trans. R. Soc. A Math. Phys. Eng. Sci.* **2007**, *365*, 539–560. [CrossRef] [PubMed]
240. Xia, Y.; Chen, B.; Weng, S.; Ni, Y.-Q.; Xu, Y.-L. Temperature Effect on Vibration Properties of Civil Structures: A Literature Review and Case Studies. *J. Civ. Struct. Health Monit.* **2012**, *2*, 29–46. [CrossRef]
241. Magalhães, F.; Cunha, A.; Caetano, E. Vibration Based Structural Health Monitoring of an Arch Bridge: From Automated OMA to Damage Detection. *Mech. Syst. Signal Process.* **2012**, *28*, 212–228. [CrossRef]
242. Peeters, B.; Maeck, J.; Roeck, G. De Vibration-Based Damage Detection in Civil Engineering: Excitation Sources and Temperature Effects. *Smart Mater. Struct.* **2001**, *10*, 518–527. [CrossRef]
243. Liu, C.; DeWolf, J.T.; Kim, J.-H. Development of a Baseline for Structural Health Monitoring for a Curved Post-Tensioned Concrete Box–Girder Bridge. *Eng. Struct.* **2009**, *31*, 3107–3115. [CrossRef]
244. Sohn, H.; Dzwonczyk, M.; Straser, E.G.; Kiremidjian, A.S.; Law, K.H.; Meng, T. An Experimental Study of Temperature Effect on Modal Parameters of the Alamosa Canyon Bridge. *Earthq. Eng. Struct. Dyn.* **1999**, *28*, 879–897. [CrossRef]
245. Ding, Y.; Li, A. Temperature-Induced Variations of Measured Modal Frequencies of Steel Box Girder for a Long-Span Suspension Bridge. *Int. J. Steel Struct.* **2011**, *11*, 145–155. [CrossRef]
246. Hassan, R.; Hekmati Athar, S.P.; Taheri, M.; Cesmeci, S.; Taheri, H. Regression Model for Structural Health Monitoring of a Lab Scaled Bridge. In *NDE 4.0 and Smart Structures for Industry, Smart Cities, Communication, and Energy, Proceedings of the SPIE Smart Structures + Nondestructive Evaluation, Online, 21 March 2021*; Meyendorf, N.G., Niezrecki, C., Farhangdoust, S., Eds.; SPIE: Bellingham, DC, USA, 2021; p. 18.
247. Mangalathu, S.; Jeon, J.-S.; DesRoches, R. Critical Uncertainty Parameters Influencing Seismic Performance of Bridges Using Lasso Regression. *Earthq. Eng. Struct. Dyn.* **2018**, *47*, 784–801. [CrossRef]
248. Kim, C.W.; Morita, T.; Oshima, Y.; Sugiura, K. A Bayesian Approach for Vibration-Based Long-Term Bridge Monitoring to Consider Environmental and Operational Changes. *Smart Struct. Syst.* **2015**, *15*, 395–408. [CrossRef]
249. Laory, I.; Trinh, T.N.; Smith, I.F.C.; Brownjohn, J.M.W. Methodologies for Predicting Natural Frequency Variation of a Suspension Bridge. *Eng. Struct.* **2014**, *80*, 211–221. [CrossRef]
250. Entezami, A.; Sarmadi, H.; Behkamal, B.; Mariani, S. Big Data Analytics and Structural Health Monitoring: A Statistical Pattern Recognition-Based Approach. *Sensors* **2020**, *20*, 2328. [CrossRef]
251. Figueiredo, E.; Brownjohn, J. Three Decades of Statistical Pattern Recognition Paradigm for SHM of Bridges. *Struct. Health Monit.* **2022**, *21*, 3018–3054. [CrossRef]
252. Gao, Z.; Li, F.; Yu, X.; Yan, J.; Chen, Z. Research and Application of Pattern Recognition LSTM Based Bridge Data Anomaly Detection. In Proceedings of the 2022 China Automation Congress (CAC), Xiamen, China, 25 November 2022; IEEE: Piscataway, NJ, USA; pp. 4630–4635.
253. Farrar, C.R.; Worden, K. An Introduction to Structural Health Monitoring. *Philos. Trans. R. Soc. A Math. Phys. Eng. Sci.* **2007**, *365*, 303–315. [CrossRef]
254. Hu, W.-H.; Tang, D.-H.; Teng, J.; Said, S.; Rohrmann, R.G. Structural Health Monitoring of a Prestressed Concrete Bridge Based on Statistical Pattern Recognition of Continuous Dynamic Measurements over 14 Years. *Sensors* **2018**, *18*, 4117. [CrossRef]

255. Haritos, N.; Owen, J.S. The Use of Vibration Data for Damage Detection in Bridges: A Comparison of System Identification and Pattern Recognition Approaches. *Struct. Health Monit.* **2004**, *3*, 141–163. [CrossRef]
256. Datteo, A.; Lucà, F.; Busca, G. Statistical Pattern Recognition Approach for Long-Time Monitoring of the G.Meazza Stadium by Means of AR Models and PCA. *Eng. Struct.* **2017**, *153*, 317–333. [CrossRef]
257. Cheung, A.; Cabrera, C.; Sarabandi, P.; Nair, K.K.; Kiremidjian, A.; Wenzel, H. The Application of Statistical Pattern Recognition Methods for Damage Detection to Field Data. *Smart Mater. Struct.* **2008**, *17*, 065023. [CrossRef]
258. Bao, Y.; Chen, Z.; Wei, S.; Xu, Y.; Tang, Z.; Li, H. The State of the Art of Data Science and Engineering in Structural Health Monitoring. *Engineering* **2019**, *5*, 234–242. [CrossRef]
259. Malekloo, A.; Ozer, E.; Alhamaydeh, M.; Girolami, M. Machine Learning and Structural Health Monitoring Overview with Emerging Technology and High-Dimensional Data Source Highlights. *Struct. Health Monit.* **2022**, *21*, 1906–1955. [CrossRef]
260. Hu, X.; Shentu, J.; Xie, N.; Huang, Y.; Lei, G.; Hu, H.; Guo, P.; Gong, X. Predicting Triaxial Compressive Strength of High-Temperature Treated Rock Using Machine Learning Techniques. *J. Rock Mech. Geotech. Eng.* **2023**, *15*, 2072–2082. [CrossRef]
261. Wedel, F.; Marx, S. Application of Machine Learning Methods on Real Bridge Monitoring Data. *Eng. Struct.* **2022**, *250*, 113365. [CrossRef]
262. Naderpour, H.; Rafiean, A.H.; Fakharian, P. Compressive Strength Prediction of Environmentally Friendly Concrete Using Artificial Neural Networks. *J. Build. Eng.* **2018**, *16*, 213–219. [CrossRef]
263. Rezazadeh Eidgahee, D.; Jahangir, H.; Solatifar, N.; Fakharian, P.; Rezaeemanesh, M. Data-Driven Estimation Models of Asphalt Mixtures Dynamic Modulus Using ANN, GP and Combinatorial GMDH Approaches. *Neural Comput. Appl.* **2022**, *34*, 17289–17314. [CrossRef]
264. Rizzo, F.; Caracoglia, L. Artificial Neural Network Model to Predict the Flutter Velocity of Suspension Bridges. *Comput. Struct.* **2020**, *233*, 106236. [CrossRef]
265. Naseri Nasab, M.; Jahangir, H.; Hasani, H.; Majidi, M.-H.; Khorashadizadeh, S. Estimating the Punching Shear Capacities of Concrete Slabs Reinforced by Steel and FRP Rebars with ANN-Based GUI Toolbox. *Structures* **2023**, *50*, 1204–1221. [CrossRef]
266. Farhangi, V.; Jahangir, H.; Eidgahee, D.R.; Karimipour, A.; Javan, S.A.N.; Hasani, H.; Fasihihour, N.; Karakouzian, M. Behaviour Investigation of SMA-Equipped Bar Hysteretic Dampers Using Machine Learning Techniques. *Appl. Sci.* **2021**, *11*, 10057. [CrossRef]
267. Kwon, T.H.; Park, S.H.; Park, S.I.; Lee, S.-H. Building Information Modeling-Based Bridge Health Monitoring for Anomaly Detection under Complex Loading Conditions Using Artificial Neural Networks. *J. Civ. Struct. Health Monit.* **2022**, *12*, 483. [CrossRef]
268. Huang, Y.-H. Artificial Neural Network Model of Bridge Deterioration. *J. Perform. Constr. Facil.* **2010**, *24*, 597–602. [CrossRef]
269. Mehrjoo, M.; Khaji, N.; Moharrami, H.; Bahreininejad, A. Damage Detection of Truss Bridge Joints Using Artificial Neural Networks. *Expert Syst. Appl.* **2008**, *35*, 1122–1131. [CrossRef]
270. Favarelli, E.; Giorgetti, A. Machine Learning for Automatic Processing of Modal Analysis in Damage Detection of Bridges. *IEEE Trans. Instrum. Meas.* **2021**, *70*, 1–13. [CrossRef]
271. Weinstein, J.C.; Sanayei, M.; Brenner, B.R. Bridge Damage Identification Using Artificial Neural Networks. *J. Bridge Eng.* **2018**, *23*, 04018084. [CrossRef]
272. Nguyen, V.V. Damage Identification of a Bridge Component Using Model Updating and Artificial Neural Network Techniques Based on Frequency Response Functions. Ph.D. Thesis, University of Technology Sydney, Ultimo, Australia, 2018.
273. Jayasundara, W.K.N. Damage Detection of Arch Bridges Using Vibration Characteristics and Artificial Neural Network. Ph.D. Thesis, Queensland University of Technology, Brisbane, Australia, 2020.
274. Xu, H.; Humar, J. Damage Detection in a Girder Bridge by Artificial Neural Network Technique. *Comput.-Aided Civ. Infrastruct. Eng.* **2006**, *21*, 450–464. [CrossRef]
275. Eftekhar Azam, S.; Rageh, A.; Linzell, D. Damage Detection in Structural Systems Utilizing Artificial Neural Networks and Proper Orthogonal Decomposition. *Struct. Control Health Monit.* **2019**, *26*, e2288. [CrossRef]
276. Nguyen, V.V.; Dackermann, U.; Li, J.; Alamdari, M.M.; Mustapha, S.; Runcie, P.; Ye, L. Damage Identification of a Concrete Arch Beam Based on Frequency Response Functions and Artificial Neural Networks. *Electron. J. Struct. Eng.* **2015**, *14*, 75–84. [CrossRef]
277. Tran, N.H. Structural Health Monitoring for Bridges Using Metaheuristic Optimization Algorithms Combined with Artificial Neural Network. Ph.D. Thesis, Ghent University, Ghent, Belgium, 2022.
278. Cortes, C.; Vapnik, V. Support-Vector Networks. *Mach. Learn.* **1995**, *20*, 273–297. [CrossRef]
279. Pan, H.; Azimi, M.; Gui, G.; Yan, F.; Lin, Z. Vibration-Based Support Vector Machine for Structural Health Monitoring. In *Experimental Vibration Analysis for Civil Structures*; Springer: Cham, Switzerland, 2018; pp. 167–178.
280. Alamdari, M.M.; Dang Khoa, N.L.; Wang, Y.; Samali, B.; Zhu, X. A Multi-Way Data Analysis Approach for Structural Health Monitoring of a Cable-Stayed Bridge. *Struct. Health Monit.* **2019**, *18*, 35–48. [CrossRef]
281. Pan, H.; Azimi, M.; Yan, F.; Lin, Z. Time-Frequency-Based Data-Driven Structural Diagnosis and Damage Detection for Cable-Stayed Bridges. *J. Bridge Eng.* **2018**, *23*, 04018033. [CrossRef]
282. Fernández-Navamuel, A.; Zamora-Sánchez, D.; Varona-Poncela, T.; Jiménez-Fernández, C.; Díez-Hernández, J.; García-Sánchez, D.; Pardo, D. Vibration-Based SHM Strategy for a Real Time Alert System with Damage Location and Quantification. In *European Workshop on Structural Health Monitoring*; Springer: Cham, Switzerland, 2021; pp. 245–255.

283. Kim, S.; Spencer, B.F.; Kim, H.K. Automated Long-Term Damping Estimation of the Cable-Stayed Bridge Using Faulty Data in Wireless Sensor Network. In *Bridge Maintenance, Safety, Management, Life-Cycle Sustainability and Innovations*; CRC Press: Boca Raton, FL, USA, 2021.
284. Gui, G.; Pan, H.; Lin, Z.; Li, Y.; Yuan, Z. Data-Driven Support Vector Machine with Optimization Techniques for Structural Health Monitoring and Damage Detection. *KSCE J. Civ. Eng.* **2017**, *21*, 523–534. [CrossRef]
285. Tang, H.; Tang, G.; Meng, L. Prediction of the Bridge Monitoring Data Based on Support Vector Machine. In Proceedings of the 2015 11th International Conference on Natural Computation (ICNC), Zhangjiajie, China, 15–17 August 2015; IEEE: Piscataway, NJ, USA, 2015; pp. 781–785.
286. Feng, C.-W.; Huang, H.-Y. Hybridization of Finite-Element Method and Support Vector Machine to Determine Scour Bridge Safety Level. *J. Perform. Constr. Facil.* **2015**, *29*, 04014079. [CrossRef]
287. Bulut, A.; Singh, A.K.; Shin, P.; Fountain, T.; Jasso, H.; Yan, L.; Elgamal, A. Real-Time Nondestructive Structural Health Monitoring Using Support Vector Machines and Wavelets. *Adv. Sens. Technol. Nondestruct. Eval. Struct. Health Monit.* **2005**, *5770*, 180. [CrossRef]
288. Li, X.; Guo, Y.; Li, Y. Particle Swarm Optimization-Based SVM for Classification of Cable Surface Defects of the Cable-Stayed Bridges. *IEEE Access* **2020**, *8*, 44485–44492. [CrossRef]
289. Breiman, L. Bagging Predictors. *Mach. Learn.* **1996**, *24*, 123–140. [CrossRef]
290. Zhou, J.; Shi, X.; Du, K.; Qiu, X.; Li, X.; Mitri, H.S. Feasibility of Random-Forest Approach for Prediction of Ground Settlements Induced by the Construction of a Shield-Driven Tunnel. *Int. J. Geomech.* **2017**, *17*, 04016129. [CrossRef]
291. Zhou, J.; Li, E.; Wei, H.; Li, C.; Qiao, Q.; Armaghani, D.J. Random Forests and Cubist Algorithms for Predicting Shear Strengths of Rockfill Materials. *Appl. Sci.* **2019**, *9*, 1621. [CrossRef]
292. Zhou, J.; Asteris, P.G.; Armaghani, D.J.; Pham, B.T. Prediction of Ground Vibration Induced by Blasting Operations through the Use of the Bayesian Network and Random Forest Models. *Soil Dyn. Earthq. Eng.* **2020**, *139*, 106390. [CrossRef]
293. Arnold, M.; Hoyer, M.; Keller, S. Convolutional Neural Networks for Detecting Bridge Crossing Events with Ground-Based Interferometric Radar Data. *ISPRS Ann. Photogramm. Remote Sens. Spat. Inf. Sci.* **2021**, *5*, 31–38. [CrossRef]
294. Lei, X.; Sun, L.; Xia, Y.; He, T. Vibration-Based Seismic Damage States Evaluation for Regional Concrete Beam Bridges Using Random Forest Method. *Sustainability* **2020**, *12*, 5106. [CrossRef]
295. Zhang, Y.; Xiong, Z.; Liang, Z.; She, J.; Ma, C. Structural Damage Identification System Suitable for Old Arch Bridge in Rural Regions: Random Forest Approach. *Comput. Model. Eng. Sci.* **2023**, *136*, 447–469. [CrossRef]
296. Wang, T.; Reiffsteck, P.; Chevalier, C.; Chen, C.W.; Schmidt, F. Application of Random Forest Algorithm in Bridge Scour Risk Prediction. In Proceedings of the Machine Learning & Risk Assessment in Geoengineering, Wrocław, Poland, 25–27 October 2021.
297. Jia, H.; Lin, J.; Liu, J. Bridge Seismic Damage Assessment Model Applying Artificial Neural Networks and the Random Forest Algorithm. *Adv. Civ. Eng.* **2020**, *2020*, 6548682. [CrossRef]
298. Bousquet, O.; von Luxburg, U.; Rätsch, G. (Eds.) *Advanced Lectures on Machine Learning*; Lecture Notes in Computer Science; Springer: Berlin/Heidelberg, Germany, 2004; Volume 3176, ISBN 978-3-540-23122-6.
299. Rasmussen, C.E. Gaussian Processes in Machine Learning. In *Summer School on Machine Learning*; Springer: Berlin/Heidelberg, Germany, 2004; pp. 63–71.
300. Asso, R.; Cucuzza, R.; Rosso, M.M.; Masera, D.; Marano, G.C. Bridges Monitoring: An Application of AI with Gaussian Processes. In Proceedings of the 14th International Conference on Evolutionary and Deterministic Methods for Design, Optimization and Control, Athens, Greece, 28–30 June 2021; pp. 245–261. [CrossRef]
301. Chalouhi, E.K.; Gonzalez, I.; Gentile, C.; Karoumi, R. Damage Detection in Railway Bridges Using Machine Learning: Application to a Historic Structure. *Procedia Eng.* **2017**, *199*, 1931–1936. [CrossRef]
302. Zhang, T.; Barthorpe, R.; Worden, K. A Simplified Treed Gaussian Process Approach to the Modelling of Bridge Data for Structural Health Monitoring. In *Proceedings of the Structural Health Monitoring 2017*; DEStech Publications, Inc.: Lancaster, PA, USA, 2017.
303. da Silva, S.; Figueiredo, E.; Moldovan, I. Damage Detection Approach for Bridges under Temperature Effects Using Gaussian Process Regression Trained with Hybrid Data. *J. Bridge Eng.* **2022**, *27*, 04022107. [CrossRef]
304. O'Connor, S.M.; Zhang, Y.; Lynch, J.; Ettouney, M.; van der Linden, G. Automated Analysis of Long-Term Bridge Behavior and Health Using a Cyber-Enabled Wireless Monitoring System. In *Nondestructive Characterization for Composite Materials, Aerospace Engineering, Civil Infrastructure, and Homeland Security 2014*; Wu, H.F., Yu, T.-Y., Gyekenyesi, A.L., Shull, P.J., Eds.; SPIE: Bellingham, DC, USA, 2014; p. 90630Y.
305. Moravej, H.; Chan, T.H.; Nguyen, K.-D.; Jesus, A. Vibration-Based Bayesian Model Updating of Civil Engineering Structures Applying Gaussian Process Metamodel. *Adv. Struct. Eng.* **2019**, *22*, 3487–3502. [CrossRef]
306. Zhang, H.; Lin, J.; Hua, J.; Gao, F.; Tong, T. Data Anomaly Detection for Bridge SHM Based on CNN Combined with Statistic Features. *J. Nondestr. Eval.* **2022**, *41*, 28. [CrossRef]
307. Gomez-Cabrera, A.; Escamilla-Ambrosio, P.J. Review of Machine-Learning Techniques Applied to Structural Health Monitoring Systems for Building and Bridge Structures. *Appl. Sci.* **2022**, *12*, 10754. [CrossRef]
308. Nguyen, D.H.; Nguyen, Q.B.; Bui-Tien, T.; De Roeck, G.; Abdel Wahab, M. Damage Detection in Girder Bridges Using Modal Curvatures Gapped Smoothing Method and Convolutional Neural Network: Application to Bo Nghi Bridge. *Theor. Appl. Fract. Mech.* **2020**, *109*, 102728. [CrossRef]

309. Shajihan, S.A.V.; Wang, S.; Zhai, G.; Spencer, B.F. CNN Based Data Anomaly Detection Using Multi-Channel Imagery for Structural Health Monitoring. *Smart Struct. Syst.* **2022**, *29*, 181–193. [CrossRef]
310. Duan, Y.; Chen, Q.; Zhang, H.; Yun, C.B.; Wu, S.; Zhu, Q. CNN-Based Damage Identification Method of Tied-Arch Bridge Using Spatial-Spectral Information. *Smart Struct. Syst.* **2019**, *23*, 507–520. [CrossRef]
311. Zou, J.Z.; Yang, J.X.; Wang, G.P.; Tang, Y.L.; Yu, C.S. Bridge Structural Damage Identification Based on Parallel CNN-GRU. *IOP Conf. Ser. Earth Environ. Sci.* **2021**, *626*, 012017. [CrossRef]
312. Yessoufou, F.; Zhu, J. One-Class Convolutional Neural Network (OC-CNN) Model for Rapid Bridge Damage Detection Using Bridge Response Data. *KSCE J. Civ. Eng.* **2023**, *27*, 1640–1660. [CrossRef]
313. Chamangard, M.; Ghodrati Amiri, G.; Darvishan, E.; Rastin, Z. Transfer Learning for CNN-Based Damage Detection in Civil Structures with Insufficient Data. *Shock. Vib.* **2022**, *2022*, 3635116. [CrossRef]
314. Li, Z.-J.; Adamu, K.; Yan, K.; Xu, X.-L.; Shao, P.; Li, X.-H.; Bashir, H.M. Detection of Nut–Bolt Loss in Steel Bridges Using Deep Learning Techniques. *Sustainability* **2022**, *14*, 10837. [CrossRef]
315. Khodabandehlou, H.; Pekcan, G.; Fadali, M.S. Vibration-based Structural Condition Assessment Using Convolution Neural Networks. *Struct. Control Health Monit.* **2018**, *26*, e2308. [CrossRef]
316. Pamuncak, A.P.; Salami, M.R.; Adha, A.; Budiono, B.; Laory, I. Estimation of Structural Response Using Convolutional Neural Network: Application to the Suramadu Bridge. *Eng. Comput. Swans.* **2021**, *38*, 4047–4065. [CrossRef]
317. Lee, K.; Byun, N.; Shin, D.H. A Damage Localization Approach for Rahmen Bridge Based on Convolutional Neural Network. *KSCE J. Civ. Eng.* **2020**, *24*, 1–9. [CrossRef]
318. Teng, S.; Chen, X.; Chen, G.; Cheng, L.; Bassir, D. Structural Damage Detection Based on Convolutional Neural Networks and Population of Bridges. *Measurement* **2022**, *202*, 111747. [CrossRef]
319. Hochreiter, S.; Schmidhuber, J. Long Short-Term Memory. *Neural Comput.* **1997**, *9*, 1735–1780. [CrossRef] [PubMed]
320. Li, S.; Li, S.; Laima, S.; Li, H. Data-driven Modeling of Bridge Buffeting in the Time Domain Using Long Short-term Memory Network Based on Structural Health Monitoring. *Struct. Control Health Monit.* **2021**, *28*, e2772. [CrossRef]
321. Gers, F.A.; Schmidhuber, J.; Cummins, F. Learning to Forget: Continual Prediction with LSTM. *Neural Comput.* **2000**, *12*, 2451–2471. [CrossRef]
322. Xiao, S.; Qiao, S.; Chen, H.; Liu, X.; Li, Y.; Tang, M.; Hu, H.; Li, G. Data Prediction Model Based on LSTM Neural Network in Bridge Health Monitoring System. *IOP Conf. Ser. Earth Environ. Sci.* **2021**, *668*, 012068. [CrossRef]
323. Shin, R.; Okada, Y.; Yamamoto, K. Application of C-LSTM Networks to Automatic Labeling of Vehicle Dynamic Response Data for Bridges. *Sensors* **2022**, *22*, 3486. [CrossRef]
324. Yue, Z.; Ding, Y.; Zhao, H.; Wang, Z. Ultra-High Precise Stack-LSTM-CNN Model of Temperature-Induced Deflection of a Cable-Stayed Bridge for Detecting Bridge State Driven by Monitoring Data. *Structures* **2022**, *45*, 110–125. [CrossRef]
325. Hou, G.; Li, L.; Xu, Z.; Chen, Q.; Liu, Y.; Qiu, B. A BIM-Based Visual Warning Management System for Structural Health Monitoring Integrated with LSTM Network. *KSCE J. Civ. Eng.* **2021**, *25*, 2779–2793. [CrossRef]
326. Yang, K.; Ding, Y.; Geng, F.; Jiang, H.; Zou, Z. A Multi-Sensor Mapping Bi-LSTM Model of Bridge Monitoring Data Based on Spatial-Temporal Attention Mechanism. *Measurement* **2023**, *217*, 113053. [CrossRef]
327. Guo, A.; Jiang, A.; Lin, J.; Li, X. Data Mining Algorithms for Bridge Health Monitoring: Kohonen Clustering and LSTM Prediction Approaches. *J. Supercomput.* **2020**, *76*, 932–947. [CrossRef]
328. Duan, D.-Y.; Wang, Z.-C.; Sun, X.-T.; Xin, Y. A Data Fusion Method for Bridge Displacement Reconstruction Based on LSTM Networks. *Smart Struct. Syst.* **2022**, *29*, 599.
329. Yue, Z.; Ding, Y.; Zhao, H.; Wang, Z. Mechanics-Guided Optimization of an LSTM Network for Real-Time Modeling of Temperature-Induced Deflection of a Cable-Stayed Bridge. *Eng. Struct.* **2022**, *252*, 113619. [CrossRef]
330. Zhao, H.; Ding, Y.; Li, A.; Sheng, W.; Geng, F. Digital Modeling on the Nonlinear Mapping between Multi-source Monitoring Data of In-service Bridges. *Struct. Control Health Monit.* **2020**, *27*, e2618. [CrossRef]
331. Sharma, S.; Sen, S. Real-Time Structural Damage Assessment Using LSTM Networks: Regression and Classification Approaches. *Neural Comput. Appl.* **2023**, *35*, 557–572. [CrossRef]
332. Wang, C.; Ansari, F.; Wu, B.; Li, S.; Morgese, M.; Zhou, J. LSTM Approach for Condition Assessment of Suspension Bridges Based on Time-Series Deflection and Temperature Data. *Adv. Struct. Eng.* **2022**, *25*, 3450–3463. [CrossRef]
333. Liu, H.; Ding, Y.; Zhao, H.; Wang, M.; Geng, F. Deep Learning-Based Recovery Method for Missing Structural Temperature Data Using LSTM Network. *Struct. Monit. Maint.* **2020**, *7*, 109–124.
334. Zhao, H.; Ding, Y.; Li, A.; Ren, Z.; Yang, K. Live-Load Strain Evaluation of the Prestressed Concrete Box-Girder Bridge Using Deep Learning and Clustering. *Struct Health Monit.* **2020**, *19*, 1051–1063. [CrossRef]
335. Lu, Y.; Tang, L.; Chen, C.; Zhou, L.; Liu, Z.; Liu, Y.; Jiang, Z.; Yang, B. Reconstruction of Structural Long-Term Acceleration Response Based on BiLSTM Networks. *Eng. Struct.* **2023**, *285*, 116000. [CrossRef]
336. Yang, K.; Ding, Y.; Jiang, H.; Zhao, H.; Luo, G. A Two-stage Data Cleansing Method for Bridge Global Positioning System Monitoring Data Based on Bi-direction Long and Short Term Memory Anomaly Identification and Conditional Generative Adversarial Networks Data Repair. *Struct. Control Health Monit.* **2022**, *29*, e2993. [CrossRef]
337. Spadavecchia, C.; Belcore, E.; Di Pietra, V. Preliminary Test on Structural Elements Health Monitoring with a LiDAR-Based Approach. *Int. Arch. Photogramm. Remote Sens. Spat. Inf. Sci.* **2023**, *XLVIII-2/W*, 247–253. [CrossRef]

338. Cha, G.; Park, S.; Oh, T. A Terrestrial LiDAR-Based Detection of Shape Deformation for Maintenance of Bridge Structures. *J. Constr. Eng. Manag.* **2019**, *145*, 04019075. [CrossRef]
339. Lõhmus, H.; Ellmann, A.; Märdla, S.; Idnurm, S. Terrestrial Laser Scanning for the Monitoring of Bridge Load Tests—Two Case Studies. *Surv. Rev.* **2018**, *50*, 270–284. [CrossRef]
340. Truong-Hong, L.; Laefer, D.F. Using Terrestrial Laser Scanning for Dynamic Bridge Deflection Measurement. *Angew. Chem. Int. Ed.* **2018**, *3*, 10–27.
341. Cafiso, S.; Di Graziano, A.; Goulias, D.; Mangiameli, M.; Mussumeci, G. Bridge Monitoring Combining Laser Scanning and Ground Penetrating Radar. In Proceedings of the International Conference of Numerical Analysis and Applied Mathematics (ICNAAM 2018), Rhodes, Greece, 13–18 September 2018; AIP Publishing: Melville, NY, USA, 2019; p. 280004.
342. Xi, C.X.; Zhou, Z.; Xiang, X.; He, S.; Hou, X. Monitoring of Long-Span Bridge Deformation Based on 3D Laser Scanning. *Instrum. Mes. Métrol.* **2018**, *18*, 113–130. [CrossRef]
343. Minehane, M.J.; O'Donovan, R.; Ruane, K.D.; O'Keeffe, B. The Use of 3D Laser Scanning Technology for Bridge Inspection and Assessment. *Struct. Health Monit.* **2014**, *13*, 14.
344. Zhang, H.; Zhu, Y.; Xiong, W.; Cai, C.S. Point Cloud Registration Methods for Long-Span Bridge Spatial Deformation Monitoring Using Terrestrial Laser Scanning. *Struct. Control Health Monit.* **2023**, *2023*, 2629418. [CrossRef]
345. Neitzel, F.; Niemeier, W.; Weisbrich, S.; Lehmann, M. Investigation of Low-Cost Accelerometer, Terrestrial Laser Scanner and Ground-Based Radar Interferometer for Vibration Monitoring of Bridges. In Proceedings of the 6th European Workshop—Structural Health Monitoring, EWSHM 2012, Dresden, Germany, 3–6 July 2012; Volume 1, pp. 542–551.
346. Psomas, L.; Trainor, R.; Alzraiee, H. Case Study: Assessing the Structural Condition of Steel Bridges Using Terrestrial Laser Scanner (TLS). In *Construction Research Congress 2022*; ASCE: Reston, VA, USA, 2022.
347. Mader, D.; Blaskow, R.; Westfeld, P.; Weller, C. Potential of Uav-Based Laser Scanner and Multispectral Camera Data in Building Inspection. *Int. Arch. Photogramm. Remote Sens. Spat. Inf. Sci.* **2016**, *XLI-B1*, 1135–1142. [CrossRef]
348. Mohammadi, M.; Rashidi, M.; Mousavi, V.; Karami, A.; Yu, Y.; Samali, B. Quality Evaluation of Digital Twins Generated Based on UAV Photogrammetry and TLS: Bridge Case Study. *Remote Sens.* **2021**, *13*, 3499. [CrossRef]
349. Zollini, S.; Alicandro, M.; Dominici, D.; Quaresima, R.; Giallonardo, M. UAV Photogrammetry for Concrete Bridge Inspection Using Object-Based Image Analysis (OBIA). *Remote Sens.* **2020**, *12*, 3180. [CrossRef]
350. Jiang, R.; Jáuregui, D.V.; White, K.R. Close-Range Photogrammetry Applications in Bridge Measurement: Literature Review. *Measurement* **2008**, *41*, 823–834. [CrossRef]

infrastructures

Article

The "*M* and *P*" Technique for Damage Identification in Reinforced Concrete Bridges

Athanasios Bakalis [1,*], Triantafyllos Makarios [1] and Vassilis Lekidis [2]

[1] Institute of Structural Analysis and Dynamics of Structures, School of Civil Engineering, Aristotle University of Thessaloniki, GR-54124 Thessaloniki, Greece; makariostr@civil.auth.gr
[2] Institute of Engineering Seismology and Earthquake Engineering (ITSAK-OASP), Dasiliou Street, Pylaia, GR-55535 Thessaloniki, Greece; lekidis@itsak.gr
* Correspondence: abakalis@civil.auth.gr

Abstract: The seismic damage in reinforced concrete bridges is identified in this study using the "*M* and *P*" hybrid technique initially developed for planar frames, where *M* signifies "Monitoring" and *P* denotes "Pushover analysis". The proposed methodology involves a series of pushover and instantaneous modal analyses with a progressively increasing target deck displacement along the longitudinal direction of the bridge. From the results of these analyses, the diagram of the instantaneous eigenfrequency of the bridge, ranging from the health state to near collapse, is plotted against the inelastic seismic deck displacement. By pre-determining the eigenfrequency of an existing bridge along its longitudinal direction through "monitoring and frequency identification", the target deck displacement corresponding to the damage state can directly be found from this diagram. Subsequently, the damage can be identified by examining the results of the pushover analysis at the step where the target deck displacement is indicated. The effectiveness of this proposed technique is evaluated in the context of straight multiple span bridges with unequal pier heights, illustrated through an example of a four-span bridge. The findings demonstrate that the damage potential in bridge piers can be successfully identified by combining the results of a monitoring process and pushover analysis.

Keywords: reinforced concrete bridges; damage identification; instantaneous eigenfrequency diagram; pushover capacity curve; seismic target deck displacement; bridge plastic mechanism

Citation: Bakalis, A.; Makarios, T.; Lekidis, V. The "*M* and *P*" Technique for Damage Identification in Reinforced Concrete Bridges. *Infrastructures* **2024**, *9*, 18. https://doi.org/10.3390/infrastructures9020018

Academic Editors: Seyed Saman Khedmatgozar Dolati and Armin Mehrabi

Received: 26 December 2023
Revised: 15 January 2024
Accepted: 23 January 2024
Published: 25 January 2024

1. Introduction

Bridges play a crucial role in infrastructure, emphasizing the need for health monitoring processes to extend their lifespan and ensure safety in the face of environmental factors like earthquakes. Detecting damage in reinforced concrete (RC) structures is a key aspect of these processes, with a primary focus on monitoring alterations in dynamic characteristics to assess the health status. Beyond damage identification, this process plays a pivotal role in establishing dependable structural models. These models, in turn, serve as a foundation for conducting advanced nonlinear analyses, providing insights into the intrinsic seismic capacity of bridges.

The identification of eigenfrequency (and mode shape) along the longitudinal axis of an existing reinforced concrete (RC) bridge can be achieved through instrumental monitoring. This involves the installation of a local multichannel network system of accelerometers to gauge acceleration forces in the longitudinal direction. Notably, fiber optic sensors have gained prominence in recent years for measuring various parameters, including natural frequencies, accelerations, stresses, and strains. The subsequent analytical processing of the recorded response necessitates the application of various stochastic and deterministic procedures. Several techniques developed in the past can be employed for this purpose, such as: (a) the "frequency domain decomposition" technique, which is used in "operational modal analysis" [1–4]; details on ambient vibration monitoring can be found in the book authored by Wenzel and Pichler [5]; (b) the "stochastic sub-space identification" techniques in which the measured responses directly fit to the parametric models; details can be found in the books authored by Overschee and De Moor [6]. Three distinct algorithms are employed in stochastic subspace techniques: principal component, canonical variate analysis algorithms, and the unweighted principal component; in all cases, random data analysis and operational modal analysis constitute the primary fields of investigation for the recorded accelerograms [7–9]; (c) the "modal time-histories method" [10], which is based on the aforementioned techniques; this method is well suited for structures exposed to earthquake ground excitation or structures experiencing significant wind pressure. Using the "modal time-histories method", eigenfrequencies, mode shapes, and modal damping ratios have been calculated within the linear domain for a variety structures [11]; (d) the "minimum rank perturbation theory" (MRPT), as proposed by Zimmerman and Kaouk [12,13], interprets a non-zero entry in the damage vector as an indicator of the damage location; (e) a technique developed by Domaneschi et al. [14,15], which involves utilizing the discontinuity of mode shape forms; (f) the concept of the damage stiffness matrix is explored in notable works, including those by Peeters [3], Amani et al. [16], and Zhang et al. [17]; (g) techniques that integrate structural health monitoring with pushover analysis are employed for the detection of damage in both individual structural elements [18] and frame structures [19]; (h) several artificial neural network techniques that were developed by Nazari and Baghalian [20] for simple symmetric beams. It is noteworthy to mention the recent research contributions of Reuland et al. [21], who conducted an extensive review of data-driven damage indicators for rapid seismic structural health monitoring. Additionally, Martakis et al. [22] explored the integration of traditional structural health monitoring techniques with innovative machine learning tools, offering a comprehensive perspective. Moreover, [23,24] provide an extensive review of available methods and case studies related to damage identification in bridge structures. Also, details regarding the damage of bearing devices and the abutment–backfill system, along with their impact on bridge behavior, can be found in [25,26]. These works focus on the estimation of seismic fragility curves for common bridges, offering a framework for efficient risk assessment and valuable information on nonlinear modeling techniques and analyses of bridges with various geometrical forms. Some other interesting works offer insights into bridge damage under environmental and operational conditions [27] or to stiffness loss due to heavy vehicle vibrations [28], combining machine learning tools and numerical simulations. In another study [29], the interpretation of dynamic tests on a single-span concrete bridge is presented to explain the observed trend in natural frequencies and is justified by a finite element model analysis.

The current work proposes an alternative and hybrid procedure for identifying damage in existing reinforced concrete (RC) bridges, recognizing that the development of new techniques for this purpose remains an open area of investigation. The technique introduced in this study is based on the "*M* and *P*" technique, where "*M*" stands for "Monitoring" and "*P*" for "Pushover," originally developed by Makarios [30,31] for the identification of damage in planar multistorey RC frames, primarily for seismic loading or wind loading cases.

The primary focus of this study is on damage identification in straight, ductile, RC bridges subjected to seismic loading, specifically along the longitudinal direction of the bridge axis. These bridges feature multiple spans of varying lengths and multiple piers of different heights while the deck spans are simply supported on the various piers. The deck is considered rigid due to the incorporation of continuation plates at the deck level. Adopting the "*M* and *P*" technique for damage identification in RC bridges, along the longitudinal axis, involves the following steps: (i) perform a series of pushover analyses and instantaneous modal analyses in a suitable nonlinear model of the bridge by gradually targeting the Near Collapse state. From the results of these analyses, the eigenfrequency curve (key diagram) is plotted against deck displacement; (ii) utilize instrumental monitoring to identify the fundamental eigenfrequency of the existing (damaged) bridge in the nonlinear regime; (iii) insert the identified fundamental eigenfrequency of the existing bridge into the key diagram. This process reveals the seismic (target) deck displacement corresponding to the monitored eigenfrequency; (iv) from the final step of a pushover analysis targeting the previously found deck displacement, recognize the damage state at the base of the piers. This involves determining the location and severity of damage. Calculate the damage stiffness of the bridge, considering both the developed plastic hinges at the piers and the stiffness degradation of the piers. Additionally, following the final step of the pushover analysis where the deck displacement is attained, perform an instantaneous modal analysis to determine the mode shape of the damaged bridge. This comprehensive procedure provides a systematic approach to identifying and characterizing seismic damage in ductile bridges, considering multiple spans, various pier heights, and the influence of nonlinear behavior.

To validate the effectiveness of the "*M* and *P*" technique for damage identification along the longitudinal direction of RC bridges, an extensive parametric analysis is conducted. This analysis involves the investigation of a group of existing, straight, ductile, multi-span, RC bridges with varying spans and pier heights. A numerical example of an RC bridge with four spans and five piers (each with two columns) of different heights is presented herein, outlining all the steps of the "*M* and *P*" technique in detail. The primary objectives of this study are twofold: (a) calculation of the eigenfrequency curve of the existing (damaged) RC bridge along the longitudinal direction as a function of seismic deck displacement; (b) evaluation of the damage stiffness and the damage image (location and severity of damage) of the RC bridge.

The application of the hybrid "*M* and *P*" technique yields successful results in identifying damage along the longitudinal direction of ductile, RC bridges by combining monitoring methods with pushover analysis. This presents an alternative technique for detecting damage in ductile RC bridges, ensuring accuracy through the "monitoring and identification of frequency". In this study, it is demonstrated that for a given damage pattern in an existing, ductile, RC bridge, the bridge stiffness at the health state undergoes changes. This alteration results in an elongation of the bridge's eigenfrequency, which is experimentally identified through the monitoring procedure. The key diagram of the method is then utilized to determine the seismic deck displacement of the bridge. This displacement corresponds to the identified damage image and ensures the same eigenfrequency value measured in the field. Consequently, the "*M* and *P*" technique stands as a self-evident process, demonstrating its reliability in practical applications.

2. Methodology

The focus of this study revolves around a typical scenario featuring an existing straight, multiple-span, reinforced concrete (RC) bridge. The bridge configuration includes piers of varying heights and the deck, which is rigid and is simply supported at the piers. The longitudinal rigidity of the deck is ensured by the presence of continuation plates at the deck level. It is also assumed that the piers exhibit highly ductile behavior, designed in accordance with the High Ductility (DCH) class of EN 1998-1 [32]. Each pier may consist of one or more columns. In an earthquake event, the inelastic behavior of this bridge type along the longitudinal direction is anticipated to manifest at the base of the piers, which exhibit a cantilever behavior. Plastic hinges will develop in this region, leading to damage occurrences at these specific locations.

In order to assess the seismic response of an existing reinforced concrete (RC) bridge with a rigid deck along its longitudinal direction, the bridge can be effectively modeled as a Single Degree of Freedom system (SDOF) [25]. The differential equation of motion for the free vibration of an SDOF, without damping and subject to an initial forced displacement or velocity, is given by:

$$m\,\ddot{u}(t) + k_o\,u(t) = 0 \tag{1}$$

Here, m represents the mass of the bridge, and k_o denotes the stiffness of the SDOF. The variables $u(t)$ and $\ddot{u}(t)$ correspond to the time-varying displacement and acceleration of the SDOF mass, respectively.

If an existing SDOF system presents a damage image during its operational life, attributed to any cause, the stiffness at any time step i will undergo a change by Δk_i:

$$k_i = k_o - \Delta k_i \tag{2}$$

where Δk_i is the change of stiffness due to damage.

Furthermore, the instantaneous mode shape at each inelastic i-step of the analysis can be defined by conducting a modal linear analysis using the instantaneous stiffness k_i, which incorporates the effects of damage on stiffness. As a result, the equation of motion, Equation (1), is reformulated as follows:

$$m\,\ddot{u}(t) + (k_o - \Delta k_i)\,u(t) = 0 \tag{3}$$

Therefore, given that the mass m remains constant, the eigenvalue problem at the inelastic i-step is expressed as:

$$\left[(k_o - \Delta k_i) - \omega^2 m\right]\,\varphi = 0 \tag{4}$$

Here, ω (rad/s) represents the instantaneous eigenvalue, and φ is the instantaneous mode shape of the SDOF system at the inelastic i-step of the analysis. The solution of the eigenvalue problem is obtained by equating the expression within the brackets to zero and determining the ω^2 value from the resulting algebraic equation:

$$\left[(k_o - \Delta k_i) - \omega^2 m\right] = 0 \tag{5}$$

Subsequently, the instantaneous mode shape φ can be computed using Equation (4) for the known value of ω^2. Moreover, with the known eigenvalue ω^2, Equation (4) can be reformulated as:

$$\Delta k_i\;\varphi = k_o\;\varphi - \omega^2\;m\;\varphi \tag{6}$$

It is essential to highlight that the identification of the instantaneous frequency ω and the instantaneous mode shape φ of the SDOF structure at the inelastic i-step cannot be achieved through the analysis of the records resulted by a time-history analysis (with accelerograms) using random data processing. These procedures necessitate a sufficient time window where the eigenfrequency remains constant, a condition often challenging in seismic scenarios. Instead, the records obtained through a monitoring multichannel network system of accelerometers should originate from the ambient vibration of an existing (damaged) bridge, without any induced motion. This underlines the significance of the ambient vibration data from a stationary (calm) damaged bridge in accurately determining key dynamic parameters for subsequent verification and advanced analytical procedures. Therefore, if ω, φ, k_o, Δk_i, m are known through the proposed methodology, Equation (6) can be employed at the end for verification purposes (e.g., for advanced optimization and probabilistic analysis).

The identification of damage along the longitudinal direction of multi-span, ductile, RC bridges, with piers of various height, can be achieved through the proposed hybrid "*M* and *P*" technique (where *M* signifies "Monitoring" and *P* denotes "Pushover") that integrates an identification system with a numerical model, according to the following phases:

(a) The eigenfrequency f of the existing (damaged) RC bridge along the longitudinal direction is identified through monitoring. This involves utilizing a local network of uniaxial accelerometers placed at characteristic positions along the degree of freedom of the system.

(b) A suitable numerical nonlinear model of the RC bridge is established, and a sequence of separate pushover analyses are performed along the longitudinal direction, targeting each time at a gradual increasing deck displacement $u_{deck,i}$. For each target displacement, one pushover analysis is performed leading to the drawing of the capacity curve of the bridge in terms of base shear V_o and deck displacement u_{deck}. The base shear of the bridge is computed as the sum of the base shears of the various bridge piers, where each pier can consist of several columns. Figure 1 illustrates the general form of the capacity curve of the bridge along the longitudinal direction. Additionally, the figure features an idealized elasto-perfectly plastic force–displacement relationship, defining the idealized yield point (u_y, V_{oy}) of the bridge and the effective (secant) stiffness at the yield point. Various performance levels of the bridge corresponding to different deck displacements u_i are presented in the figure. The Near Collapse state of the bridge is characterized by the ultimate target (deck) displacement, u_{ult}. It is noted that P-D effects should always be considered in pushover analysis, especially for more flexible bridge structures. However, caution is advised, as spurious results may arise from the above consideration in the instantaneous modal analyses that follow. This potential issue can be caused from the possible appearance of negative stiffness in the capacity curve due to the P-D effects. Regarding the effective stiffness of the RC bridge piers in the pushover analysis, a proposed stiffness scenario is outlined in the subsequent step.

(c) A sequence of instantaneous modal analysis of the bridge is performed, following the final step of each pushover analysis with an increasing deck (target) displacement $u_{deck,i}$. To be more specific, the stiffness of the existing (damaged) bridge obtained at the final step of each pushover analysis serves as the initial condition for the subsequent instantaneous modal analysis. Utilizing the results from the instantaneous modal analyses, a diagram of the instantaneous (step) cyclic eigenfrequency f_i (in Hz) of the damaged bridge is plotted against the deck displacement $u_{deck,i}$ of the bridge along the longitudinal direction. Figure 2 illustrates the general form of such a diagram, which serves as the key diagram of the "*M* and *P*" technique. In this figure, the inelastic deck displacement $u_{deck,i}$ is represented on the abscissa, while the eigenfrequency f_i of the damaged bridge is on the ordinate. By incorporating the fundamental eigenfrequency f_1, identified through the monitoring procedure in phase (a), into this diagram, the seismic inelastic deck displacement $u_{deck,i}$ of an existing damaged RC bridge can be determined.

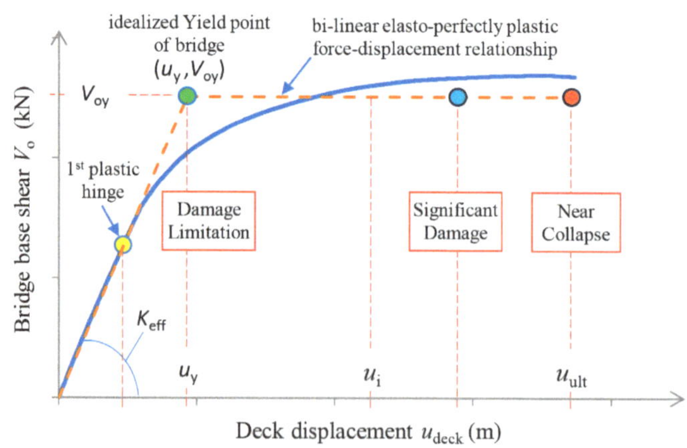

Figure 1. Pushover curve of a multispan RC bridge along the longitudinal direction (target displacement at the Near Collapse (NC) state).

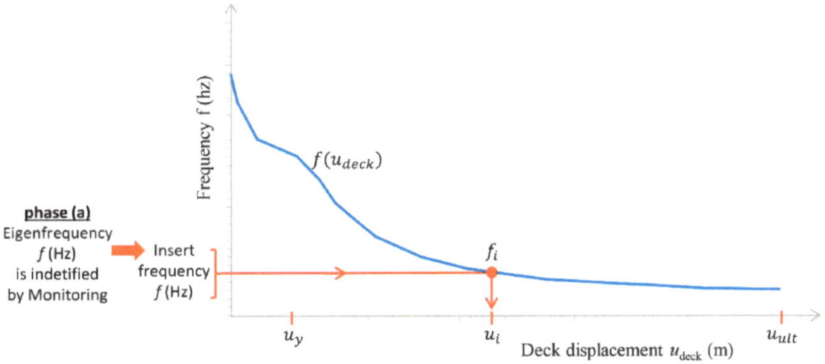

Figure 2. Instantaneous eigenfrequency diagram of the RC bridge, in the nonlinear area.

It is important to note that all pushover and instantaneous modal analyses of the reinforced concrete (RC) bridge, targeting each gradual increase in deck displacement $u_{deck,i}$, should be conducted within a suitable nonlinear model of the bridge. This model should account for discrete values $E_c I_{eff,i}$ (E_c is the elastic modulus of concrete) representing the effective bending stiffness of the various piers. This consideration is crucial due to varying levels of stiffness degradation at each target displacement $u_{deck,i}$ corresponding to different damage states (extent and magnitude of cracking along the entire critical plastic region length of the piers). These damage states align with various performance levels, including undamaged (health) state, 1st hinge (1st yield), Damage Limitation (DL), Significant Damage (SD), Near Collapse (NC), and all the intermediate ones (Figure 1).

To address this, an effective stiffness scenario, expressed in terms of the effective moment of inertia ratio $I_{eff,i}/I_g$ of the piers, should be established before performing pushover and modal analyses, as a function of the chord rotation of the bridge piers, $\theta_{c,i}$ (in rad). In this ratio, I_g is the moment of inertia of the geometric sections of the piers. Moreover, the chord rotation of the piers at the examined deck (target) displacement $u_{deck,i}$ is given by $\theta_{c,i} = u_{deck,i}/h_c$, where h_c is the pier height (Figure 3). Consequently, piers of various heights will exhibit different values of chord rotation since the lateral displacement of the rigid deck along the longitudinal direction of the bridge remains the same for all piers at their tops. This approach ensures a comprehensive consideration of stiffness variations associated with different damage states and pier characteristics during the analyses.

The main assumption regarding the effective moment of inertia $I_{eff,i}$ of the RC piers, which exhibit a cantilever bending deformation with potential plastic hinges located at the base-section of the piers, is grounded in the considerations outlined in EN 1998-3 [33]:

$$E_c I_{\text{eff}} = \frac{M_p \cdot L_v}{3 \cdot \theta_y} \quad (7)$$

where M_p is the plastic moment of the base-section of the piers, which is calculated through a section analysis using an elastoplastic idealization of the moment–curvature diagram M-φ; L_v signifies the shear span of the piers, which is equal to the height of the piers from the foundation level to the bottom of the deck girders. Additionally, θ_y denotes the chord rotation of the shear span of the piers at the yield state, approximately given by Equation (A.10) of Eurocode EN 1998-3 [33].

Equation (7) is explicitly derived from elasticity theory, and its rationale is presented in Figure 4. When a cantilever pier undergoes yielding (i.e., when its bending moment reaches the plastic moment M_p of the section), then the effective bending stiffness $E_c I_{\text{eff}}$ of the shear span L_v of the pier ($L_v = M/V$, where M is the bending moment and V is the shear force of the pier) is equal to the secant stiffness of the shear span to the yield point. EN 1998-3 [33] imposes these low values of the secant (effective) bending stiffness at yield on each pier of the RC bridge in order to perform nonlinear analysis that targets all performance levels, ranging from DL to NC. The adoption of these low values of secant stiffness at yield ensures a conservative approach in displacement calculations, contributing to an overall more flexible behavior of the bridge in the analyses.

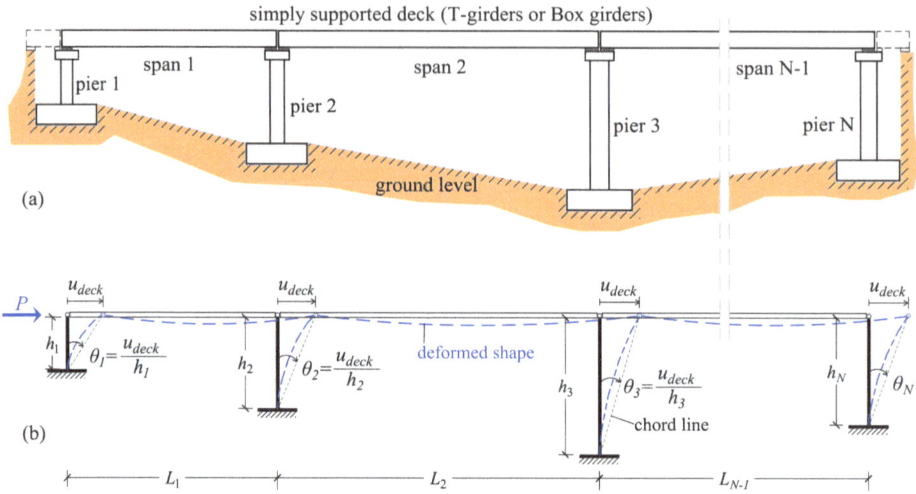

Figure 3. (**a**) Simply supported straight bridge with N-1 spans of various lengths and N-piers of various heights; (**b**) chord rotation θ_c of the bridge piers in pushover analysis with lateral force at the deck level.

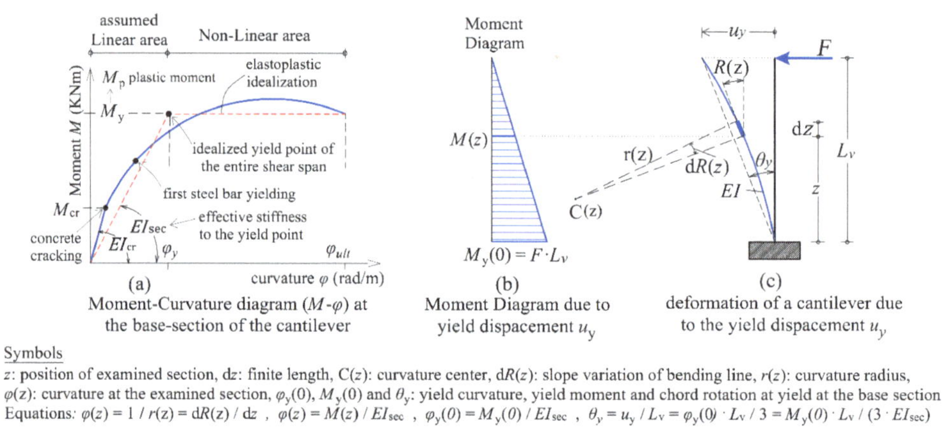

Figure 4. Base section analysis of a cantilever pier in the RC bridge, with a height equal to the shear span length L_v. This includes the calculation of curvature φ_y (rad/m), chord rotation θ_y (rad), and secant stiffness at yield EI_{sec} (kN·m^2) according to EN1998-3 [33].

To mitigate this inherent conservatism in the stiffness scenario depicted in Figure 5, a scaling approach is proposed when the pushover analysis targets higher seismic performance levels, including DL, SD, or any intermediate levels between DL to NC. According to [34], which addresses RC buildings, at the DL state, the suggested value for $E_c I_{eff}$ is equal to $2 \cdot E_c I_{eff,NC}$ and lies between $0.25 \cdot E_c I_g$ and $0.5 \cdot E_c I_g$. At the SD state, $E_c I_{eff}$ is recommended to be equal to the average of the corresponding values at the NC and DL states. Moreover, to simplify the stiffness scenario in order to encompass all piers of the RC bridge, which have various sections, mean values of the ratio of effective stiffness I_{eff}/I_g

are proposed as a function of the chord rotation θ_{pr} (where the subscript "*pr*" denotes profile) for all performance levels. These values are derived from an extended parametric investigation of various ductile RC piers with columns of cyclic cross-sections (D = 0.6 to 2 m) and with geometric longitudinal reinforcement ratios ranging from 1% to 4%. This investigation involved consecutive pushover analyses with gradually increasing target displacements, ranging from the health state to the NC state. The analyses incorporated suitable values of effective stiffness determined through a trial-and-error process. The objective was to achieve convergence, aligning the observed chord rotation at the base section of the column piers (performance level) with the assigned percentages of reduction of the moment of inertia. In these analyses, the scaling approach remains within two limits: at the NC state, the effective moment of inertia ratio I_{eff}/I_g of an RC column pier is calculated using Equation (7) of EN 1998-3 [33], while just before DL, at 1st hinge formation (1st yield), the effective moment of inertia ratio is practically equal to the 50% reduction rate proposed by EN 1998-1 [32] for the design of new RC buildings. Similarly, the effective moment of inertia corresponding to achieving the SD performance level must align with the chord rotation at the base of the columns, which, as per EN 1998-3 [33], is approximately 75% of that at the NC level. From the health state to the formation of the 1st hinge, the scaling approach for reducing the effective moment of inertia is constrained within the limits of the geometric moment of inertia I_g and the 50% reduction rate proposed by EN 1998-1 [32], respectively. In the nonlinear model of each cantilever pier, mean values of strengths were employed, and the plastic hinge at the base-section was modeled using the fiber hinge approach with a plastic hinge length in accordance with EN 1998-3 [33]:

$$L_{\mathrm{pl}} = \frac{L_{\mathrm{v}}}{30} + 0.2 \cdot h + 0.11 \cdot \frac{d_{\mathrm{bL}} \cdot f_{\mathrm{ym}}}{\sqrt{f_{\mathrm{cm}}}} \tag{8}$$

where f_{cm} is the mean concrete compressive strength, f_{ym} is the mean yield stress of steel, d_{bL} is the mean diameter of the tension reinforcement, h is the depth of the cross-section, and L_{v} is the shear span. This approach ensures a more refined and representative stiffness scenario for the entire bridge, accounting for the variability in pier sections and offering a comprehensive representation of stiffness under different performance levels.

It was found that the mean values of $\theta_{pr,i}$ between all examined RC piers (with different circular sections) at the 1st yield, DL, SD, and NC states were approximately equal to 0.0095, 0.016, 0.068, and 0.091 rad, respectively. The corresponding mean values of $I_{eff,i}/I_g$ were approximately equal to 0.5. 0.4, 0.28, and 0.22. Furthermore, the effective stiffness scenario depicted in Figure 5 introduces discrete $I_{eff,i}/I_g$ values starting from the uncracked (health) state towards the 1st hinge (indicating the 1st yield), and then progressing to DL. At the 1st hinge, the value $I_{eff,i}/I_g = 0.50$ is considered, as suggested in EN 1998-1 [32] for elastic analysis. Additionally, three straight lines for the effective stiffness ratio $I_{eff,i}/I_g$ of the piers within the linear and nonlinear areas are presented in Figure 5. These lines align with the abovementioned proposed θ_{pr} and $I_{eff,i}/I_g$ values for the 1st yield, DL, SD, and NC states:

For the linear area (health state to 1st hinge), $0 \leq \theta_{pr} \leq 0.00946$ (rad):

$$I_{eff}/I_g = 1 - 52.847 \cdot \theta_{pr} \tag{9}$$

For the assumed linear area (1st hinge to DL), $0.00946 < \theta_{pr} \leq 0.01605$ (rad):

$$I_{eff}/I_g = 0.6436 - 15.174 \cdot \theta_{pr} \tag{10}$$

For the nonlinear area (DL to NC), $0.01605 < \theta_{pr} \leq 0.0913$:

$$I_{eff}/I_g = 0.4384 - 2.391 \cdot \theta_{pr} \tag{11}$$

To utilize this figure, the known chord rotation $\theta_{c,i}$ of the base section of a pier at a discrete target deck displacement $u_{deck,i}$ should be inserted into Figure 5 for θ_{pr}, and the corresponding effective stiffness ratio $I_{eff,i}/I_g$ of the pier can be obtained.

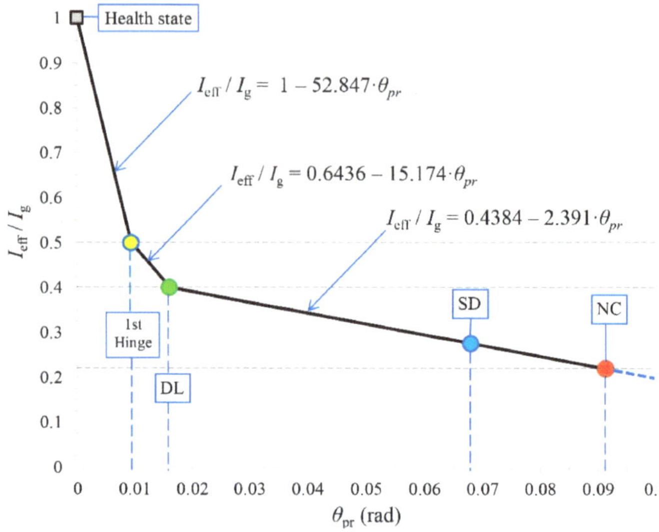

Figure 5. Effective moment of inertia ratio I_{eff}/I_g of the ductile piers of the RC bridge at various performance levels (damage states) as a function of the chord rotation θ_{pr} (rad) at their base section.

(d) The known eigenfrequency f_i (in Hz) of the existing bridge, obtained through the monitoring procedure in phase (a), is incorporated into the instantaneous eigenfrequency diagram (Figure 2). Consequently, the corresponding inelastic seismic (target) deck displacement $u_{deck,i}$ of the bridge is determined from the diagram.

(e) The bridge damage state can be identified through the results of a pushover analysis, specifically targeting the previously determined seismic deck displacement $u_{deck,i}$ from the preceding phase. The location and state of potential plastic hinges that may develop at the base-section of the piers at the last step of pushover analysis provide an estimation of the damaged state of the existing RC bridge. It is important to note that even if no plastic hinge appears at the base of the piers, the magnitude of damage at the base of the piers can still be estimated in terms of stiffness degradation relative to the health state.

(f) Additionally, a linear modal analysis is performed at the final step of pushover analysis in phase (e). The initial conditions for this analysis are derived from the instantaneous stiffness of the RC bridge along the longitudinal direction at this last pushover analysis *i*-step. From the results of this instantaneous modal analysis, both the circular eigenfrequency ω_i and the mode shape φ_i of the damaged bridge are calculated.

(g) Finally, the instantaneous stiffness k_i of the RC bridge along the longitudinal direction is computed at the examined inelastic *i*-step where the deck displacement $u_{deck,i}$ occurs. This calculation is facilitated after determining the flexibility of the damaged bridge at the same i-step. To achieve this, a linear analysis is performed with a lateral force F_i applied at the deck level, specifically along the dynamic degree of freedom of the bridge along the longitudinal direction. This analysis is conducted at the last step of the pushover analysis in phase (e), resulting in the calculation of the corresponding static displacement $u_{st,i}$. Subsequently, the stiffness k_i of the damaged bridge is computed by the ratio $F_i/u_{st,i}$. Thus, the damage stiffness Δk_i at the same inelastic *i*-step is calculated using the general equation $\Delta k_i = k_o - k_i$, where

k_o represents the known initial stiffness of the undamaged bridge, determined at the health state as mentioned above. In the preceding linear analysis, the base shear $V_{c,i}$ of the piers is also recorded. Consequently, the stiffness $k_{c,i}$ of the damaged piers can be calculated by the ratio $V_{c,i}/u_{st,i}$, and their stiffness degradation $\Delta k_{c,i} = k_{c,o} - k_{c,i}$ relative to the health state can be determined. Utilizing this information, the location and the magnitude of the damage on the bridge piers can be identified.

In Figure 6, a flowchart illustrating the application of the "*M* and *P*" technique for the identification of damage in RC bridges is presented.

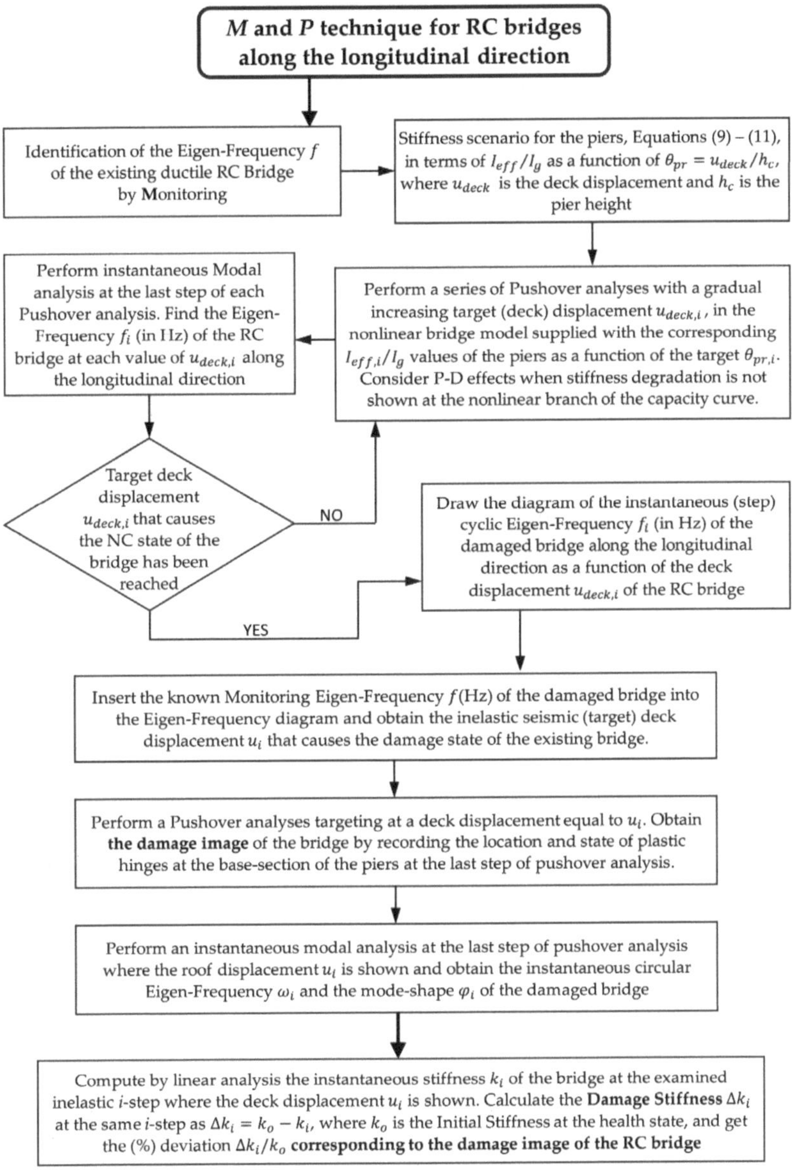

Figure 6. Flowchart for the application of the "*M* and *P*" technique in RC bridges.

3. Numerical Example

The existing, straight, ductile, RC bridge depicted in Figure 7a comprises four spans with dimensions $L_1 = 15$ m, $L_2 = L_3 = 25$ m, and $L_4 = 20$ m. It features five piers with varying heights: $h_1 = 6$ m, $h_2 = 10$ m, $h_3 = 15$ m, $h_4 = 12$ m, and $h_5 = 8$ m. There are no abutments at the bridge ends due to steep vertical rock slopes in those locations. Each pier consists of two identical circular columns that connect at the top with a cup beam of rectangular section 1.5 × 1.0 m. The cap beam extends beyond the piers with a reduced cross-section at the free end. The two circular columns of each pier are assumed fixed into a common slab foundation at the ground level. The column diameters of the five piers are $D_1 = 0.8$ m, $D_2 = 1.2$ m, $D_3 = 1.5$ m, $D_4 = 1.3$ m, and $D_5 = 1$ m. The cast in place bridge deck, with a thickness of 30 cm, is simply supported on the cap beams along each span through precast prestressed girders of a rectangular section 0.5 × 1.5 m. The bearings at the top of the cap beams (which are usually of elastomeric type in bridges) are assumed as simple support areas for the girders without altering the dynamic behavior of the bridge. The superstructure of the bridge (deck system) is considered rigid along the longitudinal direction due to the presence of continuation plates above the piers at the deck level. As a result, the dynamic simulation of the RC bridge shown in Figure 7b can be considered as an SDOF system, where the piers exhibit cantilever behavior. A detailed section of the bridge at the location of pier 5 is illustrated in Figure 8a.

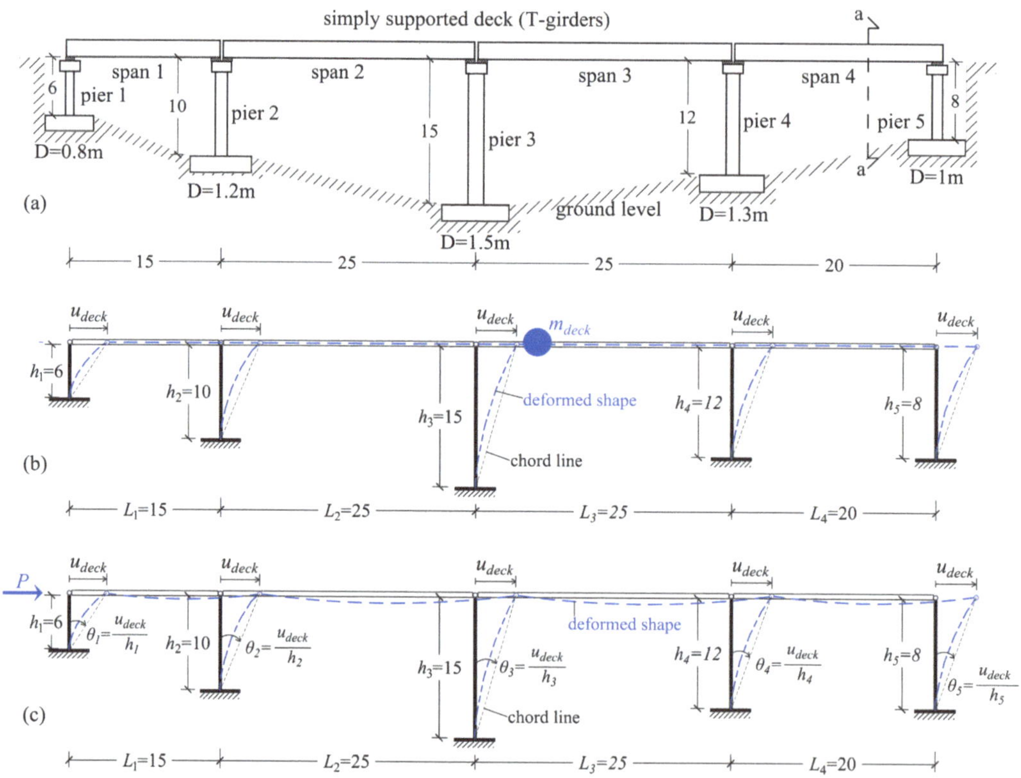

Figure 7. (**a**) Straight RC bridge with four spans and with (2 column) piers of various heights; (**b**) dynamic model of RC bridge (SDOF) for modal analysis; (**c**) static model of RC bridge for pushover analysis.

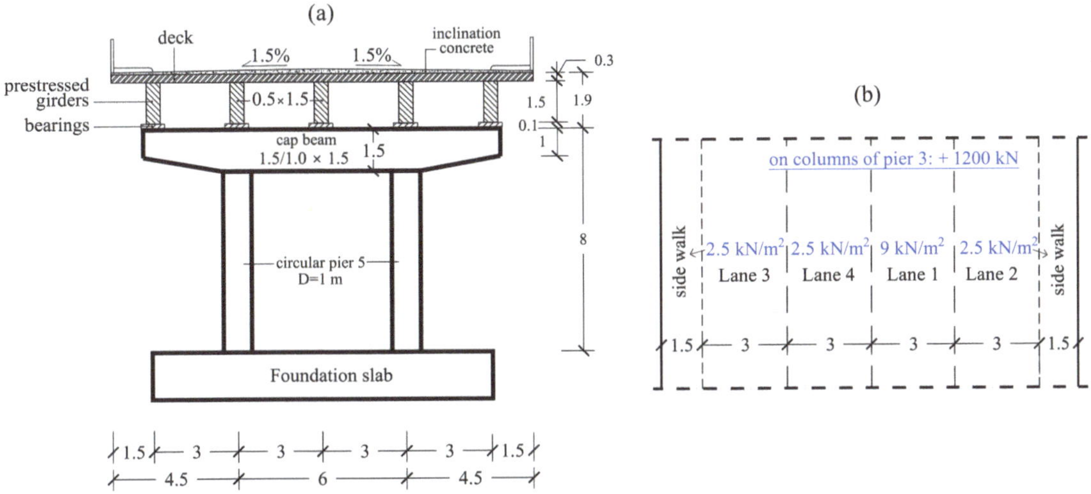

Figure 8. (**a**) Section aa at pier 5 of the RC bridge; (**b**) lane set-up and live loads on lanes.

The bridge accommodates four traffic lanes each with a width of 3 m and two sidewalks of 1.5 m (Figure 8b). The lane numbering and the live traffic loads on the lanes are detailed in Figure 8, following the LM1 load model of EN 1991-2 [35]. This load model encompasses uniformly distributed loads and includes a concentrated heavy vehicle load positioned to act at the most slender columns, specifically at the top of the two columns of pier 3. The total vertical loads $p = g + \psi_2 q$ in the seismic combination, where g is the dead load, $\psi_E q$ is the quasi-live load, with $\psi_2 = 0.2$ for bridges [36], act at the top of the columns of the five piers and are outlined in Table 1. These loads contribute to a total bridge mass of approximately 2875 tons, assumed to be concentrated at the center of gravity of the deck system. The bridge mass at the deck and the degree of freedom along the longitudinal direction of the RC bridge for modal analysis are depicted in Figure 7b. Additionally, Figure 7c illustrates the static simulation of the RC bridge with the lateral force at the deck level for the pushover analysis.

Table 1. Longitudinal (L) and shear (h) reinforcement of circular columns.

Pier Column	D (m)	Number of Bars	d_{bL} (mm)	Geom. Ratio (%)	d_{bh} (mm)	s_h (cm)
1	0.8	24	28	2.94	16	5
2	1.2	28	25	1.22	16	6
3	1.5	30	26	0.90	16	6
4	1.3	30	26	1.20	16	6
5	1	25	28	1.96	16	6

The construction materials of the RC bridge include concrete grade C35/40 with a mean compressive strength f_{cm} of 43 MPa and steel grade B500c with a mean tensile strength f_{ym} of 550 MPa. The modulus of elasticity for concrete is $E_c = 34$ GPa, while for steel, it is $E_s = 200$ GPa.

All circular column sections are symmetrically reinforced, with geometrical ratios ranging from 1% to 3% for the longitudinal reinforcement. The columns of the shortest pier 1 are the most reinforced. Confinement reinforcement in all columns consists of closed circular hoops with a diameter of 16 mm, evenly spaced at critical end sections with an axial spacing of 5 or 6 cm. The concrete cover is 5 cm. The steel reinforcement details of a typical circular column section are depicted in Figure 9. The longitudinal and shear reinforcements of the circular columns are outlined in Table 1. The reinforcement details of the beam caps and the deck are omitted here as they are not relevant. The seismic behavior of the RC bridge relies exclusively on the behavior of the piers acting as cantilevers with a common top displacement, ensured by the rigid deck system. It is noteworthy that the RC bridge has been designed in accordance with EN 1998-1 [32] and EN 1998-2 [36] for the high ductility class (DCH). Consequently, it is expected to exhibit highly ductile behavior in the nonlinear domain, developing plastic hinges at the base of the columns.

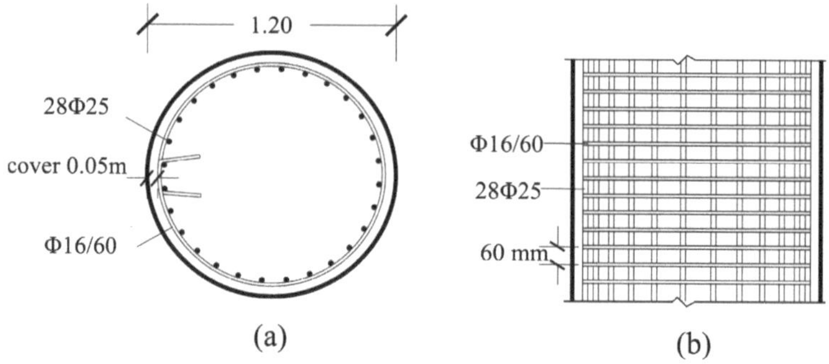

Figure 9. Pier 2: (**a**) section; (**b**) elevation.

Next, a series of pushover and instantaneous modal analyses is conducted to implement the "*M* and *P*" technique for structural damage identification in the existing RC bridge. The FEM analysis software SAP2000 [37] is utilized for this purpose. Plastic hinges of the fiber type are incorporated at the base of the piers in the nonlinear model of the bridge to simulate potential locations of inelastic behavior. The constitutive relationships used to characterize the behavior of construction materials in both the linear and nonlinear domains are as follows: (a) the model proposed by Mander, Priestley, and Park (1988) [38] for uniaxial unconfined and confined concrete (Figure 10); and (b) the model proposed by Park and Paulay (1975) [39] for steel reinforcement, which is parabolic in the strain hardening region (Figure 11). All necessary data for the nonlinear simulation of the bridge are derived from the results of a section analysis. The data for the base section of the piers are presented in Table 2, where the effective stiffness ratio I_{eff}/I_g is calculated using Equation (7) in accordance with EN 1998-3 [33]. The plastic hinge length of the fiber hinges is calculated using Equation (8) and is shown in the last column of Table 2.

Next, as outlined in phase (c) of the "*M* and *P*" technique, Equations (9)–(11) from Figure 5 are employed to establish the stiffness scenario (I_{eff}/I_g) for the piers. This information is then incorporated into a series of pushover and instantaneous modal analyses of the bridge along the longitudinal direction, with a progressively increasing (target) deck displacement $u_{deck,i}$. The stiffness scenario is outlined in Table 3 as a function of the profile angle θ_{pr} at the base of the piers (Figure 7c). The discrete values of the effective moment of inertia I_{eff} assigned to the two RC columns of each pier in the nonlinear model of the bridge depend on both the seismic (target) deck displacement of the pushover analysis and the pier height, i.e., on the target performance level of each pier (Figure 4). This is because the developed chord rotation at the base section of each pier $\theta_{c,i} = \theta_{pr,i} = u_{deck,i}/h_c$ at a target deck displacement u_i depends on both these parameters and is different for each pier (Figure 7c). For example, if the target deck displacement $u_{deck,i}$ is equal to 0.3 m, then the developed chord rotations at the base sections of the two columns of each pier in rad units are: $\theta_1 = 0.3/6 = 0.05$, $\theta_2 = 0.3/10 = 0.03$, $\theta_3 = 0.3/15 = 0.02$, $\theta_4 = 0.3/12 = 0.025$, and $\theta_5 = 0.3/8 = 0.038$. Hence, different I_{eff}/I_g values for the two columns of each pier for $u_{deck,i} = 0.3$ m are derived from Table 3, which are, respectively, equal to 0.32, 0.37, 0.39, 0.38, and 0.35.

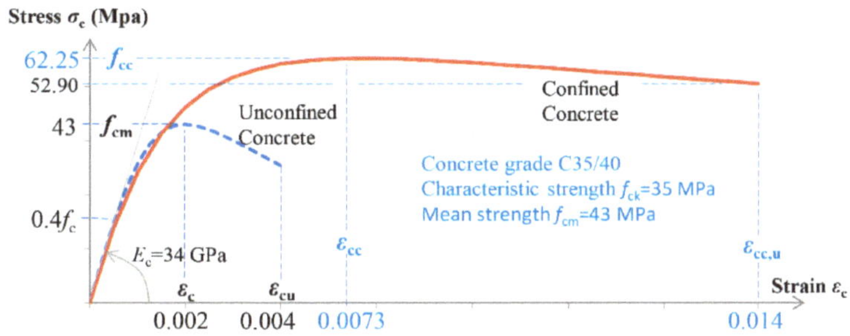

Figure 10. Stress–strain diagrams for both unconfined and confined concrete.

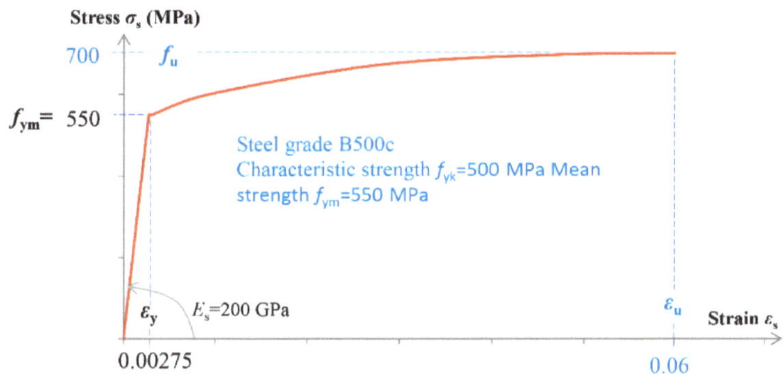

Figure 11. Stress–strain diagram for steel reinforcement bars.

Table 2. Base section analysis of RC pier columns to calculate the ratio (I_{eff}/I_g) from Equation (7), $E_c I_{\text{eff}} = M_p \cdot L_v / (3 \cdot \theta_y)$.

Pier/ Column	Axial Force N (kN)	$\nu = N/(A \cdot f_c)$	φ_y (rad/m)	M_p (kNm)	L_v (m)	θ_y (rad)	$E_c \cdot I_{eff}$ (kNm^2)	$E_c \cdot I_g$ (kNm^2)	I_{eff}/I_g	L_{pl} (m)
1	1350	0.062	0.0098	2909	6	0.0264	220,531	683,611	0.32	0.62
2	3205	0.066	0.0053	5231	10	0.0226	773,128	3,460,778	0.22	0.80
3	4240	0.056	0.0040	8150	15	0.0243	1,679,135	8,449,166	0.20	1.04
4	3545	0.062	0.0049	6534	12	0.0243	1,074,144	4,766,748	0.23	0.90
5	1760	0.052	0.0069	4017	8	0.0241	444,408	1,668,971	0.27	0.72

Table 3. Effective stiffness scenario (I_{eff}/I_g) for the ductile RC pier columns as a function of base chord rotation θ_{pr} (in radians).

θ_{pr}	I_{eff}/I_g	θ_{pr}	I_{eff}/I_g	θ_{pr}	I_{eff}/I_g	θ_{pr}	I_{eff}/I_g	θ_{pr}	I_{eff}/I_g
0	1.00	0.010	0.49	0.020	0.39	0.045	0.33	0.070	0.27
0.001	0.95	0.011	0.48	0.023	0.38	0.048	0.32	0.072	0.27
0.002	0.89	0.012	0.46	0.025	0.38	0.050	0.32	0.075	0.26
0.003	0.84	0.013	0.45	0.027	0.37	0.053	0.31	0.077	0.25
0.004	0.79	0.014	0.43	0.030	0.37	0.055	0.31	0.080	0.25
0.005	0.74	0.015	0.42	0.032	0.36	0.057	0.30	0.083	0.24
0.006	0.68	0.016	0.40	0.035	0.35	0.060	0.29	0.085	0.24
0.007	0.63	0.017	0.40	0.038	0.35	0.062	0.29	0.087	0.23
0.008	0.58	0.018	0.40	0.040	0.34	0.065	0.28	0.090	0.22
0.009	0.52	0.019	0.39	0.043	0.34	0.068	0.28	0.0903	0.22

Next, a pushover analysis is conducted to target the NC state, aiming to derive the capacity curve of the RC bridge along its longitudinal direction. The target deck displacement causing the bridge NC state is set to 0.54 m. In this analysis, the effective moment of inertia ratio I_{eff}/I_g of the columns of the various piers is determined from Table 3 (or from Figure 5) based on the corresponding values of θ_{pr} (in radians) that develop at their base sections at $u_{deck} = 0.54$ m. The resulting capacity curve of the RC bridge using the proposed stiffness scenario for the NC state is illustrated in Figure 12, along with the elastic-perfectly plastic idealization line, from which the yield displacement $u_y = 0.17$ m of the bridge results.

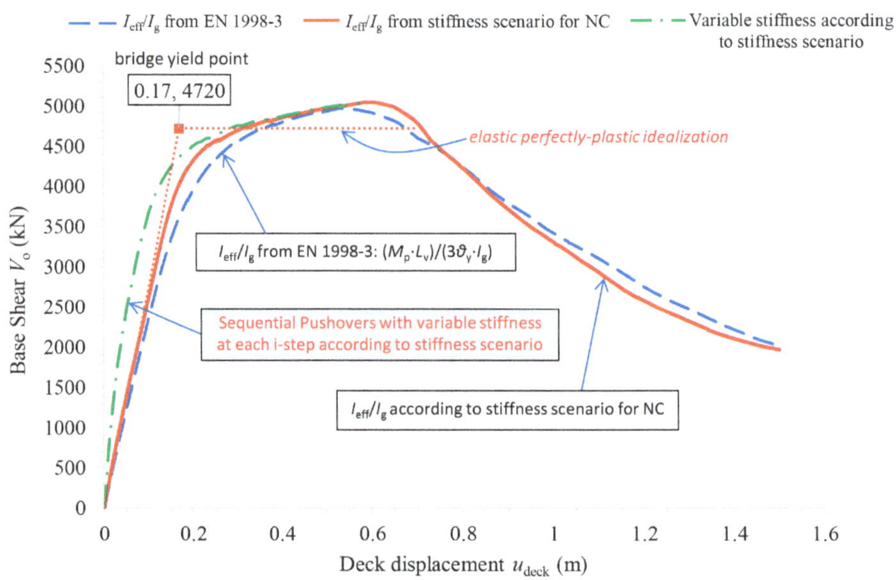

Figure 12. Capacity curve of the RC bridge along the longitudinal direction according to various stiffness scenarios. Ref. [33].

Additionally, Figure 12 includes the capacity curve of the bridge obtained from a pushover analysis using the stiffness scenario proposed in EN 1998-3 [33] (Equation (7)). Furthermore, Figure 12 displays the capacity curve of the bridge derived from sequential pushover analyses, targeting each time an increasing deck displacement from the health state to NC state, using the stiffness scenario proposed in Figure 5 for the discrete values of the target deck displacement. As observed in Figure 12, the influence of the effective stiffness scenario on the results of pushover analysis is more pronounced for higher performance levels due to the delayed onset of the yield state of the bridge. In Figure 13, the capacity curve of the bridge is presented again for the case of the proposed stiffness scenario at the NC state, along with the capacity curves of the various piers (each having two columns). The contribution of the various piers to the capacity of the bridge, in terms of base shear and deck displacement, is evident in the figure. The yield points of the piers, derived from the bilinear idealization of their curves, are also displayed in Figure 13, illustrating the yield sequence. The NC state of the bridge is induced by the failure of the shortest columns of pier 1, which develop a chord rotation θ_{pr} at their base sections equal to $\theta_1 = 0.54/6 = 0.09$ rad. At the failure of the columns of pier 1, the columns of the other piers develop the following chord rotations at their base sections: $\theta_2 = 0.54/10 = 0.054$, $\theta_3 = 0.54/15 = 0.036$, $\theta_4 = 0.54/12 = 0.045$, and $\theta_5 = 0.54/8 = 0.068$ in rad units. It is noted that for lower values of the target deck displacement driving the RC bridge to higher performance levels (for example, DL or SD), the corresponding values of I_{eff}/I_g from Table 3 should be assigned to the piers. Therefore, the capacity curve resulting from the pushover analysis of the bridge targeting a higher performance level will not exhibit exactly the same characteristics as that for the NC state (see Figure 14).

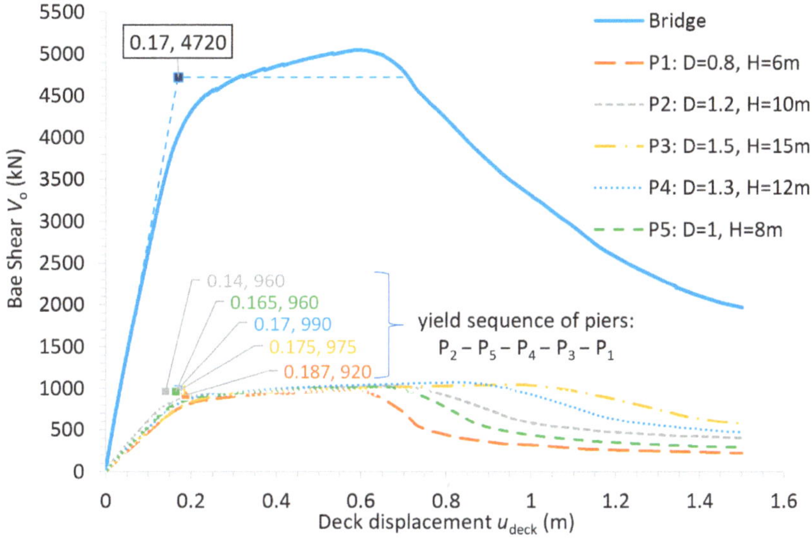

Figure 13. Capacity curve of the RC bridge and of the various (2 columns) piers along the longitudinal direction according to the stiffness scenario for the NC state.

Next, a series of pushover analyses is performed on the existing RC bridge along its longitudinal direction. Each analysis targets an increasing deck displacement corresponding to the specific values of the profile angle θ_{pr} of Table 3, ranging from 0 to 0.09 rad. The initial pushover analysis in the series targets the health (uncracked) state of the bridge. Therefore, it is conducted with a lateral (target) deck displacement set to zero and considers only the effect of the vertical loads in the seismic combination, $g + \psi_E q$. For this analysis, all piers in the nonlinear model of the RC bridge are provided with the value $I_{eff} = I_g$. In all subsequent pushover analyses in the series, the various piers of the bridge are provided with the effective moments of inertia $I_{eff,i}$ from Table 3 (or from Figure 5), corresponding to the developed values of the profile angle $\theta_{pr,i}$ at the base section of the pier columns at the target deck displacement u_i.

Subsequently, a series of instantaneous modal analyses along the longitudinal direction of the RC bridge is performed following the final step of each one of the separate pushover analyses in the series. These modal analyses have as initial condition the damage state of the bridge at the final step of the separate pushover analyses with target deck displacement u_i. In other words, the stiffness of the damaged bridge at the final step of each pushover analysis is used. From this series of modal analyses on the RC bridge, the instantaneous cyclic eigenfrequency f_i (in Hz) of the SDOF system along the longitudinal direction is recorded at each i-step. This information is then utilized to generate the diagram of the instantaneous cyclic eigenfrequency (in Hz) of the RC bridge across the linear and nonlinear domains as a function of the deck displacement, $u_{deck,i}$. Figure 14 depicts this diagram along with an approximate line that encompasses all performance levels.

The cyclic eigenfrequency f_i (in Hz) resulting from the instantaneous modal analyses at the final step of each separate pushover analysis are presented in Table 4 for indicative values of the target displacement $u_{deck,i}$. These data correspond to the information depicted in Figure 14.

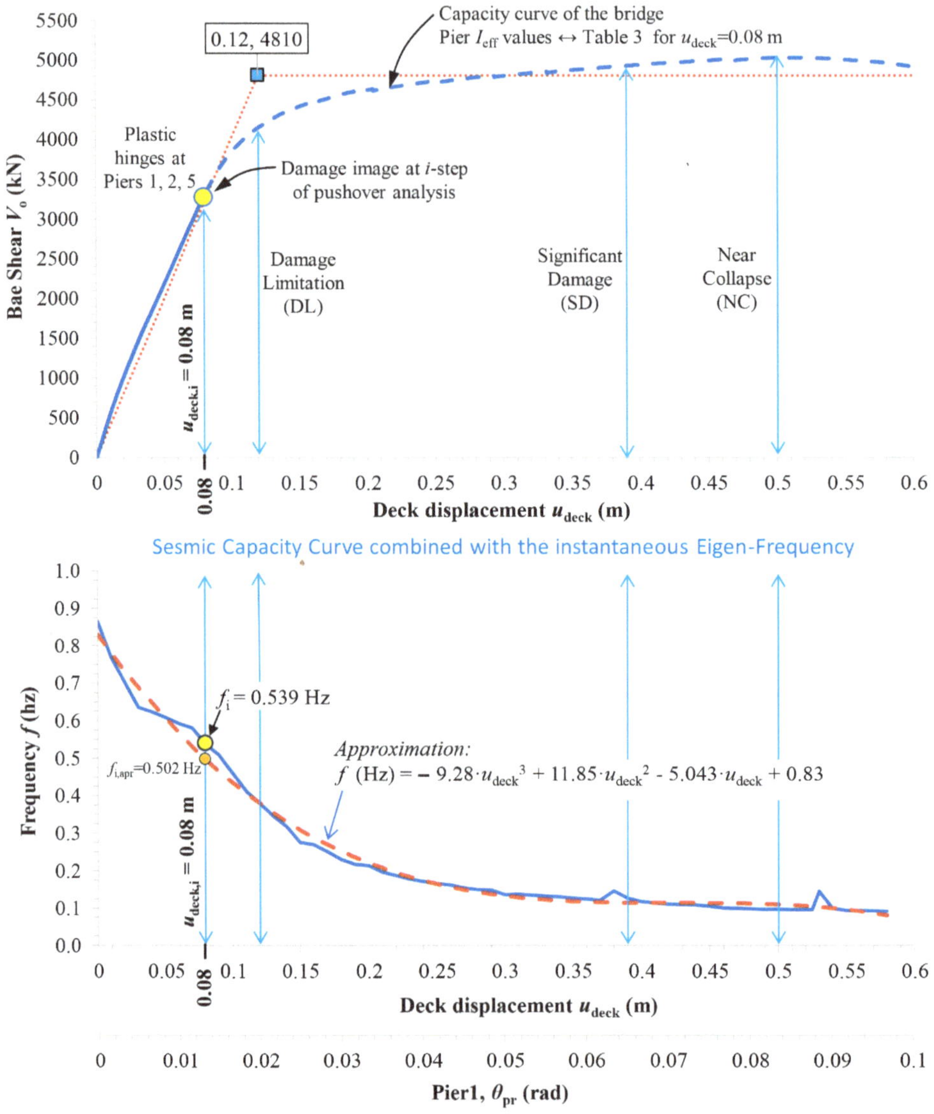

Figure 14. Instantaneous eigenfrequency diagram combined with the seismic capacity curve of the RC bridge along the longitudinal direction (key diagram).

As evident from Table 4, the instantaneous period T_i (in seconds) of the existing (damaged) RC bridge—corresponding to the inverse of the cyclic eigenfrequency f_i—is elongated with increasing deck displacement. This indicates that the RC bridge gradually becomes more flexible due to the advancing damage. The elongation of the period between the health state and the DL state is observed to be more than two times. Figure 14 presents the diagram of the instantaneous cyclic eigenfrequency f_i (in Hz) in conjunction with the capacity curve of the RC bridge along the longitudinal direction. These two diagrams are interconnected and form the *key-diagram* of the proposed "*M* and *P*" technique for bridges. This key-diagram serves as a crucial tool in the "*M* and *P*" technique, facilitating the identification and assessment of damage in the bridge structure. In this key-diagram:

(a) The monitoring fundamental frequency of the existing bridge is inserted (see Figure 2), and the deck seismic displacement $u_{deck,i}$ (or the corresponding θ_{pr} value of the piers) of the existing RC bridge is determined.

(b) At this seismic (target) deck displacement $u_{deck,i}$, we ascend to the capacity curve of the RC bridge, allowing us to visually observe the damage state.

Table 4. Instantaneous cyclic eigenfrequency f_i (Hz) of the RC bridge (Figure 14), from the health state to the Near Collapse (NC) state.

u_{deck} (m)	θ_{pr} of the Column of Piers in Rad					I_{eff}/I_g					State of Damage	f (hz)	T (s)
	P_1	P_2	P_3	P_4	P_5	P_1	P_2	P_3	P_4	P_5			
0	0.000	0.000	0.000	0.000	0.000	1	1	1	1	1	Health	0.865	1.16
0.04	0.007	0.004	0.003	0.003	0.005	0.65	0.79	0.86	0.82	0.74		0.623	1.61
0.07	0.012	0.007	0.005	0.006	0.009	0.47	0.63	0.75	0.69	0.54	1st Hinge	0.581	1.72
0.08	0.013	0.008	0.005	0.007	0.010	0.44	0.58	0.72	0.65	0.49		0.539	1.86
0.09	0.015	0.009	0.006	0.008	0.011	0.42	0.52	0.68	0.60	0.47		0.509	1.97
0.1	0.017	0.010	0.007	0.008	0.013	0.40	0.49	0.65	0.56	0.45		0.461	2.17
0.12	0.020	0.012	0.008	0.010	0.015	0.39	0.46	0.58	0.49	0.42	DL	0.379	2.64
0.14	0.023	0.014	0.009	0.012	0.018	0.38	0.43	0.51	0.47	0.40		0.318	3.15
0.18	0.030	0.018	0.012	0.015	0.023	0.37	0.40	0.46	0.42	0.38		0.229	4.36
0.2	0.033	0.020	0.013	0.017	0.025	0.36	0.39	0.44	0.40	0.38		0.213	4.69
0.22	0.037	0.022	0.015	0.018	0.028	0.35	0.39	0.42	0.39	0.37		0.188	5.33
0.25	0.042	0.025	0.017	0.021	0.031	0.34	0.38	0.40	0.39	0.36		0.166	6.03
0.27	0.045	0.027	0.018	0.023	0.034	0.33	0.37	0.40	0.38	0.36		0.154	6.48
0.3	0.050	0.030	0.020	0.025	0.038	0.32	0.37	0.39	0.38	0.35		0.137	7.33
0.33	0.055	0.033	0.022	0.028	0.041	0.31	0.36	0.39	0.37	0.34		0.133	7.54
0.35	0.058	0.035	0.023	0.029	0.044	0.30	0.35	0.38	0.37	0.33		0.127	7.87
0.37	0.062	0.037	0.025	0.031	0.046	0.29	0.35	0.38	0.36	0.33		0.121	8.28
0.4	0.067	0.040	0.027	0.033	0.050	0.28	0.34	0.37	0.36	0.32		0.118	8.49
0.42	0.070	0.042	0.028	0.035	0.053	0.27	0.34	0.37	0.35	0.31	SD	0.111	9.03
0.45	0.075	0.045	0.030	0.038	0.056	0.26	0.33	0.37	0.35	0.30		0.105	9.49
0.47	0.078	0.047	0.031	0.039	0.059	0.25	0.33	0.36	0.34	0.30		0.100	10.00
0.5	0.083	0.050	0.033	0.042	0.063	0.24	0.32	0.36	0.34	0.29	NC	0.097	10.32

The eigenvalue curve of Figure 14 was generated using a nonlinear model of the RC bridge in which the effective moment of inertia values I_{eff} for the piers were obtained from Table 1 (or from Figure 5) for a deck target displacement $u_{deck,i} = 0.08$ m. At this target displacement, the resulting chord rotations θ_{pr} at the base section of the columns for piers 1 to 5 are as follows: $\theta_1 = 0.08/6 = 0.0133$, $\theta_2 = 0.08/10 = 0.008$, $\theta_3 = 0.08/15 = 0.00533$, $\theta_4 = 0.08/12 = 0.0067$, and $\theta_5 = 0.08/8 = 0.01$ rad. Therefore, the corresponding I_{eff}/I_g values for piers 1 to 5 are 0.44, 0.58, 0.72, 0.65, and 0.49, respectively (Table 4). These values contribute to the accurate representation of the bridge's dynamic behavior as depicted in the eigenvalue curve.

According to phase (a) of the "*M* and *P*" technique for existing bridges, an identification monitoring system is installed in the RC bridge, and response accelerations along the longitudinal direction are recorded when the bridge is in a quasi-calm state. The analysis of the records reveals an identified eigenfrequency $f_i = 0.539$ Hz for the *i*-step of the pushover analysis. Subsequently, in accordance with phase (d) of the "*M* and *P*" technique, the identified eigenfrequency f_i is inserted into the eigenfrequency diagram of Figure 14, enabling the determination of the corresponding displacement $u_i = 0.08$ m of the bridge deck. This displacement corresponds to specific θ_{pr} values for each pier, as mentioned above. Additionally, at the final step of the pushover analysis targeting the deck displacement $u_i = 0.08$ m, a visual representation of the damage state in the RC bridge is acquired (damage image). As illustrated in Figure 15, the columns of piers 1, 2, and 5 have just yielded at approximately the same displacement. Figure 16 displays the capacity curves of the bridge piers resulting from a pushover analysis targeting a deck displacement $u_{deck} = 0.08$ m, in terms of base shear and θ_{pr}.

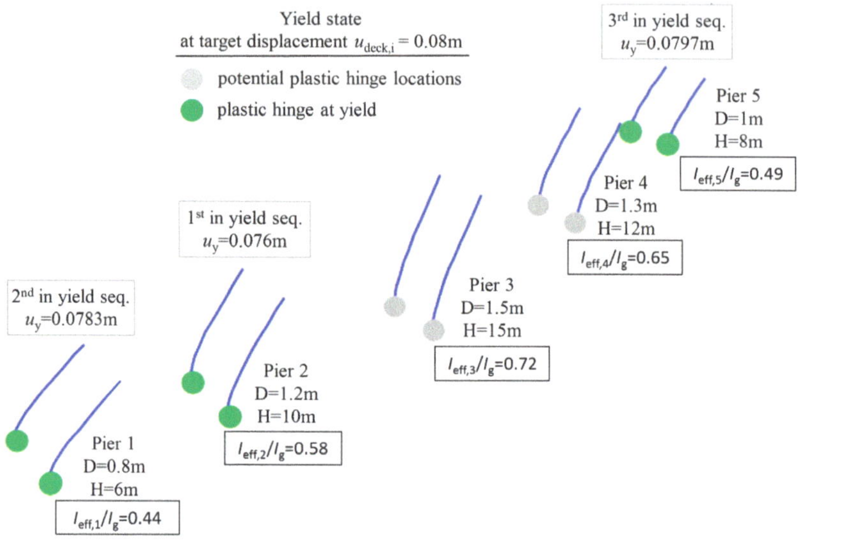

Figure 15. Yield state and sequence of yield of the bridge piers at deck target displacement $u_i = 0.08$ m.

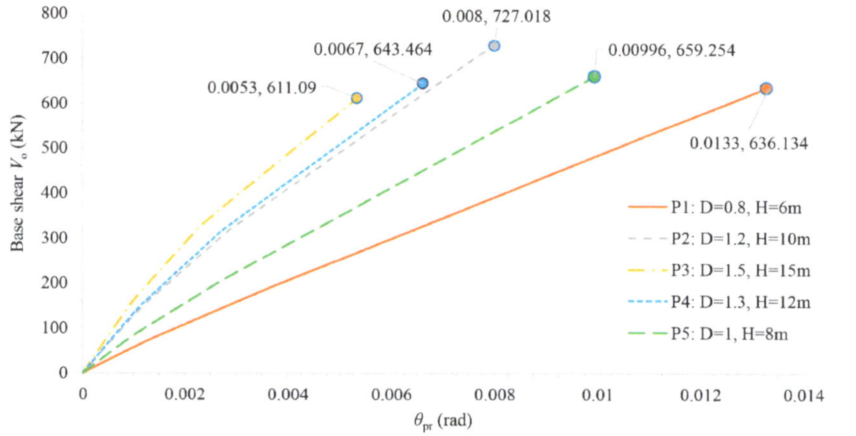

Figure 16. Capacity curve of the RC bridge piers for a deck target displacement $u_{deck} = 0.08$ m, in terms of base shear and θ_{pr}.

The eigenfrequency of the bridge $f_i = 0.539$ Hz, corresponding to a deck seismic displacement $u_{deck,i} = 0.08$ m, is approximately 63% of the eigenfrequency $f_{Health} = 0.865$ Hz at the health (uncracked) state of the bridge. This presents an elongation in period values from $T_{Health} = 1.156$ s to $T_i = 1.856$ s at this damage state.

It is emphasized that in a seismic event, the actual seismic load on the bridge along the longitudinal direction is different and varies at each time step. As a result, the damage distribution on the bridge and the cyclic degradation of stiffness, closely linked to the duration and number of cycles of the ground excitation, may slightly differ from that obtained through pushover analysis. However, the critical parameter in the current proposed methodology for bridges is the eigenfrequency f_i, identified by the monitoring procedure with a local network of uniaxial accelerometers under calm conditions. With the knowledge of the eigenfrequency f_i of the bridge, the seismic lateral deck displacement $u_{deck,i}$ along the longitudinal direction of the bridge can be estimated, as illustrated in Figure 14. Subsequently, the capacity curve is utilized to identify the damage state of the bridge piers.

In the final phase (g) of the "*M* and *P*" technique, the instantaneous stiffness k_i along the longitudinal direction of the RC bridge is computed at the examined inelastic *i*-step where the deck displacement is equal to 0.08 m. To achieve this, a linear analysis is conducted with a prescribed lateral force F_i applied at the deck level, following the final step of the pushover analysis where the deck displacement $u_{deck,i} = 0.08$ m is reached. The corresponding static displacement $u_{st,i}$ is then calculated. Next, the stiffness k_i of the damaged bridge is calculated using the ratio $F_i / u_{st,i}$. Therefore, the damage stiffness Δk_i of the bridge at the same inelastic *i*-step is derived from the general relationship $\Delta k_i = k_o - k_i$, where k_o represents the known initial stiffness of the undamaged bridge, which is calculated at the health state using the same procedure. Additionally, the damage stiffness $\Delta k_{c,i}$ of the various RC piers of the bridge can be obtained from the preceding linear analysis by recording their base shear. Consequently, the location and magnitude of the damage at the examined seismic deck displacement $u_{deck,i} = 0.08$ m can be identified.

The damage stiffness of the RC bridge and that of the bridge piers at the seismic deck displacement $u_{deck,i} = 0.08$ m is provided in Table 5. By knowing the damage stiffness Δk_i and $\Delta k_{c,i}$ of the RC bridge and the various RC piers along the longitudinal direction of the bridge, the final percentage deviation terms of Δk_i and $\Delta k_{c,i}$ can be calculated with respect to the initial stiffness k_o and $k_{c,o}$, respectively. These deviation terms are presented in the last column of Table 5. A visual representation of this table is depicted in Figure 17.

Table 5. Percentage deviation of the damage stiffness Δk_i for the RC bridge and the bridge piers at the inelastic *i*-step corresponding to the seismic deck displacement $u_{deck,i} = 0.08$ m.

Pier with 2 Columns	Health State k_o (kN/m)	Damage State at $u_{deck,i}$ = 0.08 m k_i (kN/m)	$\Delta k_i = k_o - k_i$ (kN/m)	$\Delta k_i / k_o$ (%)
C1a	8932	3330	5602	63
C1b	8932	3330	5602	63
C2a	9625	2819	6806	71
C2b	9625	2819	6806	71
C3a	7025	3107	3919	56
C3b	7025	3107	3919	56
C4a	7724	3498	4226	55
C4b	7724	3498	4226	55
C5a	9141	3711	5430	59
C5b	9141	3711	5430	59
Bridge	84,895	32,930	51,965	61

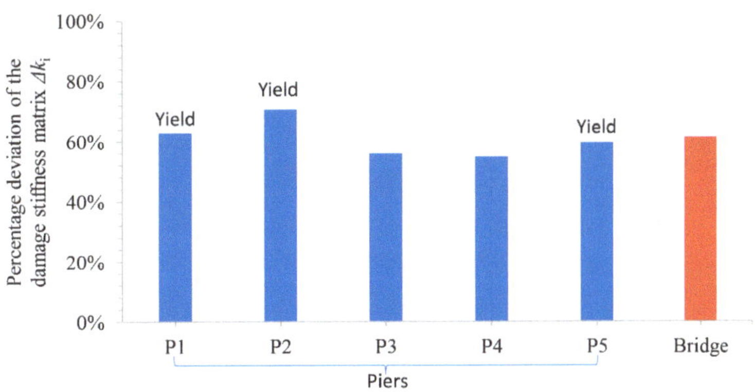

Figure 17. Percentage deviation of the damage stiffness Δk_i for the RC bridge and the bridge piers at the inelastic *i*-step corresponding to the seismic deck displacement $u_{deck,i} = 0.08$ m.

4. Discussion

As indicated by the results presented in Table 4 for the eigenfrequency, in Figure 14 for the deck (target) displacement, in Figures 15 and 17 for the plastic mechanism, as well as in Table 5 and Figure 17 for the damage stiffness of the existing RC bridge, the location and severity of the damage in the RC bridge can now be confirmed in relation to the health state. This confirmation is feasible due to the interconnected nature of all the aforementioned parameters. This achievement aligns with the objectives of the proposed "*M* and *P*" hybrid technique for bridges.

Indeed, by computing the damage stiffness Δk_i of the RC bridge along the longitudinal direction, and, additionally, the damage stiffness $\Delta k_{c,i}$ of the various bridge piers, the final percentage deviation terms of Δk_i and $\Delta k_{c,i}$ can be determined with respect to the initial stiffness k_o and $k_{c,o}$ at the health state (Table 5 and Figure 17). These deviations in damage stiffness express the extent of damage in the RC bridge as a whole and at the base section of the bridge piers, at the final step of pushover analysis where the seismic deck displacement $u_{deck,i} = 0.08$ m occurs. This aligns seamlessly with the damage image in Figure 15. This article addresses the identification of damage along the longitudinal direction of straight, ductile, RC bridges with rigid decks. In this type of existing bridges, damage always occurs at the base sections of the columns of the various bridge piers. It is important to note that the stiffness terms of the various piers correspond to the lateral dynamic degree of freedom of the SDOF RC bridge along the longitudinal direction (Figure 7). Therefore, the deviation of these stiffness terms $\Delta k_{c,i}$ relative to the health state corresponds to the overall damage of the piers at their critical base region, which occurs at a discrete seismic target deck displacement. From Table 5 and Figure 17, it is observed that the circular columns of the piers 1, 2, and 5 exhibit higher deviation terms in damage stiffness $\Delta k_{c,i}$ compared to piers 3 and 4. Hence, the stiffness terms $\Delta k_{c,i}$ of the piers, as well as their magnitude, which constitute the total stiffness degradation Δk_i of the RC bridge along the longitudinal direction, are fully consistent with the damage image in Figure 15. This image illustrates that the base sections of these piers present the greatest damage.

Hence, it is demonstrated that, given a specific damage image in an existing, ductile, RC bridge, the stiffness of the bridge undergoes changes relative to the health state. This alteration in stiffness leads to a shift in the eigenfrequency of the bridge, which is experimentally identified through the monitoring procedure. Subsequently, utilizing the key diagram of the proposed "*M* and *P*" technique for bridges, the seismic deck displacement of the bridge along the longitudinal direction is determined. This displacement, on the one hand, aligns with the observed damage image and, on the other hand, it ensures the eigenfrequency matches the field-measured value. All the parameters utilized in the hybrid "*M* and *P*" technique for bridges are interconnected, establishing a self-evident and accurate methodology.

5. Conclusions

A new proposed hybrid technique for the identification of damage in ductile, RC bridges is evaluated in the current paper by investigating a group of straight, multiple-span bridges with rigid decks and piers of various heights, from which a specific numerical example was presented here. This is a four-span RC bridge with five piers, each consisting of two circular columns behaving as cantilevers.

The newly introduced hybrid technique for bridges, referred to as the "*M* and *P* technique" (where *M* stands for "Monitoring" and *P* for "Pushover"), integrates the pushover capacity curve of the bridge along its longitudinal direction with the diagram illustrating the instantaneous eigenfrequency of the bridge in relation to the inelastic seismic (target) deck displacement. This key diagram was generated through a series of pushover and instantaneous modal analyses, where the target deck displacement $u_{deck,i}$ gradually increased, corresponding to specific values of the chord rotations $\theta_{pr,i}$ of the bridge piers. In each analysis, the circular columns of the bridge piers in the nonlinear model are assigned suitable values of the effective bending stiffness $E_c I_{eff,i}$ corresponding to the target deck displacement. This is accomplished following a proposed stiffness scenario based on the principles outlined in EN1998-3. By incorporating the eigenfrequency of the existing (damaged) RC bridge in this diagram, initially identified through a network of accelerograms in the monitoring phase, the target deck displacement $u_{deck,i}$ of the bridge arises. Consequently, the damage image of the bridge at the final step of the pushover analysis targeting this specific deck displacement is revealed. Moreover, the instantaneous stiffness and the damage stiffness of the entire RC bridge, as well as the corresponding values of the individual bridge piers are computed at this final step of pushover analysis where the target displacement $u_{deck,i}$ is observed. The damage stiffness of the various bridge piers is fully compatible with the extent of damage at the base section of the corresponding piers.

The interconnected nature of the parameters involved in the proposed hybrid "*M* and *P*" technique for bridges, including deck displacement, bridge stiffness, and eigenvalue, enables a highly accurate estimation of the damage pattern in an existing RC bridge. This is achieved because all the interconnected parameters are derived from the eigenfrequency measured in the field by the monitoring procedure. The technique consistently predicts the location and severity of damage among the various bridge piers, providing a comprehensive understanding of the bridge's health.

While the "*M* and *P*" technique is currently well-suited for damage identification along the longitudinal direction of a straight RC bridge, ongoing research is exploring its applicability for detecting damage along the direction perpendicular to the bridge axis. Furthermore, investigations are underway to assess its effectiveness in other bridge geometric forms, including skewed or curved bridges. To broaden the scope of the current research methodology, further exploration into the method's applicability should consider investigating the vertical component of earthquakes and examining the complexities introduced by soil–structure interaction. Additionally, future investigations will focus on integrating the hybrid "*M* and *P*" technique into health monitoring procedures and exploring its potential use in neural networks.

Author Contributions: Conceptualization, A.B. and T.M.; methodology, A.B. and T.M.; software, A.B.; validation, A.B., T.M. and V.L.; formal analysis, A.B. and T.M.; investigation, A.B. and T.M.; resources, A.B., T.M. and V.L.; data curation, A.B., T.M. and V.L.; writing—original draft preparation, A.B.; writing—review and editing, A.B., T.M. and V.L.; visualization, A.B. and T.M.; supervision, A.B., T.M. and V.L.; project administration, A.B. and T.M.; funding acquisition, None. All authors have read and agreed to the published version of the manuscript.

Funding: This research received no external funding.

Data Availability Statement: Data presented in this study are available in the article.

Conflicts of Interest: The authors declare no conflicts of interest.

References

1. Basseville, M.; Benveniste, A.; Goursat, M.; Hermans, L.; Mevel, L.; Auweraer, H. Output-only Subspace-based structural identification: From theory to industrial testing practice. *J. Dyn. Syst. Meas. Control* **2001**, *123*, 668–676. [CrossRef]
2. Brincker, R.; Zhang, L.; Andersen, P. Modal identification of output-only systems using frequency domain decomposition. *Smart Mater. Struct. J.* **2001**, *10*, 441–445. [CrossRef]
3. Peeters, B. Identification and Damage Detection in Civil Engineering. Ph.D. Thesis, Katholieke Universiteit, Leuven, Belgium, 2000. Available online: https://www.researchgate.net/publication/238331491_System_Identification_and_Damage_Detection_in_Civil_Engineering (accessed on 1 September 2023).
4. Peeters, B.; Roek, G. Stochastic System Identification for Operational Modal Analysis: A Review. *J. Dyn. Syst. Meas. Control.* **2001**, *123*, 659–667. [CrossRef]
5. Wenzel, H.; Pichler, D. *Ambient Vibration Monitoring*; John Wiley & Sons, Ltd.: Chichester, UK, 2005.
6. Overschee, P.; De Moor, B. *Subspace Identification for Linear Systems: Theory-Implemented-Applications*; Kluwer Academic Pub-lishers: Dordrecht, The Netherlands, 1996. [CrossRef]
7. Papoulis, A. *Signal Analysis*; International Student Edition; McGraw-Hill Book: New York, NY, USA, 1985.
8. Oppenheim, A.; Schafer, R. *Digital Signal Processing*; Prentice-Hall, Inc.: Englewood Cliffs, NJ, USA, 1999. Available online: https://research.iaun.ac.ir/pd/naghsh/pdfs/UploadFile_2230.pdf (accessed on 1 September 2023).

9. Bendat, J.; Piersol, A. *Random Data. Analysis and Measurement Procedures*, 4th ed.; John Wiley & Sons, Inc.: Hoboken, NJ, USA, 2010. Available online: https://www.wiley.com/en-gb/Random+Data:+Analysis+and+Measurement+Procedures,+4th+Edition-p-9780470248775 (accessed on 1 September 2023).
10. Makarios, T. Identification of the mode shapes of spatial tall multi-storey buildings due to earthquakes. The new "modal time-histories" method. *J. Struct. Des. Tall Spec. Build.* **2012**, *21*, 621–641. [CrossRef]
11. Dragos, K.; Makarios, T.K.; Karetsou, I.; Manolis, G.D.; Smarsly, K. Detection and Correction of Synchronization-induced Errors in Operational Modal Analysis. *Arch. Appl. Mech.* **2020**, *90*, 1547–1567. [CrossRef]
12. Zimmerman, D.C.; Kaouk, M. Structural damage detection using a minimum rank update theory. *J. Vib. Acoust.* **1994**, *116*, 222–231. [CrossRef]
13. Kaouk, M.; Zimmerman, D.C. Structural damage assessment using a generalized minimum rank perturbation theory. *AIAA J.* **1994**, *32*, 836–842. [CrossRef]
14. Domaneschi, M.; Limongelli, M.P.; Martinelli, L. Damage Identification in a benchmark Cable-Stayed Bridge using the Interpolation Method. In Proceedings of the 7th European Workshop on Structural Health Monitoring, Nantes, France, 8–11 July 2014. Available online: https://www.researchgate.net/publication/281947001_Damage_Identification_in_a_Benchmark_Cable-Stayed_Bridge_Us-ing_the_Interpolation_Method (accessed on 1 September 2023).
15. Domaneschi, M.; Limongelli, M.P.; Martinelli, L. Damage detection and localization on a Cable-Stayed Bridge. *Earthq. Struct.* **2015**, *5*, 1113–1126. [CrossRef]
16. Amani, M.G.; Riera, J.D.; Curadelli, R.O. Identification of changes in the stiffness and damping matrices of linear structures through ambient vibrations. *Struct. Control Health Monit. J.* **2007**, *14*, 1155–1169. [CrossRef]
17. Zhang, S.; Wang, H.; Wang, W.; Chen, S. Damage Detection in Structures Using Artificial Neural Networks. In Proceedings of the International Conference on Artificial Intelligence and Computational Intelligence, Sanya, China, 23–24 October 2010; pp. 207–210. Available online: https://ieeexplore.ieee.org/document/5656766 (accessed on 1 September 2023). [CrossRef]
18. Lubrano Lobianco, A.; Del Zoppo, M.; Di Ludovico, M. Seismic Damage Quantification for the SHM of Existing RC Structures. In *Civil Structural Health Monitoring. CSHM 2021. Lecture Notes in Civil Engineering*; Rainieri, C., Fabbrocino, G., Caterino, N., Ceroni, F., Notarangelo, M.A., Eds.; Springer: Cham, Switzerland, 2021; Volume 156, pp. 177–195, ISBN 978-3-030-74257-7. [CrossRef]
19. Pepe, V.; De Angelis, A.; Pecce, M.R. Damage assessment of an existing RC infilled structure by numerical simulation of the dynamic response. *J. Civil. Struct. Health Monit.* **2019**, *9*, 385–395. [CrossRef]
20. Nazari, F.; Baghalian, S. A new Method for Damage Detection in symmetric beams using artificial neural network and Finite Element Method. *Int. J. Eng. Appl. Sci. (IJEAS)* **2011**, *3*, 30–36. Available online: https://dergipark.org.tr/en/pub/ijeas/issue/23578/251156 (accessed on 1 September 2023).
21. Reuland, Y.; Martakis, P.; Chatzi, E. A Comparative Study of Damage-Sensitive Features for Rapid Data-Driven Seismic Structural Health Monitoring. *Appl. Sci.* **2023**, *13*, 2708. [CrossRef]
22. Martakis, P.; Reuland, Y.; Stavridis, A.; Chatzi, E. Fusing damage-sensitive features and domain adaptation towards robust damage classification in real buildings. *Soil Dyn. Earthq. Eng.* **2023**, *166*, 107739. [CrossRef]
23. Sun, L.; Shang, Z.; Xia, Y.; Bhowmick, S.; Nagarajaiah, S. Review of Bridge Structural Health Monitoring Aided by Big Data and Artificial Intelligence: From Condition Assessment to Damage Detection. *J. Struct. Eng.* **2020**, *146*, 04020073. [CrossRef]
24. Karimi, S.; Mirza, O. Damage identification in bridge structures: Review of available methods and case studies. *Aust. J. Struct. Eng.* **2023**, *24*, 89–119. [CrossRef]
25. Nettis, A.; Raffaele, D.; Uva, G. Seismic risk-informed prioritisation of multi-span RC girder bridges considering knowledge-based uncertainty. *Bull. Earthq. Eng.* **2023**. [CrossRef]
26. Moschonas, I.F.; Kappos, A.J.; Panetsos, P.; Papadopoulos, V.; Makarios, T.; Thanopoulos, P. Seismic fragility curves for greek bridges: Methodology and case studies. *Bull. Earthq. Eng.* **2009**, *7*, 439–468. [CrossRef]
27. Fernandez-Navamuel, A.; Pardo, D.; Magalhães, F.; Zamora-Sánchez, D.; Omella, Á.J.; Garcia-Sanchez, D. Bridge damage identification under varying environmental and operational conditions combining Deep Learning and numerical simulations. *Mech. Syst. Signal Process.* **2023**, *200*, 110471. [CrossRef]
28. Feng, K.; González, A.; Casero, M. A kNN algorithm for locating and quantifying stiffness loss in a bridge from the forced vibration due to a truck crossing at low speed. *Mech. Syst. Signal Process.* **2021**, *154*, 107599. [CrossRef]
29. Dilena, M.; Morassi, A.; Perin, M. Dynamic identification of a reinforced concrete damaged bridge. *Mech. Syst. Signal Process.* **2011**, *25*, 2990–3009. [CrossRef]
30. Makarios, T. Damage Identification in plane multi-storey reinforced concrete frame. *Open Constr. Build. Technol. J.* **2023**, *17*, e187483682302070. [CrossRef]
31. Makarios, T.K.; Bakalis, A.P. Identification of Damage in Planar Multistory Reinforced Concrete Frames Developing a Beam-Sway Plastic Mechanism Using the "M and P" Technique. *Buildings* **2023**, *13*, 2316. [CrossRef]
32. *EN 1998-1*; Eurocode 8: Design of Structures for Earthquake Resistance—Part 1: General Rules, Seismic Actions and Rules for Buildings. European Committee for Standardization: Brussels, Belgium, 2004.
33. *EN 1998-3*; Eurocode 8: Design of Structures for Earthquake Resistance—Part 3: Assessment and Retrofitting of Buildings. European Committee for Standardization: Brussels, Belgium, 2005.
34. Bakalis, A.; Makarios, T. Seismic Enforced-Displacement pushover procedure on multistorey R/C buildings. *Eng. Struct.* **2021**, *229*, 111631. [CrossRef]

35. *EN 1991-2*; Eurocode 1: Actions on Structures—Part 2: Traffic Loads on Bridges. European Committee for Standardization: Brussels, Belgium, 2003.
36. *EN 1998-2*; Eurocode 8: Design of Structures for Earthquake Resistance—Part 2: Bridges. European Committee for Standardization: Brussels, Belgium, 2005.
37. *Structural Analysis Program SAP2000v23*; Computers and Structures, Inc.: Walnut Creek, CA, USA, 2021. Available online: www.csiamerica.com (accessed on 1 September 2023).
38. Mander, J.B.; Priestley, M.J.N.; Park, R. Theoretical Stress-Strain Model for Confined Concrete. *J. Struct. Eng.* **1988**, *114*, 1804–1826. [CrossRef]
39. Park, R.; Paulay, T. *Reinforced Concrete Structures*; John Wiley & Sons, Inc.: New York, NY, USA, 1975.

infrastructures

Article

Inertial Methodology for the Monitoring of Structures in Motion Caused by Seismic Vibrations

Julio C. Rodríguez-Quiñonez [1,†], Jorge Alejandro Valdez-Rodríguez [2,*,†], Moises J. Castro-Toscano [1], Wendy Flores-Fuentes [1] and Oleg Sergiyenko [3]

[1] Facultad de Ingeniería, Universidad Autónoma de Baja California, Mexicali 21280, Mexico; julio.rodriguez81@uabc.edu.mx (J.C.R.-Q.); moises.castro@uabc.edu.mx (M.J.C.-T.); flores.wendy@uabc.edu.mx (W.F.-F.)
[2] School of Engineering, CETYS Universidad, Mexicali 21259, Mexico
[3] Instituto de Ingeniería, Universidad Autónoma de Baja California, Mexicali 21280, Mexico; srgnk@uabc.edu.mx
* Correspondence: jorge.valdez@cetys.mx
† These authors contributed equally to this work.

Abstract: This paper presents a non-invasive methodology for structural health monitoring (SHM) integrated with inertial sensors and signal conditioning techniques. The proposal uses the signal of an IMU (inertial measurement unit) tri-axial accelerometer and gyroscope to continuously measure the displacements of a structure in motion due to seismic vibrations. A system, called the "Inertial Displacement Monitoring System" or "IDMS", is implemented to attenuate the signal error of the IMU with methodologies such as a Kalman filter to diminish the influence of white noise, a Chebyshev filter to isolate the frequency values of a seismic motion, and a correction algorithm called zero velocity observation update (ZVOB) to detect seismic vibrations and diminish the influence of external perturbances. As a result, the IDMS is a methodology developed to measure displacements when a structure is in motion due to seismic vibration and provides information to detect failures opportunely.

Keywords: structural health monitoring; inertial sensors; signal conditioning; drift attenuation; correction algorithm; filtering; seismic vibration

Citation: Rodríguez-Quiñonez, J.C.; Valdez-Rodríguez, J.A.; Castro-Toscano, M.J.; Flores-Fuentes, W.; Sergiyenko, O. Inertial Methodology for the Monitoring of Structures in Motion Caused by Seismic Vibrations. *Infrastructures* **2024**, *9*, 116. https://doi.org/10.3390/infrastructures9070116

Academic Editors: Seyed Saman Khedmatgozar Dolati and Armin Mehrabi

Received: 8 June 2024
Revised: 12 July 2024
Accepted: 16 July 2024
Published: 19 July 2024

1. Introduction

Implementing structural health monitoring (SHM) technologies to measure the condition of a structure when subjected to seismic motion is a topic of interest for public safety, particularly in seismic regions. Proper failure detection generates opportunities for adequate maintenance and the understanding of what generates a fracture or collapse in structures such as buildings, tunnels, bridges, or the included mechanical structural systems such as tanks, wind turbines, and aerospace composites [1,2].

Currently, there are different approaches to performing SHM and obtaining information about a structure's condition when it is subjected to an earthquake. Research such as implementing cameras to obtain images and video is a proposal complemented by training neural networks, where the information from databases is used to detect irregularities [3–5]. These techniques implement pattern recognition techniques that seek to imitate how the human brain works [6,7]. Additionally, they are trained with vibration signals and register the information at different levels or structural states. However, neural networks and other non-deterministic models require high computational resources, and they are susceptible to noise generated in the environment, which can make it difficult to monitor the process for structures in monitoring networks [8–10].

Other research has suggested the use of optical sensors in an effort to detect cracks, delamination, or fractures [11–13], or implementing laser scanning techniques capable of detecting displacements and structural deformations [14,15]. These technologies are non-invasive and provide monitoring of structures, where digitalized images detect specified parameters that serve as a reference for the system. Hence, as optical sensors and laser scanning techniques provide useful information to monitor structural health, light reflection and the illumination intensity can cause measurement errors due to the changing nature of the environment. In addition, optical sensors are required to be isolated from the vibrations affecting the structure, and it is necessary to place them under proper light conditions to avoid measurement errors [16–18].

Furthermore, some researchers have considered accelerometers to perform SHM, because they can measure under dynamic conditions and provide vibration measurements, and this information is used to calculate displacements in structures. However, some systems have problems synchronizing the measured data when using digital accelerometers; there is an inherent drift in inertial sensors [19,20]. Hence, some of the proposals have used probabilistic or machine learning techniques. Consequently, most inertial systems are used as a complement to a more robust system to compensate for measurement errors [21,22].

Although the mentioned techniques provide valuable data on a structure's condition, they are susceptible to environmental interferences and have present difficulties monitoring when atypical behavior influences the structure. As a result, suitable detection of fractures in structures due to the displacements produced by seismic motion is still a challenge.

This paper presents the use of an accelerometer in conjunction with a gyroscope in an "Inertial Measurement Unit" IMU inertial sensor, where the acquired signal of the movements in the structure is analyzed and used to calculate the displacement with the signal conditioning phase [23–25]. As microelectromechanical systems (MEMS) can be placed at different points on a structure in a non-invasive configuration, the system provides information about the structure to perform a non-destructive evaluation.

The methodology "Inertial Displacement Monitoring System" or IDMS measures the displacements in a two-degrees-of-freedom (DoF) structural system placed on a shaking table that simulates the seismic motion of primary waves (P-waves) in a horizontal plane. The system implements the IMU's accelerometer and gyroscope placed at the top of one of the links of the structure and performs signal conditioning using a Kalman filter and a proposed zero velocity observation update (ZVOB) correction algorithm. The principle of the system, called the inertial displacement monitoring system "IDMS", is shown in Figure 1 and will be explained in Section 4.

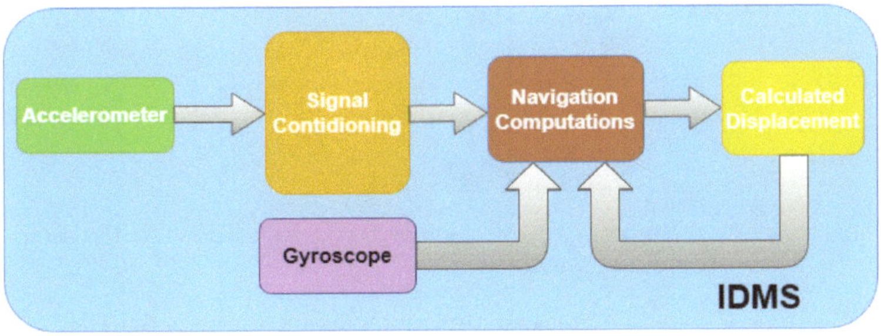

Figure 1. IDMS: A SHM system for the measurement of displacements through an inertial system methodology.

The main contributions of this paper are summarized as follows:

- Continuous monitoring and measurement of structure displacement caused by seismic vibration.
- An inertial system is implemented to perform SHM under seismic vibrations and provide information to perform a non-destructive evaluation.
- Correction algorithm ZVOB: ZVOB is a robust technique that evaluates whether a body is still in motion or in a steady-state phase.

- Drift reduction: error caused by drift due to the accelerometer is attenuated to increase the precision of measurements.
- Frequency analysis of a structure to measure the displacement when it is excited by seismic vibrations.

This paper is organized as follows: Section 2, Inertial Sensor Computations, presents equations and principles of an inertial system, describing the reference frame and how its position is calculated. Section 3, Signal Conditioning, presents the filter and correction algorithms implemented to attenuate the error and measure under seismic vibratory conditions. Section 4, Inertial Displacement Monitoring System, describes the principle of the IDMS and contains an description of the methodology; additionally, it includes a block diagram with the elements that compose the system presented in this paper. Section 5, Experimentation, explains the developed experiments and the conditions that demonstrate the system's effectiveness. Section 6, Results, gives an analysis of the obtained results and a statistical interpretation of the data. Section 7, Discussion, is where the system results are compared to other works. Section 8, Conclusions, summarizes the main points and presents future research.

2. Inertial Sensor Computations

2.1. Local Level Reference Frame

The IDMS methodology processes MEMS inertial sensor signals acquired by a data acquisition system connected to a computer and calculates the structural displacement through mathematical methods. The implemented IMU is in a strapdown configuration, where the three accelerometers are placed on a printed circut board (PCB) with their center of gravity (cg) aligned [26]. Therefore, the IMU is mounted on one of the sides of the structure and aligns with of the axes of the simulated movement. In Figure 2 is shown a diagram of the IMU with the orientation of three accelerometers mounted on the structure.

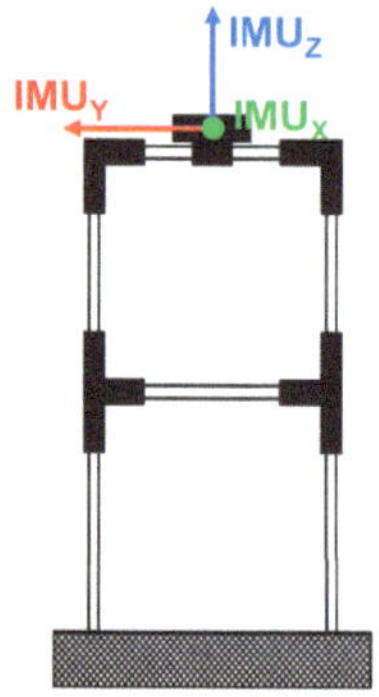

Figure 2. The IMU position in one of the links at the top of the structure to measure the displacements realized.

All the measurements and experiments of the IDMS are located in a local level reference frame *l-frame*. The *l-frame* presents the object of interest in a non-accelerating and non-rotating set of axes configuration known as ENU (East, North, Up), where the *z-axis* points up and is aligned with the Earth's gravity [27,28]. The IMU sensor is placed to measure the different displacements caused by seismic vibrations and so that it will maintain the same orientation in the same reference frame.

2.2. Velocity and Position

In an inertial system where accelerometer measurements are used to calculate a body's position, it is required to condition the sensor's information considering the *Earth's gravity* g and *Coriolis force* $\mathbf{\Omega}$ [29].

The Coriolis effect $\mathbf{\Omega}$ is a deflective force created by the Earth's rotation, affecting bodies on its surface. An inertial system can measure this with a gyroscope contained in an IMU. The IDMS system includes a gyroscope for measuring the angular velocity caused by the Coriolis force [30]. Hence, the signal acquired by the gyroscope $\boldsymbol{\omega}^l$ is a combination of the Coriolis effect $\mathbf{\Omega}$ and the sensor's angular velocity in ω_x^l, ω_y^l and ω_z^l.

$$\boldsymbol{\omega}^l = \begin{bmatrix} \omega_x^l \\ \omega_y^l \\ \omega_z^l \end{bmatrix} + \begin{bmatrix} \Omega_x \\ \Omega_y \\ \Omega_z \end{bmatrix} = (\boldsymbol{\omega} + \mathbf{\Omega}) \tag{1}$$

In a navigation body, the acceleration is affected by a radial force produced by the angular velocities contained in the $\boldsymbol{\omega}^l$ cross product with the *sensor's velocity* $\mathbf{v}^l$ and by the *gravitational acceleration* $\mathbf{g}^l$ [31]. Therefore, to calculate the inertial acceleration of the sensor, it is necessary to isolate the accelerometer's signal from the mentioned external forces, as demonstrated below in Equation (2).

$$\mathbf{a}_I^l = \begin{bmatrix} a_x^l \\ a_y^l \\ a_z^l \end{bmatrix} = \mathbf{a}^l - (\boldsymbol{\omega}^l + \mathbf{\Omega}) \times \mathbf{v}^l - \mathbf{g}^l \tag{2}$$

Afterward, integration is applied to the data in order to obtain the displacement, once to compute from acceleration to velocity and again to calculate position with velocity.

$$\mathbf{v}^l(n) = \int_{n-1}^{n} \mathbf{a}_I^l dt \approx \mathbf{v}^l[n-1] + \mathbf{a}_I^l[n]\Delta t \tag{3}$$

$$\mathbf{p}^l(n) = \int_{n-1}^{n} \mathbf{v}^l dt \approx \mathbf{p}^l[n-1] + \mathbf{v}^l[n]\Delta t \tag{4}$$

where $\boldsymbol{v}^l$ and $\boldsymbol{p}^l$ are vectors presenting the velocity and position in the three axes. Hence, calculating the position only with the equations presented in this section is insufficient to provide a precise measurement of the shaking table movement. Therefore, a signal conditioning phase is implemented to diminish the error and attenuate drift.

3. Signal Conditioning

Inertial sensor computations are a fundamental part of the system for estimating displacement. Thus, as in most inertial systems, there is an inherent bias in the sensors and a multi-factorial drift in the system. Consequently, inconsistencies exist between the obtained measurements and the calculations performed, even in ideal conditions.

Therefore, a filtering phase and a correction algorithm must be integrated to attenuate the noise signals present in the sensors arrangement.

3.1. Kalman Filter

In the IDMS, a Kalman filter attenuates the white noise in accelerometer-acquired signals, along with the estimated velocity and position [32,33]. These data are contained in the measurement vector $\boldsymbol{X}$ as follows:

$$\mathbf{X} = \begin{bmatrix} \mathbf{v}^l \\ \mathbf{p}^l \end{bmatrix} \tag{5}$$

It is important to remember that $\boldsymbol{v}^l$ and $\boldsymbol{p}^l$ are vectors with data from the three axes, as Equations (3) and (4) show.

Subsequently, the Kalman filter initiates with an a priori state, calculating the prediction *state correction vector* $\hat{\boldsymbol{x}_n^-}$, as shown below:

$$\hat{\mathbf{x}_n^-} = \mathbf{F}_{n-1}\hat{\mathbf{x}_{n-1}} \tag{6}$$

where $\mathbf{F}_{n-1}$ is the *state's transition matrix* and $\mathbf{x}_{n-1}^{\hat{}}$ represents the previous *state's correction vector*. Afterward, the error matrix covariance $\boldsymbol{P}$ is calculated, which considers the *noise covariance* $\boldsymbol{Q}_{n-1}$ of the system and $\mathbf{F}_{n-1}$. Hence, $\boldsymbol{P}$ is computed as

$$\mathbf{P}_n^- = \mathbf{F}_{n-1}\mathbf{P}_{n-1}(\mathbf{F}_{n-1})^T + \mathbf{Q}_{n-1} \tag{7}$$

Hence, the Kalman gain state is calculated, which serves as a correction value for $\mathbf{x}_{n-1}^{\hat{}}$. Therefore, the Kalman gain $\boldsymbol{K}$ is calculated using $\boldsymbol{P}$ from Equation (7).

$$\mathbf{K}_n = \mathbf{P}_n^-(\mathbf{H}_n)^T[\mathbf{P}_n^-\mathbf{H}_n(\mathbf{H}_n)^T + \mathbf{R}_n] \tag{8}$$

where matrix $\boldsymbol{H}_n$ creates a relation between the state vector, and $\boldsymbol{R}_n$ is the *measurement noise covariance matrix*, which has to be tuned up for the system.

Therefore, the gain state is followed by the a posteriori state, which estimates a new $\hat{\mathbf{x}_n}$ employing the gain estimated in Equation (8) and the previously calculated $\hat{x_n^-}$.

$$\hat{\mathbf{x}_n} = \hat{\mathbf{x}_n^-} + \mathbf{K}_n[\mathbf{X} - \mathbf{H}_n\hat{\mathbf{x}_n^-}] \tag{9}$$

The second part of Equation (9) is referred to as *Kalman's innovation*. Kalman's innovation measures the discrepancy between the actual measured $\boldsymbol{X}$ and the predicted measurement $\boldsymbol{H}_n\hat{\mathbf{x}_n}$.

Afterward, a new covariance matrix for the following sample is calculated in Equation (7).

$$\mathbf{P}_n = [\mathbf{I} - \mathbf{K}_n\mathbf{H}_n]\mathbf{P}_n^- \tag{10}$$

As the Kalman filter attenuates the white noise in the accelerometer, it is necessary to complement this with other filters to attenuate the signal's more specific characteristics.

3.2. Gravity Filter

As the Earth's gravity is presented as a low frequency, the data obtained with the accelerometer passes through a filter to obtain a precise value of the current gravity affecting the system. Hence, the accelerometer data $\boldsymbol{a}^l$ are used in the filter, as shown below:

$$\mathbf{g}^l = [\Lambda \times g_{n-1}^l] + [(1-\Lambda) \times \mathbf{a}_n^l] \tag{11}$$

where Λ is the filter coefficient; thus, the calculus provides a new value of the gravity divided into the three axes of the accelerometer.

3.3. Chebyshev Filter

In the arrangement, the movements produced by seismic vibrations are at a higher frequency than the Earth's gravity. Hence, the IDMS seeks to identify the movements caused by these vibrations and calculate the total displacements acquired from the inertial sensor. Thus, to compute only these movements, a 6th-order Chebyshev filter is designed to attenuate the signals at a higher frequency than the seismic vibrations. It is important to note that as a consequence of the higher order, acquiring more samples of the previous iterations is required to implement the filter. As a consequence, a delay is produced in the system, which is imperceptible to a user but considerably improves the precision.

$$H_p(z) = \frac{q_1 + q_2 + q_3 + q_4 + q_5 + q_6}{k_1 + k_2 + k_3 + k_4 + k_5 + k_6 + k_7} \tag{12}$$

where k and q represent the filter coefficients for the input and output.

Simplified as the function of the signal in any axis of accelerometer inputs $\boldsymbol{a}_I^l$, and filtered in any axis accelerometer signal outputs $\boldsymbol{a}_{I/f}^l$, the equation is described below.

$$\mathbf{a}^l_{I/f}(n) = k_1\mathbf{a}^l_I(n) + k_2\mathbf{a}^l_I(n-1) + \ldots + k_6\mathbf{a}^l_I(n-5) + k_7\mathbf{a}^l_I(n-6) + q_1\mathbf{a}^l_{I/f}(n-1) + q_2\mathbf{a}^l_{I/f}(n-2)\ldots + q_6\mathbf{a}^l_{I/f}(n-6) \quad (13)$$

The described Equation (13) is the *filtered inertial acceleration in a local frame*, also called $\mathbf{a}^l_{I/f}$, which is implemented in Equation (4).

3.4. Zero Velocity Observation Update (ZVOB)

As a complement to the system's filtering phase, a new methodology based on zero velocity update (ZUPT) is implemented, called zero velocity observation update (ZVOB). The algorithm is based on a pedestrian walking, where there is a standing phase with minimal or no change during the body's navigation [34,35].

ZVOB is an algorithm that analyzes if a body is still in motion or is in a steady-state phase; thus, it differs from ZUPT, because it considers a calculated coefficient for each Euler axis according to the behavior of the acceleration, instead of filtering the information with a low-pass and high-pass filter [36,37]. The purpose is to avoid the restrictions in motion detection that could occur due to the loss of information during the filter phase in the conventional ZUPT algorithm. The algorithm presented in this section uses the sensor's acceleration in *l-frame*. Therefore, an $\boldsymbol{a}^l$ vector for the three axes is compared with a designated vector for every axis constant ζ, called the "Zero Velocity Coefficient".

ζ is a constant that is estimated in a steady-state phase. First, the difference between the current sample of the acceleration vector in a local frame and the previous acceleration vector in a local frame is calculated, as Equation (14) presents.

$$\left|\mathbf{a}^{l*}\right| = \begin{bmatrix} a^l_{x,n} \\ a^l_{y,n} \\ a^l_{z,n} \end{bmatrix} - \begin{bmatrix} a^l_{x,n-1} \\ a^l_{y,n-1} \\ a^l_{z,n-1} \end{bmatrix} \quad (14)$$

Hence, the magnitude of the vector $|\mathbf{a}^{l*}|$ presented in Equation (14) is calculated, as shown below.

$$\left|\mathbf{a}^{l*}\right| = \sqrt{(a^{l*}_x)^2 + (a^{l*}_y)^2 + (a^{l*}_z)^2} \quad (15)$$

Therefore, to calculate ζ, it is necessary to take a sample of data in a steady-state phase of $\mathbf{a}^{l*}$ to estimate the mean μ_{an} of the magnitude $|\mathbf{a}^{l*}|$, and with the acceleration in x, y and *z-axis*, the standard deviation σ_{ax} is calculated, σ_{ay} and σ_{az} of the sample taken. However, when the observed signal drifts towards infinity, the sample cannot be considered for the calculus. Hence, ζ is computed as presented below.

$$\zeta = \begin{bmatrix} \mu_{an} + \sigma_{ax} \\ \mu_{an} + \sigma_{ay} \\ \mu_{an} + \sigma_{az} \end{bmatrix} \quad (16)$$

Finally, Figure 3 shows a flowchart diagram of ZVOB implementing ζ, $\boldsymbol{a}^l$, and $\boldsymbol{v}^l$.

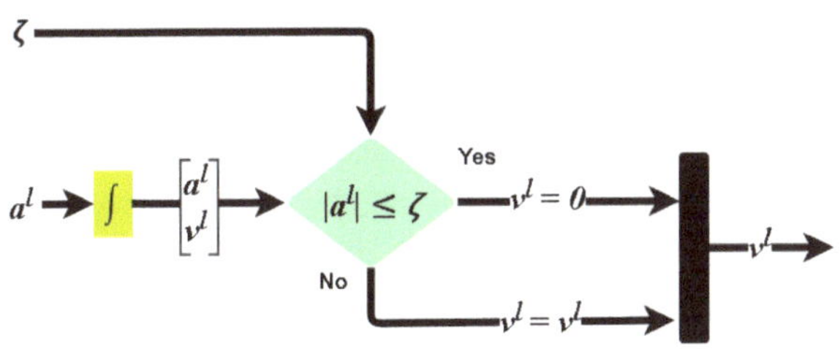

Figure 3. ZVOB flowchart.

The diagram shows that once v^l is computed, a^l is compared with ζ; if ζ is greater or equal to a^l, v^l is zero. Otherwise, it is considered that there is a v^l velocity in the sensor.

4. Inertial Displacement Monitoring System (IDMS) Methodology

The IDMS is a methodology that implements an inertial system for monitoring the displacement in diverse structures under vibration. The system considers the inherent drift of inertial sensors. It also implements the correction algorithm ZVOB to calculate the position and measure the displacements when the structure is under movement, to diminish the vibrations caused by external factors. The correction algorithm works in conjunction with a Kalman filter and a Chebyshev filter to attenuate white noise and identify the oscillations in the structure; it also uses a Gravity filter for the inertial configuration ENU to avoid considering the gravity as a movement. In Figure 4, a block diagram with the IDMS methodology is shown.

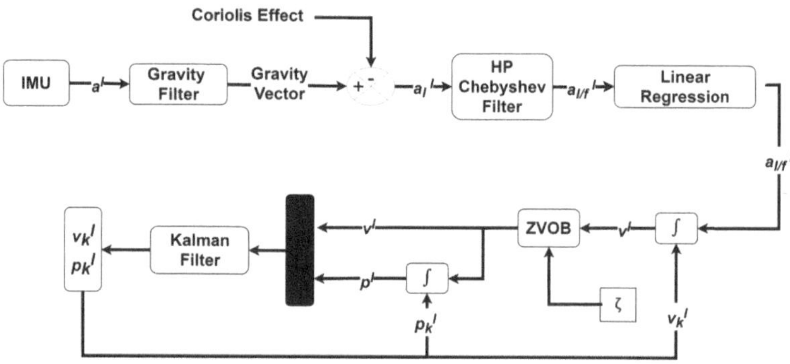

Figure 4. IDMS methodology: the block diagram shows the structure of the system. Monitoring is performed implementing the correction algorithm ZVOB, the filters of Kalman, Chebyshev, and Gravity, complemented by inertial equations to calculate the displacement of a structure.

The system initiates sample acquisition from the IMU's accelerometer and gyroscope. First, the accelerometer data passes through the gravity filter to obtain the gravity vector as shown in Equation (11). Then, the inertial acceleration a_I^l is calculated by implementing the sensor velocity a^l, which is 0 for an initial value, in conjunction with the information from the gyroscope, which includes the Ω and ω^l. As the system measures the displacement produced by seismic vibrations, it uses the accelerometer signal to determine the motion frequency, which requires proper signal conditioning. Hence, the data are filtered through a Chebyshev 6th-order filter to attenuate the low frequencies and preserve the information corresponding to the movements caused by the seismic vibrations. Afterward, a linear regression equation is calculated to estimate the frequency of the oscillation displacements. The equation implemented is shown in Equation (17) below, where f_m is the frequency measured, c_k are the linear regression coefficients and Df is the dominant frequency.

$$Df(f_m) = c_1 f_m^3 + c_2 f_m^2 + c_3 f_m + c_4 \tag{17}$$

Afterwards, the acceleration $a_{I/f}^l$, is integrated to estimate v^l, which is the sensor-measured velocity. The information from $a_{I/f}^l$ and v^l is used in the correction algorithm ZVOB in conjunction with the calculated tuning value ζ to define v^l for the current iteration. Subsequently, the v^l information is integrated to obtain p^l.

Therefore, the white noise in the calculated values v_{n-1}^l and ${p^l}_{n-1}$ is attenuated through a Kalman filter. These values are considered as the past sample in the calculus of v_n^l and ${p^l}_n$.

5. Experimentation

As mentioned, the IDMS measures the displacement of a structure under seismic conditions. Thus, to demonstrate the system's capacity to measure displacement in seismic vibratory conditions, a shaking table was implemented that performed a seismic movement oscillating forwards and backward, simulating the wave motions of primary waves (P-wave). Furthermore, the shaking table performed the movements in one axis, and the IMU was aligned in that axis direction to measure the displacements. The experiment was performed on a shaking table that moved at defined frequencies:

1. 6.25 Hz movement in *x-axis*.
2. 6.94 Hz movement in *x-axis*.
3. 7.81 Hz movement in *x-axis*.
4. 8.92 Hz movement in *x-axis*.

The selection of frequencies followed the range of the most common observed frequencies in seismic movements, and their average length time of 10 s was considered for the experiments [38–40]. The authors developed the shaking table; which includes such elements as fully threaded screws, a stepped motor 42HS4D-1704JA, and the PLA printed components of the structure used in the experiment; moreover, a program was created in Arduino, which runs on an ELEGOO UNO R3 controller board. The shaking table and the position of the IMU in the structure are shown in Figure 5.

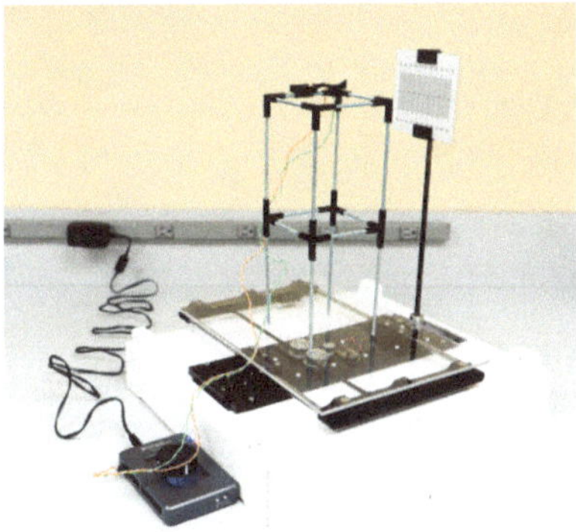

Figure 5. Shaking table and IMU position for the experiments.

The accelerometers were contained in an Invensense IMU MPU-9250 tracking device. The starting value in all the experiments was zero. The movement was performed over 5 s, and the other 5 s were in steady-state for comparison.

The inertial sensor movement at the top of the structure was measured with a graduated chart, as shown in Figure 6.

To ensure the accuracy of our results, we compared the measured movements with the actual displacement performed by the shaking table. This comparison was made possible by recording the movements with a high-quality slow-motion camera, with 12 MP, an aperture of f/1.6, and a 26 mm wide video setup of 1080p@120fps.

Therefore, experimentation data measured by the IMU (acceleration) were acquired with a MyRio of NI, an FPGA for data acquisition, and connected to an Asus TUF505 portable personal computer for processing and visualization. The sample rate of the captured read data was 1 ms. In the computer, a virtual instrument program was developed in order to implement the methodology and simultaneously take samples of data from the NIMyRio.

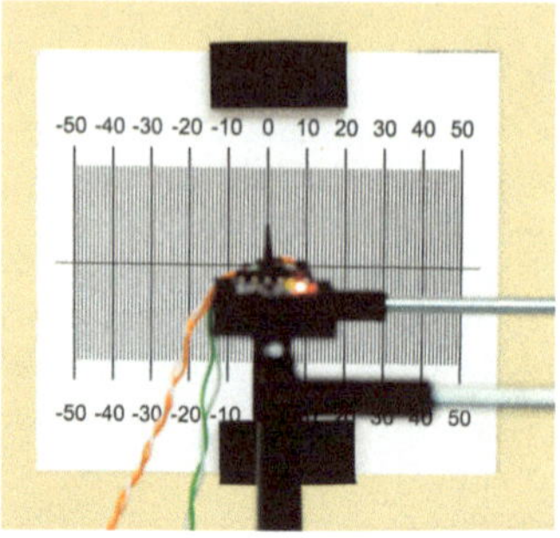

Figure 6. The inertial sensor movement was measured with a graduated chart in millimeters. In order to verify the precision of the displacement, the movements were recorded with a slow-motion camera.

6. Results

The experiment results are presented in two phases: signal frequency analysis and displacement analysis. As different frequencies were employed in the experiment, verifying that the system was measuring at the same frequency at which the movements took place was required. Thus, the displacements calculated with the methodology were compared with the real displacements performed by the shaking table.

6.1. Signal Analysis

With the data obtained through the experiment, it was possible to compute the dominant frequency of the movements in each experiment realized. The IDMS identified a frequency accordant with the structure motion and posteriorly attenuated non-related frequencies resulting from the shaking table and other external sources [41,42]. In Figure 7, all the frequencies measured in the experimentation are presented; this includes the dominant frequency measured in every experiment and the external detected frequencies in the experimentation. As a Chebyshev filter was applied, the system only use the data at the mentioned frequency to calculate the displacements performed by the shaking table.

The data presented in Table 1 show the relative percentage error calculated in the frequency measured. Furthermore, the system frequency column shows the mean of all the experiment-measured frequencies for each of the four proposed experiment types. Finally, the last column displays the relative percentage error from all the experiments for each case. In the four experiments, a low relative percentage error existed in the measurement of the frequency and a "Root Mean Square Error" or "RMSE" of 0.088, which presents the system's precision in determining the motion frequency at which the structure was performing the displacements. Therefore, the IDMS monitored the different excitations in the structure produced by seismic vibrations and attenuated those produced by the ambient elements. As a result, the system was capable of following the movements produced by the shaking table with precision, demonstrating that the displacements measured corresponded with the movements that occurred in the structure.

Table 1. Frequency analysis relative percentage error and RMSE

Experiment Frequency (Hz)	System Frequency (Hz)	Relative Percentage Error
6.25	6.2546	1.33%
6.94	6.9527	1.74%
7.81	7.88367	0.49%
8.92	8.9375	1.004%
	Overall	**1.14%**
	RMSE	**0.088**

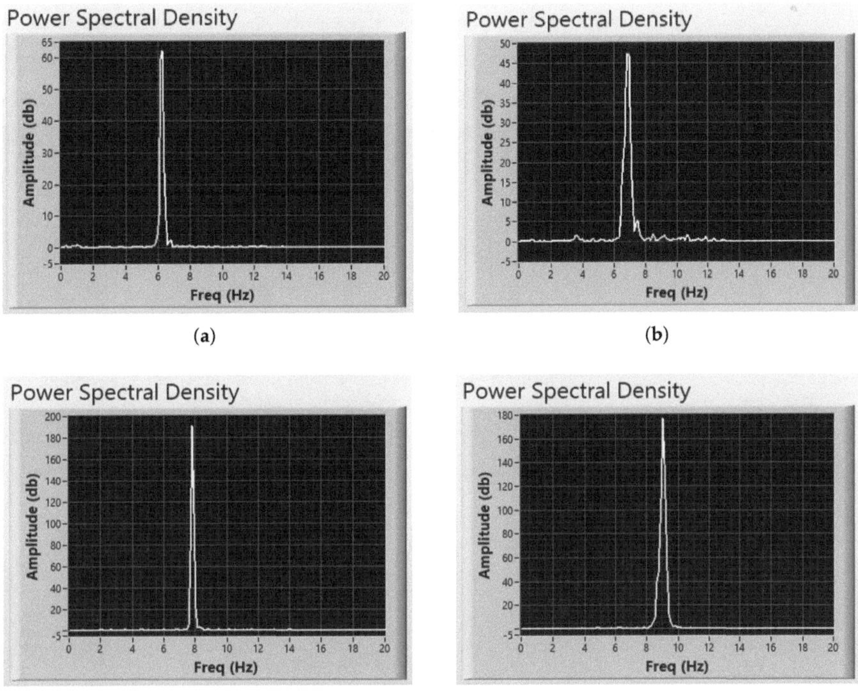

Figure 7. Power Spectral density chart showing the dominant frequency in each of the experiments. The graphics present the dominant frequency of the oscillation movements. (**a**) Frequency measured for a seismic vibratory movement that oscillated at 6.25 Hz. (**b**) Frequency measured for a seismic vibratory movement that oscillated at 6.94 Hz. (**c**) Frequency measured for a seismic vibratory movement that oscillated at 7.81 Hz. (**d**) Frequency measured for a seismic vibratory movement that oscillated at 8.92 Hz.

6.2. Displacement Measurements

As the IMDS calculated the frequency of the movements performed by the shaking table, the dominant frequency data were used to estimate and normalize the displacements. Posteriorly, the computed displacements were compared with the real displacements. In Figure 8, the four types of motion frequency performed by the shaking table are presented. Since the experiment simulated the seismic p-wave movement in a structure, the system measurement considered total displacement as the amplitude in a complete cycle.

It must be noted in Table 2 that as the frequency of the oscillation increments, the total displacement diminishes. The reason for this is that the shaking table had to adapt to the selected movements, which reduced the displacement to perform the required oscillations for each experiment frequency.

In Table 2 is shown how the system followed the movement behavior with an RMSE of 0.4092, demonstrating the precision of most of the movements. Therefore, an increment in the relative percentage error of the frequency of 7.81 hz was noted. This was due to the difference in displacement between the experiment at frequencies of 6.25 and 6.94 in comparison with the other experiments.

Thus, in all four experiments, the IDMS presented a relative percentage error below 7%, demonstrating the system's capacity to measure the displacement of structures under seismic motion. Additionally, as the system presented signal conditioning of the IMU's accelerometer data to obtain a measurable distance in the *x-axis*, the drift presented in an inertial system is still a subject of interest for future works.

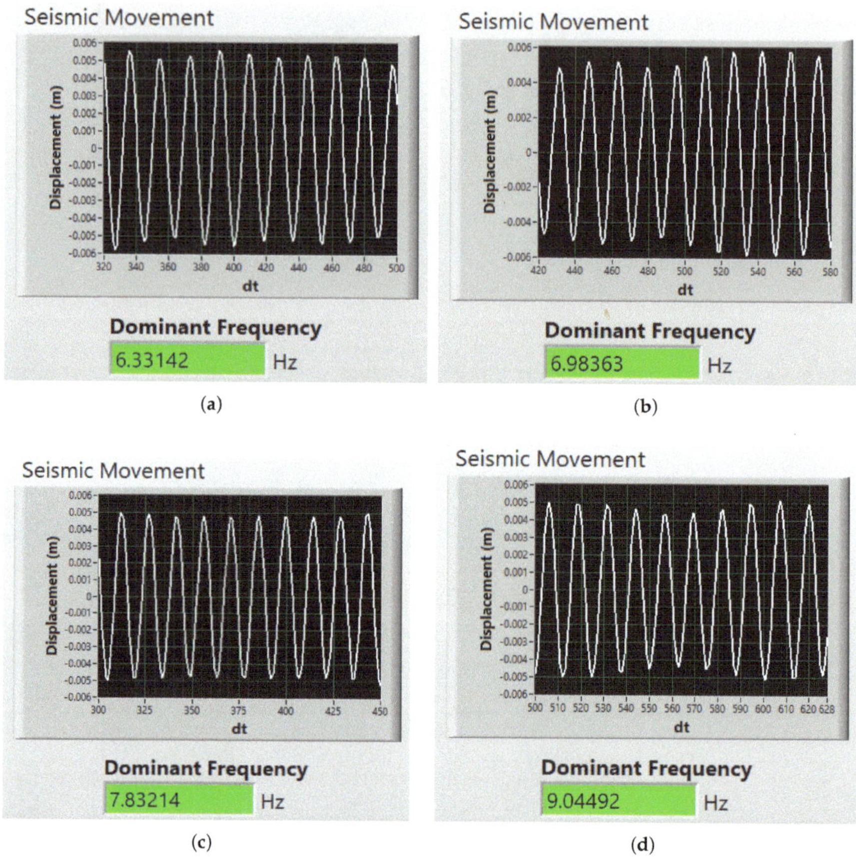

Figure 8. Total displacements and measured frequency. (**a**) Total displacement and frequency measured for the movement at 6.25 Hz. (**b**) Total displacement and frequency measured for the movement at 6.94 Hz. (**c**) Total displacement and frequency measured for the movement at 7.81 Hz. (**d**) Total displacement and frequency measured for the movement at 8.92 Hz.

Table 2. Relative Percentage Error Analysis of the Measured Displacement.

Experiment Frequency (Hz)	Structure Displacement (mm)	System Measured Displacement (mm)	Relative Percentage Error
6.25	11.6000	11.8851	**2.0132%**
6.94	11.3000	11.6923	**2.6201%**
7.81	10.9000	11.6555	**6.1587%**
8.92	10.4000	10.4712	**0.5712%**
		RMSE	**0.4092**

7. Discussion of the Results

The IDMS methodology is an inertial system that uses an IMU's accelerometer to determine displacements when a structure is in motion due to seismic vibrations.

The experiments provided data to measure the motion frequency and measure the structural displacements when subjected to simulated p-wave seismic motion. Therefore, it was compared with other methodologies that also measure the displacement in structures under the influence of seismic motion.

In [43], a binocular vision system approach based on deep learning was proposed to measure a structure's displacement. As the system measures tiny vibrations of a structure, performing image magnification of the motion is required, particularly where the displacement is smaller than one pixel.

However, ref. [44] presented a system for long-term monitoring, implementing four fiber-optic sensors to measure the structural displacements. This system considers the information of all the sensors in a structure; therefore, it is necessary to calibrate the current state of the structure's components to prevent a reduction in the correlation between measurement of the sensors.

As a result, the mean errors in the displacement measurement of Shao's and Garcia's works were 9.02% and 4.44%, respectively, compared with the IDMS system, which presented a mean error of 2.84%. Additionally, the IDMS relies on the IMU's accelerometer, which does not continuously re-evaluate the conditions of the structural components.

The work in [45] presented a multi-camera 3D DIC (digital image correlation) method using four high-speed cameras in a 1/8 scaled concrete structure. The multi-camera approach provides stereo vision measurements and the field of view cannot be overlapped. The RMSE in their *x-axis* displacement measurements was 0.43, compared to the IDMS's RMSE of 0.409.

The system's capacity to measure the displacement of structures demonstrates precision and efficiency under seismic vibratory conditions when monitoring the displacement produced in a structure.

8. Conclusions

The IDMS is an SHM methodology for performing continuous non-destructive measurements of a structure's displacements when it is subjected to seismic motions. As with all inertial systems, the measurement signal of an accelerometer and a gyroscope contained in an IMU requires conditioning to correct the influence of drift and to properly predict the displacements of a structure under motion.

The conditioning phase involves Kalman and Chebyshev filters to reduce the influence of external perturbations and measure the motion frequency, to determine that the displacement measurement that corresponds to the seismic motion that the structure is subjected to. An algorithm called zero velocity observation update (ZVOB) was included to determine if a body is in motion or steady-state; this provides the system with adaptability to distinctive seismic movements and avoids the influence of external perturbations.

The IDMS methodology is designed to continuously monitor and measure displacements during seismic motion in a structure. The system was tested with experiments simulating seismic p-waves in one of the axes and aligning this with the accelerometers contained in the IMU, where the selection of the frequencies considered a range of typical frequencies of seismic movements in different parts of the world. Additionally, the time of the vibration motion in the experiments was performed according to the usual time length of seismic motions.

As a result, the IDMS measured frequencies with an RMSE of 0.088 hz overall, identifying the frequencies of the seismic vibrations and avoiding frequencies created by machines, human walking, or other external elements that could create errors in the measurement of the displacement. Furthermore, the system measured the displacement with an RMSE of 0.4092 mm, monitoring the different motions to which the structure was subjected.

Implementing the IDMS in different structures, such as multiple-floor buildings, bridges, or linear machines that are usually in a single line, could provides information to evaluate the structural conditions and prevent potential hazards due to structural failures. As the system monitors displacements, it is possible to detect separations in structural elements, the size of cracks, and delamination, or to elaborate studies to prevent structures from colliding with adjacent buildings.

The efficacy of the implemented ZVOB and analysis of the developed methodology demonstrated an inertial SHM system capable of measuring displacements under seismic vibratory conditions with precision. Furthermore, if the IDMS is implemented with multiple sensors at different points of the structure, adapting the current methodology could allow improving the precision, performing a better analysis to monitor displacements and vibration, and obtaining more data to complement the presented IDMS methodology.

For further research, multiple inertial sensors are being contemplated to build a nested IDMS and improve the obtained capabilities. Additionally, fusion with other technologies, such as laser scanning techniques, as mentioned by Castro-Toscano [46], could provide a wider range of measurements and detection of deformation in structures.

Additionally, it could be connected with other inertial sensors or with a network of similar sensors to improve measurement precision.

Author Contributions: Conceptualization, J.C.R.-Q. and J.A.V.-R.; Methodology, J.C.R.-Q., J.A.V.-R., M.J.C.-T. and O.S.; Software, J.C.R.-Q. and J.A.V.-R.; Validation, M.J.C.-T., W.F.-F. and O.S.; Formal analysis, W.F.-F.; Investigation, J.C.R.-Q. and J.A.V.-R.; Writing—original draft, J.C.R.-Q. and J.A.V.-R. All authors have read and agreed to the published version of the manuscript.

Funding: This research received no external funding.

Data Availability Statement: The original data presented in the study are openly available in Zenodo at https://zenodo.org/doi/10.5281/zenodo.11583394 (accessed on 11 June 2024).

Acknowledgments: We gratefully acknowledge the support of Universidad Autónoma de Baja California and CONACYT.

Conflicts of Interest: The authors declare no conflicts of interest.

References

1. Zhang, C.; Mousavi, A.A.; Masri, S.F.; Gholipour, G.; Yan, K.; Li, X. Vibration feature extraction using signal processing techniques for structural health monitoring: A review. *Mech. Syst. Signal Process.* **2022**, *177*, 109175. [CrossRef]
2. Deng, Z.; Huang, M.; Wan, N.; Zhang, J. The Current Development of Structural Health Monitoring for Bridges: A Review. *Buildings* **2023**, *13*, 1360. [CrossRef]
3. Dang, H.V.; Tran-Ngoc, H.; Nguyen, T.V.; Bui-Tien, T.; De Roeck, G.; Nguyen, H.X. Data-Driven Structural Health Monitoring Using Feature Fusion and Hybrid Deep Learning. *IEEE Trans. Autom. Sci. Eng.* **2021**, *18*, 2087–2103. [CrossRef]
4. Shao, Y.; Li, L.; Li, J.; An, S.; Hao, H. Computer vision based target-free 3D vibration displacement measurement of structures. *Eng. Struct.* **2021**, *246*, 113040. [CrossRef]
5. Dong, C.Z.; Bas, S.; Catbas, F.N. Investigation of vibration serviceability of a footbridge using computer vision-based methods. *Eng. Struct.* **2020**, *224*, 111224. [CrossRef]
6. Zinno, R.; Haghshenas, S.S.; Guido, G.; VItale, A. Artificial Intelligence and Structural Health Monitoring of Bridges: A Review of the State-of-the-Art. *IEEE Access* **2022**, *10*, 88058–88078. [CrossRef]
7. Luan, L.; Zheng, J.; Wang, M.L.; Yang, Y.; Rizzo, P.; Sun, H. Extracting full-field subpixel structural displacements from videos via deep learning. *J. Sound Vib.* **2021**, *505*, 116142. [CrossRef]
8. Boscato, G.; Fragonara, L.Z.; Cecchi, A.; Reccia, E.; Baraldi, D. Structural Health Monitoring through Vibration-Based Approaches. *Shock Vib.* **2019**, *2019*, 2380616. [CrossRef]
9. Moradi, M.; Broer, A.; Chiachío, J.; Benedictus, R.; Loutas, T.H.; Zarouchas, D. Intelligent health indicator construction for prognostics of composite structures utilizing a semi-supervised deep neural network and SHM data. *Eng. Appl. Artif. Intell.* **2023**, *117*, 105502. [CrossRef]
10. Torzoni, M.; Manzoni, A.; Mariani, S. A Deep Neural Network, Multi-fidelity Surrogate Model Approach for Bayesian Model Updating in SHM. In *European Workshop on Structural Health Monitoring*; Rizzo, P., Milazzo, A., Eds.; Springer: Cham, Switzerland, 2023; pp. 1076–1086.
11. Lopez-Higuera, J.M.; Rodriguez Cobo, L.; Quintela Incera, A.; Cobo, A. Fiber Optic Sensors in Structural Health Monitoring. *J. Light. Technol.* **2011**, *29*, 587–608. [CrossRef]

12. Carani, L.B.; Martin, T.D.; Eze, V.O.; Okoli, O.I. Impact sensing and localization in composites structures with embedded mechanoluminescence-perovskite sensors. *Sens. Actuators A Phys.* **2022**, *346*, 113843. [CrossRef]
13. Rao, A.S.; Radanovic, M.; Liu, Y.; Hu, S.; Fang, Y.; Khoshelham, K.; Palaniswami, M.; Ngo, T. Real-time monitoring of construction sites: Sensors, methods, and applications. *Autom. Constr.* **2022**, *136*, 104099. [CrossRef]
14. Jeon, J.Y.; Miao, Y.; Park, G.; Flynn, E. Compressive laser scanning with full steady state wavefield for structural damage detection. *Mech. Syst. Signal Process.* **2022**, *169*, 108626. [CrossRef]
15. Choi, Y.; Abbas, S.H.; Lee, J.R. Aircraft integrated structural health monitoring using lasers, piezoelectricity, and fiber optics. *Measurement* **2018**, *125*, 294–302. [CrossRef]
16. Real-Moreno, O.; Rodríguez-Quiñonez, J.C.; Sergiyenko, O.; Flores-Fuentes, W.; Castro-Toscano, M.J.; Miranda-Vega, J.E.; Mercorelli, P.; Valdez-Rodríguez, J.A.; Trujillo-Hernández, G.; Sanchez-Castro, J.J. A Quadrant Approach of Camera Calibration Method for Depth Estimation Using a Stereo Vision System. In Proceedings of the IECON 2022—48th Annual Conference of the IEEE Industrial Electronics Society, Brussels, Belgium, 17–20 October 2022; pp. 1–6. [CrossRef]
17. Feng, D.; Feng, M.Q. Identification of structural stiffness and excitation forces in time domain using noncontact vision-based displacement measurement. *J. Sound Vib.* **2017**, *406*, 15–28. [CrossRef]
18. Trujillo-Hernández, G.; Rodríguez-Quiñonez, J.C.; Flores-Fuentes, W.; Sergiyenko, O.; Ontiveros-Reyes, E.; Real-Moreno, O.; Hernández-Balbuena, D.; Murrieta-Rico, F.N.; Rascón, R. Development of an integrated podometry system for mechanical load measurement and visual inspection. *Measurement* **2022**, *203*, 111866. [CrossRef]
19. Veluthedath Shajihan, S.A.; Chow, R.; Mechitov, K.; Fu, Y.; Hoang, T.; Spencer, B.F. Development of Synchronized High-Sensitivity Wireless Accelerometer for Structural Health Monitoring. *Sensors* **2020**, *20*, 4169. [CrossRef] [PubMed]
20. Shih, J.Y.; Weston, P.; Entezami, M.; Roberts, C.; O'Callaghan, M. Experiences Using MEMS Accelerometers on Railway Bearers at Switches and Crossings to Obtain Displacement—Awkward Situations. *Infrastructures* **2024**, *9*, 91. [CrossRef]
21. Hassani, S.; Dackermann, U.; Mousavi, M.; Li, J. A systematic review of data fusion techniques for optimized structural health monitoring. *Inf. Fusion* **2024**, *103*, 102136. [CrossRef]
22. Qu, X.; Ding, X.; Xu, Y.L.; Yu, W. Real-time outlier detection in integrated GNSS and accelerometer structural health monitoring systems based on a robust multi-rate Kalman filter. *J. Geod.* **2023**, *97*, 38. [CrossRef]
23. Castro-Toscano, M.; Rodríguez-Quiñonez, J.; Hernández-Balbuena, D.; Rivas-Lopez, M.; Sergiyenko, O.; Flores-Fuentes, W. Obtención de Trayectorias Empleando el Marco Strapdown INS/KF: Propuesta Metodológica. *Rev. Iberoam. Autom. Inform. Ind.* **2018**, *15*, 391–403. [CrossRef]
24. Rodríguez-Quiñonez, J.C. Intelligent Automatic Object Tracking Method by Integration of Laser Scanner System and INS. *Program. Comput. Softw.* **2020**, *46*, 619–625. [CrossRef]
25. Castro-Toscano, M.J.; Valdez-Rodríguez, J.A.; Rodríguez-Quiñonez, J.C.; Flores-Fuentes, W.; Sergiyenko, O.; Trujillo-Hernandez, G.; Real-Moreno, O. Determination of Trajectories Using IKZ/CF Inertial Navigation: Methodological Proposal/CF Inertial Navigation: Methodological Proposal. *Heliyon* **2023**, *9*, e13863. [CrossRef] [PubMed]
26. Sokolov, S.; Pogorelov, V.; Shatalov, A. Solving the Autonomous Initial Navigation Task for Strapdown Inertial Navigation System on the Perturbed Basis Using Rodriguez—Hamilton Parameters. *Russ. Aeronaut.* **2019**, *62*, 42–51. [CrossRef]
27. Bose, A.; Bhat, K.; Kurian, T. *Fundamentals of Navigation and Inertial Sensors*; Eastern Economy Edition; PHI Learning: Delhi, India, 2014.
28. Valdez-Rodríguez, J.A.; Rodríguez-Quiñonez, J.C.; Flores-Fuentes, W.; Ramírez-Hernández, L.R.; Trujillo-Hernández, G.; Real-Moreno, O.; Castro-Toscano, M.J.; Miranda-Vega, J.E.; Mercorelli, P. Visual-Inertial Navigation Systems and Technologies. In *Optoelectronic Devices in Robotic Systems*; Sergiyenko, O., Ed.; Springer International Publishing: Cham, Switzerland, 2022; pp. 137–166. [CrossRef]
29. Lin, Y.; Zhang, W.; Xiong, J. Specific force integration algorithm with high accuracy for strapdown inertial navigation system. *Aerosp. Sci. Technol.* **2015**, *42*, 25–30. [CrossRef]
30. Acar, C.; Shkel, A. Introduction; MEMS Vibratory Gyroscopes: Structural Approaches to Improve Robustness. In *MEMS Vibratory Gyroscopes: Structural Approaches to Improve Robustness*; Springer US: Boston, MA, USA, 2009; pp. 1–14. [CrossRef]
31. Klančar, G.; Zdešar, A.; Blažič, S.; Škrjanc, I. Chapter 5—Sensors Used in Mobile Systems. In *Wheeled Mobile Robotics*; Klančar, G., Zdešar, A., Blažič, S., Škrjanc, I., Eds.; Butterworth-Heinemann: Oxford, UK, 2017; pp. 207–288. [CrossRef]
32. Kumari, N.; Kulkarni, R.; Ahmed, M.R.; Kumar, N. Use of Kalman Filter and Its Variants in State Estimation: A Review. In *Artificial Intelligence for a Sustainable Industry 4.0*; Awasthi, S., Travieso-González, C.M., Sanyal, G., Kumar Singh, D., Eds.; Springer International Publishing: Cham, Switzerland, 2021; pp. 213–230. [CrossRef]
33. Quan, W.; Gong, X.; Fang, J.; Li, J. INS/GNSS Integrated Navigation Method. In *INS/CNS/GNSS Integrated Navigation Technology*; Springer Berlin Heidelberg: Berlin/Heidelberg, Germany, 2015; pp. 185–235. [CrossRef]
34. Wang, Y.; Shkel, A.M. Adaptive Threshold for Zero-Velocity Detector in ZUPT-Aided Pedestrian Inertial Navigation. *IEEE Sens. Lett.* **2019**, *3*, 1–4. [CrossRef]
35. Tong, X.; Su, Y.; Li, Z.; Si, C.; Han, G.; Ning, J.; Yang, F. A Double-Step Unscented Kalman Filter and HMM-Based Zero-Velocity Update for Pedestrian Dead Reckoning Using MEMS Sensors. *IEEE Trans. Ind. Electron.* **2020**, *67*, 581–591. [CrossRef]
36. Benzerrouk, H.; Nebylov, A.V. Robust IMU/UWB integration for indoor pedestrian navigation. In Proceedings of the 2018 25th Saint Petersburg International Conference on Integrated Navigation Systems (ICINS), Saint Petersburg, Russia, 28–30 May 2018; pp. 1–5. [CrossRef]

37. Suresh, R.P.; Sridhar, V.; Pramod, J.; Talasila, V. Zero Velocity Potential Update (ZUPT) as a Correction Technique. In Proceedings of the 2018 3rd International Conference On Internet of Things: Smart Innovation and Usages (IoT-SIU), Bhimtal, India, 23–24 February 2018; pp. 1–8. [CrossRef]
38. Schwardt, M.; Pilger, C.; Gaebler, P.; Hupe, P.; Ceranna, L. Natural and Anthropogenic Sources of Seismic, Hydroacoustic, and Infrasonic Waves: Waveforms and Spectral Characteristics (and Their Applicability for Sensor Calibration). *Surv. Geophys.* **2022**, *43*, 1265–1361. [CrossRef] [PubMed]
39. Aydan, Ö. *Earthquake Science and Engineering*; CRC Press: London, UK, 2022.
40. Towhata, I. *Geotechnical Earthquake Engineering*; Springer Series in Geomechanics and Geoengineering; Springer: Berlin/Heidelberg, Germany, 2008.
41. Chen, Y.; Li, H.; Hou, L.; Bu, X. Feature extraction using dominant frequency bands and time-frequency image analysis for chatter detection in milling. *Precis. Eng.* **2019**, *56*, 235–245. [CrossRef]
42. Shi, Y.; Mahr, F.; von Wagner, U.; Uhlmann, E. Chatter frequencies of micromilling processes: Influencing factors and online detection via piezoactuators. *Int. J. Mach. Tools Manuf.* **2012**, *56*, 10–16. [CrossRef]
43. Shao, Y.; Li, L.; Li, J.; An, S.; Hao, H. Target-free 3D tiny structural vibration measurement based on deep learning and motion magnification. *J. Sound Vib.* **2022**, *538*, 117244. [CrossRef]
44. Garcia-Sanchez, D.; Fernandez-Navamuel, A.; Sánchez, D.Z.; Alvear, D.; Pardo, D. Bearing assessment tool for longitudinal bridge performance. *J. Civ. Struct. Health Monit.* **2020**, *10*, 1023–1036. [CrossRef]
45. Wei, K.; Yuan, F.; Shao, X.; Chen, Z.; Wu, G.; He, X. High-speed multi-camera 3D DIC measurement of the deformation of cassette structure with large shaking table. *Mech. Syst. Signal Process.* **2022**, *177*, 109273. [CrossRef]
46. Castro-Toscano, M.J.; Rodríguez-Quiñonez, J.C.; Sergiyenko, O.; Flores-Fuentes, W.; Ramirez-Hernandez, L.R.; Hernández-Balbuena, D.; Lindner, L.; Rascón, R. Novel sensing approaches for structural deformation monitoring and 3D measurements. *IEEE Sens. J.* **2020**, *21*, 11318–11328. [CrossRef]

infrastructures MDPI

Article

Fractality–Autoencoder-Based Methodology to Detect Corrosion Damage in a Truss-Type Bridge

Martin Valtierra-Rodriguez [1], Jose M. Machorro-Lopez [2], Jesus J. Yanez-Borjas [1], Jose T. Perez-Quiroz [3], Jesus R. Rivera-Guillen [1] and Juan P. Amezquita-Sanchez [1,*]

1 ENAP-RG, CA Sistemas Dinámicos y Control, Facultad de Ingeniería, Departamento de Electromecánica, Universidad Autónoma de Querétaro, Campus San Juan del Río, San Juan del Río 76807, Querétaro, Mexico; martin.valtierra@enap-rg.org (M.V.-R.); jesus.yanez@enap-rg.org (J.J.Y.-B.); jesus.rooney.rivera@uaq.mx (J.R.R.-G.)
2 Investigador CONAHCYT—Instituto Mexicano del Transporte, Km 12 Carretera Estatal No. 431 "El Colorado-Galindo" San Fandila, Pedro Escobedo 76703, Querétaro, Mexico; jmachorro@imt.mx
3 Instituto Mexicano del Transporte, Km 12 Carretera Estatal No. 431 "El Colorado-Galindo" San Fandila, Pedro Escobedo 76703, Querétaro, Mexico; jtperez@imt.mx
* Correspondence: jamezquita@uaq.mx or juan.amezquita@enap-rg.org

Abstract: Corrosion negatively impacts the functionality of civil structures. This paper introduces a new methodology that combines the fractality of vibration signals with a data processing stage utilizing autoencoders to detect corrosion damage in a truss-type bridge. Firstly, the acquired vibration signals are analyzed using six fractal dimension (FD) algorithms (Katz, Higuchi, Petrosian, Sevcik, Castiglioni, and Box dimension). The obtained FD values are then used to generate a gray-scale image. Then, autoencoders analyze these images to generate a damage indicator based on the reconstruction error between input and output images. These indicators estimate the damage probability in specific locations within the structure. The methodology was tested on a truss-type bridge model placed at the Vibrations Laboratory from the Autonomous University of Queretaro, Mexico, where three damage corrosion levels were evaluated, namely incipient, moderate, and severe, as well as healthy conditions. The results demonstrate that the proposal is a reliable tool to evaluate the condition of truss-type bridges, achieving an accuracy of 99.8% in detecting various levels of corrosion, including incipient stages, within the elements of truss-type structures regardless of their location.

Keywords: autoencoders; corrosion detection; fractal dimension; images; truss-type bridge; vibration analysis

check for updates

Citation: Valtierra-Rodriguez, M.; Machorro-Lopez, J.M.; Yanez-Borjas, J.J.; Perez-Quiroz, J.T.; Rivera-Guillen, J.R.; Amezquita-Sanchez, J.P. Fractality–Autoencoder-Based Methodology to Detect Corrosion Damage in a Truss-Type Bridge. *Infrastructures* **2024**, 9, 145. https://doi.org/10.3390/infrastructures9090145

Academic Editors: Armin Mehrabi and Seyed Saman Khedmatgozar Dolati

Received: 27 July 2024
Revised: 21 August 2024
Accepted: 28 August 2024
Published: 29 August 2024

1. Introduction

Civil engineering infrastructure, e.g., buildings, bridges, towers, and other facilities, is a vital component of modern infrastructure, supporting essential societal functions such as work, residence, and communication across cities and nations [1]. However, these structures inevitably undergo aging processes due to environmental factors (e.g., environmental temperature, humidity, aggressive agents, etc.) and dynamic loads (e.g., forceful winds, seismic events, human-induced perils, etc.), leading to deterioration that compromises their operational integrity [2]. Among various degradation mechanisms, corrosion is a gradual deterioration process that can diminish structural component thickness, potentially resulting in critical structural failures [3]. It is produced by a chemical or electrochemical reaction with the environment, affecting the physical and mechanical properties of the elements that integrate a civil construction, and therefore, the risk of failure increases [4,5]. Consequently, addressing corrosion requires significant financial investment for its prevention and repair [6]. Therefore, early corrosion detection is crucial in civil structures [7].

Recent studies have confirmed the critical importance of developing advanced methodologies for detecting and assessing corrosion in civil structures. For instance, Melchers contributed with empirical models and phenomenological approaches aimed at providing reliable frameworks for predicting corrosion behavior and assessing structural reliability in marine environments [8,9]. These models are particularly relevant for offshore structures, where corrosion can significantly compromise structural integrity over time [10]. On the other hand, Huras et al. [11] introduced a novel method for quantifying local stiffness loss in beams using rotation rate sensors, emphasizing the need for precise and localized monitoring techniques to ensure structural safety. These contributions highlight the ongoing efforts to enhance the understanding of corrosion and its impact on structural health, yet there remains a need for methodologies that can effectively utilize vibration signals for early detection. In this regard, the stochastic reliability-based design optimization framework proposed by Sokołowski and Kamiński [12] demonstrated the application of probabilistic methods to account for the uncertainties in corrosion effects on structural components. This approach, combined with the uncertainty modeling techniques discussed by Melchers [13], provides a comprehensive strategy for managing the risks associated with corrosion in steel structures. Despite these advancements, recent studies have concentrated on using experimental vibration signals for structural health monitoring, with a particular emphasis on applying these methodologies in real-world scenarios, where vibration-based indicators could provide an alternative for assessing civil structures exposed to or affected by corrosion-induced damage.

Furthermore, the identification of structural damage and the implementation of SHM systems have become pivotal subjects within engineering fields such as civil, structural, and construction engineering [14–16]. These disciplines seek effective systems or methods to assess structural health by analyzing physical variables [17,18]. A vibration-based SHM system involves three essential steps for assessing structural condition: acquisition, processing, and interpretation. The processing step is especially complex due to low signal-to-noise ratios (SNR) and non-stationary characteristics in civil structures' vibrational responses. Structures with early- or incipient-stage damage exhibit subtle changes in their vibration responses, further complicating analyses [19]. These complexities can vary with damage levels and challenge the efficacy of signal processing techniques [20]. Therefore, the advancement and implementation of methodologies or signal processing techniques adept at discerning and extracting indicators of structural damage, particularly at its incipient stages, are imperative for evaluating structural integrity.

In the last decade, a variety of parametric and non-parametric machine learning algorithms have emerged to evaluate the condition of civil structures by using vibration signals [21–24]. These algorithms typically involve two primary steps: pattern identification and categorization [25]. In the pattern identification step, vibration signals are analyzed using linear or non-linear techniques to identify patterns based on their statistical characteristics, energy, geometric complexity, and other features indicative of the structure health condition. In the categorization step, the extracted characteristics are used to train classifiers that automatically assess the health of the structure [26]. Parametric features such as modal properties (e.g., modal shapes, damping ratios, and natural frequencies) [27] and non-parametric features such as statistical methods [28], autoregressive models [29], fractal dimensions [30], entropy methods [31], wavelet transforms [32], empirical mode decomposition [33], the multiple signal classification method [34], and principal component analysis [35] have been widely explored. Various classifiers, including decision trees [36], artificial neural networks [37,38], support vector machines [39], and fuzzy logic classifiers [40], have been employed to associate these features with the structural condition. Despite promising results, these methods face limitations such as environmental sensitivity, the need for fine calibration, and the reliance on expert knowledge for feature and classifier selection [41].

Recent advances in artificial intelligence, particularly deep learning, have provided new opportunities for overcoming these limitations [42,43]. Autoencoders, a type of unsupervised learning model, have shown significant promise in this regard [44–48]. Autoencoders consist of an encoder, which compresses and encodes the data into a different representation space, and a decoder, which reconstructs the original data from this encoded representation. This architecture allows autoencoders to effectively extract and select features from data, minimizing reconstruction error [46,49]. When machine learning algorithms are combined with fractal analysis, which captures the complexity and self-similarity of vibration signals, they can provide a robust framework for structural health monitoring [50–52]. Fractal dimensions offer a detailed characterization of vibration signals, and their integration with autoencoders enhances the ability of the model for detecting subtle signs of structural damage. Despite some challenges, such as long training times and the interpretability of extracted features, ongoing research and improvements in autoencoder models hold the potential for more accurate and efficient damage detection in civil structures.

This paper proposes a novel methodology that combines fractal analysis of vibration signals with autoencoders for corrosion detection in a truss-type bridge. Initially, vibration signals were processed using six different fractal dimension algorithms, i.e., Katz [53], Higuchi [54], Petrosian [55], Sevcik [56], Box dimension [57], and Castiglioni [58], in order to generate images that capture the fractal characteristics of the vibration signals. These images were then analyzed using an autoencoder, which generates a damage indicator based on the image reconstruction error to estimate the probability of detecting corrosion damage, regardless of the structural damage location. The proposed procedure was proven using a truss-type bridge lab model exposed to different levels of corrosion damage ranging from incipient to severe. The experimental setup, located at the Autonomous University of Queretaro, Mexico, provides a comprehensive testbed for evaluating the effectiveness of the methodology under various damage conditions. The results demonstrate the potential of combining fractal analysis with autoencoders for enhancing the accuracy of corrosion detection in civil structures, thereby contributing to more effective and efficient SHM practices.

2. Theoretical Background

This section introduces the main concepts and mathematical definitions utilized in the proposed fractality–autoencoder-based methodology.

2.1. Fractal Dimension

As previously mentioned, the vibration signals generated by a civil structure contain information about its physical constitution; therefore, being able to quantitatively characterize the changes in this information will allow the criteria establishment to determine the structure's condition, such as a healthy structure or a structure with corrosion damage. However, characterizing these changes is not trivial because of the high nonlinearities and low SNR of vibration signals.

Over the years and across different fields of science, fractal dimension, a nonlinear measurement of the complexity or fractality of a signal, has been successfully used to quantify these changes, encompassing applications from the diagnosis of electric motors [59] to the diagnosis of heart diseases [60]. This success promotes its use in this work in the topic of structural diagnosis by corrosion.

The measurement of the fractality of a signal has been proposed with different approaches, the most common being Petrosian (PFD), Higuchi (HFD), Sevcik (SFD), Box counting (BFD), Katz (KFD), and Castiglioni (CFD) fractal dimensions [53–58]. As each of these indicators quantifies the fractality of the signal in a different manner, it is essential that all of them are analyzed and taken into account to characterize the changes in the fractality of the vibration signals. In particular, the ability of Petrosian's method to identify abrupt changes makes it ideal for signals experiencing peaks or sudden discontinuities. The Sevcik method is simple to implement and provides good discrimination for signals with varying levels of complexity. Higuchi's method is highly effective for analyzing non-stationary signals and is sensitive to changes in signal roughness. Box counting offers a clear visualization of the spatial distribution of the signal and is suitable for signals with complex geometric structures. Katz's method is particularly useful for capturing the complexity of signals that exhibit repetitive patterns or cycles. Finally, Castiglioni's method offers a robust measure of signal complexity, particularly useful for signals with noise. Therefore, the simultaneous use of these methods allows a more comprehensive characterization of the geometric structure of vibration signals. Since each method emphasizes different aspects of signal complexity, their combined application may maximize the ability to detect structural damage, providing a robust and multifaceted characterization of changes in vibration signals. In the next subsections, the six fractals used in this work are detailly described.

2.1.1. Katz Fractal Dimension

The KFD of a signal sampled from the time domain (x_i, y_i) with N samples is computed by the following [53]:

$$\mathrm{KFD} = \frac{\log(N)}{\log(N) + \log\left(\frac{d}{L}\right)} \tag{1}$$

where d represents the maximum distance from the first sample to the others, and L is the Euclidian distance separating adjacent samples. These values are computed as follows:

$$L = \sum_{i=2}^{N} \sqrt{(x_i - x_{i-1})^2 + (y_i - y_{i-1})^2} \tag{2}$$

$$d = max\left((x_i - x_1)^2 + (y_i - y_1)^2\right) \tag{3}$$

2.1.2. Higuchi Fractal Dimension

For calculating the fractality according to HFD, the input signal $x(t)$ must first be decomposed into new signal sequences according to the following [54]:

$$S_k^m = \{S_m, S_{m+k}, S_{m+2k}, \ldots, S_{m+ak}\} \tag{4}$$

where $k = 1, 2, \ldots, k_{max}$, and $a = (n - m)/k$ for $m = 1, 2, \ldots, k$. Once the new sequences are obtained, their length must be calculated as follows:

$$L_k^m = \frac{\sum_{i=1}^{a} \left|S_{m+ik} - S_{m+(i-1)k}\right| (N-1)}{ak} \tag{5}$$

where N represents the number of samples for S_k^m. With these values, the average value is obtained through the following:

$$L_k = \frac{1}{k} \sum_{m=1}^{k} L_k^m \tag{6}$$

To conclude, create a log-log plot of $L(k)$ against $(1/k)$ and determine the best-fit line for the slope of the curve. Thus, the slope value of this line is the HFD.

2.1.3. Box Counting Fractal Dimension

To calculate BFD of the 1-dimensional signal $x(t)$ with N samples in the time domain, it is necessary to consider a grid of boxes that enclose the signal. In this way, the fractality will correspond to the boxes that are occupied. The BFD of a signal can be approximately calculated [57]:

$$BFD \cong -\frac{ln\left(\sum_{i=1}^{N-1} |x_{i+1} - x_i| / \Delta t\right)}{ln(\Delta t)} \tag{7}$$

where Δt is the sampling time.

2.1.4. Petrosian Fractal Dimension

PFD must be computed over a thresholder signal; i.e., the input signal $x(n) = x_1, x_2, \ldots, x_N$ with N samples has to be converted through the following [55]:

$$z_i = \begin{cases} 1\ x_i > mean(x) \\ -1\ x_i \leq mean(x) \end{cases} for\ i = 1,\ 2,\ \ldots,\ N \tag{8}$$

Thus, the PFD value is computed by the following:

$$PFD = \frac{log(N)}{log(N) + log(N/(N + 0.4N_\Delta))} \tag{9}$$

where N_Δ represents the total changes of z_i and is computed as follows:

$$N_\Delta = \sum_{i=1}^{N-1} \left| \frac{z_{i+1} - z_i}{2} \right| \tag{10}$$

2.1.5. Sevcik Fractal Dimension

In general, the SFD index is considered a variation of KFD since it also depends on the Euclidian distance; however, for SFD, the input signal (x_i, y_i) with N samples must be initially normalized [56]:

$$x_i^* = \frac{x_i - min(x)}{max(x) - min\ (x)} \tag{11}$$

$$y_i^* = \frac{y_i - min(y)}{max(y) - min\ (y)} \tag{12}$$

where the functions $min(\cdot)$ and $max(\cdot)$ compute the minimum and maximum values of x and y, respectively, for the sample i.

Once the input signal is normalized, the SFD is determined as follows:

$$SFD = 1 + \frac{log(L) + log(2)}{log(2 \cdot (N + 1))} \tag{13}$$

where L is computed by the following:

$$L = \sum_{i=2}^{N} \sqrt{\left(x_i^* - x_{i-1}^*\right)^2 + \left(y_i^* - y_{i-1}^*\right)^2} \tag{14}$$

2.1.6. Castiglioni Fractal Dimension

The CFD is also a modified version of KFD, where L, the total distance between adjacent points, and d, the non-stable range of the analyzed time-series signal, are computed as follows [58]:

$$L = \sum_{i=1}^{N-1} \sqrt{(y_{i+1} - y_i)^2 + (x_{i+1} - x_i)^2} \tag{15}$$

$$d = max(y_i) - min(y_i) \tag{16}$$

Once these values are obtained, the CFD value is computed:

$$\text{CFD} = \frac{\log(N-1)}{\log(N-1) + \log\left(\frac{d}{L}\right)} \tag{17}$$

2.2. *Autencoder*

An autoencoder, a tool used in machine learning, is a neural network designed to learn in an unsupervised way to reconstruct the input data at the output as shown in Figure 1 [61]; i.e., the output image of an autoencoder is a reconstructed version of the input image. This neural network architecture comprises two main components: an encoder and a decoder.

The encoder is tasked with converting the input data $a(n)$ into the compressed/encoded representation $b(m)$ through the function $b(m) = enc(a(n))$. In this work, the input data are the fractal dimension values of the analyzed signal. Conversely, the decoder reconstructs the original input from this encoded representation by applying the function $g(n) = dec(b(m))$. Given that $b(m)$ serves as a compact representation of $a(n)$, autoencoders are commonly employed for dimensionality minimization tasks. Based on these points, an autoencoder is defined as follows [62]:

$$dec(enc(a(n))) = c(n) \tag{18}$$

where $c(n)$ is as similar as possible to $a(n)$.

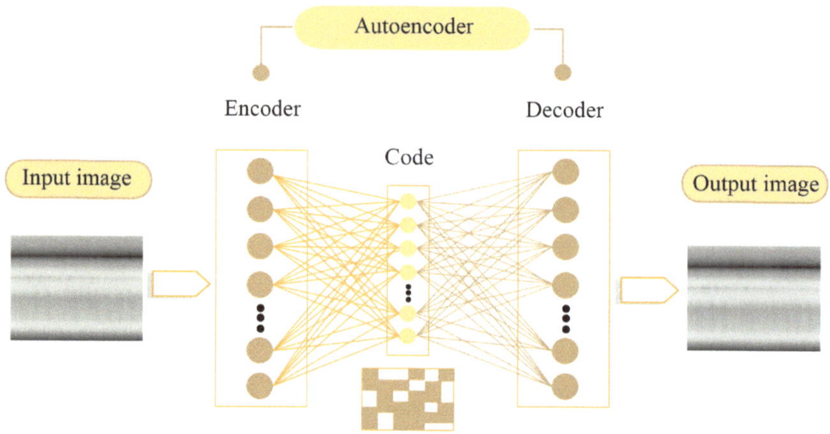

Figure 1. General structure of an autoencoder.

To be more specific regarding the functions $enc(\cdot)$ and $dec(\cdot)$, the following vector operations have to be carried out:

$$\mathrm{b} = h\left(\mathrm{W}^{\mathrm{E}}\mathrm{a} + \mathrm{B}^{\mathrm{E}}\right) \tag{19}$$

where h, W, and B are the transfer function, the weight matrix, and the bias vector, respectively. As transfer function, the *logsig* function is utilized in this work [63]. The superscript E denotes the numbers associated with the encoder stage. After encoding the input data, the decoder generates an approximation of the input data through the following:

$$\mathrm{c} = h\left(\mathrm{W}^{\mathrm{D}}\mathrm{b} + \mathrm{B}^{\mathrm{D}}\right) \tag{20}$$

where the superscript D refers to the numbers related with the decoder stage. To adjust the bias and weight values during the training, the loss function is evaluated utilizing the cross-entropy function and the scaled conjugate gradient backpropagation method [63].

3. Methodology

Civil infrastructure, e.g., bridges, buildings, and towers, among others, are constantly experiencing dynamic or changing forces or excitations, i.e., earthquakes, wind, and traffic, among others. These dynamic excitations provide an opportunity to assess the structural behavior through vibrational responses recorded or acquired during these events [64]. Vibration signals play a crucial role in structural evaluation, as they can capture changes or patterns indicative of damage within the structure. The extent of these changes varies with the damage severity level; for instance, an initial damage known as incipient or light damage may cause subtle alterations in the signals [65]. Hence, the capability of a method or technique to identify relevant patterns in the measured signals is essential for accurate structural condition assessment.

Figure 2 illustrates the steps of the proposed method for determining the condition of a truss structure in bridge configuration. It consists of four main steps. As a first step, the vibration signals are measured in three axes, represented by Vx, Vy, and Vz, under two conditions: healthy and damaged (with elements affected by corrosion). These signals are acquired from a 3D nine-bay truss structure configurated as a bridge, which is subjected to low-amplitude dynamic excitations (comparable to ambient vibrations) imposed by an electromechanical system known as electrodynamic shaker [66]. Then, the acquired vibration responses are evaluated by mean of six FD methods: SFD, KFD, CFD, BFD, HFD, and PFD, with the aim of calculating features or patterns that can effectively determine the bridge's condition. It is worth mentioning that the monitored vibration signals of the truss-type bridge, which are collected for 20 s during each test (this time window was chosen to ensure that the electromechanical system can reliably generate an excitation range from 0 to 100 Hz, allowing capturing the relevant natural frequencies of the structure), are segmented into time windows of 1 s. Each of these segmented signals is analyzed by the proposed method, leading to a total of 360 values (60 values per FD method, considering the three axes).

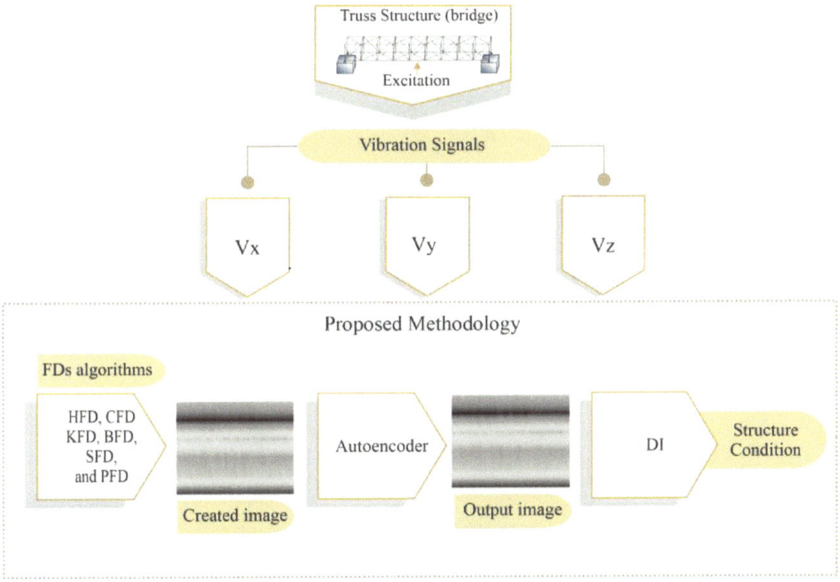

Figure 2. Proposed methodology.

The third step involves the creation of images through the FD values obtained in the previous stage (see Figure 3a), which can be processed by an unsupervised algorithm, specifically an autoencoder. The obtained 360 FD values for each test are visualized using a surface plot, where the X and Y axes represent the number of FD values estimated and the FD methods employed, respectively, while the Z-axis indicates the amplitude or magnitude of these FD values (see Figure 3b). The "turbo" color map was chosen for its smooth color transition. To create the image, a top-down view is utilized, where variations in color depict the estimated FD values, and the dimensions of the image indicate the number of FD values estimated and the FD methods employed, respectively, ensuring clarity and minimal overlap between data elements (as illustrated in Figure 3c). The generated image is converted to grayscale to reduce computational load compared to an RGB format (Figure 3d) [67]. The generated images have dimensions of 666 by 531 pixels.

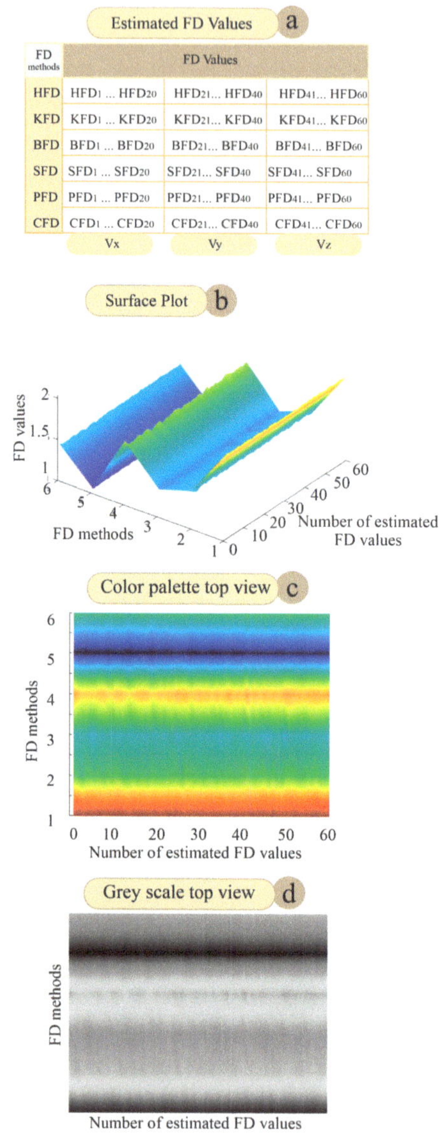

Figure 3. Process for creating the image: (**a**) Estimation of FD values, (**b**) surface plot generated based on the estimated FD values, (**c**) top view of the surface plot, and (**d**) transformation of the top view to grayscale.

Finally, in the fourth step, an autoencoder model is trained for reconstructing input images as outputs, making its training an unsupervised learning task. It is crucial to note that the autoencoder is trained exclusively on non-damaged FD values to establish a baseline model. Therefore, a significant reconstruction error indicates the model's difficulty in reconstructing images generated by FD values obtained from a damaged structure. This error serves as a damage indicator, DI, to identify or determine the bridge's condition.

4. Experimental Setup

The experimental test rig used to evaluate the integrity condition of the bridge studied in this article is shown in Figure 4 and is situated at the Autonomous University of Queretaro, Campus San Juan del Rio, Mexico. This experimental prototype contains 162 bar elements, creating 9 bays (see Figure 4a), similar to the model studied in [68]. The material utilized for this truss-type bridge is aluminum grade 6061-T6 (Mg~0.8–1.2%, Al~95–98%, Cu~0.15–0.40%, and Si~0.4–0.8%) [69], and its general dimensions are length of 6.40 m, height of 0.71 m, and width of 0.71 m. On the other hand, the diameter of the bars is 19 mm, the horizontal and vertical elements have 0.70 m length, and the diagonal bars have corresponding lengths of $0.70\sqrt{2}$ m. With the aim of assessing the structural state of this experimental platform (see Figure 4b), an electrodynamic shaker (Labworks model ET-126B, Labworks, Costa Mesa, CA, USA) along with a linear amplifier (Labworks model PA-138, Labworks, Costa Mesa, CA, USA) were used to generate continuous dynamic excitations of low-intensity/frequency into the structure, simulating, in this way, the vibratory behavior of real bridges [70]. So as to produce the corresponding dynamic signals, 200 Hz sampling frequency white Gaussian noise was used; hence, a 0–100 Hz bandwidth with a frequency content in the whole range was generated. Taking into account that the central region of this experimental bridge is the one that registers the highest vibration amplitudes [71,72], the fourth bay, which is located in this zone, was selected for acquiring the respective dynamic signals; in this way, valuable data were monitored and analyzed. A Kistler tri-axial accelerometer (Vx, Vy, and Vz) model 8395A with ± 10 g of measuring range, 400 mV/g of resolution, and 0–1000 Hz bandwidth, was utilized for measuring the structure dynamic response. This sensor was placed on the fourth bay, as can be seen Figure 5, indicated by a blue dot. Lastly, a National Instruments data acquisition system (DAS) model NIUSB-6002 with a 16-bit resolution analog-to-digital converter was employed to store and transfer the measured vibrational signals into a computer.

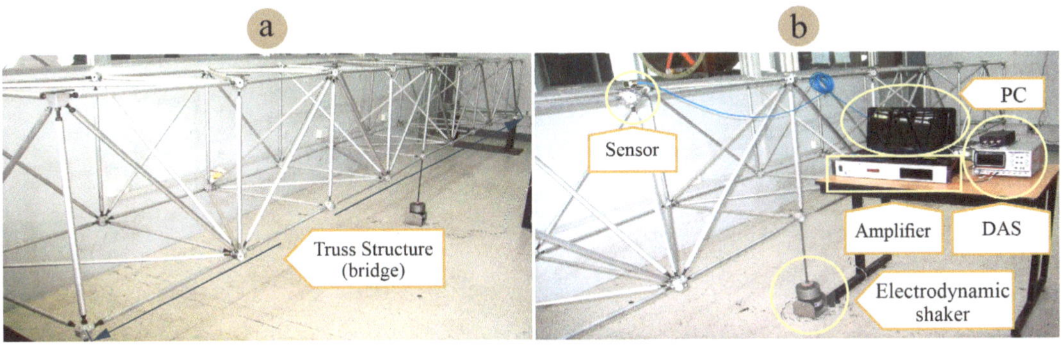

Figure 4. Experimental configuration. (**a**) Truss-type bridge; (**b**) experimental arrangement.

As a means to simulate corrosion damage in the bar elements at different levels, a diameter reduction of 1 mm was induced for incipient damage (Figure 5b), 4 mm for moderate damage (Figure 5c), and 8 mm for severe damage (Figure 5d). The selected sampling frequency for each experiment was 200 Hz, resulting in a total of 4000 samples considering the 20 s duration per test. Likewise, the experiments were performed 100 times for each condition; that is, 100 times for the healthy case and 100 times for each damage scenario (corrosion level) at each of the nine bays, generating a total of 2800 tests. It should be noted that each corrosion condition (incipient, moderate, and severe corrosion level) was applied one at a time into the structure and one after the other, replacing a healthy element with a corroded bar. Thus, the experimental procedure consisted of substituting a healthy bar at a certain location by another damaged bar until completing the three damaged scenarios, beginning in the first bay and, subsequently, repeating the process for the other eight bays of the structure. For each bay, the selection of the corroded element's location was random (see Figure 5) so that the efficacy of the proposed methodology to detect damage in the experimental bridge could be verified regardless of the damage position.

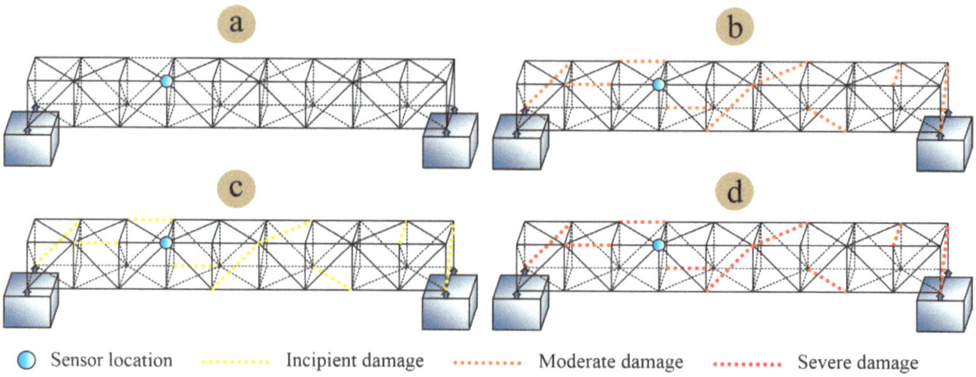

Figure 5. Location of sensor (blue dot) and corroded elements mounted on the bridge (colored lines). (**a**) Healthy condition, (**b**) incipient damage, (**c**) moderate damage, and (**d**) severe damage.

Damaged Elements

The natural corrosion process can take a long time [73]. Hence, for the purpose of this research, the corrosion process was accelerated in order to assess the effect of this kind of damage at different levels in the behavior of the experimental truss-type structure; thus, in order to produce external corrosion [74–76], the bars' ends were submerged in hydrochloric acid (HCl). In this way, the three damaged levels obtained in the bars included diameter reductions of 1 mm, 4 mm, and 8 mm for incipient, moderate, and severe damage, respectively, with the corresponding mass reductions of 19.7 g, 49.7 g, and 84.4 g with respect to the healthy element. The controlled diameter reductions were chosen to ensure that the integrity of the elements was not compromised [77]. In Figure 6, pictures of a healthy bar (19 mm of diameter and 445 g of mass) and the three damaged bars with the different corrosion levels (incipient, moderate, and severe) can be observed.

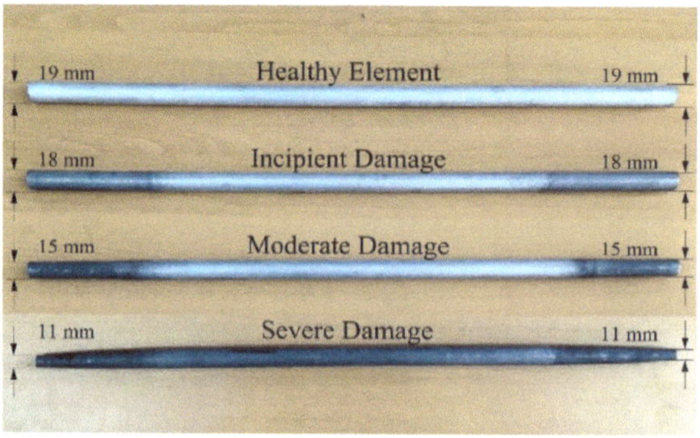

Figure 6. Picture of the different conditions of the bars: healthy condition (diameter of 19 mm in all sections of the bar) and damage conditions considering different corrosion levels (diameter reductions of 1 mm, 4 mm, and 8 mm at the ends of the bars).

5. Results

Figure 7 illustrates the vibrational responses of the Vx, Vy, and Vz axes monitored from the fourth bay location for both the healthy state and the incipient damage condition.

As depicted, both vibrational responses exposed similar visual patterns, underscoring the necessity of employing robust algorithms like FD methods to discover reliable patterns or differences into the monitored signals for accurate structural condition assessment.

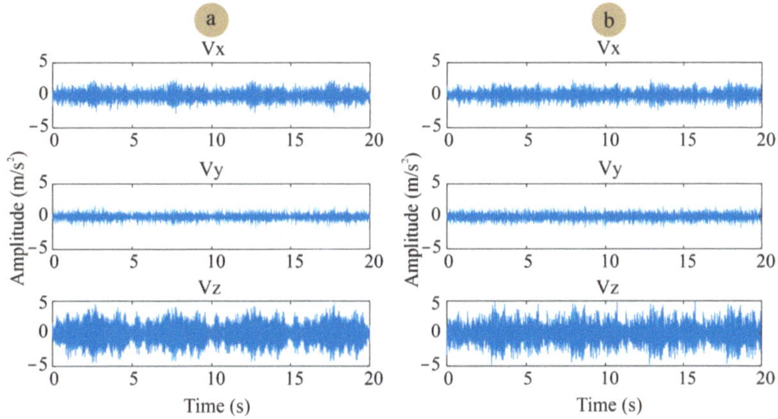

Figure 7. Acquired vibrational responses in the Vx, Vy, and Vz axes for (**a**) the healthy condition and (**b**) incipient corrosion in the 4th bay.

Once the different FD methods were applied to diverse conditions, the values obtained from these FD methods for each condition were used to create images that were employed to determine the bridge's condition. For illustrative purposes only, Figure 8 presents the images obtained for the conditions of a healthy structure and a structure exhibiting an incipient corrosion damage located in the fourth bay, respectively. As depicted, both images exhibit similar visual patterns, emphasizing the importance of using robust algorithms such as autoencoders to detect subtle differences or patterns in the signals to accurately assess the bridge's reliability.

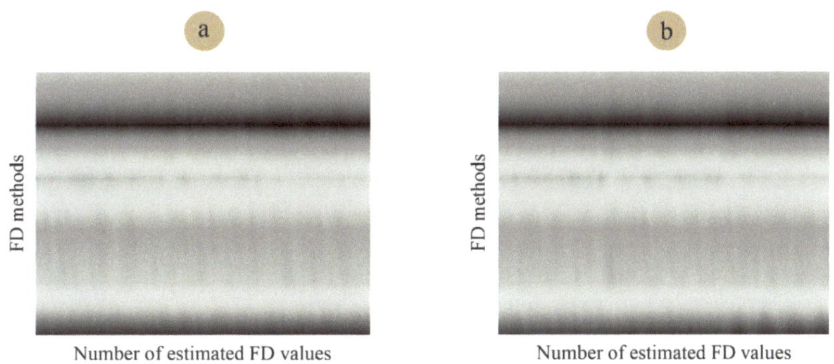

Figure 8. Images obtained for (**a**) healthy bridge and (**b**) corrosion damage with incipient level.

With the purpose of establishing a baseline model, the autoencoder was exclusively trained with images obtained by a healthy or initial condition. It is essential to investigate the adequate number of hidden neurons in the autoencoder to minimize computational costs while maintaining high performance. To perform this, we conducted research, spanning from 10 to 100 neurons, incremented by 10. The obtained results indicated that using 30 neurons yields high performance and a low error rate, suggesting that the reconstructed images closely match the input images generated from FD values obtained by a healthy structure. Table 1 summarizes the mean values obtained for the loss function (formulated as the mean squared error combined with a regularization term) during the training phase and the reconstruction error (calculated as the mean squared error between the input data and the reconstructed output) during the testing phase of the autoencoder. It highlights that using 30 neurons results in the lowest loss and error rates (see Figure 9). These results are derived from training with 70 images and testing with 30 images. Additionally, five-fold cross-validation was employed for mitigating a bias and assessing the methodology's efficacy [78].

Table 1. Estimated mean values obtained for loss function and reconstruction error for different number of neurons.

No. Neurons	Loss Value	Estimated Error
100	1.8456×10^{-4}	1.7902×10^{-4}
90	2.3020×10^{-4}	1.9357×10^{-4}
80	3.9727×10^{-4}	3.2216×10^{-4}
70	4.2435×10^{-4}	4.0953×10^{-4}
60	2.2616×10^{-4}	2.1101×10^{-4}
50	2.6931×10^{-4}	2.5435×10^{-4}
40	2.6117×10^{-4}	2.3685×10^{-4}
30	1.6913×10^{-4}	1.5919×10^{-4}
20	2.2099×10^{-4}	2.0457×10^{-4}
10	2.4064×10^{-4}	2.0065×10^{-4}

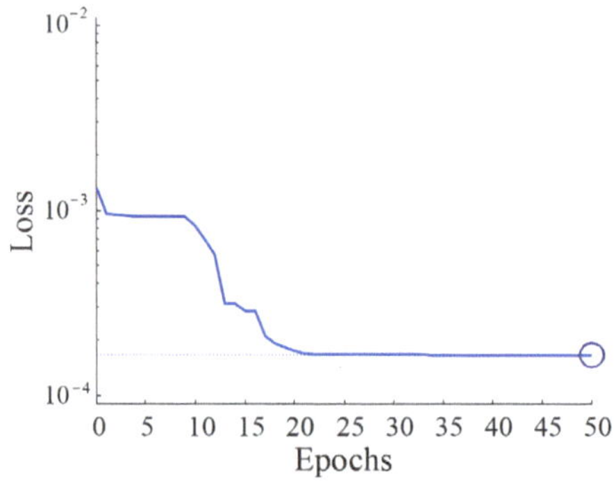

Figure 9. Obtained loss function with 30 neurons.

Once the most appropriate number of neurons for the autoencoder method used in evaluating the bridge condition was identified, it helped to establish a baseline model for determining whether the truss bridge exhibits a damage. In this context, a DI is defined based on the extent of deviations between predicted and actual structural responses, where higher DI values suggest potential damage or deterioration. Mathematically, it is defined as follows:

$$DI = \frac{1}{n}\sum_{i=1}^{n}(y_i - \hat{y}_i)^2 \tag{21}$$

where n is the total number of values or observations, and $\hat{y}_i$ and y_i denote the predicted values and the actual observed values, respectively. A lower DI value implies a better fit of the predictive model to the observed data, suggesting minimal discrepancies and, consequently, less structural damage.

In structural applications, the DI can be employed to identify changes in the vibrational responses of civil structures over time, providing insights into potential damage progression. By establishing baseline DI values during initial assessments, engineers can monitor deviations that may signal structural deterioration or failure. Therefore, DI serves not only as a measure of model performance but also as a vital tool for assessing the overall health of structures. A baseline or threshold is established by comparing DI values obtained from a healthy bridge with those from a slightly damaged one (i.e., incipient damage condition).

Figure 10 displays a box plot of the calculated DI values for both the healthy bridge condition, denoted by the letter H, and for incipient, moderate, and severe levels of corrosion in each bay, represented as Bi (where I = 1, . . ., 9 corresponds to the bay number). The obtained results in Figure 10 demonstrate that the DI effectively identifies the bridge's condition across all three levels of corrosion, particularly the last two levels of damage, regardless of the damage location. However, it is important to note that the DI values for bays B1 to B6 are higher than those for bays B7 to B9 due to their proximity to the sensor. The farther bays show a decrease in DI values, particularly for incipient damage, while the DI values remain higher for moderate and severe damage, indicating that these more severe conditions are detected with greater accuracy. Despite this variation, the DI, estimated through fractality and the autoencoder, can successfully identify the presence of initial or incipient damage. Based on the results obtained, a threshold value, represented by a dotted blue line in Figure 10, was established according to the maximum DI value obtained for the undamaged structure in order to determine the bridge condition automatically. A DI value greater than 1.788×10^{-4} is established as an indicative of damage, while a DI value less than or equal to 1.788×10^{-4} is established as a healthy condition.

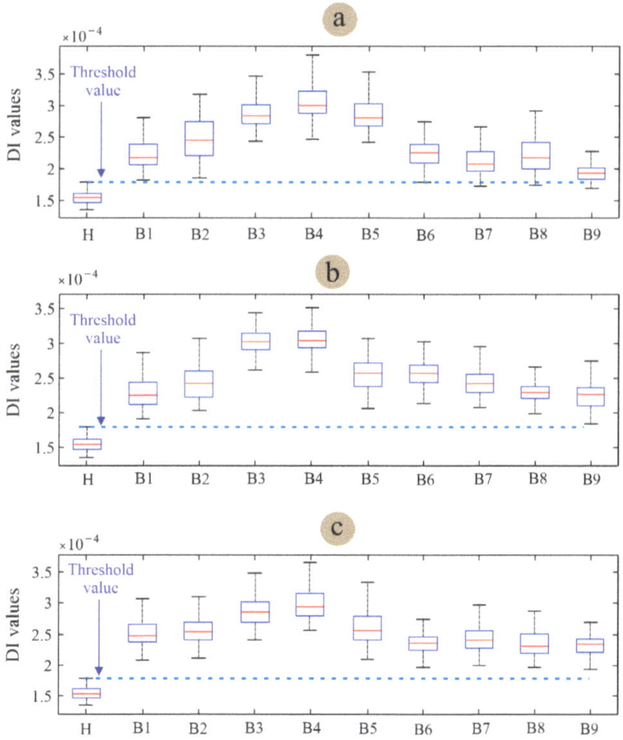

Figure 10. Distribution of calculated DI values for (**a**) incipient, (**b**) moderate, and (**c**) severe corrosion in each bay (indicated by Bi, where i = 1, . . ., 9) and the healthy bridge (indicated by H). Threshold value is indicated by the dotted blue line.

Figure 11 shows a box plot diagram with the data distribution of the calculated DI values for the three levels of corrosion and healthy bridge investigated in this study. It should be noted that the estimated DI values from all nine bays for each damage level were combined to demonstrate the method's capability for determining the truss structure condition regardless of the corrosion location. Along with this figure, the computed DI values for a healthy bridge and the moderate and severe corrosion levels do not overlap, implying that the proposal can recognize these two levels of corrosion irrespective of its location. However, the incipient corrosion condition shows a slight overlap. This overlap does not impede the identification of this condition, as it is shown in the confusion matrix (Table 2). Nevertheless, is noteworthy, as it may result from minimal changes in the structure's dynamics since only one element is damaged, and it can be far away from sensor location when the vibration responses are measured.

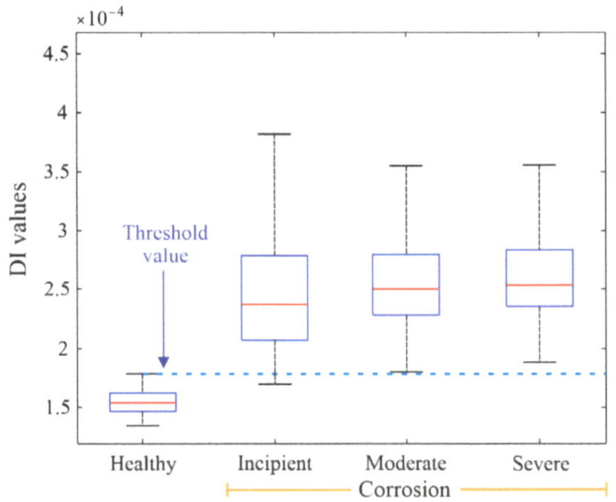

Figure 11. Distribution of calculated DI values for a healthy bridge and three corrosion levels (incipient, moderate, and severe), with DI values consolidated or joined from all bays for each respective condition. Threshold value is indicated by the dotted blue line.

Table 2. Obtained results for determining the truss condition based on the calculated DI values and threshold selected.

Truss Structure Condition		**Predicted**		**Effectiveness (%)**
		Healthy	**Corrosion**	
Actual	Healthy	30	0	100
	Corrosion	10	2690	99.6
		Total accuracy (%)		99.8

Table 2 exhibits a confusion matrix with the estimated classification results. It can be observed that the proposed DI accurately evaluates the bridge condition, achieving a 99.8% accuracy for distinguishing between a healthy truss-type bridge and one with corrosion damage. It is important to note that 30 tests were considered for the healthy condition (as 70 tests were used to train the autoencoder) and 900 tests for each damage condition, resulting in 2700 (incipient, moderate, and severe corrosion).

6. Discussion

Although that the results of this research were obtained from a lab model under controlled conditions (without load or environmental variations), it is significant to point out that the method proposed in this work is promissory and could be a useful SHM tool for evaluating the structural condition of civil constructions such as bridges for three main reasons: (a) Only one sensor is required to determine the structure global condition regardless of the defect level and location, which is an important advantage and even more considering that other methodologies need multiple sensors to detect damage, and in some cases, the instrumentation is installed close to the damage region, which would not be helpful for real cases [79,80]. (b) The complexity as well as the computational load are low, due to the fact that neither pre-processing nor domain transformation are required for the measured signals, whereas other methods need to carry out diverse signals' pre-processing and obtain non-linear features to define the integrity condition of a structure [70,77,81–83]. (c) Considering the previous advantages, the method proposed in this article could be implemented on site to assess the structural state of bridges in real time. Therefore, in order to avoid tragic collapses of bridges, this methodology could evaluate the health condition of civil structures permanently, and once a damage is detected, more detailed analyses will have to be performed to define the damage location and establish/execute appropriate maintenance plans.

In future works, the operational variations as well as acquisition time window, which can generate uncertainties that affect the methodology performance, will be modeled and considered for the analyses; in this way, the proposed methodology can help avoid tragic collapses of bridges by determining the integrity condition of real-life civil structures permanently, and once a damage is detected, more detailed analyses will have to be performed to define the damage location and establish/execute appropriate maintenance plans. In addition, with the aim of optimizing the computational resources of the proposed methodology and gaining a clear understanding of the contribution of each fractal method and each vibration axis to the proposal, statistical tests such as ANOVA or Kruskal–Wallis, will be applied to rank these indicators according to their discrimination ability between classes (e.g., healthy condition and damage condition) as well as some metaheuristic optimization algorithms for the optimal selection of the number and type of indicators.

7. Conclusions

Corrosion negatively impacts the functionality of civil structures. This paper investigates the integration of fractal dimension with autoencoders for determining the health condition of a tridimensional truss-type structure of nine bays configurated as a bridge exposed to forced excitations. The proposed DI revealed the model's difficulty in reconstructing images generated from FD values obtained from a damaged structure, allowing an effective assessment of the bridge's condition.

The estimated results prove that integrating FD and autoencoders produces a DI capable of accurately detecting a structural damage, regardless of its location within the bridge. This indicates the model's proficiency in identifying reliable features from vibration signals, achieving a 99.8% accuracy in determining the bridge condition. Consequently, the effectiveness of the proposed method in corrosion detection, compared to the healthy condition, is reliable and outperforms existing methodologies. Notably, unlike other approaches, the proposal utilizes a single sensor for damage detection, eliminating the need for pre-processing vibration signals to extract features. These attributes indicate that it can be a tool for real-time SHM systems, offering simplicity in classifier usage and feature extraction. Furthermore, future evaluations will explore the application of these FD algorithms integrated with autoencoders or other deep learning methods for detecting other types of damage (e.g., loosened bolts, cracks, etc.) as well as varying the quantity of sensors employed to avoid a sensitivity loss of extracted features across various civil infrastructures (e.g., towers, houses, buildings, etc.) in order to determine the most reliable FD-based features for different damage scenarios.

Author Contributions: Visualization, resources, and investigation, J.J.Y.-B., J.T.P.-Q., J.M.M.-L., M.V.-R., J.R.R.-G. and J.P.A.-S.; writing—review, writing—original draft preparation, and editing, all authors; conceptualization, J.P.A.-S. and M.V.-R.; data curation, J.R.R.-G., M.V.-R., J.T.P.-Q., J.M.M.-L. and J.P.A.-S. All authors have read and agreed to the published version of the manuscript.

Funding: This work was supported by the "Fondo para el Fortalecimiento de la Investigación, Vinculación y Extensión (FONFIVE-UAQ 2024)" project; the National Council of Humanities, Sciences, and Technologies (Consejo Nacional de Humanidades, Ciencias y Tecnologías, CONAHCYT-México) through project 34/2018 of the Program "Investigadoras e Investigadores por México" del CONAHCYT (Cátedras CONAHCYT); and the scholarships 296574, 161138, 91805, 160267, and 253652 awarded to Martin Valtierra-Rodriguez, Jose M. Machorro-Lopez, Jose T. Perez-Quiroz, Jesus R. Rivera-Guillen, and Juan P. Amezquita-Sanchez, respectively, through the "Sistema Nacional de Investigadoras e Investigadores" (SNII)-CONAHCYT-México.

Institutional Review Board Statement: Not applicable.

Informed Consent Statement: Not applicable.

Data Availability Statement: The data presented in this study are not publicly available due to privacy issues.

Conflicts of Interest: The authors declare no conflicts of interest.

References

1. Gonzalez, A.; Schorr, M.; Valdez, B.; Mungaray, A.; Gonzalez, A.; Schorr, M.; Valdez, B.; Mungaray, A. Bridges: Structures and Materials, Ancient and Modern. In *Infrastructure Management and Construction*; IntechOpen: London, UK, 2020; ISBN 978-1-78984-549-5.
2. Hao, H.; Bi, K.; Chen, W.; Pham, T.M.; Li, J. Towards next Generation Design of Sustainable, Durable, Multi-Hazard Resistant, Resilient, and Smart Civil Engineering Structures. *Eng. Struct.* **2023**, *277*, 115477. [CrossRef]
3. van de Velde, M.; Vandecruys, E.; Verstrynge, E.; Reynders, E.; Lombaert, G. Vibration Monitoring and Acoustic Emission Sensing during Progressive Load Tests of Corroded Reinforced Concrete Beams. *Eng. Struct.* **2024**, *306*, 117851. [CrossRef]
4. Schofield, M.J. 33—Corrosion. In *Plant Engineer's Reference Book*, 2nd ed.; Snow, D.A., Ed.; Butterworth-Heinemann: Oxford, UK, 2002; pp. 33-1–33-25; ISBN 978-0-7506-4452-5.
5. Kruger, J.; Begum, S. Corrosion of Metals: Overview. In *Reference Module in Materials Science and Materials Engineering*; Elsevier: Amsterdam, The Netherlands, 2016; ISBN 978-0-12-803581-8.
6. Ameli, Z.; Nesheli, S.J.; Landis, E.N. Deep Learning-Based Steel Bridge Corrosion Segmentation and Condition Rating Using Mask RCNN and YOLOv8. *Infrastructures* **2024**, *9*, 3. [CrossRef]
7. Amezquita-Sanchez, J.P.; Adeli, H. Signal Processing Techniques for Vibration-Based Health Monitoring of Smart Structures. *Arch. Computat. Methods Eng.* **2016**, *23*, 1–15. [CrossRef]
8. Melchers, R.E. Probabilistic Models for Corrosion in Structural Reliability Assessment—Part 1: Empirical Models. *J. Offshore Mech. Arct. Eng.* **2003**, *125*, 264–271. [CrossRef]
9. Melchers, R.E. Modeling of Marine Immersion Corrosion for Mild and Low-Alloy Steels—Part 1: Phenomenological Model. *Corrosion* **2003**, *59*, 319–334. [CrossRef]
10. Melchers, R.E. The Effect of Corrosion on the Structural Reliability of Steel Offshore Structures. *Corros. Sci.* **2005**, *47*, 2391–2410. [CrossRef]
11. Huras, L.; Zembaty, Z.; Bońkowski, P.A.; Bobra, P. Quantifying Local Stiffness Loss in Beams Using Rotation Rate Sensors. *Mech. Syst. Signal Process.* **2021**, *151*, 107396. [CrossRef]
12. Sokołowski, D.; Kamiński, M. Stochastic Reliability-Based Design Optimization Framework for the Steel Plate Girder with Corrugated Web Subjected to Corrosion. *Materials* **2022**, *15*, 7170. [CrossRef]
13. Melchers, R.E. Corrosion Uncertainty Modelling for Steel Structures. *J. Constr. Steel Res.* **1999**, *52*, 3–19. [CrossRef]
14. Mishra, M.; Lourenço, P.B.; Ramana, G.V. Structural Health Monitoring of Civil Engineering Structures by Using the Internet of Things: A Review. *J. Build. Eng.* **2022**, *48*, 103954. [CrossRef]
15. Agyemang, I.O.; Zhang, X.; Adjei-Mensah, I.; Acheampong, D.; Fiasam, L.D.; Sey, C.; Yussif, S.B.; Effah, D. Automated Vision-Based Structural Health Inspection and Assessment for Post-Construction Civil Infrastructure. *Autom. Constr.* **2023**, *156*, 105153. [CrossRef]
16. Smarsly, K.; Dragos, K.; Stührenberg, J.; Worm, M. Mobile Structural Health Monitoring Based on Legged Robots. *Infrastructures* **2023**, *8*, 136. [CrossRef]
17. Avci, O.; Abdeljaber, O.; Kiranyaz, S.; Hussein, M.; Gabbouj, M.; Inman, D.J. A Review of Vibration-Based Damage Detection in Civil Structures: From Traditional Methods to Machine Learning and Deep Learning Applications. *Mech. Syst. Signal Process.* **2021**, *147*, 107077. [CrossRef]
18. Hasani, H.; Freddi, F. Operational Modal Analysis on Bridges: A Comprehensive Review. *Infrastructures* **2023**, *8*, 172. [CrossRef]
19. Amezquita-Sanchez, J.P. Entropy Algorithms for Detecting Incipient Damage in High-Rise Buildings Subjected to Dynamic Vibrations. *J. Vib. Control.* **2021**, *27*, 426–436. [CrossRef]
20. Deng, Z.; Huang, M.; Wan, N.; Zhang, J. The Current Development of Structural Health Monitoring for Bridges: A Review. *Buildings* **2023**, *13*, 1360. [CrossRef]
21. Luo, J.; Zheng, F.; Sun, S. A Few-Shot Learning Method for Vibration-Based Damage Detection in Civil Structures. *Structures* **2024**, *61*, 106026. [CrossRef]
22. Poorghasem, S.; Bao, Y. Review of Robot-Based Automated Measurement of Vibration for Civil Engineering Structures. *Measurement* **2023**, *207*, 112382. [CrossRef]

23. Hou, R.; Xia, Y. Review on the New Development of Vibration-Based Damage Identification for Civil Engineering Structures: 2010–2019. *J. Sound Vib.* **2021**, *491*, 115741. [CrossRef]
24. Yu, X.; Fu, Y.; Li, J.; Mao, J.; Hoang, T.; Wang, H. Recent Advances in Wireless Sensor Networks for Structural Health Monitoring of Civil Infrastructure. *J. Infrastruct. Intell. Resil.* **2024**, *3*, 100066. [CrossRef]
25. Avci, O.; Abdeljaber, O.; Kiranyaz, S.; Hussein, M.; Inman, D.J. Wireless and Real-Time Structural Damage Detection: A Novel Decentralized Method for Wireless Sensor Networks. *J. Sound Vib.* **2018**, *424*, 158–172. [CrossRef]
26. Amezquita-Sanchez, J.P.; Adeli, H. Nonlinear Measurements for Feature Extraction in Structural Health Monitoring. *Sci. Iran.* **2019**, *26*, 3051–3059. [CrossRef]
27. Vafaei, M.; Alih, S.C. Adequacy of First Mode Shape Differences for Damage Identification of Cantilever Structures Using Neural Networks. *Neural Comput. Appl.* **2018**, *30*, 2509–2518. [CrossRef]
28. Datteo, A.; Lucà, F.; Busca, G. Statistical Pattern Recognition Approach for Long-Time Monitoring of the G.Meazza Stadium by Means of AR Models and PCA. *Eng. Struct.* **2017**, *153*, 317–333. [CrossRef]
29. Datteo, A.; Busca, G.; Quattromani, G.; Cigada, A. On the Use of AR Models for SHM: A Global Sensitivity and Uncertainty Analysis Framework. *Reliab. Eng. Syst. Saf.* **2018**, *170*, 99–115. [CrossRef]
30. Shi, B.; Qiao, P. A New Surface Fractal Dimension for Displacement Mode Shape-Based Damage Identification of Plate-Type Structures. *Mech. Syst. Signal Process.* **2018**, *103*, 139–161. [CrossRef]
31. Li, H.; Bao, Y.; Ou, J. Structural Damage Identification Based on Integration of Information Fusion and Shannon Entropy. *Mech. Syst. Signal Process.* **2008**, *22*, 1427–1440. [CrossRef]
32. Kankanamge, Y.; Hu, Y.; Shao, X. Application of Wavelet Transform in Structural Health Monitoring. *Earthq. Eng. Eng. Vib.* **2020**, *19*, 515–532. [CrossRef]
33. Azami, M.; Salehi, M. Response-Based Multiple Structural Damage Localization through Multi-Channel Empirical Mode Decomposition. *J. Struct. Integr. Maint.* **2019**, *4*, 195–206. [CrossRef]
34. Jiang, X.; Adeli, H. Pseudospectra, MUSIC, and Dynamic Wavelet Neural Network for Damage Detection of Highrise Buildings. *Int. J. Numer. Methods Eng.* **2007**, *71*, 606–629. [CrossRef]
35. Tibaduiza, D.; Torres-Arredondo, M.Á.; Vitola, J.; Anaya, M.; Pozo, F. A Damage Classification Approach for Structural Health Monitoring Using Machine Learning. *Complexity* **2018**, *2018*, 5081283. [CrossRef]
36. Rabcan, J.; Levashenko, V.; Zaitseva, E.; Kvassay, M.; Subbotin, S. Non-Destructive Diagnostic of Aircraft Engine Blades by Fuzzy Decision Tree. *Eng. Struct.* **2019**, *197*, 109396. [CrossRef]
37. Padil, K.H.; Bakhary, N.; Abdulkareem, M.; Li, J.; Hao, H. Non-Probabilistic Method to Consider Uncertainties in Frequency Response Function for Vibration-Based Damage Detection Using Artificial Neural Network. *J. Sound Vib.* **2020**, *467*, 115069. [CrossRef]
38. Lin, T.-K.; Chen, Y.-C. Integration of Refined Composite Multiscale Cross-Sample Entropy and Backpropagation Neural Networks for Structural Health Monitoring. *Appl. Sci.* **2020**, *10*, 839. [CrossRef]
39. Ruocci, G.; Cumunel, G.; Le, T.; Argoul, P.; Point, N.; Dieng, L. Damage Assessment of Pre-Stressed Structures: A SVD-Based Approach to Deal with Time-Varying Loading. *Mech. Syst. Signal Process.* **2014**, *47*, 50–65. [CrossRef]
40. Abu-Mahfouz, I.; Banerjee, A. Crack Detection and Identification Using Vibration Signals and Fuzzy Clustering. *Procedia Comput. Sci.* **2017**, *114*, 266–274. [CrossRef]
41. Valtierra-Rodriguez, M.; Rivera-Guillen, J.R.; Basurto-Hurtado, J.A.; De-Santiago-Perez, J.J.; Granados-Lieberman, D.; Amezquita-Sanchez, J.P. Convolutional Neural Network and Motor Current Signature Analysis during the Transient State for Detection of Broken Rotor Bars in Induction Motors. *Sensors* **2020**, *20*, 3721. [CrossRef]
42. Cha, Y.-J.; Ali, R.; Lewis, J.; Büyüköztürk, O. Deep Learning-Based Structural Health Monitoring. *Autom. Constr.* **2024**, *161*, 105328. [CrossRef]
43. Eltouny, K.; Gomaa, M.; Liang, X. Unsupervised Learning Methods for Data-Driven Vibration-Based Structural Health Monitoring: A Review. *Sensors* **2023**, *23*, 3290. [CrossRef]
44. Ghazimoghadam, S.; Hosseinzadeh, S.A.A. A Novel Unsupervised Deep Learning Approach for Vibration-Based Damage Diagnosis Using a Multi-Head Self-Attention LSTM Autoencoder. *Measurement* **2024**, *229*, 114410. [CrossRef]
45. Sarwar, M.Z.; Cantero, D. Probabilistic Autoencoder-Based Bridge Damage Assessment Using Train-Induced Responses. *Mech. Syst. Signal Process.* **2024**, *208*, 111046. [CrossRef]
46. Giglioni, V.; Venanzi, I.; Poggioni, V.; Milani, A.; Ubertini, F. Autoencoders for Unsupervised Real-Time Bridge Health Assessment. *Comput.-Aided Civ. Infrastruct. Eng.* **2023**, *38*, 959–974. [CrossRef]
47. Coraça, E.M.; Ferreira, J.V.; Nóbrega, E.G.O. An Unsupervised Structural Health Monitoring Framework Based on Variational Autoencoders and Hidden Markov Models. *Reliab. Eng. Syst. Saf.* **2023**, *231*, 109025. [CrossRef]
48. Junges, R.; Rastin, Z.; Lomazzi, L.; Giglio, M.; Cadini, F. Convolutional Autoencoders and CGANs for Unsupervised Structural Damage Localization. *Mech. Syst. Signal Process.* **2024**, *220*, 111645. [CrossRef]
49. Li, P.; Pei, Y.; Li, J. A Comprehensive Survey on Design and Application of Autoencoder in Deep Learning. *Appl. Soft Comput.* **2023**, *138*, 110176. [CrossRef]
50. Hoxha, E.; Vidal, Y.; Pozo, F. Damage Diagnosis for Offshore Wind Turbine Foundations Based on the Fractal Dimension. *Appl. Sci.* **2020**, *10*, 6972. [CrossRef]
51. Bao, Y.; Li, H. Machine Learning Paradigm for Structural Health Monitoring. *Struct. Health Monit.* **2021**, *20*, 1353–1372. [CrossRef]

52. Xu, N.; Zhang, Z.; Liu, Y. 14—Spatiotemporal Fractal Manifold Learning for Vibration-Based Structural Health Monitoring. In *Structural Health Monitoring/Management (SHM) in Aerospace Structures*; Yuan, F.-G., Ed.; Woodhead Publishing Series in Composites Science and Engineering; Woodhead Publishing: Sawston, UK, 2024; pp. 409–426; ISBN 978-0-443-15476-8.
53. Katz, M.J. Fractals and the Analysis of Waveforms. *Comput. Biol. Med.* **1988**, *18*, 145–156. [CrossRef]
54. Higuchi, T. Approach to an Irregular Time Series on the Basis of the Fractal Theory. *Phys. D Nonlinear Phenom.* **1988**, *31*, 277–283. [CrossRef]
55. Petrosian, A. Kolmogorov Complexity of Finite Sequences and Recognition of Different Preictal EEG Patterns. In Proceedings of the Proceedings Eighth IEEE Symposium on Computer-Based Medical Systems, Lubbock, TX, USA, 9–10 June 1995; pp. 212–217.
56. Sevcik, C. A Procedure to Estimate the Fractal Dimension of Waveforms. *Complex Int.* **1998**, *5*, 1–19. Available online: https://arxiv.org/pdf/1003.5266 (accessed on 23 July 2024).
57. Wang, B. Detection of Structural Damage Using Fractal Dimension Technique. *Zhendong Yu Chongji (J. Vibr. Shock)* **2005**, *24*, 87–88.
58. Castiglioni, P. What Is Wrong in Katz's Method? Comments on: "A Note on Fractal Dimensions of Biomedical Waveforms". *Comput. Biol. Med.* **2010**, *40*, 950–952. [CrossRef]
59. Medina, R.; Sánchez, R.-V.; Cabrera, D.; Cerrada, M.; Estupiñan, E.; Ao, W.; Vásquez, R.E. Scale-Fractal Detrended Fluctuation Analysis for Fault Diagnosis of a Centrifugal Pump and a Reciprocating Compressor. *Sensors* **2024**, *24*, 461. [CrossRef] [PubMed]
60. Captur, G.; Karperien, A.L.; Hughes, A.D.; Francis, D.P.; Moon, J.C. The Fractal Heart—Embracing Mathematics in the Cardiology Clinic. *Nat. Rev. Cardiol.* **2017**, *14*, 56–64. [CrossRef] [PubMed]
61. Du, B.; Xiong, W.; Wu, J.; Zhang, L.; Zhang, L.; Tao, D. Stacked Convolutional Denoising Auto-Encoders for Feature Representation. *IEEE Trans. Cybern.* **2017**, *47*, 1017–1027. [CrossRef] [PubMed]
62. Liu, P.; Zheng, P.; Chen, Z. Deep Learning with Stacked Denoising Auto-Encoder for Short-Term Electric Load Forecasting. *Energies* **2019**, *12*, 2445. [CrossRef]
63. Agarwala, A.; Pennington, J.; Dauphin, Y.; Schoenholz, S. Temperature Check: Theory and Practice for Training Models with Softmax-Cross-Entropy Losses. *arXiv* **2020**, arXiv:2010.07344.
64. Tabiatnejad, D.; Tabiatnejad, B.; Khedmatgozar Dolati, S.S.; Mehrabi, A. Damage Detection in External Tendons of Post-Tensioned Bridges. *Infrastructures* **2024**, *9*, 103. [CrossRef]
65. Moghadam, A.; Melhem, H.G.; Esmaeily, A. A Proof-of-Concept Study on a Proposed Ambient-Vibration-Based Approach to Extract Pseudo-Free-Vibration Response. *Eng. Struct.* **2020**, *212*, 110517. [CrossRef]
66. Abdeljaber, O.; Avci, O.; Kiranyaz, S.; Gabbouj, M.; Inman, D.J. Real-Time Vibration-Based Structural Damage Detection Using One-Dimensional Convolutional Neural Networks. *J. Sound Vib.* **2017**, *388*, 154–170. [CrossRef]
67. Hagara, M.; Stojanović, R.; Bagala, T.; Kubinec, P.; Ondráček, O. Grayscale Image Formats for Edge Detection and for Its FPGA Implementation. *Microprocess. Microsyst.* **2020**, *75*, 103056. [CrossRef]
68. Blachowski, B.; An, Y.; Spencer, B.F., Jr.; Ou, J. Axial Strain Accelerations Approach for Damage Localization in Statically Determinate Truss Structures. *Comput.-Aided Civ. Infrastruct. Eng.* **2017**, *32*, 304–318. [CrossRef]
69. Khodabandehlou, H.; Pekcan, G.; Fadali, M.S. Vibration-Based Structural Condition Assessment Using Convolution Neural Networks. *Struct. Control. Health Monit.* **2019**, *26*, e2308. [CrossRef]
70. Rafiei, M.H.; Adeli, H. A Novel Machine Learning-based Algorithm to Detect Damage in High-rise Building Structures. *Struct. Des. Tall Spec. Build.* **2017**, *26*, e1400. Available online: https://onlinelibrary.wiley.com/doi/abs/10.1002/tal.1400 (accessed on 23 July 2024). [CrossRef]
71. Sydenham, P.H. Chapter 11—Vibration. In *Instrumentation Reference Book*, 4th ed.; Boyes, W., Ed.; Butterworth-Heinemann: Boston, MA, USA, 2010; pp. 113–125; ISBN 978-0-7506-8308-1.
72. Lin, J.-F.; Xu, Y.-L.; Law, S.-S. Structural Damage Detection-Oriented Multi-Type Sensor Placement with Multi-Objective Optimization. *J. Sound Vib.* **2018**, *422*, 568–589. [CrossRef]
73. Affonso, L.O.A. 7—Corrosion. In *Machinery Failure Analysis Handbook*; Affonso, L.O.A., Ed.; Gulf Publishing Company: Houston, TX, USA, 2006; pp. 83–99; ISBN 978-1-933762-08-1.
74. Liu, D.; Qiu, X.; Shao, M.; Gao, J.; Xu, J.; Liu, Q.; Zhou, H.; Wang, Z. Synthesis and Evaluation of Hexamethylenetetramine Quaternary Ammonium Salt as Corrosion Inhibitor. *Mater. Corros.* **2019**, *70*, 1907–1916. [CrossRef]
75. Pang, B.; Qian, J.; Zhang, Y.; Jia, Y.; Ni, H.; Pang, S.D.; Liu, G.; Qian, R.; She, W.; Yang, L.; et al. 5S Multifunctional Intelligent Coating with Superdurable, Superhydrophobic, Self-Monitoring, Self-Heating, and Self-Healing Properties for Existing Construction Application. *ACS Appl. Mater. Interfaces* **2019**, *11*, 29242–29254. [CrossRef]
76. Park, S.; Park, S.-K. Quantitative Corrosion Monitoring Using Wireless Electromechanical Impedance Measurements. *Res. Nondestruct. Eval.* **2010**, *21*, 184–192. [CrossRef]
77. Moreno-Gomez, A.; Amezquita-Sanchez, J.P.; Valtierra-Rodriguez, M.; Perez-Ramirez, C.A.; Dominguez-Gonzalez, A.; Chavez-Alegria, O. EMD-Shannon Entropy-Based Methodology to Detect Incipient Damages in a Truss Structure. *Appl. Sci.* **2018**, *8*, 2068. [CrossRef]
78. Marcot, B.G.; Hanea, A.M. What Is an Optimal Value of k in K-Fold Cross-Validation in Discrete Bayesian Network Analysis? *Comput. Stat.* **2021**, *36*, 2009–2031. [CrossRef]
79. Yang, L.; Fu, C.; Li, Y.; Su, L. Survey and Study on Intelligent Monitoring and Health Management for Large Civil Structure. *Int. J. Intell. Robot. Appl.* **2019**, *3*, 239–254. [CrossRef]

80. Babajanian Bisheh, H.; Ghodrati Amiri, G.; Nekooei, M.; Darvishan, E. Damage Detection of a Cable-Stayed Bridge Using Feature Extraction and Selection Methods. *Struct. Infrastruct. Eng.* **2019**, *15*, 1165–1177. [CrossRef]
81. Yang, J.; Li, P.; Yang, Y.; Xu, D. An Improved EMD Method for Modal Identification and a Combined Static-Dynamic Method for Damage Detection. *J. Sound Vib.* **2018**, *420*, 242–260. [CrossRef]
82. Chen, Z.; Pan, C.; Yu, L. Structural Damage Detection via Adaptive Dictionary Learning and Sparse Representation of Measured Acceleration Responses. *Measurement* **2018**, *128*, 377–387. [CrossRef]
83. Zamani HosseinAbadi, H.; Amirfattahi, R.; Nazari, B.; Mirdamadi, H.R.; Atashipour, S.A. GUW-Based Structural Damage Detection Using WPT Statistical Features and Multiclass SVM. *Appl. Acoust.* **2014**, *86*, 59–70. [CrossRef]

infrastructures

Article

Improved Blob-Based Feature Detection and Refined Matching Algorithms for Seismic Structural Health Monitoring of Bridges Using a Vision-Based Sensor System

Luna Ngeljaratan [1,*], Mohamed A. Moustafa [2,3,*], Agung Sumarno [1], Agus Mudo Prasetyo [1], Dany Perwita Sari [1] and Maidina Maidina [1]

1 Research Center for Structural Strength Technology, National Research and Innovation Agency, BJ Habibie Science and Technology Center, Gedung 220, Setu, Tangerang Selatan 15314, Indonesia; agun025@brin.go.id (A.S.); agus108@brin.go.id (A.M.P.); dany003@brin.go.id (D.P.S.); maid002@brin.go.id (M.M.)
2 Department of Civil and Environmental Engineering, University of Nevada, Reno, NV 89557, USA
3 Division of Engineering, New York University Abu Dhabi, Abu Dhabi P.O. Box 129188, United Arab Emirates
* Correspondence: luna001@brin.go.id (L.N.); mohamed.moustafa@nyu.edu (M.A.M.)

Abstract: The condition and hazard monitoring of bridges play important roles in ensuring their service continuity not only throughout their entire lifespan but also under extreme conditions such as those of earthquakes. Advanced structural health monitoring (SHM) systems using vision-based technology, such as surveillance, traffic, or drone cameras, may assist in preventing future impacts due to structural deficiency and are critical to the emergence of sustainable and smart transportation infrastructure. This study evaluates several feature detection and tracking algorithms and implements them in the vision-based SHM of bridges along with their systematic procedures. The proposed procedures are implemented via a two-span accelerated bridge construction (ABC) system undergoing a large-scale shake-table test. The research objectives are to explore the effect of refined matching algorithms on blob-based features in improving their accuracies and to implement the proposed algorithms on large-scale bridges tested under seismic loads using vision-based SHM. The procedure begins by adopting blob-based feature detectors, i.e., the scale-invariant feature transform (SIFT), speeded-up robust features (SURF), and KAZE algorithms, and their stability is compared. The least medium square (LMEDS), least trimmed square (LTS), random sample consensus (RANSAC), and its generalization maximum sample consensus (MSAC) algorithms are applied for model fitting, and their sensitivity for removing outliers is analyzed. The raw data are corrected using mathematical models and scaled to generate displacement data. Finally, seismic vibrations of the bridge are generated, and the seismic responses are compared. The data are validated using target-tracking methods and mechanical sensors, i.e., string potentiometers. The results show a good agreement between the proposed blob feature detection and matching algorithms and target-tracking data and reference data obtained using mechanical sensors.

Keywords: blob-based features; bridge; seismic structural health monitoring; SIFT; SURF; KAZE; LMEDS; LTS; RANSAC; MSAC; matching accuracy

Citation: Ngeljaratan, L.; Moustafa, M.A.; Sumarno, A.; Prasetyo, A.M.; Sari, D.P.; Maidina, M. Improved Blob-Based Feature Detection and Refined Matching Algorithms for Seismic Structural Health Monitoring of Bridges Using a Vision-Based Sensor System. *Infrastructures* **2024**, *9*, 97. https://doi.org/10.3390/infrastructures9060097

Academic Editors: Armin Mehrabi and Seyed Saman Khedmatgozar Dolati

Received: 13 May 2024
Revised: 3 June 2024
Accepted: 8 June 2024
Published: 14 June 2024

1. Introduction

Structural health monitoring (SHM) has emerged as a beneficial system for assessing structural performance under ambient and forced vibrations either in laboratory or field environments, especially in monitoring civil infrastructure. Key civil infrastructure systems like bridges are mostly equipped with SHM systems to allow monitoring of traffic, wind, and other environmental loading as well as natural hazard incidents such as earthquakes. In bridge monitoring, most SHM systems use wireless network systems or even mechanical sensors with wires that may interfere with traffic and practically provide data at limited locations. Their installment and implementation are also challenging, especially for some locations such as under bridges, bridges over water, or highly elevated bridges. High sampling rates that are mostly used by these types of sensors also require more data processing to capture the low-vibration frequency of flexible-type bridges or long-span bridges. Moreover, when considering SHM for reconnaissance efforts, rapid bridge evaluation that will reveal whether to open or close it after the earthquake; this decision strongly depends on the bridge's condition in the aftermath. These quick assessments are challenging using current monitoring inspection procedures since they are unable to capture the global behavior of the bridge during earthquakes, and they are further incapable of monitoring the progress of more permanent forms of damage such as plastic deformation, rotation, and drifts.

Advanced SHM technologies using cameras, i.e., vision-based technology, offer advantages over their SHM counterparts [1–5]. A larger field of view that depends on the camera type and setting enables the monitoring of a broader bridge area and provides options for assessing the deck and bent of the bridge all in one setting. Lower camera video sampling rates also allow for the dynamic characterization of a bridge under a very low to high excitation, from low ambient vibration to the high amplitude of earthquakes. The global response of bridges recorded at the bent, deck, base, and abutment levels generate valuable data for estimating permanent damage to bridges. Moreover, an intelligent robotic system provided by an unmanned aerial vehicle (UAV) or drone, which is also integrated with camera technology, is capable of recording bridges and other infrastructure during earthquakes or any vibration events; it does so without disturbances from ground movement due to its airborne operation [6–10].

The vast development of computer vision (CV) algorithms greatly supports either laboratory or field implementations of vision-based SHM. Tracking algorithms using well-known template matching techniques such as digital image correlation (DIC) have been proven one of the most effective procedures in generating structural displacement, strain mapping, and other structural dynamic characteristics [3,4,11,12]. As this technique requires stamping on the surface area of the specimen in the form of spray-painting or artificial targets, it is challenging to implement in real-life structures, especially for long-term monitoring. Therefore, advancement in CV feature detection, extraction, and tracking algorithms offers advantages in object recognition using the natural features of these structures. Multiscale feature detection and description algorithms such as the scale-invariant feature transform (SIFT) [13], speeded-up robust features (SURF) [14], and KAZE [15] algorithms are among the most popular algorithms that demonstrate high repeatability and distinctiveness with various forms of image transformation such as noise and blurring. Matching algorithms, such as the greedy nearest neighbor, optimal fair or full, or exact algorithm, are selected to match features between images depending on the matching goal. The returned matching features may not be exact; therefore, it is necessary to filter the outliers using an optimization algorithm such as random sample consensus (RANSAC) or its variants, namely, the M-estimator SAC (MSAC) [16], progressive SAC (PROSAC) [17], or maximum likelihood estimator SAC (MLESAC) [18] algorithm, of which selection is based on its accuracy, speed, robustness, or optimality [19].

Performing feature detection and description on digital images is primarily to extract specific features based on which descriptor is used. Among the available detectors like edge or corner detectors, the blob-based detector is the simplest method that aims to analyze the shape features of an object in the image that contrast their backgrounds in color, brightness, or other properties. The SURF and SIFT algorithms are among the frequently used operators that can extract blob-based features in the image and have been implemented in bridge SHM [20,21]. Meanwhile, the KAZE algorithm is not commonly selected in bridge monitoring; however, previous work has reported its implementation in wind turbine monitoring [22]. Their matching accuracies, i.e., the number of tracked and matched features after the detection and extraction steps, have barely been reported in previous works. Therefore, the objectives of this study are to explore the effect of refined matching algorithms on blob-based features in improving their accuracies and to implement the proposed algorithms on large-scale bridges tested under seismic loads using vision-based SHM. The major contributions of this study are a detailed procedure that exploits the impact of selecting detection and matching operators to improve blob-based feature performance and a CV-oriented procedure for seismic SHM using vision-based sensors. The remainder of this paper is organized as follows. In Section 2, selected blob-based feature detectors and matching and refined matching algorithms are explained. The testing setup of two-span accelerated bridge construction (ABC) bridge for implementing the proposed algorithms is given in Section 3. Their results and important constraints are presented and their accuracies verified in Section 4, and conclusions are drawn in Section 5.

2. Methodology

Computer vision algorithms cover a wide range of operators that can detect, extract, and match features within image sequences recorded from tests. Their general procedures are adopted in this study, as shown in Figure 1, together with the selected operators. They start by detecting blob-based features and continue with the matching procedure. Then, the proposed refined matching algorithm is applied to improve the number of correctly matched pairs. In Figure 1, three blob-based feature detectors, i.e., SURF, SIFT, and KAZE, are used to detect blob features in test images. These features are then matched with the nearest neighbour (NN) algorithm. Four operators are proposed to return more exact matching, i.e., the least median of squares (LMEDS), least trimmed square (LTS), random sample consensus (RANSAC), and M-estimator sample consensus (MSAC) algorithms. These algorithms have commonly been used in distance or matching problems in previous works [23–28]. The details of the proposed procedure are provided in the next subsections.

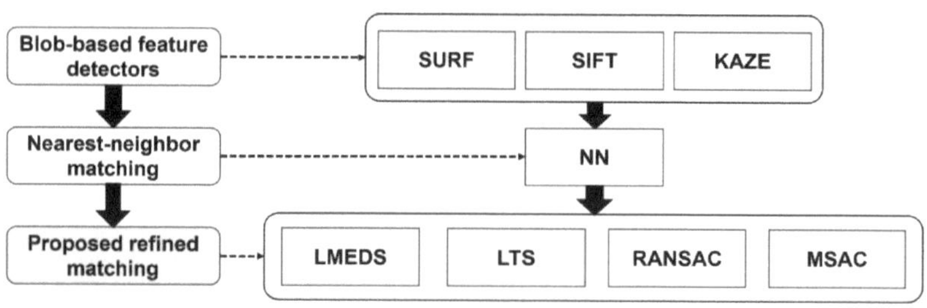

Figure 1. Computer vision procedure using blob-based feature detectors and nearest neighbor matching with the proposed refined matching algorithms.

2.1. Blob-Based Feature Detection Algorithm

In CV image registration, there are five general stages of relating image sequence characteristics so the image datasets can be transformed into a single unified coordinate system. They are feature detection and description, matching, refined matching to filter outliers, image transformation, and reconstruction. Feature detector operators are algorithms that detect features in the image that can be in the form of edges, corners, lines, blobs, etc. The three blob feature descriptors selected in this study are SURF, SIFT, and KAZE. SURF [14] detectors are based on the determinant of the Hessian matrix $\mathcal{H}$, as shown in Equation (1) in point $x = (x, y)$ at scale σ. They use integral images to improve the algorithm speed, relying on Gaussian space analysis where $L_{xx}(x, \sigma)$ is the convolution of the Gaussian second-order derivative of image I in point x and for $L_{xy}(x, \sigma)$ and $L_{yy}(x, \sigma)$. Specific bin dimensions are used to describe the detected blob features with Haar wavelet distribution within certain regions. The dimensions can be extended to 64-D or even 128-D depending on the change of image perspective.

$$\mathcal{H}(x,\sigma) = \begin{bmatrix} L_{xx}(x,\sigma) & L_{xy}(x,\sigma) \\ L_{xy}(x,\sigma) & L_{yy}(x,\sigma) \end{bmatrix} \quad (1)$$

The most well-known feature descriptor is SIFT [13], which is based on the difference of Gaussian (DoG) operator, i.e., an approximation of Laplacian of Gaussian. Local maxima on an image are searched using DoG to detect feature points at various scales. It is strongly invariant to image scale and rotations with affine variations; however, it has a high computational cost. SURF, on the other hand, has a low computational cost compared to SIFT. Equation (2) describes the convolution of difference D between two Gaussians at different scales within image $I(x, y)$, in which G corresponds to the Gaussian function.

$$D(x, y, \sigma) = (G(x, y, k\sigma) - G(x, y, \sigma)) * I(x, y) \tag{2}$$

The KAZE detector is also computed at multiple scale levels based on the normalized determinant of the Hessian matrix. It uses non-linear diffusion filtering that benefits blurred image processing as it reduces noise and maintains the boundaries of regions in subject images simultaneously. A moving window is used to select the maxima of the detector. Feature points are detected by discovering major orientations in circular regions around each detected feature. Similar to SIFT and SURF, KAZE features are also invariant to rotation, scale, and limited affine. This detector has more uniqueness at varying scales; therefore, it also increases the computational time. Equation (3) shows the typical nonlinear diffusion formula with divergence div, conductivity function c, gradient operator ∇, and image luminance L.

$$\frac{\partial L}{\partial t} = div(c(x, y, t) \cdot \nabla L \tag{3}$$

2.2. Matching Algorithm

Following the blob-based feature detection, the next step is to search for the most similar matches of blob-based features within image sequences. This is the most computationally expensive segment of y computer vision algorithms as it involves searching and tracking for the most similar matches to high-dimensional vectors. Therefore, a robust yet efficient algorithm is required to perform fast searching and tracking in such large datasets. Nearest neighbor (NN) is selected in this study as it provides speedy computation in several orders of magnitude. The NN problem consists of pre-processing a set of points X such that the operation in Equation (4) can be performed efficiently. Equation (4) describes the NN search in a metric space. It is defined as follows: for a set of feature points $X = \{x_1, x_2, x_3, \ldots, x_n\}$ in a metric space M with a query point $q \in M$, the element NN $(q, X) \in X$ is searched such that the match is the closest to q with respect to a metric distance $d : M \times M \rightarrow \mathbb{R}$.

$$NN(q, X) = argmin_{x \in X}\, d(q, x) \tag{4}$$

2.3. Proposed Refined Matching Algorithm

It is known that working with high-dimensional features will mostly produce incorrect searches and matches. Many practical applications of the NN algorithm return approximate matches with some outliers, meaning that some results are estimated yet still close to the exact matches. Therefore, in image registration, it is common that NN is just a part of complete procedures combined with applications of other CV algorithms that contain other approximations. The refined matching procedures used to filter the outliers selected in this study are the least median of squares (LMEDS) [29], least trimmed square (LTS) [30], random sample consensus (RANSAC) [31], and M-estimator sample consensus (MSAC) [18] algorithms. Both the LMEDS and LTS estimators are common regression estimators. Considering a sample with n observations consisting of inliers and outliers, the sample with the least maximal squared residual is called LMEDS, while the sample with the highest residual sum of squares is defined as LTS. LMEDS estimator estimates the parameters by solving the nonlinear minimization problem and reduces the median of squared standardized residuals for the entire dataset. Meanwhile, LTS consists of finding a subset of cases whose deletion from the dataset would lead to the regression with the smallest residual sum of squares. It is regression-, scale-, and affine-equivariant and is used as a general-purpose high-breakdown method.

The RANSAC algorithm is widely used to detect unique transformation by random sampling. The corresponding transformation is estimated, and the adequacy of the transformation to the rest of the data is then tested. The transformation that yields the most effective consensus is then kept. RANSAC is a robust and fast algorithm, yet several important parameters need to be set in the analysis. The RANSAC step relates to the theory of a minimal sample set, which the initial set is randomly selected from the input. It is followed by computing the model parameters, after which RANSAC requires the setting of a certain threshold that iteratively checks which observations of the entire dataset are consistent with the hypothetical model. It specifies the maximum distance from a point of interest to a hypothetical model. When it fits the criterion, the point is treated as a hypothetical inlier; otherwise, it is treated as an outlier. The estimated model is exact when an adequate amount of points which are classified as exact observations (inliers) are reached. MSAC is a generalization of the RANSAC estimator that is principally used to robustly estimate multiple new relations from point correspondences. MSAC implements the same sampling approach as RANSAC to generate exact solutions. However, it selects the solution that maximizes the likelihood rather than just the number of inliers.

3. Testing Setup

The proposed methods in Section 2 are experimentally evaluated on a two-span accelerated bridge construction (ABC) through a shake-table test. ABC is a bridge construction that uses state-of-the-art design, materials, and construction methods safely and cost-effectively. ABC not only improves site constructability and projects' construction schedule but also reduces impacts on traffic and project delays due to weather conditions. The tested ABC bridge shown in Figure 2 is a one-third-scale two-span reinforced concrete bridge with seismic connections. The length of each span (in the longitudinal direction) is 10.4 m with a 3.4 m width in the transverse direction. It has a two-column pier in the middle, which is spaced at 1.8 m with a height of 2.1 m. The bridge sits on a seat-type abutment at both ends. The bidirectional ground motion that was recorded during the 1994 Northridge earthquake was scaled according to a design-level (DE) seismic demand. Three tests with increasing seismic intensity from 20% and up to 75% of the DE level are selected to investigate the effect of each applied algorithm on the feature matching as well as the seismic response of the ABC bridge system. More details about the design, construction, and studies of the tested ABC bridge can be found in [32,33]. The bridge is placed on three biaxial shake tables manufactured by MTS. Each table measures 4.3×4.5 m with a stroke of ± 300 mm. It can reach a peak velocity of 1000 mm/s) and an acceleration of 1 g with a 50-ton payload (about 445 kN). All three tables are constrained to act together as a single large table, with the option to be operated individually with independent motions depending on the loading requirements. The specification of the SHM system is given in Table 1. The camera lens is 35 mm with a CMOS sensor. The image is set to monochrome to accelerate the later image processing using the proposed algorithms. The camera's full ROI is used, with 2560×2048 image pixels and a record duration of 30 s with 30 Hz sampling rates or 30 frames-per-second rates. The reference sensor R, as shown in Figure 2c, is a string potentiometer recorded at 256 Hz; therefore, adjustment of displacement is required later to enable the comparison.

Vision-based SHM of the ABC bridge has been previously reported in [3]. The difference between this study and previous works lies in its methods, i.e., the CV algorithms applied to image sequences that later affect the seismic data. Prior work needed a stereophotogrammetry technique to generate the three-dimensional coordinates of the features shown in Figure 2d. Then, these features were tracked using tracking algorithms to detect the change in each feature coordinate within the image sequences. Therefore, previous work required a set of photogrammetry images and test images, also involving two steps in processing the vision-based data. Meanwhile, this work simply uses test data without the prerequisite of taking photogrammetric images. The proposed algorithms are directly applied to the test images, and the displacement data are generated using the scale factor method [34].

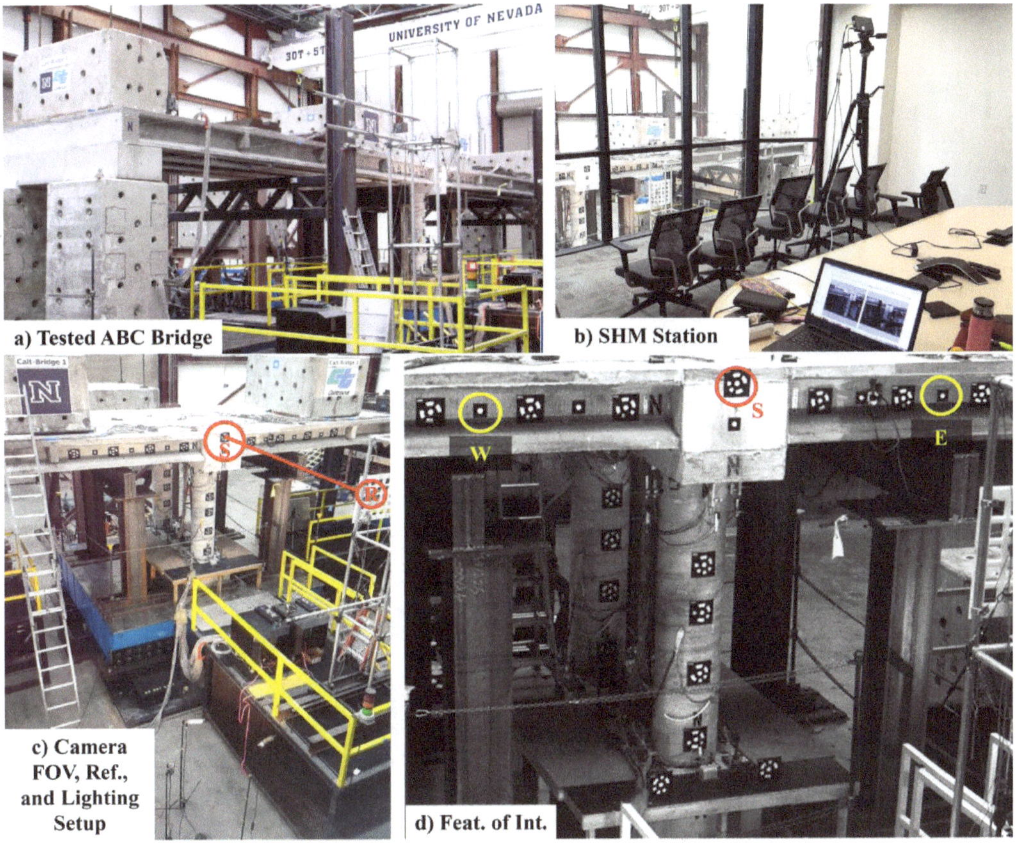

Figure 2. Tested ABC bridge (**a**), vision-based SHM setup with camera field of view (FOV) (**b**,**c**), and the feature of interest to be detected using proposed algorithms (**d**). Comparison is made between target-tracking methods S, E, W, and reference string potentiometer R to propose refined matching.

Table 1. SHM system specifications.

Sensor	CMOS	**Shutter**	3–41.654 μs	**Format**	.tiff
Lens	35 mm	Image resolution (pix.)	2560 × 2048	Image type	Monochrome

4. Results and Discussions

4.1. Blob-Based Feature Detection and Matching Results

Following the method described in Section 2, this section explores the results of the selected feature detection and matching algorithms using the images of the tested ABC bridge in Figure 2. The selected algorithms for blob-based feature detections are SURF, SIFT, and KAZE. Their thresholding is given in Table 2, i.e., with the matching NN algorithm. Two figures of each test are taken to visualize the results of each algorithm and the bridge's condition before and after testing, i.e., an image taken at the beginning of the test at $t = 0$, $I_{1,t=0}$ and at the end of the test at $t = T$, $I_{N,t=T}$. Figure 3 shows the results of blob-based features and NN matching using each algorithm. The red and blue circles indicate detected blob features on $I_{1,t=0}$ and $I_{N,t=T}$, respectively. Magenta lines show the matching results for both correct and incorrect matchings. Straight lines connecting red and blue blob features show correct matching, while crossing lines demonstrate an incorrect match between images. Algorithm selections and matching feature results are given in Table 3. Table 3 shows that the number of detected blob features on each image varies along with the selection of the algorithm. More blobs are detected using KAZE; however, less than 60% of matching accuracies are computed. It is clear that NN matching does not return 100% exact matches; rather, some features are paired incorrectly and identified as outliers, as shown in Figure 3.

Table 2. Selected algorithm for blob-based feature detections.

Parameters	Algorithm		
	SURF	**SIFT**	**KAZE**
Threshold	6000	0.05	0.005
Matching	Nearest neighbour (NN)	Nearest neighbour (NN)	Nearest neighbour (NN)

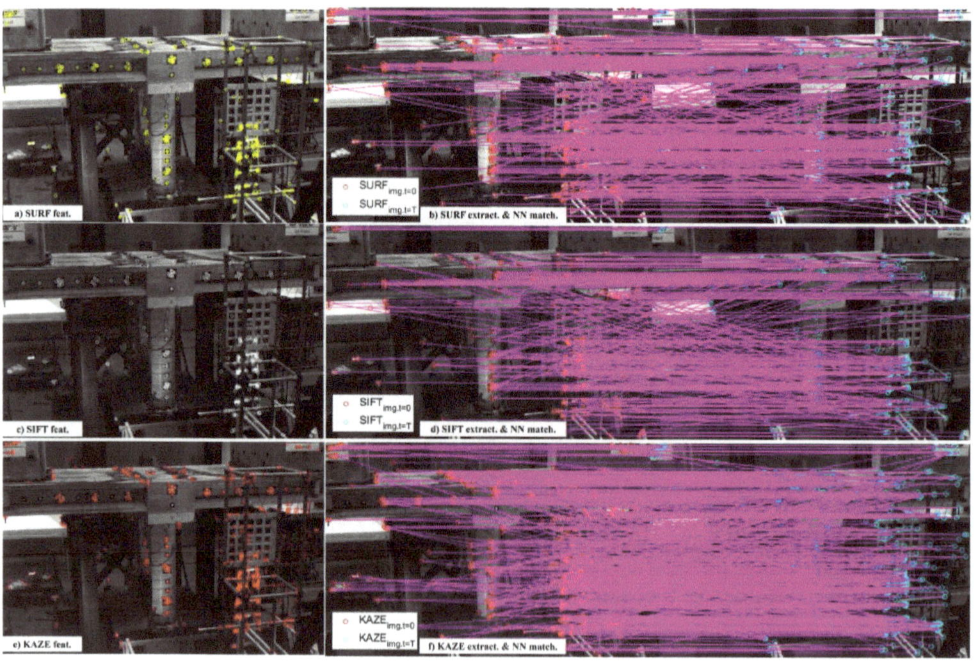

Figure 3. Feature detection, extraction, and NN matching using (**a**,**b**) SURF, (**c**,**d**) SIFT, and (**e**,**f**) KAZE algorithms.

Table 3. Results of algorithm selection and matching features.

Algorithm	**SURF**	**SIFT**	**KAZE**
$I_{1,\ t=0}$	387	448	909
$I_{N,\ t=T}$	632	499	1610
Matching feat.	229	254	533
NN Accuracy (%)	59.17	56.69	58.63

The next step is to remove the outliers and refine the matching using the LMEDS, LTS, RANSAC, and MSAC algorithms, which is proposed to improve the matching accuracy of the NN algorithm. The selected thresholds (TH) as parameters are 0.001, 0.01, 0.1, and 1. The results are given in Table 4 with the example of matching using threshold = 0.1 given in Figure 4. Similar to Figure 3, Figure 4 also shows an example of images taken at the beginning of the test at $t = 0$, $I_{1,\ t=0}$, and at the end of the test at $t = T$, $I_{N,\ t=T}$. Figure 4a, f, and k show SURF, SIFT and KAZE detections, respectively, in which blob features are matched using only the NN algorithm. Incorrect pairs are shown as crossed lines between images, which are improved further using the selected refined matching algorithms. Correct pairs are shown for all blob features after applying refined matching algorithms in Figure 4b–e for the SURF detector, Figure 4g–j for the SIFT detector, and Figure 4l–o for the KAZE detector. LMEDS and LTS return similar matching results at each increment in the threshold (193 features and 194 features associated with 49.87% and 50.15% matching accuracies for LMEDS and LTS, respectively). Meanwhile, both RANSAC and MSAC results show their effectiveness throughout each selected threshold. The highest improvement is obtained from the smallest threshold of 0.001 using the KAZE detection algorithm: 96.48% and 96.92% accuracies for RANSAC and MSAC, respectively. As the threshold increases, the number of returned exact matches decreases and drops to 44.44% and 43.15% using RANSAC and MSAC detected by SURF.

Table 4. Exact feature matching methods with their respective accuracies using the LMEDS, LTS, RANSAC, and MSAC refined matching algorithms.

				SURF Detection				
Ref. Alg.	**LMEDS**		**LTS**		**RANSAC**		**MSAC**	
TH	**Exact Match**	**Acc. (%)**	**Exact Match**	**Acc. (%)**	**Exact Match**	**Acc. (%)**	**Exact Match**	**Acc. (%)**
0.001	193	49.87	194	50.13	364	94.06	363	93.80
0.01	193	49.87	194	50.13	328	84.75	329	85.01
0.1	193	49.87	194	50.13	254	65.63	252	65.12
1	193	49.87	194	50.13	172	44.44	167	43.15
				SIFT Detection				
Ref. Alg.	**LMEDS**		**LTS**		**RANSAC**		**MSAC**	
TH	**Exact Match**	**Acc. (%)**	**Exact Match**	**Acc. (%)**	**Exact Match**	**Acc. (%)**	**Exact Match**	**Acc. (%)**
0.001	224	50	224	50	421	93.97	409	91.29
0.01	224	50	224	50	372	83.04	388	86.61
0.1	224	50	224	50	310	69.20	289	64.51
1	224	50	224	50	229	51.12	241	53.79
				KAZE Detection				
Ref. Alg.	**LMEDS**		**LTS**		**RANSAC**		**MSAC**	
TH	**Exact Match**	**Acc. (%)**	**Exact Match**	**Acc. (%)**	**Exact Match**	**Acc. (%)**	**Exact Match**	**Acc. (%)**
0.001	454	49.94	455	50	877	96.48	881	96.92
0.01	454	49.94	455	50	828	91.09	831	91.42
0.1	454	49.94	455	50	651	71.62	702	77.23
1	454	49.94	455	50	440	48.40	446	49.06

Table 5 shows the change in matching accuracies of each blob-based feature detection algorithm with their respective refined matching algorithm and threshold. LMEDS and LTS have no effect in improving the matching accuracies of blob features detected by SURF, SIFT, and KAZE. The applied algorithms even impair some features, resulting in a decrease of 9.30% and 9.04% in matching accuracies for LMEDS and LTS, respectively. RANSAC and MSAC, on the other hand, are proven to improve the matching accuracies by 37.28% and 34.60% using the SIFT detector and by 37.84% and 38.28% using the KAZE detector at the lowest threshold of 0.001. However, increasing threshold intensifies the incorrect matching as at threshold = 1, RANSAC and MSAC return false matching rates of 5.58% and 2.9% using the SIFT detector and 10.23% and 9.57% using the KAZE detector.

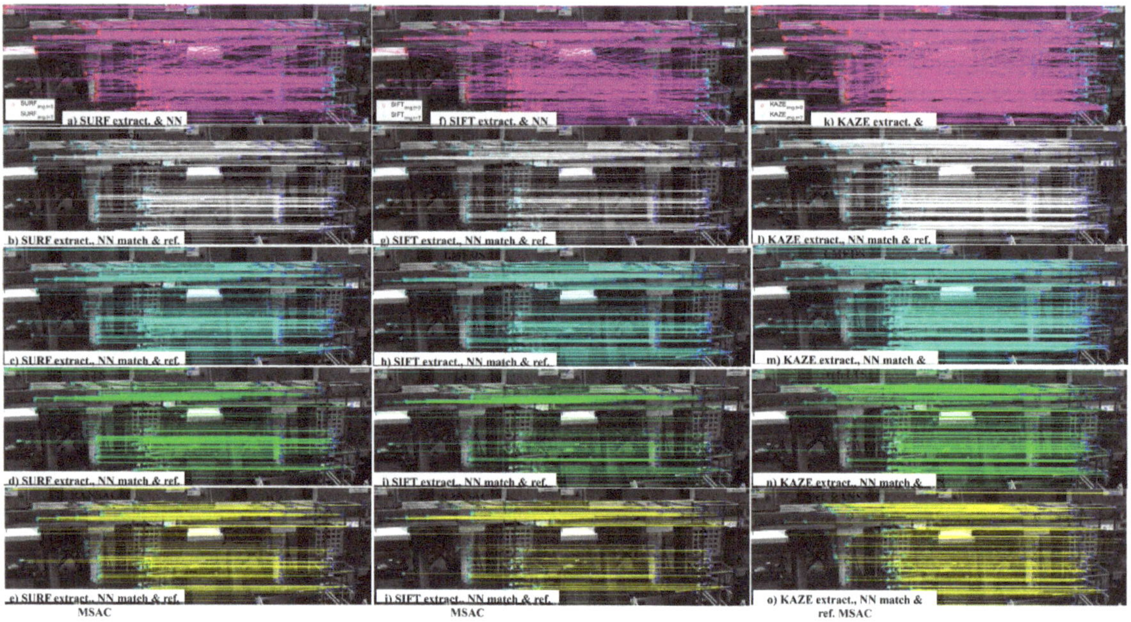

Figure 4. LMEDS, LTS, RANSAC, and MSAC refined matching using threshold = 0.01. (**a**,**f**,**k**) SURF, SIFT, KAZE with NN matching only; (**b**,**g**,**l**) SURF, SIFT, KAZE with NN matching and refined LMDES matching; (**c**,**h**,**m**) SURF, SIFT, KAZE with NN matching and refined LTS matching; (**d**,**i**,**n**) SURF, SIFT, KAZE with NN matching and refined RANSAC matching; (**e**,**j**,**o**) SURF, SIFT, KAZE with NN matching and refined MSAC matching.

Table 5. Changes in matching accuracies ($\Delta_{acc.}(\%)$) with their respective algorithms.

Alg.	**SURF Detection**				**SIFT Detection**				**KAZE Detection**			
Ref. Alg.	**(1)**	**(2)**	**(3)**	**(4)**	**(1)**	**(2)**	**(3)**	**(4)**	**(1)**	**(2)**	**(3)**	**(4)**
TH	$\Delta_{acc.}$ (%)	$\Delta_{acc.}$ (%)	$\Delta_{acc.}$ (%)	$\Delta_{acc.}$ (%)	$\Delta_{acc.}$ (%)	$\Delta_{acc.}$ (%)	$\Delta_{acc.}$ (%)	$\Delta_{acc.}$ (%)	$\Delta_{acc.}$ (%)	$\Delta_{acc.}$ (%)	$\Delta_{acc.}$ (%)	$\Delta_{acc.}$ (%)
0.001	−9.30	−9.04	34.88	34.63	−6.70	−6.70	37.28	34.60	−8.69	−8.64	37.84	38.28
0.01	−9.30	−9.04	25.58	25.84	−6.70	−6.70	26.34	29.91	−8.69	−8.64	32.45	32.78
0.1	−9.30	−9.04	6.46	5.94	−6.70	−6.70	12.50	7.81	−8.69	−8.64	12.98	18.59
1	−9.30	−9.04	−14.73	−16.02	−6.70	−6.70	−5.58	−2.90	−8.69	−8.64	−10.23	−9.57

(1) LMEDS (2) LTS (3) RANSAC (4) MSAC

4.2. Seismic Response of the ABC Bridge

The blob-based feature detection and matching results in the previous section suggest that RANSAC and MSAC returned more exact matches as compared to their counterparts, i.e., LMEDS and LTS. Careful selection should be considered, however, when choosing the threshold parameters, since higher thresholds may decrease the accuracy of NN matching. The next procedure is to generate the seismic response of the tested bridge and to compare the results from prior work using target-tracking and reference sensors to the results of the proposed blob-based feature refined matching. Based on Figure 2, two comparisons are made: the first at the bent cap level and the second at the west and east sides of the bridge girder. At the bent cap level, the results of target-tracking method S and reference string potentiometer R are compared to the blob-based feature KAZE and RANSAC refined matching BK results. At the girder level, only the target-tracking method of the feature at the east (E) and west (W) sides is compared to blob-based feature matching using KAZE combined with RANSAC, BK, since the reference sensors are not available at the girder level. KAZE combined with RANSAC is selected as an example of refined matching in this section because it generates the highest accuracies as compared to LMEDS, LTS, and MSAC.

The seismic response of the ABC bridge at the bent cap level is given in Figure 5 and at the girder level is shown in Figure 6. The displacement time histories are given for each level of design earthquake, from 20% to 75% design earthquake, DE. Since the recordings of vision-based measurement and reference sensor types have different time reference and sampling rates, the displacement responses from the reference sensor R are adjusted to have a similar starting time as the target-tracking S and KAZE combined with RANSAC, i.e., the BK measurements in Figure 5. As the sampling rates are also different between both systems, the vision-based system uses a practical 30Hz or 30 image-per-second rate, while the reference sensor R uses 256 Hz; resampling is conducted on vision-based results. Therefore, to accurately compare the sensor data, the results in Figure 5 are after resampling of vision-based results, i.e., the resampling of the S and BK displacement results. The tolerance is 0.0001 in the signal processing. Meanwhile, the results at girder level do not require any further signal processing since the results in Figure 6 are taken from vision-based data. Only the result of features E and W is generated using the target-tracking method, while the BK result uses the proposed blob-based feature detector KAZE and refined matching RANSAC algorithms.

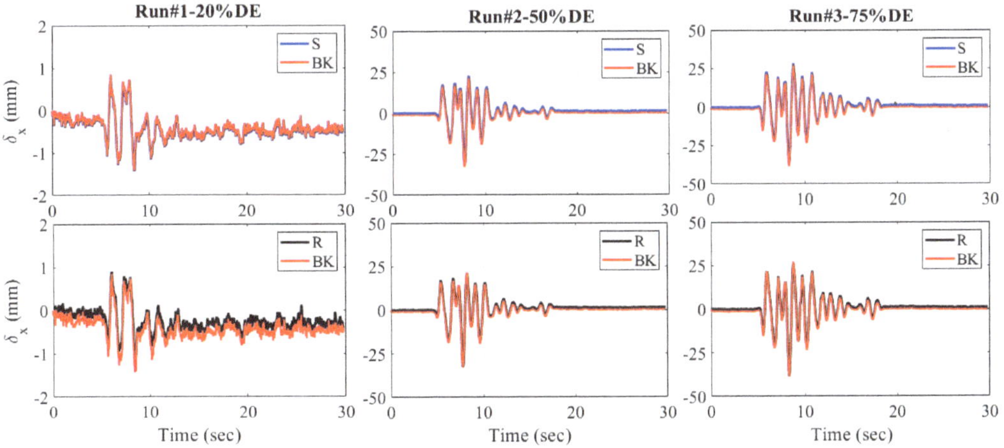

Figure 5. Displacement comparison at bent cap level between target-tracking method S, reference string potentiometer R, and blob feature KAZE combined with RANSAC matching BK.

To measure the accuracy of the proposed blob-based feature KAZE detector and RANSAC matching the target-tracking and reference sensor results, an error analysis is performed using the root mean square error (RMSE). A summary of the RMSE computed in millimeter (mm) units is given in Table 6. As shown in the table, the normalized RMSE is very low, with a lowest result of 0.02 mm and a highest result of 1.15 mm. This confirms the accuracy of blob-based feature KAZE and RANSAC matching and also justifies the technique and the camera's practical 30 fps sampling rates. A slight difference is shown between the proposed algorithm and string potentiometer measurements, from R to BK at 75% DE. This is reasonable because as the seismic load increases, the error also becomes larger. Furthermore, this is because no triangulation is considered for the reference sensor R data, even though the impact of seismic shake on the bridge should observe the possibility that the string potentiometer is no longer aligned after the shakes. Therefore, based on the equivalent response shown in Figures 5 and 6 together with the RMSE discussion above and the results in Table 6, the validity of the proposed algorithms is proven.

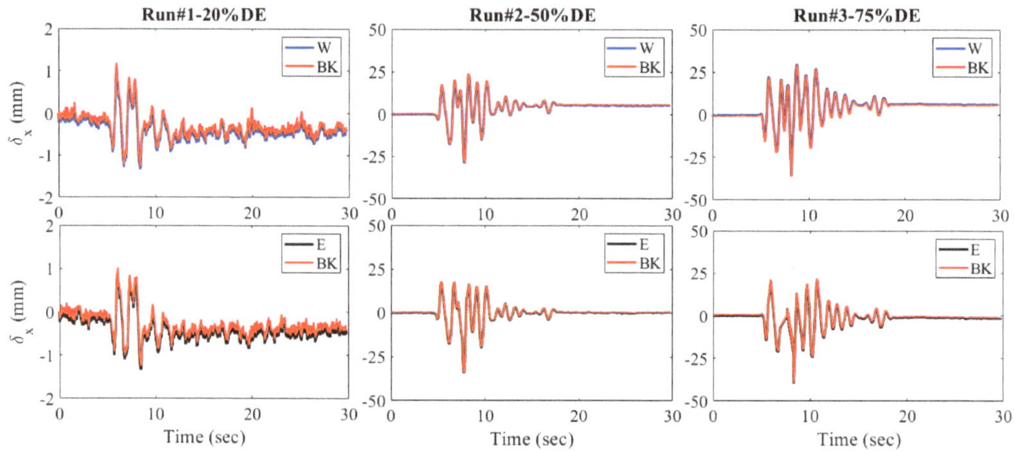

Figure 6. Displacement comparison at girder level between target-tracking method E, W, and blob feature KAZE combined with RANSAC matching BK.

Table 6. Measurement differences from the shake-table tests between the target-tracking method S, E, W and reference string potentiometer R and blob feature KAZE combined with RANSAC matching BK.

EQ Test	RMSE (mm)			
	S to BK	R to BK	E to BK	W to BK
RUN#1-25%DE	0.02	0.17	0.08	0.06
RUN#2-50%DE	1.09	0.96	0.16	0.22
RUN#3-75%DE	1.02	1.15	0.50	0.38

5. Conclusions

This study explores the effect of blob-based feature detectors combined with refined matching algorithms in improving their matching accuracies and then implements them on a two-span ABC bridge tested under seismic loads. The SHM is conducted using vision-based systems. The major contributions of this study are a detailed procedure that explores the impact of selecting detection and matching operators with their parameters to improve blob-based feature performance and a CV-oriented procedure implemented on seismic SHM using vision-based sensors. The study uses three design earthquake tests to quantify the effect of each selected algorithm on the matching accuracy. The main conclusions and key findings of this study are as follows:

1. Using only the NN algorithm to return the exact match is not efficient, as only less than 60% of correct matches are returned for SURF, SIFT, and KAZE detectors. This confirms the results of previous works stating that using only the NN algorithm to find exact matches is challenging; combining it with a refined matching algorithm is necessary to improve the results.

2. The LMEDS and LTS algorithms have no effect on refining the false matching; rather, they decrease the number of correct matches as well as the accuracies from NN matching. The change in their thresholds also has no impact; they show a constant number of matching features either at low- or high-threshold selection.
3. RANSAC and MSAC, however, are proven to increase the number of correct matches as well as the matching accuracies of blob-based features using SURF, SIFT, and KAZE. Careful selection of threshold is mandatory because the matching accuracies decrease following the increase in the threshold. There is a chance that with selection of a higher threshold, the number of correct feature matches will even decrease from NN matching numbers, similar to when using LMEDS or LTS.
4. The ABC bridge seismic test verifies the potential and accuracy of the proposed blob-based feature and refined matching algorithms, especially for bridge displacement responses. The displacement responses of the proposed algorithms show a similar trend to previous studies using target-tracking and string potentiometers. A minimum error of 0.02 mm is computed, while a slightly higher RMSE of 1.15 mm is calculated for the response to the higher-magnitude design earthquake.
5. Overall, it is concluded that the proposed blob-based feature detection and refined matching algorithms have the potential for implementation in vision-based SHM of bridges. The combination of the blob-based feature detector SURF, SIFT, or KAZE with NN matching then refined with RANSAC and MSAC is proven to generate more accurate matching results. Therefore, they are recommended for implementation in detecting, extracting, and matching blob-based features of interest on bridge structures. Additionally, further study about detecting natural features on bridges or other civil infrastructures using the proposed algorithms is recommended.

Author Contributions: L.N. and M.A.M. conceptualization, methodology, review, and editing; L.N. formal analysis, investigation, visualization; M.A.M. supervision, project administration, funding acquisition; L.N., A.S., A.M.P., D.P.S. and M.M. writing—original draft preparation. All authors have read and agreed to the published version of the manuscript.

Funding: This work was funded in part by the US Department of Transportation Accelerated Bridge Construction University Transportation Center (ABC-UTC) 2016 Grant.

Data Availability Statement: The data presented in this study are available on request from the authors.

Acknowledgments: We would like to thank The Indonesia Endowment Funds for Education (LPDP) and the National Research and Innovation Agency (BRIN) through Riset and Inovasi untuk Indonesia Maju (RIIM) program, which also supported the first author in this research.

Conflicts of Interest: The authors declare no conflict of interest.

References

1. Ngeljaratan, L.; Moustafa, M.A.; Pekcan, G. A compressive sensing method for processing and improving vision-based target-tracking signals for structural health monitoring. *Comput. Civ. Infrastruct. Eng.* **2021**, *36*, 1203–1223. [CrossRef]
2. Ngeljaratan, L.; Moustafa, M.A. Implementation and evaluation of vision-based sensor image compression for close-range photogrammetry and structural health monitoring. *Sensors* **2020**, *20*, 6844. [CrossRef] [PubMed]
3. Ngeljaratan, L.; Moustafa, M.A. Structural health monitoring and seismic response assessment of bridge structures using target-tracking digital image correlation. *Eng. Struct.* **2020**, *213*, 110551. [CrossRef]
4. Ngeljaratan, L.; Moustafa, M.A. System identification of large-scale bridges using target-tracking digital image correlation. *Front. Built Environ.* **2019**, *5*, 85. [CrossRef]
5. Ngeljaratan, L.; Moustafa, M.A. Underexposed Vision-Based Sensors' Image Enhancement for Feature Identification in Close-Range Photogrammetry and Structural Health Monitoring. *Appl. Sci.* **2021**, *11*, 11086. [CrossRef]
6. Wang, X.; Wittich, C.; Hutchinson, T.; Bock, Y.; Goldberg, D.; Lo, E.; Kuester, F. Methodology and validation of UAV-based video analysis approach for tracking earthquake-induced building displacements. *J. Comput. Civil Eng.* **2020**, *34*, 04020045. [CrossRef]
7. Wang, X.; Lo, E.; De Vivo, L.; Hutchinson, T.C.; Kuester, F. Monitoring the earthquake response of full-scale structures using UAV vision-based techniques. *Struct. Control Health Monit.* **2022**, *29*, e2862. [CrossRef]

8. Bai, X.; Yang, M. UAV based accurate displacement monitoring through automatic filtering out its camera's translations and rotations. *J. Build. Eng.* **2021**, *44*, 102992. [CrossRef]
9. Asadzadeh, S.; de Oliveira, W.J.; de Souza Filho, C.R. UAV-based remote sensing for the petroleum industry and environmental monitoring: State-of-the-art and perspectives. *J. Pet. Sci. Eng.* **2022**, *208*, 109633. [CrossRef]
10. Ngeljaratan, L.; Moustafa, M.A.; Sumarno, A.; Prasetyo, A.M.; Sari, D.P.S.; Maidina. Exploratory Study of Drone Data Stabilization with Implications in Vibration-based Structural Health Monitoring. *Evergreen* **2023**, *10*, 1776–1783. [CrossRef]
11. Janeliukstis, R.; Chen, X. Review of digital image correlation application to large-scale composite structure testing. *Compos. Struct.* **2021**, *271*, 114143. [CrossRef]
12. Pagani, A.; Azzara, R.; Carrera, E.; Zappino, E. Static and dynamic testing of a full-composite VLA by using digital image correlation and output-only ground vibration testing. *Aerosp. Sci. Technol.* **2021**, *112*, 106632. [CrossRef]
13. Lowe, D.G. Distinctive Image Features from Scale-Invariant Keypoints. *Int. J. Comput. Vis.* **2004**, *60*, 91–110. [CrossRef]
14. Bay, H.; Tuytelaars, T.; Van Gool, L. *SURF: Speeded Up Robust Features*; Springer: Berlin/Heidelberg, Germany, 2006; pp. 404–417.
15. Alcantarilla, P.F.; Bartoli, A.; Davison, A.J. KAZE features. In Proceedings of the Computer Vision–ECCV 2012: 12th European Conference on Computer Vision, Florence, Italy, 7–13 October 2012; Part VI 12. pp. 214–227.
16. Torr, P.H.; Zisserman, A. Robust parameterization and computation of the trifocal tensor. *Image Vis. Comput.* **1997**, *15*, 591–605. [CrossRef]
17. Chum, O.; Matas, J. Matching with PROSAC-progressive sample consensus. In Proceedings of the 2005 IEEE computer society conference on computer vision and pattern recognition (CVPR'05), San Diego, CA, USA, 20–25 June 2005; pp. 220–226.
18. Torr, P.H.; Zisserman, A. MLESAC: A new robust estimator with application to estimating image geometry. *Comput. Vis. Image Underst.* **2000**, *78*, 138–156. [CrossRef]
19. Martínez-Otzeta, J.M.; Rodríguez-Moreno, I.; Mendialdua, I.; Sierra, B. Ransac for robotic applications: A survey. *Sensors* **2022**, *23*, 327. [CrossRef] [PubMed]
20. Yu, S.; Zhang, J. Fast bridge deflection monitoring through an improved feature tracing algorithm. *Comput. Civ. Infrastruct. Eng.* **2020**, *35*, 292–302. [CrossRef]
21. Duan, X.; Chu, X.; Zhu, W.; Zhou, Z.; Luo, R.; Meng, J. Novel method for bridge structural full-field displacement monitoring and damage identification. *Appl. Sci.* **2023**, *13*, 1756. [CrossRef]
22. Prasad, S.; Chiang, C.-H.; Kumar, D.; Kalra, S.; Khandelwal, A. Robust and efficient feature-based method for structural health monitoring of large structures. *J. Civil Struct. Health Monit.* **2023**, *13*, 961–982. [CrossRef]
23. Nugraha, G.D.; Sudiarto, B.; Ramli, K. Machine learning-based energy management system for prosumer. *Evergreen* **2020**, *7*, 309–313. [CrossRef]
24. Srivastava, A.K.; Singh, B.K.; Gupta, S. Prediction of Tool Wear Using Machine Learning Approaches for Machining on Lathe Machine. *Evergreen* **2023**, *10*, 1357–1365. [CrossRef]
25. Nugroho, A.C.P.T.; Al Hakim, B.; Hendrik, D.; Sasmito, C.; Muttaqie, T.; Tjolleng, A.; Kurniawan, M.A.; Komariyah, S. Mission Analysis of Small-Scale LNG Carrier as Feeder for East Indonesia: Ambon City as the Hub Terminal. *Evergreen* **2023**, *10*, 1938–1950. [CrossRef]
26. Prasetyo, A. Gait Analysis Parameter Study Using Xbox Kinect Aimed at Medical Rehabilitation Tool. *Evergreen* **2022**, *9*, 511–518.
27. Kamra, M.M.; Hu, C. Implementation of unstructured multi-dimensional THINC for practical multi-phase flow simulations. *Evergreen* **2017**, *4*, 52–57. [CrossRef]
28. Hesty, N.W.; Supriatna, N.K.; Akhmad, K.; Kuncoro, A.H.; Nurliyanti, V.; Rahardja, M.B.; Sudarto, S.; Mulyadi, W.; Utama, P.A. Promoting Wind Energy by Robust Wind Speed Forecasting Using Machine Learning Algorithms Optimization. *Evergreen* **2024**, *11*, 354–370.
29. Rousseeuw, P.J.; Leroy, A.M. *Robust Regression and Outlier Detection*; John Wiley & Sons: New York, NY, USA, 2005.
30. Ruppert, D.; Carroll, R.J. Trimmed least squares estimation in the linear model. *J. Am. Stat. Assoc.* **1980**, *75*, 828–838. [CrossRef]
31. Fischler, M.A.; Bolles, R.C. Random sample consensus: A paradigm for model fitting with applications to image analysis and automated cartography. *Commun. ACM* **1981**, *24*, 381–395. [CrossRef]
32. Benjumea, J.; Saiidi, M.S.; Itani, A. Biaxial Seismic Performance of a Two-Span Concrete Bridge Model with Six ABC Connections. *J. Bridge Eng.* **2021**, *26*, 04021056. [CrossRef]
33. Benjumea, J.; Saiidi, M.S.; Itani, A. Large-Scale Biaxial Shake-Table Studies of a Precast Bridge Model. *J. Struct. Eng.* **2021**, *147*, 04021104. [CrossRef]
34. Ngeljaratan, L.; Bas, E.E.; Moustafa, M.A. Unmanned Aerial Vehicle-Based Structural Health Monitoring and Computer Vision-Aided Procedure for Seismic Safety Measures of Linear Infrastructures. *Sensors* **2024**, *24*, 1450. [CrossRef]

infrastructures

MDPI

Article

Machine Learning and Signal Processing for Bridge Traffic Classification with Radar Displacement Time-Series Data

Matthias Arnold [1,2,*] and Sina Keller [2]

[1] ci-tec GmbH, 76139 Karlsruhe, Germany
[2] Institute of Photogrammetry and Remote Sensing, Karlsruhe Institute of Technology, 76131 Karlsruhe, Germany; sina.keller@kit.edu
* Correspondence: matthias.arnold@partner.kit.edu

Abstract: This paper introduces a novel nothing-on-road (NOR) bridge weigh-in-motion (BWIM) approach with deep learning (DL) and non-invasive ground-based radar (GBR) time-series data. BWIMs allow site-specific structural health monitoring (SHM) but are usually difficult to attach and maintain. GBR measures the bridge deflection contactless. In this study, GBR and an unmanned aerial vehicle (UAV) monitor a two-span bridge in Germany to gather ground-truth data. Based on the UAV data, we determine vehicle type, lane, locus, speed, axle count, and axle spacing for single-presence vehicle crossings. Since displacement is a global response, using peak detection like conventional strain-based BWIMs is challenging. Therefore, we investigate data-driven machine learning approaches to extract the vehicle configurations directly from the displacement data. Despite a small and imbalanced real-world dataset, the proposed approaches classify, e.g., the axle count for trucks with a balanced accuracy of 76.7% satisfyingly. Additionally, we demonstrate that, for the selected bridge, high-frequency vibrations can coincide with axles crossing the junction between the street and the bridge. We evaluate whether filtering approaches via bandpass filtering or wavelet transform can be exploited for axle count and axle spacing identification. Overall, we can show that GBR is a serious contender for BWIM systems.

Keywords: bridge weigh-in-motion; ground-based radar; bridge monitoring; deep learning; MiniRocket; feature extraction; wavelets

Citation: Arnold, M.; Keller, S. Machine Learning and Signal Processing for Bridge Traffic Classification with Radar Displacement Time-Series Data. *Infrastructures* **2024**, *9*, 37. https://doi.org/10.3390/infrastructures9030037

Academic Editors: Seyed Saman Khedmatgozar Dolati and Armin Mehrabi

Received: 24 January 2024
Revised: 16 February 2024
Accepted: 20 February 2024
Published: 22 February 2024

1. Introduction

Extracting vehicle configurations on bridges is essential, like detecting overweight trucks and acquiring site-specific traffic information. This information can be considered for structural health monitoring (SHM). At an early stage of weight control, static scales have been used. However, the weighing process takes a long time, and measurement stations can be easily bypassed. An alternative approach uses sensors on the pavement that acquire dynamic measurements. This approach reduces the weighing time, but the sensors are exposed to more stress, making the system maintenance difficult. Current solutions use the bridge as a weighing scale [1–3]. These systems are called bridge weigh-in-motion (BWIM). A part of the bridge, such as the girder, is usually used to measure strain or acceleration during a vehicle crossing, which is henceforth referred to as an event. Axle count, spacing, and driving speed information determine the vehicle's weight. Pavement sensors are generally exploited to acquire these three parameters. Nothing-on-road (NOR) BWIMs are entirely dispensed with pavement sensors by only using the lower side of the bridge for sensor attachment. This setup increases the challenge of acquiring data about the vehicle configuration. Also, the lower bridge side has to be accessible, which is not always the case. A remote NOR BWIM has barely been studied, except by Ojio et al. [3], who use cameras for contactless measurements.

Ground-based radar (GBR) has frequently been investigated in the context of bridge monitoring in recent years [4–8]. It achieves comparable or even better accuracy than established, conventional strain, or acceleration sensors [7] with the advantage of being remote and non-invasive. Thus, GBRs can be quickly set up for measurements. As they measure the bridge deflection and displacement, it can be directly applied to many SHM methods, like determining the dynamic input factor [9]. Furthermore, GBR has been investigated regarding the detection of bridge crossing events and, to some extent, the vehicle type classification [10,11]. Arnold and Keller [12] show that it is possible to differentiate between single- and multi-presence events from GBR deflection using machine learning (ML) despite facing challenges like variable-length time series and dataset imbalance. Yet, GBR has never been explicitly studied in the context of BWIM. In this paper, we will investigate the potential of GBR bridge displacement time series for BWIM by applying standard signal processing methods such as bandpass filtering and continuous wavelet transform (CWT) and using data-driven ML.

Our study is motivated by using remote displacement measurements with GBR in the context of BWIM. We develop an output-only method using GBR for analysis and an unmanned aerial vehicle (UAV) to gather ground-truth data. Measuring the displacement is challenging, so it has barely been studied for BWIM. Yet, it is regarded as a relevant feature for SHM [13]. To our knowledge, to this date, no study exists that tries to extract traffic information from only real-world bridge deflection. Since GBR-based bridge monitoring has many advantages, such a system would be beneficial. Therefore, we focus on two objectives.

- We analyze and implement ML approaches to determine relevant BWIM vehicle configurations, including vehicle type, lane, locus, speed, axle count, and spacing.
- We investigate signal processing techniques in the context of axle-related parameters.

An important aspect of BWIM is the determination of the traffic load. This study will not investigate load estimation since we do not have ground-truth data. However, Ojio et al. [3] showed that it is possible to extract the vehicle load from bridge displacement data. Furthermore, our approach is currently limited to single-vehicle events. Otherwise, the number of vehicles during an event has to be extracted from bridge displacement data. To date, no study in this regard has been conducted. Therefore, events during which multiple vehicles are present simultaneously have been discarded as Arnold and Keller [12] indicates that data-driven differentiation between single- and multi-presence events is possible.

First, we describe the related work relevant to this study in Section 2. In Section 3, we explain the setup for GBR measurements and describe the used dataset. In Section 4, we first cover the basics of wavelet transform. Then, the proposed feature extraction, the ML models, and the methodology for each task are detailed. The results of our approach are laid out in Section 5, and a comprehensive discussion is given in Section 6. Finally, Section 7 summarizes the main findings of this study.

2. Related Work

The relevance of time-series data in ML has never measured up to the prominence of images, although the variety in the data poses an exciting challenge. A wide range in scaling, length, sampling frequency, channels, etc., makes it challenging to develop versatile models to handle such complexity. The UCR dataset provides extensive datasets and is often used as a benchmark for new approaches [14]. However, at the date of writing this paper, only six datasets are both multi-variate and of unequal length. Moreover, the differences in signal length are minor. Oftentimes, an unequal length is circumvented by rescaling or padding in the time domain or extracting features, for instance, by fitting an autoregressive integrated moving average model [15]. Utilizing the UCR dataset, ref. [16] gives a detailed overview of the most relevant state-of-the-art methods for time-series classification. Random convolutional kernel transform (ROCKET), introduced by [17], is highlighted due to its training speed and as it has the best results, especially for multi-variate time-series classification. ROCKET uses random convolutional kernels to extract features, which are then used as input data for a linear ridge classifier. Dempster et al. [18] further refined ROCKET to a version called MiniRocket. MiniRocket is faster than ROCKET and is mostly deterministic while achieving comparable results. The authors also contributed a version of MiniRocket to the Python library sktime [19], which can handle unequal-length time series. Ref. [20] took an approach based on convolutional neural networks (CNNs) in combination with padding and masking for acoustic scene classification. To this end, they implemented a global pooling layer that supports masking to prevent the model from learning the padding. Arnold and Keller [12] used hand-crafted features for tree-based learners and MiniRocket for a bridge event classification on variable-length GBR displacement data. Additionally, they investigated the potential of data augmentation in this context. Overall, they achieve a balanced accuracy of over 90% when classifying crossings in single- or multi-presence events.

Several studies used ML or deep learning (DL) methods to extract vehicle information from NOR BWIM time series data. Kawakatsu et al. [21] attached a single strain sensor to the span of a 300 m long-concrete bridge in Japan. They used CNNs to detect vehicles and estimate the speed, locus, and axle count. As input, 8 s windows of strain data are passed to the network. They use a traffic surveillance system for their ground-truth data, resulting in up to 996,093 samples depending on the classification task. In [22], they expand their dataset by a steel bridge and investigate acceleration sensors as an alternative to strain sensors. The mean absolute error (MAE) for locus estimation is 0.097 m or 0.127 m for strain data depending on the driving direction. Kawakatsu et al. [2] extend their previous work by load estimation using multi-task CNNs. A load meter provides the necessary ground-truth data. They also increase the sequence length of the input data to 20 s. For a 74 m-long steel bridge, they achieve an MAE of 0.92 m/s for speed, and an 0.354 m for axle spacing. The effects of more than one sensor are investigated in [23]. Eleven sensors are used to input different CNN architectures for several bridges, further improving the previous results. For a two-span bridge, they achieve a balanced accuracy score of 91.92% for an imbalanced lane estimation. Other features are comparable or slightly improved compared to their previous studies.

For axle detection, simple peak detection algorithms are often used [22]. Yu et al. [24] applied a combination of wavelet transform and peak detection on strain data from finite element method (FEM) simulations. Although they only use FEM data and do not transfer to real-world data, they make several relevant findings. Firstly, it is shown that axle information can be extracted from global bridge responses. Furthermore, the sampling frequency significantly impacts the identification accuracy since it leads to sharper peaks. For example, at 200 Hz, the axle spacing identification errors for a three-axle vehicle traveling at 30 m/s are 25.3% and 97.1%, respectively. At a sampling frequency of 500 Hz, these errors decrease to 0.2% and 1.43%. Finally, they show that road surface conditions can severely impact the results. Lechner et al. [25] also use wavelets for BWIM based on crack displacement sensor data. They measure the width changes of an existing crack during traffic loading. With this local response, they can successfully obtain vehicle speed, axle count, and distances. Using the influence line, they can also compute individual axle loads. Zhao et al. [26] use free-of-axle detectors (FAD) in combination with wavelets for improved strain-based axle detection. FAD BWIM uses additional FAD sensors attached to the lower side of the bridge. They have shown that axle-induced peaks in the FAD strain signal with Daubechies wavelets are more easily distinguished. While such FAD BWIMs give good results, they need many sensors, leading to a higher probability of failure.

Concerning contactless BWIM, Ojio et al. [3] investigate the potential of cameras for bridge displacement measurements. They can extract the axle loads of a few reference vehicles with known weights from the bridge displacement using the influence line. However, regarding other aspects such as speed, lane, and axle spacing, they rely on a second camera instead of directly extracting these parameters from the displacement data. An overview of BWIM studies relevant to this work is presented in Table 1.

Table 1. Work related to the comparison of extracted vehicle configurations.

Parameter	Response, Sensor, or Model	References
Speed	Camera Displacement Strain FEM	[3] [25] [2,21,23] [24]
Lane or locus	Strain	[21,23]
Axle count	Camera Displacement Strain FEM	[3] [25] [2,21,23,26] [24,26]
Axle spacing	Strain FEM	[2,23] [24]

3. GBR Measurements and Dataset

In this section, we briefly introduce GBR measurements. We refer you to, e.g., [6] for a more detailed explanation. Next, we outline our process for acquiring ground-truth data from UAV images. Figure 1 shows our overall measurement setup. Finally, we will highlight several aspects of the dataset relevant to this study.

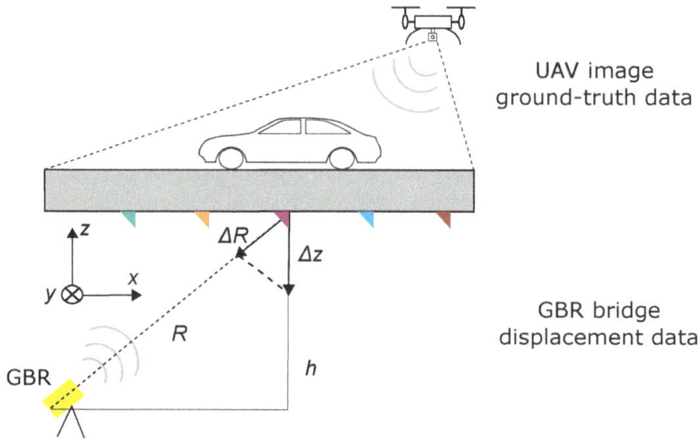

Figure 1. Measurement setup applied in this study. The GBR measures the bridge displacement from underneath it. The UAV records the bridge deck with a camera for traffic ground-truth data.

3.1. GBR Measurement Setup

Measurements have been conducted at a prestressed concrete bridge in Germany. To achieve a high signal-to-noise ratio (SNR), five corner reflectors have been attached to the lower side of one of the two spans of the bridge (see Figures 1 and 2). The monitored field has a length of 28.5 m. Both fields are loosely connected via a 0.8 m thick asphalt layer. Two lanes are present with opposing driving directions, as indicated by the black arrows. The speed limit is 100 km/h. We monitored the bridge with two GBRs, represented by yellow rectangles in the upper image of Figure 2. While the upper GBR measures the y and z components, the left GBR primarily records the x and z components. From here on out, we will primarily focus on the left GBR.

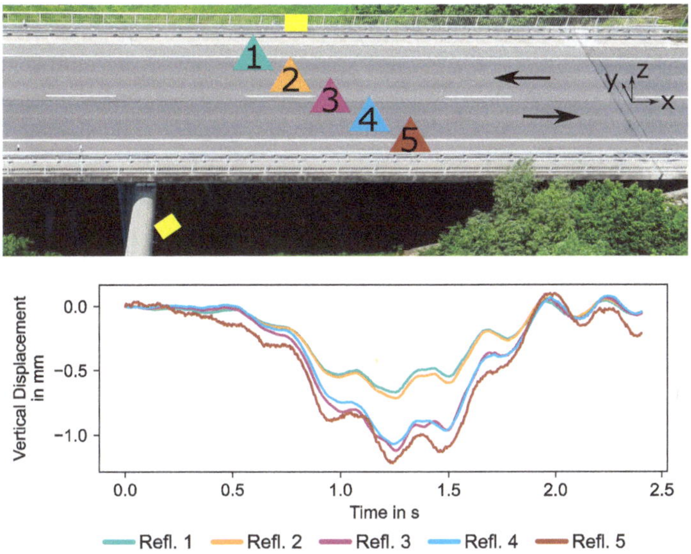

Figure 2. Overview of the bridge and an example displacement time series. The yellow rectangles in the upper image represent the GBRs, and the numbered triangles represent the five reflectors at their approximate position.

The GBR uses two different principles to acquire displacement data [27]. Firstly, using frequency modulation, the GBR can distinguish measurement points along its line of sight (LOS). The range resolution Δr can be calculated according to

$$\Delta r = \frac{c}{2 \cdot B} = 0.75\,\text{m}, \tag{1}$$

with the speed of light $c = 3 \times 10^8\,\text{m/s}^2$ and a bandwidth $B = 200\,\text{MHz}$. Every 0.75 m, the GBR measures the displacement by summing all reflected signals of a range bin. The reflectors are thus spread along the x and y plane to be distinguishable for both GBRs.

The second principle to acquire the displacement signal from each range bin is interferometry. With a sampling rate of up to 200 Hz, the phase shift $\Delta\phi$ of each point is measured and then transformed to radial displacement ΔR along LOS using

$$\Delta R = \frac{\lambda}{4\pi} \cdot \Delta\phi. \tag{2}$$

λ represents the wavelength of the GBR.

Finally, using triangulation and assuming only one dominant displacement component, the vertical displacement Δz can be calculated according to

$$\Delta z = \frac{R}{h} \cdot \Delta R, \tag{3}$$

where h is the height difference between the bridge and GBR, and R is the distance of the GBR to a measurement point (see Figure 1). The lower plot of Figure 2 shows an example displacement signal for all five reflectors. The vehicle inducing the signal is the bus shown in Figure 3.

3.2. UAV Dataset as the Ground-Truth Data

During our measurement campaigns, we deployed a UAV to monitor the bridge deck with a sampling frequency of around 25 fps. Furthermore, several control points along the bridge were measured using a tachometer. This enables us to transform image coordinates in pixels to positions in meters in our radar reference system. In Figure 3, two control points are highlighted with yellow circles. As they are visible in the UAV video data for each frame, deviations in the UAV position do not influence the transformation. Five measurement campaigns have covered various weather conditions over almost two years.

To acquire information about locus, axle spacing, and speed, we labeled each axle on three images per vehicle using a labeling tool [28]. Figure 3 shows a two-axle bus's labeling results (ground-truth data). For the position of an axle, we used the x-wise center of a bounding box, which is then transformed into the meter. The distance between consecutive axles has been calculated via the position difference. By labeling more than one image per vehicle, we determined the speed. For that, we used the positions of the first axle, the number of frames elapsed between each labeled image, and the UAV sampling frequency. Finally, the y position of the lower bound of the first axle represents the locus. Labeling three images per vehicle has allowed us to detect outliers in the extracted properties. Imprecisions during the labeling process can lead to minor variations in the ground-truth data. Furthermore, the transformation accuracy precisely depends on how the control points are recognized during the transformation process. To validate the ground-truth data, we compared datasheets for recognizable vehicles to the extracted axle spacing to verify our procedure. This provided a satisfactory match. Therefore, the margin of these errors is not significant enough to have a relevant impact. One challenge during the labeling process was posed by the darkness in combination with wet streets since it rendered the asphalt and the wheels barely distinguishable. Especially for trucks with raisable tires, it has been challenging to say whether they have touched the street or not.

Figure 3. Example of a labeled vehicle using the tool provided by Tzuta [28]. Bounding boxes have been drawn around the axles, as indicated by the orange bounding boxes. The yellow circles represent two control points for the coordination transformation. Furthermore, the lane number is shown.

Table 2 lists the axle counts distinguished by vehicle type. As expected, cars mainly have two axles, whereas trucks have varying axle configurations. Overall, 455 single cars and 131 single trucks have been registered and labeled. Additionally, Table 2 shows the distribution of events for both lanes. Lane 1 has been used more often, but all values are similar.

Table 2. Axle configuration and lane information for cars and trucks in our dataset.

Type	2 Axles	3 Axles	4 Axles	5 Axles	7 Axles	Lane 1	Lane 2
Car	443	7	5	-	-	235	220
Truck	24	12	41	53	1	78	53

Figure 4 shows the distributions for our regression tasks: speed, locus, and axle spacing estimation. The car's speed is generally higher and sometimes exceeds the speed limit. Trucks drive slower on average. The locus shows similar distributions for both vehicle types and corresponds to the existence of two lanes. The most significant difference is shown in the axle spacing, as cars have values of approximately 2.8 m, while trucks have values ranging from approximately 0.7 m to 10.3 m. At the peak of the car distribution, the truck distribution has a local minimum.

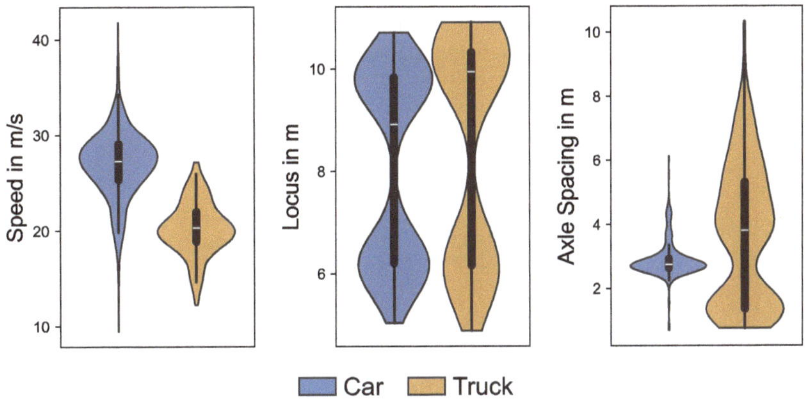

Figure 4. Distributions of speed, locus, and axle spacing in the applied dataset.

4. Methodology

In this chapter, we will shortly introduce wavelet transform. Then, we explain our preprocessing and feature extraction steps and, afterward, the data-driven approaches and methods for each vehicle parameter. First, we investigate the distinction between cars and trucks. For all other tasks, we investigate both vehicle types simultaneously to have a more extensive dataset, as well as only trucks, by disregarding cars, as trucks are more relevant to SHM.

4.1. Prerequisite Concerning Wavelets

We will only superficially explain the concept of wavelets. For a more in-depth explanation, please see, e.g., [29]. One motivation behind wavelets is to acquire local frequency information while maintaining a high temporal resolution, which is impossible using Fourier transformation. The method is analog, as the original signal is expressed as a family of functions. These functions are constructed from a so-called mother wavelet:

$$W(a,b) = \int_{-\infty}^{+\infty} f(t) \frac{1}{\sqrt{a}} \psi\left(\frac{t-b}{a}\right) dt \qquad a \in \mathbb{R}^+, b \in \mathbb{R} \tag{4}$$

Different coefficients are generated from the input signal by varying scaling a and time delay b. One example of a mother-wavelet is the Gaussian wavelet

$$\psi(t) = C \cdot exp(-t^2), \tag{5}$$

where C is an order-dependent normalization factor. Another mother-wavelet is the Gaussian derivative wavelet, the m-th order derivative of Equation (5), where m lies between 1 and 8. It is defined as

$$\psi(t) = \frac{1}{\sqrt{\Gamma(m+\frac{1}{2})}} \frac{d^m}{d\eta^m} exp(-t^2/2), \tag{6}$$

where Γ represents the gamma function [30]. The Gaussian derivative wavelet will be used in this study, as implemented by Lee et al. [31].

4.2. Preprocessing and Feature Extraction

We minimize the preprocessing, so removing each time series's offset is the only step except for axle counting and axle spacing estimation, where we use an additional high-pass filter before feature extraction. This offset comes from long-term drifts in the signal due to environmental influences [10]. Otherwise, no filtering is applied since we want to maintain high-frequency information.

Figure 5 shows the methodology for our feature-based approaches. Among others, we test various models in combination with manually crafted features. Table 3 summarizes all features used in this study and how they are calculated. Each feature is calculated for each used time series. Only a part of these features is used for a specific task to avoid making ML predictions more challenging by adding irrelevant features. This selection will be stated in the corresponding subsections. Different input features are passed to the ML models since the number of reflectors also varies depending on the task. Since tree-based models are scale-independent, no scaling is applied to the input features for random forest (RF) [32] and gradient boosting (GB) [33]. The input features are scaled in the case of *k*-NearestNeighbours (KNN) [34].

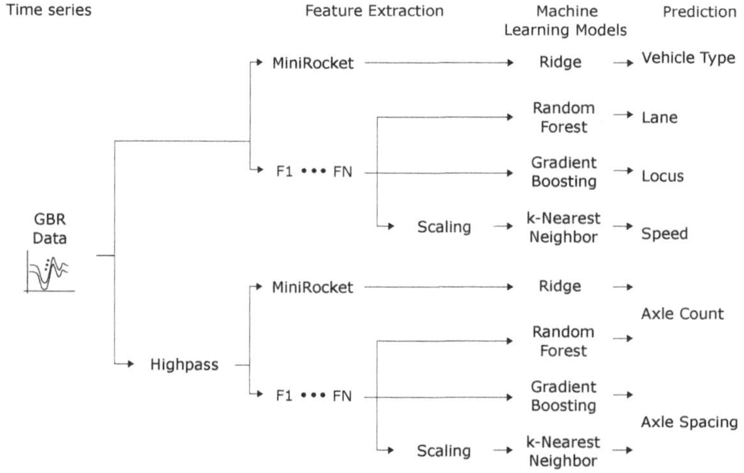

Figure 5. Methodology for our feature-based methods. $F1$–FN represent manually extracted features. We use a high-pass filter before feature extraction for the axle counting and axle spacing estimation.

Table 3. These 11 features were extracted from the GBR time-series data $\underline{x}$. The Python packages `numpy` and `scipy` were used for the calculation. All non-default values are stated.

Feature No.	Name of Feature	Basis of Calculation
1	Maximum	$\max(\underline{x})$
2	Minimum	$\min(\underline{x})$
3	Mean	$\min(\underline{x})$
4	Skewness	$\text{mean}(\underline{x}^3)$
5	Kurtosis	$\text{mean}(\underline{x}^4)$
6	Median	$\text{median}(\underline{x})$
7	Length	$\text{len}(\underline{x})$
8	xMinPosRatio	$\text{argmin}(\underline{x})/\text{len}(\underline{x})$
9	Power	$\text{sum}(\underline{x}^2)/\text{len}(\underline{x})$
10	MAD	$\text{median\_abs\_deviation}(\underline{x})$
11	GradMax	$\max(\text{gradient}(\underline{x}))$

Unlike manual feature selection, MiniRocket [18] uses random convolutional kernels to extract 9996 features. The kernels have a fixed length of 9, and their weights are restricted to two values: $\alpha = -1$ and $\beta = 2$. Dilation, or the spread of a kernel over the time series, lies within the range of $\lfloor 2^0 \rfloor$ and $\lfloor 2^{max} \rfloor$, where max $= log_2(l_{input} - 1)/8$ and with l_{input} are the input length. Padding is fixed due to the convolution and alternates between no and zero padding. Biases are calculated based on the result of the convolution of randomly selected training examples, therefore being the only non-deterministic aspect of the MiniRocket approach. The final extracted feature is the proportion of positive values (PPV), which can be calculated with

$$\mathrm{PPV}(Z) = \frac{1}{N}\sum_{i=0}^{N-1}[z_i > 0] \qquad i \in \mathbb{N}, \tag{7}$$

where Z is the convolution result between the input signal and a kernel and N represents the non-zero signal length [17].

Theoretically, the number of kernels can be regarded as a hyperparameter, but Dempster et al. [18] recommend using 10,000 kernels since they do not observe a significant impact on the accuracy for different values. Normalization is unnecessary using the PPV and the bias drawn from the convolution results.

We split our data in an 80 : 20 manner for training and testing, and we use stratified sampling for classification tasks to maintain class frequency. During training, we apply a 5-fold cross-validation grid search to find the best hyperparameters for each regression and classification task. For MiniRocket, we use the configuration recommended by the authors, which also includes a ridge regressor or classifier as the final step. During grid search, we optimize the mean squared error (MSE) for regression and the balanced accuracy for classification.

4.3. Vehicle Type

Since we distinguish between cars and trucks in the following tasks, we want to investigate whether data-driven distinction is possible for completeness. Although it seems trivial to use a threshold, the driving lane also comes into play. Therefore, we use the reflectors 2 and 4 (see Section 3.1). As input features, we use 2, 3, and 9 of Table 3. So, for each time series, we extract the minimum, mean, and power using the corresponding calculation methods. As we use two reflectors and extract each feature for each reflector, six features are extracted in this task. These features are then directly passed to RF, GB, and KNN after scaling to predict the vehicle type.

4.4. Lane and Locus

The offset of the reflectors in the y direction, as depicted in Figure 2, makes it possible to discern the vertical driving position of a vehicle. In the first step, the lane shall be determined in a classification task using reflectors 2 and 4. As a baseline, which we will refer to as POWER, we compare the signal power (Feature 9 in Table 3) for both reflectors. The lane is determined depending on which one has a higher value. The signal power of reflector 2 is greater, and the vehicle drives on lane 2. The input for the features-based models consists of the features 2, 3, and 9. In addition to the lane, the locus is estimated by regression to have a more precise vehicle localization. We use all 5 reflectors with the same features for locus regression.

4.5. Speed

Global responses, such as displacement, require a different approach than usual to calculate the speed since extracting it via peak detection from local responses is impossible. Therefore, we use data-driven ML to extract the vehicle speed. Arnold and Keller [10] and Arnold et al. [11] can extract vehicle crossings, but just using the length of such an event is not enough since the vehicle length also plays an important role. To show this, we additionally train a linear regression (LR) for speed estimation only on the signal length as a baseline. A one-dimensional LR tries to find the function of the coefficient w and the estimated intercept c, such that

$$y = w \cdot x + c. \quad (8)$$

y is the target vector and x the input vector. In our case, x represents the event length, and y represents the vehicle speed. The feature-based models are trained with the features 1 to 11 from Table 3. As the input time series, we use reflector 3.

4.6. Axle Count

We treat the determination of the axle count as a classification problem with either 2, 3, 4, or 5+ axles. With more data, a more precise classification might be possible. All predictions are made using only the signal of reflector 3.

Unlike the previous task, we do not use the displacement directly as input or for feature extraction. Instead, we filter the signal with a 50th order forward-backward Butterworth high-pass with a cut-off frequency of 45 Hz. This is performed so that the models learn high-frequency features and do not use low frequencies. For our feature-based approaches, we use the features from 1 to 10 in Table 3. We drop feature 11 since we already filter the signal with a high pass.

Apart from using ML, we also investigate the deterministic methods using a bandpass filter (BANDPASS) and CWT (WAVELET) as novel approaches. Their pipelines are depicted in Figures 6 and 7, respectively. They are similar except for the first step of the pipeline, where its corresponding filtering or transformation is applied. As input, they receive unfiltered GBR bridge displacement data from one reflector. The output consists of a list of the positions of the detected peak. For BANDPASS, we apply a forward-backward Butterworth bandpass in the first step. The order of the filter is 50 and the critical frequencies are (45 Hz, 65 Hz). We choose `gaus7` as implemented by Lee et al. [31] and the 3. coefficient for our WAVELET approach. This corresponds to $m = 7$ regarding Equation (6). These parameters have been determined as part of the parameter tuning process but outside the grid search. The bandpass-filtered signal and the wavelet transform result are then squared and smoothed by a weighted moving-average filter to obtain the distinguishable peaks. Ultimately, the signals are normalized to their highest value before searching all peaks to achieve generalization over all vehicles. We use the `find_peaks`-method of the Python-package scipy for peak detection [35]. The window size of the moving-average filter and the distance and prominence of a peak during peak detection are regarded as hyperparameters and deduced using the training dataset. Depending on the driving side and thus the driving direction, we discard peaks in half of the signal during which the vehicle does not enter or leave the bridge. As seen from Figures 8 and 9, only the entering or leaving process is relevant for peak detection. Finally, we treat the length of the list of detected peak positions as the axle count. If no peak or only one is detected, we assume two axles.

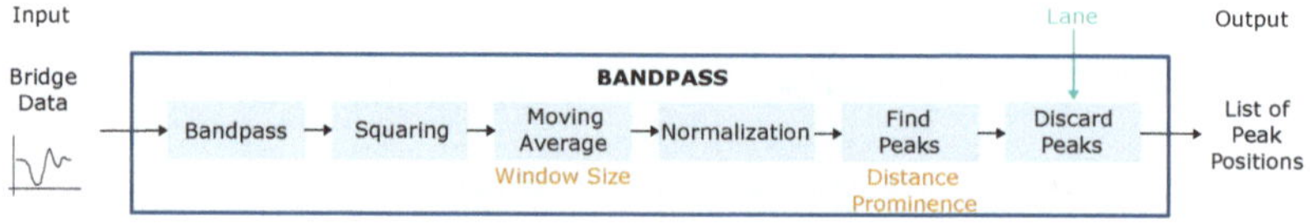

Figure 6. Pipeline for our BANDPASS approach for axle configuration extraction. The orange-colored hyperparameters are determined during the grid search. The lane information is used in the `Discard Peaks` step as an additional input parameter to filter the detected peaks.

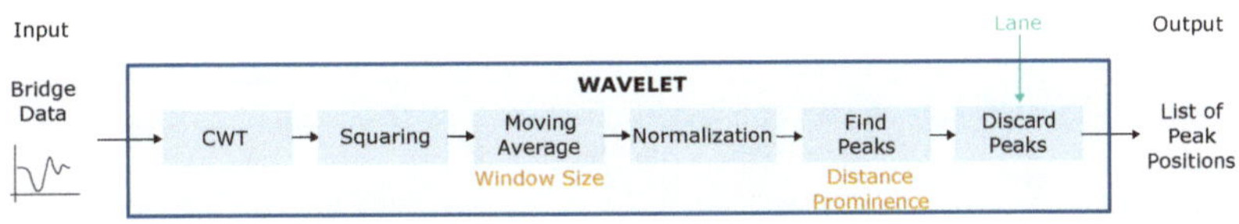

Figure 7. Pipeline for our WAVELET approach for axle configuration extraction. The orange-colored hyperparameters are determined during the grid search. The lane information is used in the `Discard Peaks` step as an additional input parameter to filter the detected peaks.

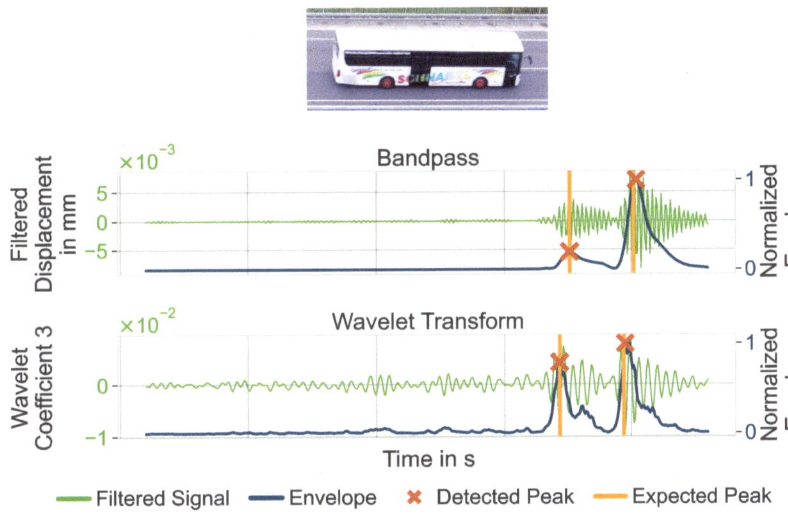

Figure 8. Example result of a bus for our filtering approaches.

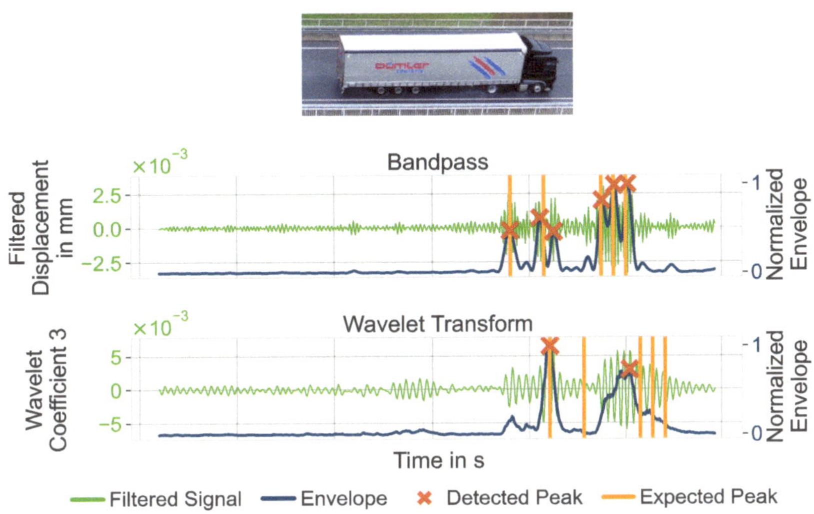

Figure 9. Example result of a truck for our filtering approaches.

4.7. Axle Spacing

Finally, we investigate how well the distance of axles can be determined. We treat this as a multi-output regression. Our BANDPASS and WAVELET procedures return a list of detected peak positions x (see Figures 6 and 7). A peak at position i in the list is interpreted as axle number i. The temporal distance between the peaks i and j can be calculated by subtracting the consecutive positions x_j and x_i and then divided by the GBR sampling rate of 200 Hz. With the speed v_{UAV} from the UAV data, this can finally be transformed into axle spacing d_{ij} between axle i and j according to

$$d_{ij} = \frac{x_j - x_i}{200\,\text{Hz}} \cdot v_{UAV}. \tag{9}$$

We assume the speed can be correctly extracted from Section 4.5. We also assume that the axle count is known and ignore additionally detected spacings during evaluation. The same goes for our ML approaches. We also use the same inputs as in Section 4.6.

5. Results

This study aims to investigate the potential of GBR for contactless NOR BWIM. In this section, we will state the results of our methods described in Section 4 in a corresponding order. Each approach except vehicle type classification is evaluated once for all vehicles and once for only trucks since they are more relevant for SHM. Classification tasks are evaluated based on balanced accuracy (BA), overall accuracy (OA), precision (P), and recall (RC). Regression performance is expressed by the coefficient of determination (R^2), mean squared error (MSE), and mean absolute error (MAE). The results of our filtering approaches will be addressed in detail in Section 5.4.

5.1. Vehicle Type

The results for the distinction between cars and trucks, which is relevant for SHM, are displayed in Table 4. All feature-based models classify the vehicle type perfectly. Using the raw time series data, only MiniRocket misclassifies one truck as a car.

Table 4. Classification results for vehicle type estimation with reflectors 2 and 4. The highlighted values represent the best results for each vehicle group.

Model	BA in%	OA in%	P in%	RC in%
MiniRocket	98.1	99.2	100	96.1
RF	**100**	100	100	100
GB	**100**	100	100	100
KNN	**100**	100	100	100

5.2. Land and Locus

We investigate both lane classification (see Table 5) as well as locus regression (see Table 6), which is the lateral position of a vehicle. MiniRocket provides the best results for both vehicle groups with 94.3% and 97.9%. Except for RF, all models have a BA of over 90%. POWER shares the first place with MiniRocket in the case of only trucks. Both approaches misclassify only one truck, leading to an RC of 100%. Yet, RF, whose concept relies on comparing input values, has the worst results for both cases. The difference between both vehicle selections is slight, but the performance for only trucks is better overall.

A more fine granular lateral position estimation via the locus using all five reflectors resulted in Table 6. Again, MiniRocket beats all other models for all vehicles, with an R^2 of 0.84. When discarding cars, KNN has an almost perfect R^2 score of 0.99, slightly surpassing the MiniRocket with 0.96. All models improve from all vehicles to only trucks, like in lane classification, except for RF, which achieves worse results.

Table 5. Classification results for lane estimation with reflectors 2 and 4. The highlighted values represent the best results for each vehicle group.

Model	All Vehicles				Trucks Only			
	BA in%	OA in%	P in%	RC in%	BA in%	OA in%	P in%	RC in%
POWER	92.7	92.4	87.1	98.2	**96.9**	96.3	91.7	100
MiniRocket	**94.3**	90.0	98.2	94.1	**96.9**	96.3	91.7	100
RF	83.2	83.1	79.7	85.4	89.2	88.9	83.3	90.1
GB	93.3	93.2	91.2	94.5	93.8	92.6	83.3	90.1
KNN	93.4	93.2	89.8	96.4	92.3	92.6	90.1	90.1

Table 6. Regression results for locus estimation with all five reflectors. The highlighted figures represent the best results for each vehicle group.

Model	All Vehicles			Trucks Only		
	R^2	MSE in m^2	MAE in m	R^2	MSE in m^2	MAE in m
LR	0.57	1.50	1.06	0.91	0.38	0.47
MiniRocket	0.84	0.57	0.54	0.96	0.18	0.36
RF	0.59	1.40	0.79	0.32	2.86	0.99
GB	0.80	0.69	0.57	0.91	0.40	0.49
KNN	0.81	0.65	**0.50**	0.99	0.04	**0.16**

5.3. Speed

Table 7 shows the results for the speed regression task. MiniRocket achieves the best performance for both vehicle sets. For all vehicles, it achieves an R^2 of 0.94 and an MAE of 0.76 m/s, which is slightly better than the 0.86 m/s of Kawakatsu et al. [23]. All feature-based methods have comparable results for R^2 and are less effective than MiniRocket. LR, on only the duration of the bridge crossing event, has the worst results but still is not far off from MAE. Its MSE, however, is significantly worse compared to the other models.

LR does improve regarding MSE and MAE when only using trucks, although R^2 decreases. The same behavior can be observed for KNN and GB. GB has the best R^2 for trucks and is slightly worse than MiniRocket in MSE and MAE. MiniRocket, with an MAE of 1.02 m/s is still only 0.16 m/s worse than Kawakatsu et al. [23]. However, they evaluate more data, making their results more robust.

Table 7. Regression results for speed estimation with reflector 3. The highlighted values represent the best MAE for each vehicle group.

Model	All Vehicles			Trucks Only		
	R^2	MSE in m^2/s^2	MAE in m/s	R^2	MSE in m^2/s^2	MAE in m/s
LR	0.71	5.08	1.81	0.67	2.22	1.22
MiniRocket	0.94	1.06	**0.76**	0.71	1.93	**1.02**
RF	0.85	2.68	1.12	0.53	3.16	1.25
GB	0.85	2.56	1.14	0.75	1.67	1.06
KNN	0.85	2.60	1.14	0.70	2.03	1.13

5.4. Axle Count

As explained in Section 4.6, we investigate the potential of bandpass filtering and wavelet transform to detect axles in GBR bridge displacement data. Three examples of these approaches can be seen in Figures 8–10. In all images, first, the vehicle is depicted, and then the results of the filtering and the smoothed time series. The filtered signal corresponds to the intermediate result after the bandpass filtering and CWT in Figures 6 and 7, respectively. The envelope represents the unitless waveform after the normalization step from which the peak positions are acquired. We have marked the peaks, which have been detected with the respective approach and the learned hyperparameters. Finally, starting from the first detected peak, positions where we would expect subsequent peaks due to the labeled UAV data are shown.

Figure 8 shows the bus with two axles from Figure 3. The bandpass filtered signal has two areas with a high amplitude vibration in the frequency range of 45 Hz to 65 Hz. Accordingly, the envelopes for both approaches have two peaks. Assuming that the first peak corresponds to the first axle, the second peak matches the expected position. A more complex example is depicted in Figure 9 since the truck has five axles, three of which are very close to each other. Again, these axles are recognizable and at their expected position in the bandpass signal and its envelope. However, one additional peak does not belong to any axle. This peak does not exist in the wavelet signal. Conversely, the first peak is missed, and the final three axles are combined into one axle group in the signal. Finally, Figure 10 shows a truck with a similar axle configuration. Seven peaks are detected using the BANDPASS approach. The final three axles are again registered as a single axle group. WAVELET clearly shows the first two axles at the expected positions, yet the peaks for the three back axles are too small to be distinguishable from noise. The behavior in the previous pictures is also evident with cars, although they are usually lighter.

We also investigated the other reflectors as well as the second GBR. While very distinct peaks, like in Figure 8, are also visible with other reflectors, reflector 3 shows the best results overall. No axle information could be extracted from the second GBR using our two methods, BANDPASS and WAVELET.

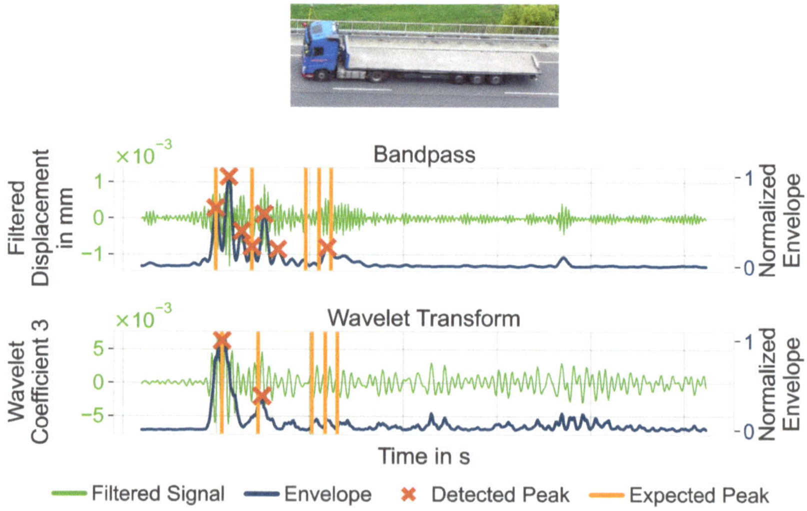

Figure 10. Second example result of a truck for our filtering approaches.

As described in Section 4.6, the hyperparameters for BANDPASS and WAVELET have been determined using the training data. To quantify the performance, Table 8 displays the results for all approaches. Again, MiniRocket has the best results for both categories. Especially for trucks, it achieves a BA of 76.7%, delivering results comparable to Kawakatsu et al. [21], who achieve BA of 85.8% using a single strain sensor. When including the cars, however, the BA decreases to only 60.7%. Only the OA improves compared to the trucks. BANDPASS and WAVELET have similar results for all vehicles, with WAVELET having a slightly better BA. Still, the BA is only 25.1%. It improves together with P and RC to 46.6% for only trucks, whereas BANDPASS achieves results of over 30%. RF, GB, and KNN tend to outperform BANDPASS and WAVELET but still fall heavily behind MiniRocket.

Table 8. Classification results for axle count estimation with reflector 3. The highlighted figures represent the best results for each vehicle group.

Model	All Vehicles				Trucks Only			
	BA in %	OA in %	P in%	RC in%	BA in%	OA in%	P in%	RC in%
BANDPASS	23.4	34.7	24.5	23.4	35.1	29.6	30.3	35.2
WAVELET	25.1	28.8	21.5	25.0	46.6	37.0	44.6	46.6
MiniRocket	**60.7**	86.4	58.1	60.7	**76.7**	70.4	74.4	76.7
RF	50.5	89.0	46.0	50.5	45.4	51.9	23.7	45.5
GB	40.8	83.9	38.0	40.9	38.6	48.1	32.0	38.6
KNN	47.0	87.3	46.0	47.0	45.3	51.8	40.9	45.3

5.5. Axle Spacing

As shown in the previous section, it is challenging to formalize a procedure for BANDPASS and WAVELET, which works for all vehicles. The same goes for axle spacing since it builds upon the peak detection. Yet, we want to indicate that the information can sometimes be extracted successfully. A more generalized analysis will be performed afterward.

In Figure 8, two peaks are easily recognizable. Using their distance, the GBR sampling frequency of 200 Hz, and the vehicle speed, we can calculate an axle spacing of 6.33 m using Equation (9). Comparing this to the UAV axle spacing of 6.19 m, we overestimate the spacing by only 0.14 m. Likewise, the results for Figure 9 would deliver close values, as the expected peaks are near the detected positions. Yet, additional peaks, such as the third peak in Figure 9 for BANDPASS, would lead to underestimating the second axle spacing, as we would also overestimate the axle count by one. These challenges are reflected in the overall results in Table 9. The R^2 is even harmful for both BANDPASS and WAVELET. While the MAE is only slightly more than 1 m for only trucks, the MSE shows that there are a lot of spacings estimated with large deviations, as previously outlined. The same tendency can be found when including cars. Interestingly, all ML approaches achieve similar results. MiniRocket has the best results for MAE with 0.64 m for only trucks. Also, the MSE is considerably smaller for the ML approaches, meaning fewer outliers exist. With all vehicles, GB achieves the best MAE of 0.21 m, although all models achieve similar results again.

Table 9. Regression results for axle spacing estimation with reflector 3. The highlighted figures represent the best results for each vehicle group.

Model	All Vehicles			Trucks Only		
	R^2	MSE in m^2	MAE in m	R^2	MSE in m^2	MAE in m
BANDPASS	−0.22	3.80	0.54	−0.45	7.78	1.30
WAVELET	−0.14	3.54	0.49	−0.16	6.25	1.19
MiniRocket	0.45	1.71	0.27	0.66	1.78	**0.64**
RF	0.53	1.45	0.23	0.63	2.00	0.71
GB	0.53	1.45	**0.21**	0.68	1.73	0.66
KNN	0.43	1.75	0.22	0.64	1.95	0.68

6. Discussion

The central focus of this study is the potential of GBR displacement signals for a remote and data-driven BWIM. This section will discuss the results stated in the previous section. First, we will analyze the potential of ML for all classification and regression tasks. Afterward, our filter approaches regarding axle configuration identification are discussed.

6.1. Machine Learning for Displacement-Based BWIM

As the results for vehicle-type classification are immaculate for feature-based methods, it is possible only to regard trucks for SHM if desired. Trucks cause a significantly greater deflection than cars. Therefore, the promising results are unsurprising. The issue of the driving lane can be avoided by using one reflector per lane as input data. Our distinction between vehicle types during vehicle configuration tasks is thus well founded.

It seems that, for all tasks except axle spacing estimation, ML models can extract vehicle configurations. MiniRocket, which uses the raw time series data and extracts features via convolutional kernels, shows especially auspicious results. The model with the best performance can vary from task to task, but MiniRocket either has the best performance or follows close behind. Interpretability is an important aspect of bridge monitoring [36–38]. The overall satisfying results of our models indicate that the extracted configurations are identifiable from global bridge displacement data using data-driven ML approaches.

To acquire the vertical position of a vehicle, we test both a lane classification and a locus regression. The lane can successfully be extracted for both vehicle categories. Trucks can be classified almost perfectly, with only one vehicle misclassified. This is unsurprising, as our reflectors are spread along the y axis (see Figure 2). Accordingly, the maximum displacement during an event correlates to the driving side. Especially for heavy vehicles, a clear distinction can be made. In Figure 2, e.g., reflectors 3 to 5 show a steeper curve than reflector 1 and 2. This suggests that the vehicle drives in lane 1, which corresponds to the bus from Figure 8, which caused the bending. Thus, it seems enough for trucks to compare the signal power of reflectors 2 and 4 to acquire the lane.

The locus regression shows a similar behavior in that the results improve when discarding cars. Due to KNN's outstanding performance, we surmise that scaling our features will help with this task. This makes sense, as the maxima ratio within one event is more relevant than their absolute values.

There is a significant difference in speed between all vehicles and only trucks. The reason for this might lie in the dataset composition. As cars are more frequent and with less variation in the vehicle configurations, like the length, the correlation between event duration and speed is more prominent. Thus, for some models, the R^2 decreases while the MAE improves. For MiniRocket, R^2 and MAE decrease when discarding cars, suggesting that it does not mainly look at the event duration.

The results for axle count classification imply that our features from Table 3 are not helpful for this task, as all models that depend upon them have a BA of less than 51%. Only MiniRocket achieves satisfactory results, especially for trucks. The high OA but low BA show that the imbalance in the dataset, when including cars, leads to more misclassified trucks.

Comparing both datasets, it can be said that, although having more data generally helps, cars drastically change the distribution of many configurations. This makes it more challenging for models to learn truck properties. Regarding BWIM, where only trucks are considered relevant, this can be seen as a disadvantage. The truck dataset is very small as we have only one truck with seven axles, for example. Since trucks come in significant axle count and spacing variations, our models naturally have generalized difficulties. The dilemma of high imbalance and a small dataset is also apparent for axle spacing regression. Here, the MSE is very high for all vehicles compared to MAE, as the axle spacing for cars is very closely distributed (see Figure 4). Conversely, models mainly learn about the car distribution, regarding trucks as outliers due to the highly imbalanced datapoints.

6.2. *Filtering for Axle Configuration Identification*

Figures 8–10 indicate that the information of axles is present in a bridge's global displacement time series. Individual examples produce promising results concerning axle count and axle spacing. To our knowledge, this has not been presented before, as displacement is barely exploited in BWIM. However, the results for BANDPASS and WAVELET in Tables 8 and 9 show that it is challenging to find a general procedure. Especially the negative R^2 values in Table 9 demonstrate this, as this means that simply using the mean of the data as a prediction is a better fit than our methods. Neither BANDPASS nor WAVELET can be described as the better approach overall, as their respective performances varied heavily between samples. Unfortunately, we can only show a small portion of all the recorded vehicles, as there are cases in which no axle-induced peaks are visible in the bandpass signal, but they exist with WAVELET.

While having a high SNR to measure these small vibrations is necessary, a good SNR alone will not lead to valuable measurements concerning axle detection. Interestingly enough, the weight of a vehicle or axle seems not to be the decisive factor, as for some cars, the peaks are easily recognizable, whereas they are not visible for trucks. Therefore, we assume that the relative axle weight within one event is relevant. The truck in Figure 10 appears heavier in the front than in the back. Accordingly, the peaks for the latter axles are less visible. However, this is only a conjecture since no axle load data have been recorded during this study. Also, the driving direction appears unrelated to how well axles can be detected. More peaks might have been detected with a higher sampling frequency, or at least their distinction might have been more straightforward for CWT. Yu et al. [24] have demonstrated that a sampling frequency of only 200 Hz leads to large errors. These vibrations seem to be caused at the junction between the bridge and the street. Consequently, this behavior might be specific to this bridge. Since we only monitored one field, we cannot say if the same vibrations can be measured for the other junction. Both fields seem too loosely coupled to transmit the signal if they exist. For one-span bridges, both leaving and entering might be observable. In such cases, our BANDPASS and WAVELET approaches can also extract the vehicle speed. In this study, however, this was not possible.

For vehicles with clearly visible peaks, like in Figure 8, the axle-induced response could also be observed in the signal of other reflectors. This suggests that the specific reflector or its attachment does not induce the vibration but is a high-frequency vibration in the bridge displacement. Also, as mentioned in Section 3.1, we monitor the bridge with two GBRs that measure different components. There have not been visible vibrations or peaks for the GBR measuring the z and y components, although the SNR lies in an adequate range. This indicates that the vibration mainly occurs in the x-component. However, a more detailed investigation is necessary.

7. Conclusions and Outlook

In this study, we discuss a novel data-driven, displacement-based BWIM approach. The data were recorded at a two-span bridge in Germany using GBR and a UAV for ground-truth data. We investigate the potential of both classic signal processing and ML to extract the vehicle configurations from bridge-crossing events. These configurations include vehicle type, speed, lane, locus, axle count, and spacing. One challenge herein is that displacement is a non-local bridge response. Furthermore, our dataset is imbalanced and consists of variable-length time series. We evaluate all approaches on all vehicles and all trucks only.

As ML approaches, we test four different models, three of which depend on manually crafted features. The fourth model, MiniRocket, uses the raw time series data. Over all configurations, MiniRocket achieves the most auspicious results, comparable to other studies such as Kawakatsu et al. [21]. While speed, lane, and locus can be extracted, axle count classification is more challenging for all models. Only MiniRocket can classify a truck axle count with a BA of 76.7%. Finally, the results for axle spacing regression suffer from a small dataset. Therefore, more bridges should be monitored. However, more complex models and using extracted values like speed as an input feature could lead to better results.

We have recorded a high-frequency vibration for this bridge that coincides with axles crossing the junction. Using bandpass filtering and wavelet transform, we could demonstrate examples of this behavior. Based on this finding, we try two approaches for axle count and axle spacing determination purely based on signal processing. While the information seems available, we have not found a comprehensive procedure for the automatic extraction of axle configurations. A more sophisticated approach exploiting CNNs might be more successful as they can learn more generalizable features. For this, however, either the dataset needs to be increased, or data augmentation needs to be applied, as it has been investigated in Arnold and Keller [12]. Furthermore, it must be investigated whether other bridges show a similar behavior in the high-frequency range. Ideally, GBRs with a higher sampling rate will be used for these measurements.

We showed that a purely data-driven BWIM exploiting GBR-displacement time series data is possible. However, these promising results must be refined using a more extensive dataset. Ojio et al. [3] indicated that determining axle loads is possible with bridge displacement signals and the vehicle configurations extracted in this study. Thus, our results work towards a fully remote and data-driven BWIM.

Author Contributions: Conceptualization, M.A. and S.K.; methodology, M.A. and S.K.; software, M.A.; validation, M.A.; formal analysis, M.A.; investigation, M.A. and S.K.; resources, M.A.; data curation, M.A.; writing—original draft preparation, M.A. and S.K.; writing—review and editing, M.A. and S.K.; visualization, M.A. and S.K.; supervision, S.K. All authors have read and agreed to the published version of the manuscript.

Funding: This research study is part of the ZEBBRA project funded by the German Federal Ministry of Education and Research (BMBF).

Data Availability Statement: The data are not publicly available as they contain sensitive information about the investigated bridge structure.

Acknowledgments: We especially thank our partners, the Institute for Automation and Applied Informatics at the Karlsruhe Institute of Technology and Büro für Strukturmechanik, Coburg, for their support during the measurement campaigns. We acknowledge support by the KIT-Publication Fund of the Karlsruhe Institute of Technology.

Conflicts of Interest: Author Matthias Arnold was employed by the company ci-tec GmbH. The remaining author declares that the research was conducted in the absence of any commercial or financial relationships that could be construed as a potential conflict of interest.

Abbreviations

The following abbreviations are used in this manuscript:

BA	Balanced accuracy
BWIM	Bridge weigh-in-motion
CNN	Convolutional neural network
CWT	Continuous wavelet transform
DL	Deep learning
FDA	Free of axle
FEM	Finite element method
GB	Gradient boosting
GBR	Ground-based radar
KNN	K-nearest neighbor
LOS	Line of sight
LR	Linear regression
ML	Machine learning
MAE	Mean absolute error
MSE	Mean squared error
NOR	Nothing-on-road
OA	Overall accuracy
P	Precision
PPV	Proportion of positive values
RC	Recall
RF	Random forest
ROCKET	Random convolutional kernel transform
SHM	Structural health monitoring
SNR	Signal-to-noise ratio
UAV	Unmanned aerial vehicle

References

1. Yu, Y.; Cai, C.; Deng, L. State-of-the-art review on bridge weigh-in-motion technology. *Adv. Struct. Eng.* **2016**, *19*, 1514–1530. [CrossRef]
2. Kawakatsu, T.; Aihara, K.; Takasu, A.; Adachi, J. Fully-Neural Approach to Heavy Vehicle Detection on Bridges Using a Single Strain Sensor. In Proceedings of the ICASSP 2020–2020 IEEE International Conference on Acoustics, Speech and Signal Processing (ICASSP), Barcelona, Spain, 4–8 May 2020; pp. 3047–3051, ISSN 2379-190X. [CrossRef]
3. Ojio, T.; Carey, C.H.; OBrien, E.J.; Doherty, C.; Taylor, S.E. Contactless Bridge Weigh-in-Motion. *J. Bridge Eng.* **2016**, *21*, 04016032. [CrossRef]
4. Gentile, C.; Bernardini, G. Output-only modal identification of a reinforced concrete bridge from radar-based measurements. *NDT E Int.* **2008**, *41*, 544–553. [CrossRef]
5. Michel, C.; Keller, S. Introducing a non-invasive monitoring approach for bridge infrastructure with ground-based interferometric radar. In Proceedings of the EUSAR 2021, 13th European Conference on Synthetic Aperture Radar, Online, 29 March–1 April 2021; pp. 1–5.
6. Michel, C.; Keller, S. Advancing Ground-Based Radar Processing for Bridge Infrastructure Monitoring. *Sensors* **2021**, *21*, 2172. [CrossRef]
7. Michel, C.; Keller, S. Determining and Investigating the Variability of Bridges' Natural Frequencies with Ground-Based Radar. *Appl. Sci.* **2022**, *12*, 5354. [CrossRef]
8. Michel, C.E. Generic Radar Processing Methods for Monitoring Tasks on Bridge Infrastructure. Ph.D. Thesis, Karlsruher Institut für Technologie (KIT), Karlsruhe, Germany, 2023. [CrossRef]
9. Pieraccini, M.; Miccinesi, L.; Abdorazzagh Nejad, A.; Naderi Nejad Fard, A. Experimental Dynamic Impact Factor Assessment of Railway Bridges through a Radar Interferometer. *Remote Sens.* **2019**, *11*, 2207. [CrossRef]
10. Arnold, M.; Keller, S. Detection and classification of bridge crossing events with ground-based interferometric radar data and machine learning approaches. *ISPRS Ann. Photogramm. Remote Sens. Spat. Inf. Sci.* **2020**, *V-1-2020*, 109–116. [CrossRef]
11. Arnold, M.; Hoyer, M.; Keller, S. Convolutional neural networks for detecting bridge crossing events with ground-based interferometric radar data. *ISPRS Ann. Photogramm. Remote Sens. Spat. Inf. Sci.* **2021**, *V-1-2021*, 31–38. [CrossRef]
12. Arnold, M.; Keller, S. Machine-learning for analyzing bridge displacement using radar data. In *Proceedings of the Bridge Maintenance, Safety, Management, Life-Cycle Sustainability and Innovations*; CRC Press: Boca Raton, FL, USA, 2024.
13. Zhao, X.; Liu, H.; Yu, Y.; Xu, X.; Hu, W.; Li, M.; Ou, J. Bridge Displacement Monitoring Method Based on Laser Projection-Sensing Technology. *Sensors* **2015**, *15*, 8444–8463. [CrossRef] [PubMed]
14. Dau, H.A.; Bagnall, A.; Kamgar, K.; Yeh, C.C.M.; Zhu, Y.; Gharghabi, S.; Ratanamahatana, C.A.; Keogh, E. The UCR Time Series Archive. *arXiv* **2019**, arXiv:1810.07758.
15. Bier, A.; Jastrzebska, A.; Olszewski, P. Variable-Length Multivariate Time Series Classification Using ROCKET: A Case Study of Incident Detection. *IEEE Access* **2022**, *10*, 95701–95715. [CrossRef]
16. Ruiz, A.P.; Flynn, M.; Large, J.; Middlehurst, M.; Bagnall, A. The great multivariate time series classification bake off: A review and experimental evaluation of recent algorithmic advances. *Data Min. Knowl. Discov.* **2021**, *35*, 401–449. [CrossRef] [PubMed]
17. Dempster, A. ROCKET: Exceptionally fast and accurate time series classification using random convolutional kernels. *Data Min. Knowl. Discov.* **2020**, *34*, 1454–1495. [CrossRef]
18. Dempster, A.; Schmidt, D.F.; Webb, G.I. MINIROCKET: A Very Fast (Almost) Deterministic Transform for Time Series Classification. *arXiv* **2021**, arXiv:2012.08791. https://doi.org/10.1145/3447548.3467231.
19. Löning, M.; Bagnall, A.; Ganesh, S.; Kazakov, V. Sktime: A Unified Interface for Machine Learning with Time Series. *arXiv* **2019**, arXiv:1909.07872.
20. Hertel, L.; Phan, H.; Mertins, A. Classifying Variable-Length Audio Files with All-Convolutional Networks and Masked Global Pooling. *arXiv* **2016**, arXiv:1607.02857.
21. Kawakatsu, T.; Aihara, K.; Takasu, A.; Adachi, J. Deep Sensing Approach to Single-Sensor Bridge Weighing in Motion. In Proceedings of the 9th European Workshop on Structural Health Monitoring, Manchester, UK, 10–13 July 2018.
22. Kawakatsu, T.; Aihara, K.; Takasu, A.; Adachi, J. Deep Sensing Approach to Single-Sensor Vehicle Weighing System on Bridges. *IEEE Sens. J.* **2019**, *19*, 243–256. [CrossRef]
23. Kawakatsu, T.; Aihara, K.; Takasu, A.; Nagayama, T.; Adachi, J. Data-Driven Bridge Weigh-In-Motion. *IEEE Sens. J.* **2023**, *23*, 17064–17077. [CrossRef]
24. Yu, Y.; Cai, C.; Deng, L. Vehicle axle identification using wavelet analysis of bridge global responses. *J. Vib. Control* **2017**, *23*, 2830–2840. [CrossRef]
25. Lechner, B.; Lieschnegg, M.; Mariani, O.; Pircher, M.; Fuchs, A. A Wavelet-Based Bridge Weigh-In-Motion System. *Int. J. Smart Sens. Intell. Syst.* **2010**, *3*, 573–591. [CrossRef]
26. Zhao, H.; Tan, C.; OBrien, E.J.; Uddin, N.; Zhang, B. Wavelet-Based Optimum Identification of Vehicle Axles Using Bridge Measurements. *Appl. Sci.* **2020**, *10*, 7485. [CrossRef]
27. Gentile, C.; Bernardini, G. An interferometric radar for non-contact measurement of deflections on civil engineering structures: Laboratory and full-scale tests. *Struct. Infrastruct. Eng.* **2010**, *6*, 521–534. [CrossRef]
28. Tzuta, L. LabelImg. 2015. Available online: https://github.com/tzutalin/labelImg (accessed on 17 July 2023).
29. Kaiser, G. *A Friendly Guide to Wavelets*; Birkhäuser: Boston, MA, USA, 2011. [CrossRef]

30. Rosyidi, S.A.P.; Taha, M.R.; Chik, Z.; Ismail, A. Signal reconstruction of surface waves on SASW measurement using Gaussian Derivative wavelet transform. *Acta Geophys.* **2009**, *57*, 616–635. [CrossRef]
31. Lee, G.R.; Gommers, R.; Waselewski, F.; Wohlfahrt, K.; O'Leary, A. PyWavelets: A Python package for wavelet analysis. *J. Open Source Softw.* **2019**, *4*, 1237. [CrossRef]
32. Breiman, L. Random Forests. *Mach. Learn.* **2001**, *45*, 5–32. [CrossRef]
33. Friedman, J.H. Greedy Function Approximation: A Gradient Boosting Machine. *Ann. Stat.* **2001**, *29*, 1189–1232. [CrossRef]
34. Cover, T.; Hart, P. Nearest neighbor pattern classification. *IEEE Trans. Inf. Theory* **1967**, *13*, 21–27. [CrossRef]
35. Virtanen, P.; Gommers, R.; Oliphant, T.E.; Haberland, M.; Reddy, T.; Cournapeau, D.; Burovski, E.; Peterson, P.; Weckesser, W.; Bright, J.; et al. SciPy 1.0: Fundamental Algorithms for Scientific Computing in Python. *Nat. Methods* **2020**, *17*, 261–272. [CrossRef]
36. Kuok, S.C.; Yuen, K.V. Investigation of modal identification and modal identifiability of a cable-stayed bridge with Bayesian framework. *Smart Struct. Syst.* **2016**, *17*, 445–470. [CrossRef]
37. Kim, S.; Kim, N.; Park, Y.S.; Jin, S.S. A Sequential Framework for Improving Identifiability of FE Model Updating Using Static and Dynamic Data. *Sensors* **2019**, *19*, 5099. [CrossRef] [PubMed]
38. Katafygiotis, L.S.; Lam, H.F.; Papadimitriou, C. Treatment of Unidentifiability in Structural Model Updating. *Sensors* **2000**, *3*, 19–39.

infrastructures

Article

Deep Learning-Based Steel Bridge Corrosion Segmentation and Condition Rating Using Mask RCNN and YOLOv8

Zahra Ameli [1,*], Shabnam Jafarpoor Nesheli [2] and Eric N. Landis [1]

[1] Department of Civil and Environmental Engineering, University of Maine, Orono, ME 04469, USA; landis@maine.edu
[2] Department of Engineering, University of Science and Culture, Tehran 1461968151, Iran; shabnamjafarpoor1371@gmail.com
* Correspondence: zahra.ameli@maine.edu

Abstract: The application of deep learning (DL) algorithms has become of great interest in recent years due to their superior performance in structural damage identification, including the detection of corrosion. There has been growing interest in the application of convolutional neural networks (CNNs) for corrosion detection and classification. However, current approaches primarily involve detecting corrosion within bounding boxes, lacking the segmentation of corrosion with irregular boundary shapes. As a result, it becomes challenging to quantify corrosion areas and severity, which is crucial for engineers to rate the condition of structural elements and assess the performance of infrastructures. Furthermore, training an efficient deep learning model requires a large number of corrosion images and the manual labeling of every single image. This process can be tedious and labor-intensive. In this project, an open-source steel bridge corrosion dataset along with corresponding annotations was generated. This database contains 514 images with various corrosion severity levels, gathered from a variety of steel bridges. A pixel-level annotation was performed according to the Bridge Inspectors Reference Manual (BIRM) and the American Association of State Highway and Transportation Officials (AASHTO) regulations for corrosion condition rating (defect #1000). Two state-of-the-art semantic segmentation algorithms, Mask RCNN and YOLOv8, were trained and validated on the dataset. These trained models were then tested on a set of test images and the results were compared. The trained Mask RCNN and YOLOv8 models demonstrated satisfactory performance in segmenting and rating corrosion, making them suitable for practical applications.

Keywords: deep learning; instance segmentation; corrosion condition rating; bridge inspection; YOLOv8; Mask RCNN

Citation: Ameli, Z.; Nesheli, S.J.; Landis, E.N. Deep Learning-Based Steel Bridge Corrosion Segmentation and Condition Rating Using Mask RCNN and YOLOv8. *Infrastructures* **2024**, *9*, 3. https://doi.org/10.3390/infrastructures9010003

Academic Editor: Armin Mehrabi

Received: 13 October 2023
Revised: 5 December 2023
Accepted: 7 December 2023
Published: 20 December 2023

1. Introduction

In the United States, more than 40 percent of all bridges are built of steel (U.S. National Bridge Inventory (NBI)), and corrosion is a major factor in the deterioration of these structures. Each year, the federal government and state Departments of Transportation (DOTs) spend billions of dollars on bridge rehabilitation and maintenance due to corrosion [1]. Corrosion reduces the effective cross-sections of critical structural components and subsequently decreases the axial, bending, and fatigue strength, which can lead to the failure of individual elements and entire structures. In order to detect and mitigate corrosion damage, regular inspection is essential. Steel structural elements must be evaluated in terms of the initiation of corrosion, section loss, or any other possible adverse effect on the strength or serviceability of structural elements. Since maintenance decisions mainly depend on the information collected during inspection, accurate bridge inspection and data analysis is important in optimizing bridge maintenance strategies and avoiding unnecessary repair costs.

The manual inspection of the bridge is always time-consuming and the result is inconsistent, subjective, and largely dependent on the experience and judgment of the inspector [2,3]. In addition, traditional bridge inspection is costly and associated with safety risks for the inspection crew and public. Unmanned aerial vehicles (UAVs) have proven to be a viable and safer solution to perform such inspections in many adverse conditions by flying up close to the structures and taking large numbers of high-resolution images from multiple angles [4,5]. Integrating UAVs with artificial intelligence (AI) in infrastructure inspections, particularly in tasks like defect detection, classification, and segmentation, offers significant benefits in terms of reduced time and costs and the enhanced accuracy of the results [6,7]. This study has focused on automated corrosion classification and condition rating to address the inefficiencies of traditional inspection.

Corrosion exhibits two main visual characteristics: a rough surface texture and a red color spectrum. These properties have been used for corrosion detection by applying feature-based classification techniques [8]. Machine learning-based corrosion detection, such as support vector machines (SVM) and random forest, requires experts to manually define relevant features. The algorithm then learns to detect corrosion based on these features. For instance, in a study by Khayatazad et al. (2020), the authors quantified and combined roughness and color for corrosion detection. The uniformity metric calculated from the gray-level co-occurrence matrix and the histogram of corrosion-representative colors were used to analyze roughness and color, respectively [9]. These techniques require manual feature engineering, which may not capture all variations in the data, and the quality of the manually selected features might affect performance of the algorithm. Image processing techniques such as histogram equalization and edge detection mitigate the need for the manual adjustment of threshold parameters, but they still require predefined rules to extract features directly from the images. The designed features may lack robustness when faced with diverse variations, leading to a significant decrease in recognition accuracy if the algorithm fails to effectively capture the advanced features present in the images [10]. Deep learning-based (DL) approaches automatically learn hierarchical features from raw data and adapt the learned features, effectively reducing the need for human intervention. DL-based techniques can handle complex, unstructured data and often achieve state-of-the-art performance. Convolutional neural networks (CNNs) are a class of deep neural networks, most commonly applied to analyze images [11]. CNNs have been used for object detection, classification, and segmentation and consistently exhibit remarkable performance compared to traditional classification algorithms. Petricca et al. (2016) presented a comparison between computer vision techniques and CNNs for automatic metal corrosion detection. The authors classified the image pixels based on the specific color components. The findings indicated that the deep learning approach achieved more uniform results and significantly higher accuracy compared to computer vision techniques [12]. Some studies have evaluated the performance of CNNs with different architectures in corrosion detection. Atha and Jahanshahi (2018) evaluated VGG and ZF Net, two different architectures of CNN, for corrosion detection. The authors also proposed two CNN architectures specifically designed for corrosion detection, namely Corrosion7 and Corrosion5. It was shown that the deeper architecture, VGG16, performed better in terms of accuracy than a smaller network such as ZF Net. The authors achieved 96.68% mean precision using 926 images. However, only clearly visible surface images were considered in this study [13]. Jin Lim et al. (2021) applied Faster RCNN with RGB and thermographic images for steel bridge corrosion inspection. They achieved an average precision (AP) value of 88% for surface and subsurface corrosion detection [14].

While corrosion detection is concerned with identifying the presence and location of corrosion in an image using bounding boxes, segmentation is focused on the pixel-level classification and delineation of object boundaries. Long et al. (2015) introduced fully convolutional networks (FCNs) as the first deep learning-based technique to perform the segmentation task, and since then, many image segmentation techniques, such as U-Net [15], DeepLab [16], PSPNet [17], SegNet [18], and Mask RCNN [19], have been developed. Each of these techniques has unique architectural features, which make them suitable for different image segmentation tasks. Recently, Ta et al. (2022) utilized the

Mask RCNN algorithm, an instance segmentation approach, for the detection of corroded bolts. The model was able to accurately detect corrosion in the tested structure. The accuracy of the model declined along with the image capturing distance [20]. Nash et al. (2018) proposed a modified VGG-16 architecture to perform the pixel-level segmentation of corrosion. Neither of the models trained on small and large datasets achieved human-level accuracy for corrosion detection [21]. In their study, Rahman et al. (2021) proposed a semantic CNN model for corrosion detection and segmentation. The authors proposed a corrosion evaluation method based on the grayscale value for the classification of each pixel of a corrosion segment into heavy, medium, and light corrosion categories [8].

For the problem of bridge inspection, inspectors are responsible for assigning corrosion condition states to the corroded structural elements according to the BIRM [22] and American Association of State Highway and Transportation Officials (AASHTO) regulations [23]. Table 1 demonstrates these condition states. The number of studies that use DL-based methods to evaluate and assign standardized condition states to the corroded areas in the image is very limited. In a recent study, Munawar et al. (2022) divided the images into four corrosion levels, no corrosion (negative class), low-level corrosion (images having less than 5% of corroded pixels), medium-level corrosion (images having less than 15% of corroded pixels), and high-level corrosion (images having more than 15% of corroded pixels) [24], based on the number of corroded pixels. Notably, this classification method differs from the approach suggested by the Bridge Inspection Reference Manual (1), which recommends assessing the corrosion condition based on the severity of the corrosion.

Table 1. Corrosion condition states (AASHTO).

Condition State	Defect Description in Steel Members
"Good"	None
"Fair"	Corrosion of the steel has initiated
"Poor"	Section loss is evident or pack rust is present but does not warrant structural review
"Severe"	The condition warrants a structural review to determine the effect on the strength or serviceability of the element or bridge, or a structural review has been completed and the defects impact the strength or serviceability of the element or bridge

To address the need for a practical corrosion condition rating approach, in the current study, we trained two DL-based corrosion segmentation algorithms to automatically rate the corroded areas and assign corrosion severity levels to them. The required corrosion images for the training of the algorithms were collected and annotated according to the AASHTO and BIRM regulations. The acquired corrosion images and corresponding annotations have been made openly accessible online and provide a resource for the training of future corrosion and condition rating models [25].

2. Materials and Methods

2.1. Image Acquisition and Annotation

A Nikon D750 camera and a DJI Mini drone were used to gather 514 high-quality corrosion images. These images were taken from steel bridge components such as girders and supports. To mitigate the risk of collision and ensure the safety of both the drone and the structure being inspected, the drone maintained a distance of more than one meter from the structure. Images were captured from multiple angles to provide a comprehensive view of the corroded areas. The blurry and dark images were removed from the dataset to maintain the overall quality of the input training data. Supervised learning algorithms such as Mask RCNN and YOLOv8 rely on labeled data for effective training; therefore, the collected images needed to be annotated. Annotation involves assigning each image with detailed labels specifying the boundaries or areas of interest. Pixel-wise annotation is time-consuming and challenging due to several factors. The smooth boundaries of corrosion make it difficult to accurately outline the damage, leading to potential inconsistencies. Additionally, the lack of distinct characteristics in corrosion severity makes it challenging to classify and annotate them correctly. Moreover, distinguishing actual corrosion from other visual artifacts, like dark backgrounds, dirt, or smudges, adds another layer of complexity to the annotation task. To address these challenges, the annotations were initially made by an experienced structural engineer and fine-tuned by another expert to ensure the consistency and accuracy. On average, approximately 12 min were dedicated to annotating each image. Corroded areas in each image were outlined and assigned labels. The three labels that were used in this study represented three different levels of corrosion severity according to the AASHTO and BIRM regulations (Table 1). These labels and the corresponding condition states are as follows.

- Class 1 ("Fair"): Mild corrosion in which corroded portions of steel were not distinguishable. Corrosion that appeared to be streaked or discolored was annotated as "Fair".
- Class 2 ("Poor"): Corrosion areas that appeared deeper than surface corrosion or were beginning to disintegrate were considered "Poor".
- Class 3 ("Severe"): Corrosion that caused section loss, deep corrosion with disintegrated portions of steel, and/or corrosion that seemed to affect the structural integrity of the bridge component was annotated as "Severe". Corrosion with complete section loss (holes or cavities) was also annotated as "Severe", with surrounding areas annotated appropriately.

Four sample images and their corresponding annotations are shown in Figure 1. The "Fair", "Poor", and "Severe" classes in the annotated images are shown with green, yellow, and red colors, respectively. As can be seen in the figure, no annotation is assigned to peeling paint, visible debris, or minimal corrosion. Freckled rust/sporadic corrosion, exposed steel, and surface corrosion are annotated as "Fair". Deeper corrosion, such as disintegrated portions of sections or pack rust, is annotated as "Poor", while any corrosion with section loss, holes, or cavities, or multiple perforations, is annotated as "Severe". The software used for the semantic mask annotations was LabelMe, an open-source annotation tool that enables users to mark objects or regions of interest in images by drawing bounding boxes, polygons, or points and assigning labels to them.

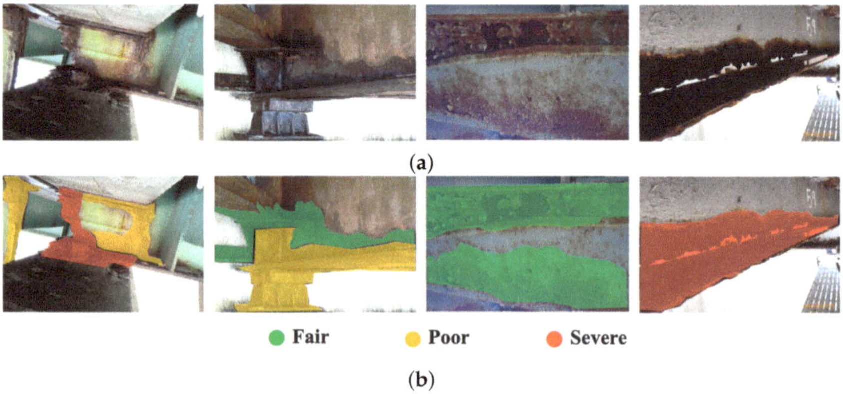

Figure 1. (**a**) Sample images and (**b**) corresponding annotations.

2.2. *Instance Segmentation Models*

2.2.1. Mask RCNN

The mask region-based convolutional neural network (Mask RCNN) [19] is a deep learning model designed for instance segmentation tasks. It is an extension of Faster RCNN, which combines object detection and image segmentation by adding a branch for the prediction of object masks. The Mask RCNN backbone is a ResNet-101 architecture that serves as a feature extractor with 101 layers. The early layers detect low-level features, while later layers detect higher-level ones. The feature map in the Mask RCNN architecture retains a spatial abstraction of the input image obtained through convolutional layers. Mask RCNN improves its backbone performance by incorporating a region proposal network (RPN) to find areas containing objects. The regions that the RPN scans over are boxes distributed over the image area called anchors. The final step in the RPN is non-maximum suppression (NMS), which suppresses overlapping object proposals and selects the best ones based on their confidence scores. The remaining object proposals are then passed to the next stage of the Mask RCNN architecture, which performs bounding box regression, classification, and mask prediction. The structure of the Mask RCNN network is shown in Figure 2.

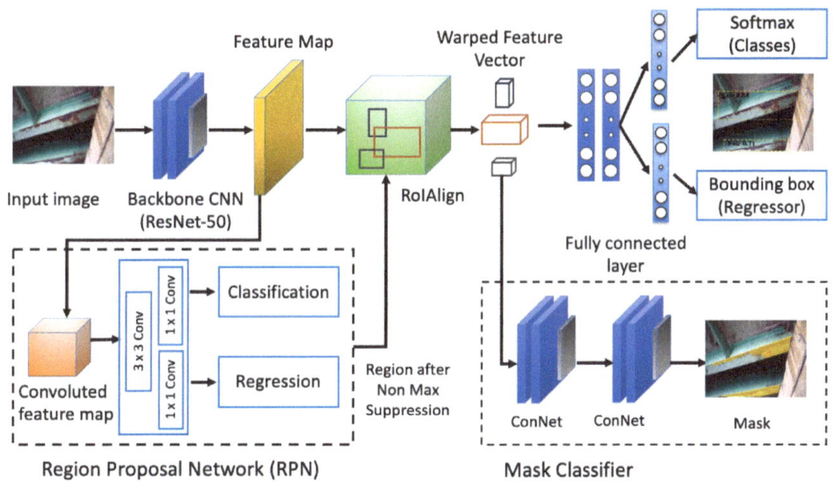

Figure 2. Structure of Mask RCNN.

The Mask RCNN loss function is calculated as follows:

$$L_{total} = L_{cls} + L_{reg} + L_{msk} \quad (1)$$

where L_{cls}, L_{reg}, and L_{msk} represent classification, regression, and mask loss, respectively. Classification loss measures the discrepancy between the predicted class probabilities and the ground truth class labels for each region proposal and is calculated as

$$L_{cls} = -(p_{true}\log(p_{pred}) + (1 - p_{true})\log(1 - p_{pred})) \quad (2)$$

In this equation, p_{true} is the ground truth label (0 or 1) indicating the true class of the sample, and p_{pred} is the predicted probability of the positive class (object presence) for the sample. Regression loss measures the difference between the predicted bounding box coordinates (e.g., top-left corner, width, height) and the ground truth bounding box coordinates for each region proposal, and it is calculated by

$$L_{\text{reg}} = \sum_i \text{R}(t_i - t_i^*) \quad (3)$$

Here, index i represents the specific instance or predicted frame, and t_i is the predicted frame coordinate vector for the i-th instance. t_i^* is the coordinate vector corresponding to the border of the real area for the i-th instance, and R is the loss function, which computes the element-wise loss between the predicted and target coordinates. Finally, L_{msk} or the mask loss quantifies the dissimilarity between the predicted instance masks and the ground truth masks for each region proposal, and it is defined as

$$L_{\text{msk}} = -\sum_{i=1}^{N}[x_i^* \cdot \log(p(x_i)) + (1 - x_i^*) \cdot \log(1 - p(x_i))] \tag{4}$$

where i iterates over the pixels from 1 to N. The x_i^* term represents the category label where the pixel is located, and $p(x_i)$ is the probability of the predicted category (x_i). The goal during training is to minimize the total loss value (Equation (1)) as it quantifies how well the model is performing.

Our implementation of Mask RCNN is a library called Detectron2, which is based upon the Mask RCNN benchmark [26]. This library was developed by Facebook AI Research, and it achieved satisfactory results in object detection and segmentation problems. Its implementation is in PyTorch and requires CUDA® due to the heavy computations involved.

2.2.2. YOLOv8

You Only Look Once version 8 (YOLOv8) is the newest state-of-the-art YOLO model that can be used for object detection, image classification, and instance segmentation tasks. YOLOv8 introduces a new backbone network, CSPDarknet53, which consists of 53 convolutional layers and employs cross-stage partial connections to improve the information flow between the different layers [27]. The head of YOLOv8 consists of multiple convolutional layers followed by a series of fully connected layers. YOLOv8 uses an anchor-free model with a decoupled head to independently perform objectness, classification, and regression tasks. This design allows each branch to focus on its task and improves the model's overall accuracy. YOLOv8 provides a semantic segmentation model called the YOLOv8-Seg model. YOLOv8 segmentation models are pretrained on the COCO dataset. These weights serve as a starting point for the training of any custom dataset. The structure of the YOLOv8 network is shown in Figure 3. One of the key features of YOLOv8 is the use of a self-attention mechanism in the head of the network. This mechanism allows the model to focus on different parts of the image and adjust the importance of different features based on their relevance to the task. Another important feature of YOLOv8 is its ability to perform multi-scale object detection. The model utilizes a feature pyramid network to detect objects of different sizes and scales within an image. This feature pyramid network consists of multiple layers that detect objects at different scales, allowing the model to detect large and small objects within an image.

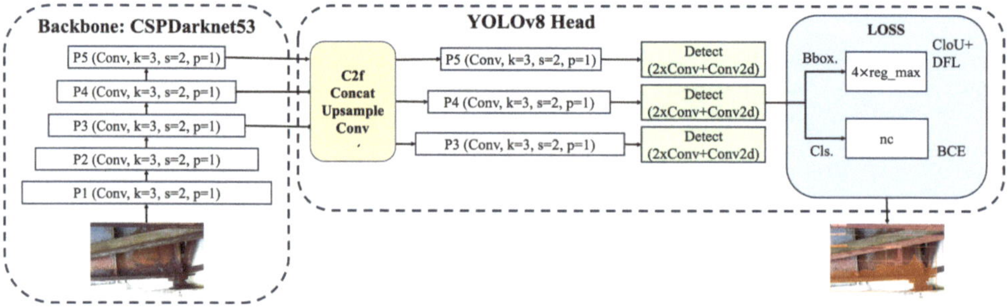

Figure 3. Structure of the YOLOv8.

YOLOv8 uses the complete intersection over union (CIoU) loss [28] and distribution focal loss (DFL) [29] functions for bounding box loss and binary cross-entropy (BCE) for classification loss. Bounding box loss (box_loss) measures the error in the predicted bounding box coordinates and dimensions compared to the ground truth. Segmentation loss (seg_loss) measures the discrepancy between the predicted segmentation masks and the ground truth masks. It encourages the model to generate accurate and precise segmentation masks that align with the ground truth annotations. Classification loss (cls_loss) measures the error in the predicted class probabilities for each object in the image compared to the ground truth. Distribution focal loss (dfl_loss) is a new addition to the YOLO architecture in YOLOv8. DFL loss considers the problem of class imbalance while training a neural network. Class imbalance occurs when one class occurs too frequently and another occurs less frequently. DFL solves this problem by placing more importance on less frequent classes. The overall loss value is typically a weighted sum of these individual losses.

Both the Mask RCNN and YOLOv8 models have demonstrated promising results in various segmentation tasks. Mask RCNN is a two-stage instance segmentation model that combines object detection and pixel-level segmentation. It first generates region proposals using a CNN and then refines these proposals by predicting the class label and generating a mask for each object instance. Mask RCNN is known for its accuracy in segmenting objects with precise boundaries. On the other hand, YOLOv8 is a one-stage object detection model that directly predicts bounding boxes and class labels for objects in an image. It divides the input image into a grid and assigns each grid cell responsibility for predicting objects. YOLOv8 is known for its real-time performance and efficiency.

3. Model Training and Validation

In this study, YOLOv8m-seg, which is a medium-sized segmentation model with over 27 million parameters, was used. The annotated files, which were in .json format originally, needed to be converted to .txt format for the training of the YOLOv8 model. For this conversion, the "json2yolo" github repository was utilized [30]. The training and validation images were randomly selected in each dataset following a ratio of 80% for training and 20% for validation and testing. A total of 412 images were considered for training, 90 images for validation, and 12 images were randomly selected to test the model's performance. The same training, validation, and testing images were used to train both the YOLOv8 and Mask RCNN models for comparison purposes.

The distribution of instances for each class is depicted in Figure 4. As can be seen in the figure, the number of "Fair" instances is significantly higher than that of "Poor" and "Severe" cases. This observation can be attributed to the relative ease of identifying "Fair" corrosion in the structural elements of the bridge. While most of the collected images contain mild corrosion, it is more challenging to find deeper and more severe corrosion in real bridges.

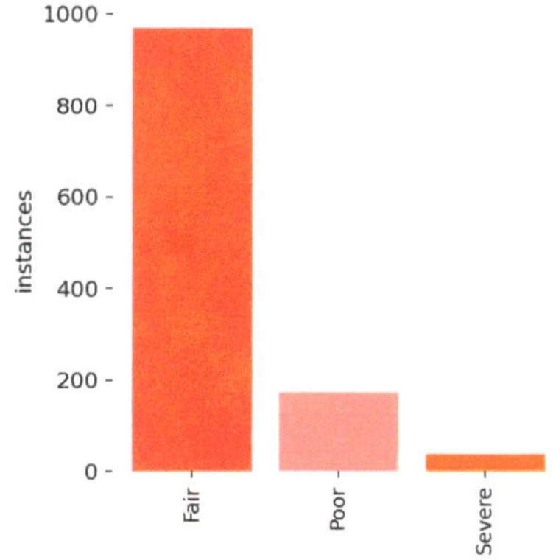

Figure 4. Number of instances for each class: "Fair", "Poor", and "Severe".

In order to perform the training task, Google Colab Pro was employed, taking advantage of a T4 GPU, the Python programming language, and the large RAM resources. Hyperparameters such as the learning rate, batch size, and number of epochs were optimized based on the performance of the models during training and validation. For the

YOLOv8 model, a learning rate of 0.0001 was obtained, along with a batch size of 8 and 250 epochs. The image size was set to 640 by 640 pixels. These values were determined through trial and error to find the optimal balance between model accuracy and training efficiency. Similarly, for the Mask RCNN model, a learning rate of 0.00025 was chosen, along with a batch size of 8 and 200 epochs. The image size was also set to 640 by 640. These values also were selected based on experimentation and previous knowledge of the model's performance. After training was completed, a file containing the custom-trained weights was saved in the training path. These weights were adjusted during the optimization process to minimize the difference between the predicted outputs and the actual outputs (labels) in the training data. The final weight values were utilized to perform predictions on the selected set of images (test images). This process involved using the trained model to classify and rate the condition of corrosion within the given images. The training loss values versus the number of epochs for the Mask RCNN and YOLOv8 models are shown in Figure 5a and b, respectively. The overall trend of the loss plot shows a consistent decrease, which indicates that the training model is learning and improving its performance over time. The validation loss values versus the number of epochs are also shown in Figure 6.

As can be seen in Figure 5, all types of loss converge to a low and stable value, indicating that both models have learned the patterns in the data. In the case of Mask RCNN, the segmentation loss has a higher value compared to the bounding and classification loss, which suggests that the model struggles with the segmentation tasks. The YOLOv8 trained model has higher segmentation and DFL loss, which shows the inherent difficulty in accurately predicting object masks. The higher DFL loss may indicate that the model is struggling with the class imbalance.

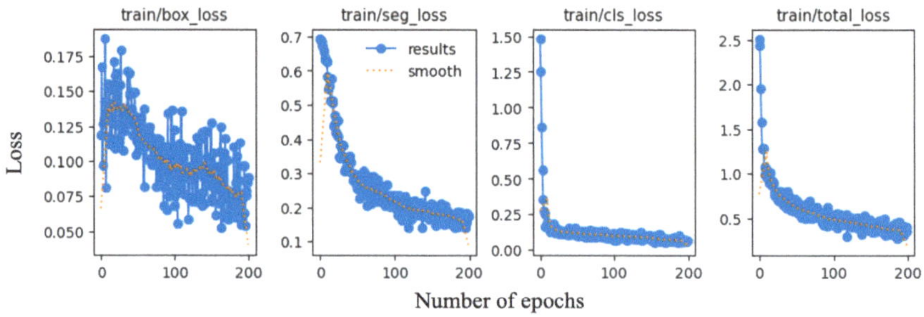

(a)

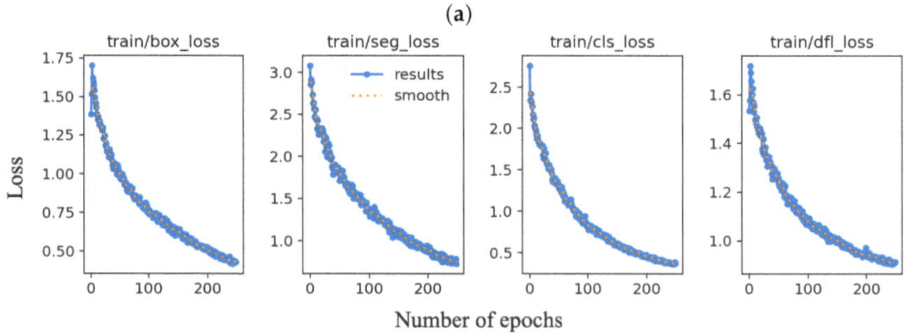

(b)

Figure 5. (**a**) Mask RCNN: bounding box, segmentation, classification, and total loss. (**b**) YOLOv8: bounding box, segmentation, classification, and distribution focal loss for the training dataset.

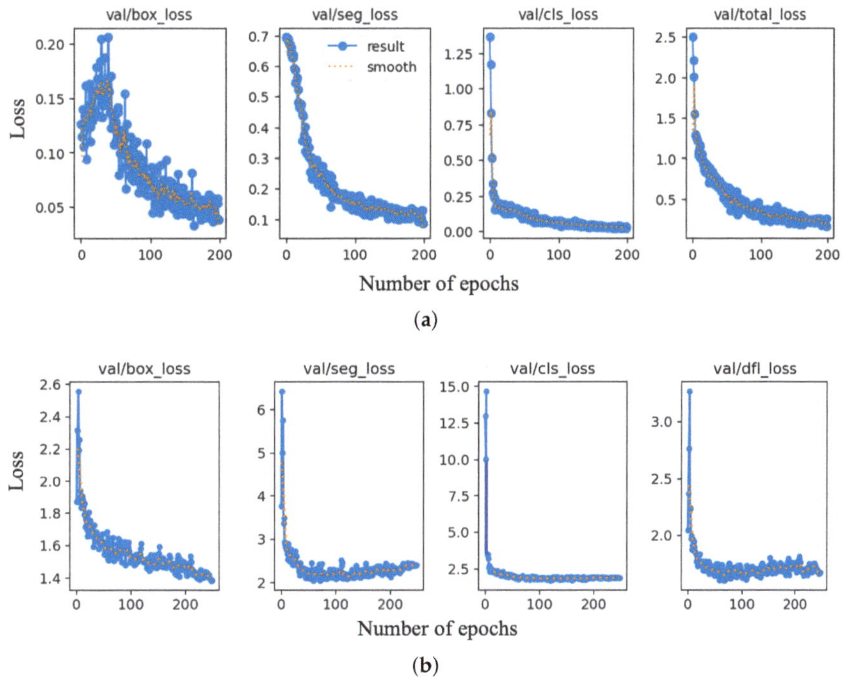

Figure 6. (**a**) Mask RCNN: bounding box (box_loss), segmentation (seg_loss), classification (cls_loss), and total loss. (**b**) YOLOv8: bounding box, segmentation, classification, and distribution focal loss (dfl_loss) for the validation dataset.

Validation loss measures the model's performance on a validation dataset, and it provides an indication of how well the model generalizes to unseen data. It was used to tune the hyperparameters, such as the learning rate, number of epochs, batch size, etc. Overfitting (when the training loss continues to decrease, but the validation loss starts to increase) and underfitting (when both the training and validation loss are high) can be avoided by comparing the validation and training loss. The validation loss of the Mask RCNN trained model is as low as its training losses (Figure 6). YOLOv8 shows higher validation loss values compared to the training losses. This could be an indication of overfitting or a lack of diversity in the training dataset.

4. Evaluation

Precision and recall values are two important metrics used to assess the performance of classification models. The precision value quantifies the accuracy of the corrosion detection algorithm, indicating the proportion of correctly identified corrosion instances. The recall value represents the fraction of all target corrosion instances that are successfully detected. The precision and recall values are obtained as follows:

$$\text{Precision} = \frac{\text{TP}}{\text{TP} + \text{FP}} \tag{5}$$

$$\text{Recall} = \frac{\text{TP}}{\text{TP} + \text{FN}} \tag{6}$$

True positive (TP) represents the number of cases that are positive and detected as positive (correctly detected corrosion). False positive (FP) represents the number of cases that are negative but detected as positive (falsely detected non-corrosion). False negative (FN) represents the number of cases that are positive but detected as negative (missed corrosion). Finally, true negative (TN) is the number of negative instances correctly predicted as negative (correctly detected non-corrosion).

The $F1$ score combines the precision and recall of a classifier into a single metric by taking their harmonic mean as follows:

$$F1\text{ Score} = \frac{2 \times (\text{Precision} \times \text{Recall})}{\text{Precision} + \text{Recall}} \tag{7}$$

Mean average precision (mAP) measures the accuracy of object localization and segmentation by considering the precision and recall values at different levels of intersection over union (IoU) thresholds. It evaluates how well the predicted bounding boxes or segmentation masks align with the ground truth annotations:

$$\text{mAP} = \frac{1}{\text{N}} \sum_{i=1}^{N} \text{AP}_i \tag{8}$$

where N is the number of classes or categories and AP_i is the average precision (AP) for class i. In terms of evaluation metrics for bounding box regression, intersection over union (IoU) is the most popular metric. The IoU for the bounding box and mask can be obtained as follows:

$$\begin{aligned} \text{Box IoU} &= \frac{|B^{Pr} \cap B^{GT}|}{|B^{Pr} \cup B^{GT}|} \\ \text{Mask IoU} &= \frac{|M^{Pr} \cap M^{GT}|}{|M^{Pr} \cup M^{GT}|} \end{aligned} \tag{9}$$

In Equation (9), (B) and (M) distinguish the metrics of segmentation. Pr and GT represent the prediction and ground truth, while (B) and (M) evaluate the detection (Box) and segmentation (Mask), respectively. mAP50 represents the average precision when the IoU threshold is set to 0.5. This threshold considers a detection as a true positive if the IoU between the predicted and the ground truth mask is greater than or equal to 0.5. This metric provides a more comprehensive evaluation of the models' performance on the validation and testing datasets. These values for both the YOLOv8 and Mask RCNN trained models were calculated and are listed in Table 2.

Table 2. Evaluation metrics for the trained Mask RCNN and YOLOv8 models.

Trained Model	Dataset	F1 Score	Precision	Recall	mAP50
YOLOv8	Validation	0.451	0.602	0.360	0.484
	Test	0.730	0.853	0.647	0.726
Mask RCNN	Validation	0.519	0.428	0.660	0.483
	Test	0.745	0.625	0.922	0.674

The F1 score values in Table 2 show that YOLOv8 achieved an F1 score of 0.730, while Mask RCNN achieved a slightly higher F1 score of 0.745. This indicates that both models exhibit reasonable overall performance, with Mask RCNN having a slight advantage in terms of F1 score and striking a better balance between precision and recall. YOLOv8 achieved higher precision (0.853) compared to Mask RCNN (0.625). This indicates that YOLOv8 tends to make fewer false positive predictions. The Mask RCNN trained model had a higher recall rate (0.922) compared to YOLOv8 (0.647), which indicates its better ability to detect and include relevant objects in its predictions. The mAP50 values of the Mask RCNN trained model were 0.483 on the validation dataset and 0.674 on the test dataset. The YOLOv8 model achieved mAP50 values of 0.484 on the validation dataset and 0.726 on the test dataset. The Mask RCNN model had slightly lower mAP50 values compared to YOLOv8, indicating that YOLOv8 performs better in terms of mean average precision at the 50% IoU threshold. The YOLOv8 model outperformed the Mask RCNN model in terms of speed, with an inference time of 0.03 s per image, compared to 0.12 s per image for Mask RCNN. Based on the available metrics in the chart, the Mask RCNN model generally outperforms the YOLOv8 model in terms of F1 score and recall but YOLOv8 achieves higher precision and mAP50. This implies that YOLOv8 tends to provide more accurate predictions concerning the relevant objects that it detects, albeit with potentially fewer true positives. It is important to note that the comparison is limited to the provided metrics, and further analysis, including additional metrics and experimental conditions, is needed to discuss the performance of these models accurately.

5. Test Results and Discussion

To evaluate the performance of the trained Mask RCNN and YOLOv8 models for corrosion segmentation and condition rating, prediction was performed on the test images that were not used for training and validation. Only objects identified with confidence of 0.5 and 0.35 or higher for Mask RCNN and YOLOv8, respectively, were considered in the test. The representative results are shown in Figures 7–9. Each figure contains four images in a row. The first row shows the original test images, the second row represents the corresponding annotations, the third row shows the prediction of the YOLOv8 trained model, and the last row depicts the prediction of Mask RCNN. A brief discussion of the results for each sample image is provided below.

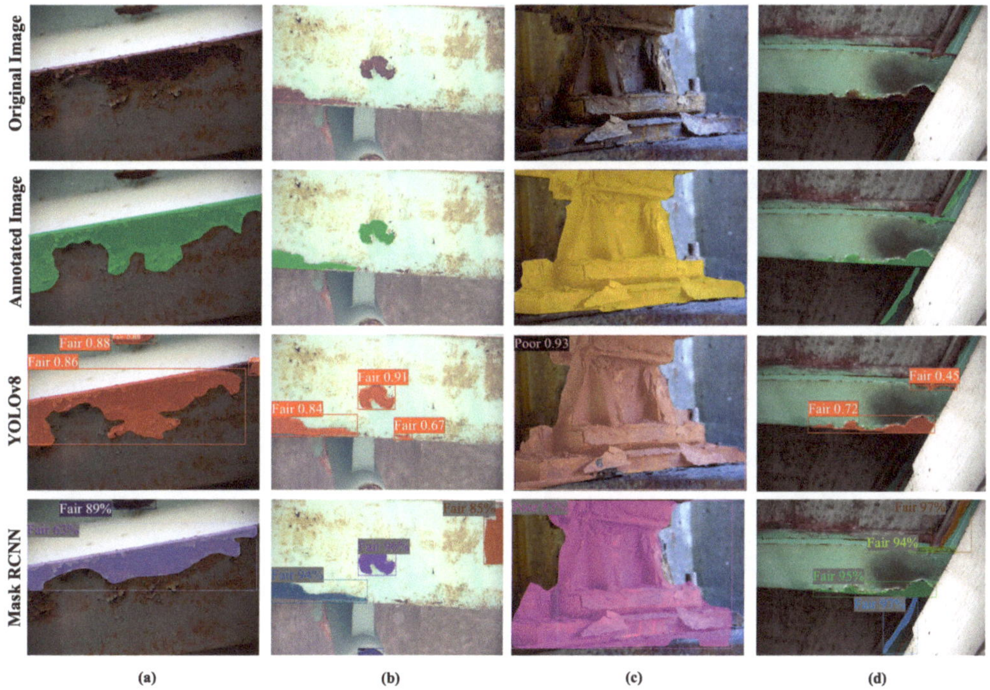

Figure 7. (**a**) Bottom flange of a steel beam (**b**) bottom flange of a main girder (**c**) Girder support (**d**) bottom flange of a cross girder.

Figure 7a shows mild corrosion in the bottom flange of a beam. The corroded portion of the flange is annotated as "Fair" and its mask is shown in the second row of Figure 7a with a green color. As can be seen in the third and fourth rows of Figure 7a, both the Mask RCNN and YOLOv8 models classified and segmented the corroded portions of the beam correctly. YOLOv8's mask area and shape is closer to the annotation and has higher confidence compared to Mask RCNN. A small corroded area in the middle of the upper side of the image is correctly detected as "Fair" and segmented by both models, but it is not considered in the annotation due to its small area.

Similar to Figure 7a, Figure 7b contains some "Fair" corrosion. There is a perfect match between the annotation and output of the YOLOv8 model. YOLOv8 also labeled a small area as "Fair", which is not labeled in the annotation. Mask RCNN labeled two additional instances in the image as "Fair" that are not included in the annotation.

The third picture from the left (Figure 7c) shows a corroded support with relatively severe corrosion, which was annotated as "Poor". Both of the trained models correctly labeled the corroded area as "Poor", while Mask RCNN labeled it with lower confidence compared to YOLOv8 (83% vs. 93%). There is some mismatch between the mask depicted by the Mask RCNN trained model and the annotation. It incorrectly detected a small portion of the concrete below the support as poor corrosion.

In Figure 7d, Mask RCNN correctly labels the entire corroded area with higher confidence compared to YOLOv8. Two corrosion instances are missing in the prediction of the YOLOv8 model, while the Mask RCNN trained model correctly classified and labeled all of them with a high confidence score (93% and 97%).

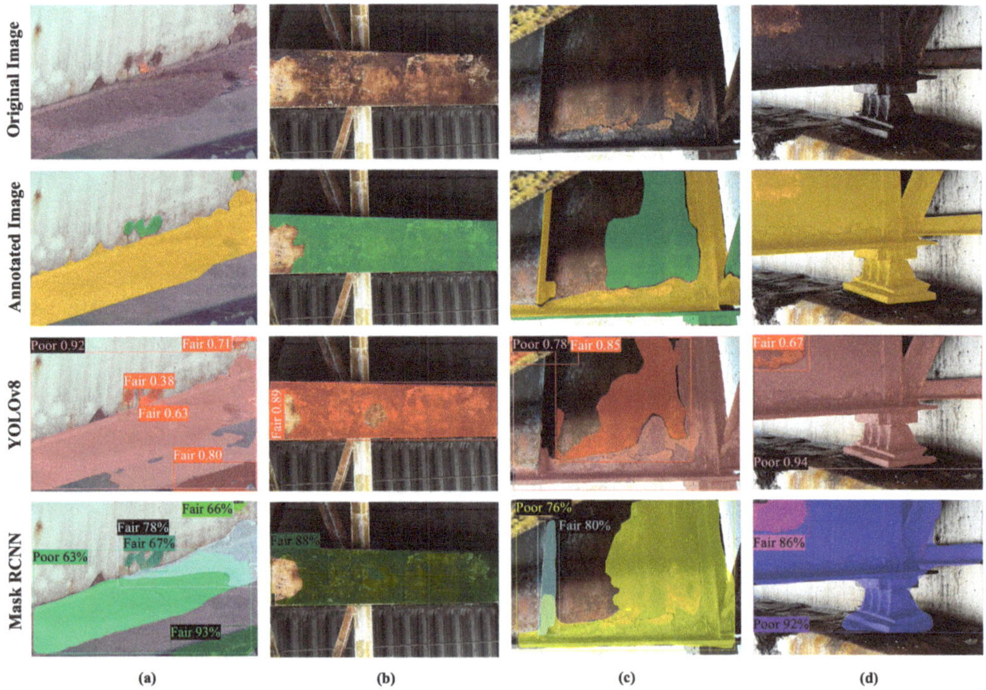

Figure 8. (**a**) A girder flange and its cover plate (**b**) bottom flange of a beam (**c**) girder web (**d**) girder web, support and bracing.

Figure 8a shows a girder flange. The lower part of the beam flange, which was annotated as "Poor", is correctly labeled and segmented by the YOLOv8 trained model. Some parts of the deck concrete are labeled as "Poor" by YOLOv8, which is not correct. In the lower right corner of the image, a corroded portion of an adjacent girder can be seen, which is not included in the annotation, but both models labeled it as "Fair". Mask RCNN labeled some parts of the flange as "Poor" and the rest as "Fair", which is not correct.

As illustrated in Figure 8b, the corroded portion of the lower flange is correctly detected and labeled by both of the models. YOLOv8 has slightly higher confidence compared to

Mask RCNN. A small part in the middle of the flange, which seems less corroded compared to the rest of the flange, is excluded from the mask provided by YOLOv8.

YOLOv8 segmented and labeled the web and the upper part of the support, the flange, and the brace as "Poor", while, in the annotation, only the girder web is considered "Poor" (Figure 8c). The middle part of the web that is less corroded is annotated as "Fair", while Mask RCNN labeled the entire web and flange as "Poor". The support and some parts of the brace are excluded from the mask depicted by the Mask RCNN trained model.

As shown in Figure 8d, the corroded support, beam flange, web, and braces are precisely detected by the YOLOv8 trained model. The dark background between the brace and the web stiffener is not included in the YOLOv8-predicted mask, while Mask RCNN detected it as "Poor" corrosion. Moreover, some small parts of the background concrete are included in the mask depicted by the Mask RCNN trained model. A small corroded area in the girder web is labeled as "Fair" by both models, which appears as less severe corrosion compared to the other parts.

Figure 9. (**a**) Bottom flange of a girder (**b**) bottom flange of a beam (**c**) lateral bracing (**d**) top lateral strut.

Figure 9a shows the bottom flange of a girder, which is delaminated due to corrosion. This corroded area is annotated as "Poor", but YOLOv8 and Mask RCNN labeled it as "Fair" with confidence of 89% and 62%, respectively. Both models provided a very precise mask of the corroded area since it is easily distinguishable from the background in this image.

As shown in Figure 9b, YOLOv8 was able to detect all instances that were annotated as "Fair", with a small mismatch in the detected mask. Both the YOLOv8 and Mask RCNN models correctly labeled the corroded parts of the girder flange. The corrosion mask provided by Mask RCNN is closer to the annotation, and it has higher confidence compared to YOLOv8.

Figure 9c illustrates a severely corroded beam with section loss, which was annotated as "Severe". The YOLOv8 trained model provided a precise mask of the severely corroded beam and labeled it correctly with high confidence (93%). A small part of the background concrete is also included in the mask. Mask RCNN mislabeled the severely corroded brace as "Poor" with confidence of 84%. A small portion of the beam is missing from the mask provided by Mask RCNN.

Figure 9d shows a severely corroded beam with section loss. The dark background of the image was not included in the annotation. The YOLOv8 trained model labeled the severely corroded area of the beam correctly with high confidence (90%). The model labeled a small portion of the dark background as "Severe", even though this area was not included in the annotation due to its invisible degree of corrosion. Similar to the previous image, the trained Mask RCNN model mislabeled the severely corroded area of the beam as "Poor" with confidence of 73%.

Analyzing the evaluation metric values on the validation and test data, as well as the predictions of the two trained algorithms on the test data, suggests that the YOLOv8 trained model excels in the precision and mAP50 values, providing accurate predictions in some cases. On the other hand, Mask RCNN demonstrates higher recall, capturing more relevant objects. If precision and the minimization of false positives are prioritized, YOLOv8 may be preferred. If the primary focus is comprehensive detection and recall, Mask RCNN might be the better choice.

6. Conclusions

In this study, 514 corrosion images with various degrees of severity were gathered from bridge structural elements, and pixel-wise annotation was performed on each image. The corroded areas were annotated into three classes: "Fair", "Poor", and "Severe". Our annotation process adhered to the guidelines outlined in the Bridge Inspectors Reference Manual (BIRM) and American Association of State Highway and Transportation Officials (AASHTO) regulations for corrosion condition rating (defect #1000). This annotation at the pixel level ensures that the trained algorithms can be directly applied to the corrosion condition rating process, offering practical application for engineers. The collected corrosion images and corresponding annotations can be found online and offer valuable support in advancing the segmentation models by providing high-quality images and corresponding annotations [25]. Two state-of-the-art segmentation algorithms, YOLOv8 and Mask RCNN, were chosen and trained on the labeled dataset to perform pixel-level corrosion segmentation and condition rating. The training process involved optimizing the hyperparameters to obtain the best weights of convolutional layers for accurate and reliable corrosion severity assessments. The trained Mask RCNN and YOLOv8 models achieved mAP50 values of 0.674 and 0.726 on the test images, which is an indication of the models' ability to perform semantic segmentation with a reasonable level of overlap with the ground truth annotations (IoU = 50%). These results indicate the efficacy of trained Mask RCNN and YOLOv8 models in segmenting and rating corrosion. The main highlights of this study can be summarized as follows.

- The applicability of these algorithms is broad and extends to various conditions. However, it is important to note that the performance of these methods may be influenced by the quality of images and factors such as the lighting conditions and the complexity of the corrosion patterns. It is advisable to ensure the quality of the images during data acquisition to optimize the performance of these segmentation algorithms.
- In terms of performance, we found that both the Mask RCNN and YOLOv8 models achieved satisfactory results in segmenting and rating corrosion. The YOLOv8 model had slightly higher mAP50 values on the validation (0.484) and test (0.726) datasets compared to Mask RCNN (0.483 and 0.674), indicating more accurate predictions while maintaining a good balance between avoiding false positives and capturing as many true positives as possible.
- Mask RCNN model achieved higher F1 score and recall values, which indicate that the model is adept at identifying most of the relevant instances in the dataset. Mask

RCNN excels in providing precise instance segmentation with detailed masks, making it suitable for objects with crucial boundaries.
- Mask RCNN has a two-stage architecture, which can be computationally intensive, impacting the real-time performance. Its training may require more computational resources compared to some single-stage approaches. YOLOv8 is designed for real-time object detection, making it efficient in scenarios where low latency is critical.
- In YOLOv8, the default task is object detection, which involves detecting and localizing objects in an image or video. Semantic segmentation, on the other hand, aims to label every pixel in an image with the corresponding class value. While YOLOv8 can be used for object detection, it is not designed specifically for semantic segmentation. It may face challenges in achieving pixel-level instance segmentation accuracy comparable to algorithms specifically designed for semantic segmentation, such as Mask RCNN. Its performance may be influenced by the sensitivity of certain hyperparameters, requiring careful tuning for optimal results.

In the future, to improve the accuracy of the corrosion condition rating model, we aim to obtain more diverse images with varying corrosion severity levels, lighting, and background conditions. Including a greater number of images depicting severe and poor corrosion instances will help to address the class imbalance problem and enhance the segmentation models' performance considerably.

Author Contributions: Conceptualization, Z.A. and E.N.L.; methodology, Z.A. and S.J.N.; software, Z.A. and S.J.N.; validation, Z.A. and S.J.N.; formal analysis, S.J.N.; investigation, Z.A.; resources, Z.A.; data curation, Z.A.; writing—original draft preparation, Z.A.; writing—review and editing, Z.A. and E.N.L.; visualization, Z.A. and S.J.N.; supervision, E.N.L.; project administration, E.N.L.; funding acquisition, E.N.L. All authors have read and agreed to the published version of the manuscript.

Funding: Funding for this research was provided by the Transportation Infrastructure Durability Center (TIDC) at the University of Maine under grant 69A3551847101 from the U.S. Department of Transportation's University Transportation Centers Program.

Data Availability Statement: The data presented in this study are available on request from the corresponding author. The data sources are publicly available and referenced in the study.

Acknowledgments: The authors are grateful to MaineDOT and VHB Consulting Engineers for providing infrastructure inspection images and for their valuable insights throughout this work. In particular, we acknowledge Robert Blunt and Benjamin Foster for their assistance with corrosion classification.

Conflicts of Interest: The authors declare no conflict of interest.

References

1. Han, X.; Yang, D.Y.; Frangopol, D.M. Risk-based life-cycle optimization of deteriorating steel bridges: Investigation on the use of novel corrosion resistant steel. *Adv. Struct. Eng.* **2021**, *24*, 1668–1686. [CrossRef]
2. Jiang, F.; Ding, Y.; Song, Y.; Geng, F.; Wang, Z. Automatic pixel-level detection and measurement of corrosion-related damages in dim steel box girders using Fusion-Attention-U-net. *J. Civ. Struct. Health Monit.* **2023**, *13*, 199–217. [CrossRef]
3. Savino, P.; Tondolo, F. Civil infrastructure defect assessment using pixel-wise segmentation based on deep learning. *J. Civ. Struct. Health Monit.* **2023**, *13*, 35–48. [CrossRef]
4. Han, Q.; Zhao, N.; Xu, J. Recognition and location of steel structure surface corrosion based on unmanned aerial vehicle images. *J. Civ. Struct. Health Monit.* **2021**, *11*, 1375–1392. [CrossRef]
5. Ameli, Z.; Aremanda, Y.; Friess, W.A.; Landis, E.N. Impact of UAV hardware options on bridge inspection mission capabilities. *Drones* **2022**, *6*, 64. [CrossRef]
6. Feroz, S.; Abu Dabous, S. UAV-based Remote Sensing Applications for Bridge Condition Assessment. *Remote Sens.* **2021**, *13*, 1809. [CrossRef]
7. Khan, M.A.-M.; Kee, S.-H.; Pathan, A.-S.K.; Nahid, A.-A. Image Processing Techniques for Concrete Crack Detection: A Scientometrics Literature Review. *Remote Sens.* **2023**, *15*, 2400. [CrossRef]
8. Rahman, A.; Wu, Z.Y.; Kalfarisi, R. Semantic deep learning integrated with RGB feature-based rule optimization for facility surface corrosion detection and evaluation. *J. Comput. Civ. Eng.* **2021**, *35*, 04021018. [CrossRef]
9. Khayatazad, M.; De Pue, L.; De Waele, W. Detection of corrosion on steel structures using automated image processing. *Dev. Built Environ.* **2020**, *3*, 100022. [CrossRef]

10. Zhou, Q.; Ding, S.; Feng, Y.; Qing, G.; Hu, J. Corrosion inspection and evaluation of crane metal structure based on UAV vision. *Signal Image Video Process.* **2022**, *16*, 1701–1709. [CrossRef]
11. Seo, H.; Raut, A.D.; Chen, C.; Zhang, C. Multi-Label Classification and Automatic Damage Detection of Masonry Heritage Building through CNN Analysis of Infrared Thermal Imaging. *Remote Sens.* **2023**, *15*, 2517. [CrossRef]
12. Petricca, L.; Moss, T.; Figueroa, G.; Broen, S. Corrosion detection using AI: A comparison of standard computer vision techniques and deep learning model. In Proceedings of the Sixth International Conference on Computer Science, Engineering and Information Technology, Qingdao, China, 29–30 May 2016; Volume 91, p. 99.
13. Atha, D.J.; Jahanshahi, M.R. Evaluation of deep learning approaches based on convolutional neural networks for corrosion detection. *Struct. Health Monit.* **2018**, *17*, 1110–1128. [CrossRef]
14. Jin Lim, H.; Hwang, S.; Kim, H.; Sohn, H. Steel bridge corrosion inspection with combined vision and thermographic images. *Struct. Health Monit.* **2021**, *20*, 3424–3435. [CrossRef]
15. Ronneberger, O.; Fischer, P.; Brox, T. U-Net: Convolutional Networks for Biomedical Image Segmentation. In Proceedings of the International Conference on Medical Image Computing and Computer-Assisted Intervention (MICCAI), Munich, Germany, 5–9 October 2015.
16. Chen, L.C.; Papandreou, G.; Kokkinos, I.; Murphy, K.; Yuille, A.L. DeepLab: Semantic image segmentation with deep convolutional nets, atrous convolution, and fully connected CRFs. *IEEE Trans. Pattern Anal. Mach. Intell. (TPAMI)* **2017**, *40*, 834–848. [CrossRef] [PubMed]
17. Zhao, H.; Shi, J.; Qi, X.; Wang, X.; Jia, J. Pyramid scene parsing network. In Proceedings of the IEEE Conference on Computer Vision and Pattern Recognition (CVPR), Honolulu, HI, USA, 21–26 July 2017.
18. Badrinarayanan, V.; Kendall, A.; Cipolla, R. SegNet: A deep convolutional encoder-decoder architecture for image segmentation. *IEEE Trans. Pattern Anal. Mach. Intell. (TPAMI)* **2017**, *39*, 2481–2495. [CrossRef] [PubMed]
19. He, K.; Gkioxari, G.; Dollár, P.; Girshick, R. Mask R-CNN. In Proceedings of the IEEE International Conference on Computer Vision, Venice, Italy, 22–29 October 2017; pp. 2961–2969.
20. Ta, Q.B.; Huynh, T.C.; Pham, Q.Q.; Kim, J.T. Corroded Bolt Identification Using Mask Region-Based Deep Learning Trained on Synthesized Data. *Sensors* **2022**, *22*, 3340. [CrossRef] [PubMed]
21. Nash, W.; Drummond, T.; Birbilis, N. Quantity beats quality for semantic segmentation of corrosion in images. *arXiv* **2018**, arXiv:1807.03138.
22. Ryan, T.W.; Mann, J.E.; Chill, Z.M.; Ott, B.T. Bridge inspector's reference manual (BIRM), Federal Highway Administration, National Highway Institute (HNHI-10), Report No.: FHWA NHI 12-050, February 2012. Available online: https://www.dot.state.mn.us/bridge/pdf/insp/birm/birmchapt0-cover.pdf, (accessed on 1 December 2023).
23. AASHTO. *Manual for Bridge Element Inspection*; American Association of State Highway and Transportation Officials: Washington, DC, USA, 2019.
24. Munawar, H.S.; Ullah, F.; Shahzad, D.; Heravi, A.; Qayyum, S.; Akram, J. Civil infrastructure damage and corrosion detection: An application of machine learning. *Buildings* **2022**, *12*, 156. [CrossRef]
25. Ameli, Z.; Lanis, E.N. Corrosion Condition Rating Database. 2023. Available online: https://digitalcommons.library.umaine.edu/student_work/30/ (accessed on 26 July 2023).
26. Facebook AI Research (FAIR). Detectron2: Next Generation Object Detection with PyTorch. 2023. Available online: https://github.com/facebookresearch/detectron2 (accessed on 15 June 2023).
27. Terven, J.; Cordova-Esparza, D. A comprehensive review of YOLO: From YOLOv1 to YOLOv8 and beyond. *arXiv* **2023**, arXiv:2304.00501.
28. Zheng, Z.; Wang, P.; Liu, W.; Li, J.; Ye, R.; Ren, D. Distance-IoU loss: Faster and better learning for bounding box regression. In Proceedings of the AAAI Conference on Artificial Intelligence, New York, NY, USA, 7–12 February 2020; Volume 34, pp. 12993–13000.
29. Li, X.; Wang, W.; Wu, L.; Chen, S.; Hu, X.; Li, J.; Tang, J.; Yang, J. Generalized focal loss: Learning qualified and distributed bounding boxes for dense object detection. *Adv. Neural Inf. Process. Syst.* **2020**, *33*, 21002–21012.
30. Jocher, G. ultralytics/COCO2YOLO: Improvements. Zenodo. 2019. Available online: https://zenodo.org/record/2738322 (accessed on 1 August 2023).

infrastructures

MDPI

Article

Application of Enhanced K-Means and Cloud Model for Structural Health Monitoring on Double-Layer Truss Arch Bridges

Chengzhong Gui [1,2,*], Dayong Han [3,4,*], Liang Gao [3,4], Yingai Zhao [3,4], Liang Wang [3,4], Xianglong Xu [3,4] and Yijun Xu [3,4]

1 Department of Civil Engineering, Institute of Disaster Prevention, Sanhe 065201, China
2 Key Laboratory of Building Failure Mechanism and Disaster Prevention, China Earthquake Administration, Sanhe 065201, China
3 Power China Road Bridge Group Co., Ltd., Beijing 100038, China; lqwangliang@powerchina.cn (L.W.)
4 Power China (Guangdong) Zhongkai Highway Co., Ltd., Jiangmen 529142, China
* Correspondence: 13115287@bjtu.edu.cn (C.G.); lqhandy@powerchina.cn (D.H.); Tel.: +86-13520890891 (D.H.)

Abstract: Bridges, as vital infrastructure, require ongoing monitoring to maintain safety and functionality. This study introduces an innovative algorithm that refines bridge component performance assessment through the integration of modified K-means clustering, silhouette coefficient optimization, and cloud model theory. The purpose is to provide a reliable method for monitoring the safety and serviceability of critical infrastructure, particularly double-layer truss arch bridges. The algorithm processes large datasets to identify patterns and manage uncertainties in structural health monitoring (SHM). It includes field monitoring techniques and a model-driven approach for establishing assessment thresholds. The main findings, validated by case studies, show the algorithm's effectiveness in enhancing clustering quality and accurately evaluating bridge performance using multiple indicators, such as statistical significance, cluster centroids, average silhouette coefficient, Davies–Bouldin index, average deviation, and Sign-Rank test *p*-values. The conclusions highlight the algorithm's utility in assessing structural integrity and aiding data-driven maintenance decisions, offering scientific support for bridge preservation efforts.

Keywords: bridge infrastructure; performance assessment; enhanced k-means clustering; silhouette coefficient; cloud model; structural health monitoring

Citation: Gui, C.; Han, D.; Gao, L.; Zhao, Y.; Wang, L.; Xu, X.; Xu, Y. Application of Enhanced K-Means and Cloud Model for Structural Health Monitoring on Double-Layer Truss Arch Bridges. *Infrastructures* **2024**, *9*, 161. https://doi.org/10.3390/infrastructures9090161

Academic Editors: Seyed Saman Khedmatgozar Dolati and Armin Mehrabi

Received: 13 August 2024
Revised: 3 September 2024
Accepted: 5 September 2024
Published: 12 September 2024

1. Introduction

1.1. Background and Significance

Bridges serve as pivotal nodes in transportation networks, with their safety being paramount to the stability of socio-economic systems. The double-layer truss arch bridge, a large-span structure often tasked with carrying significant transportation burdens, is subject to environmental erosion, traffic loads, and material aging. Therefore, regular structural performance assessments are crucial to ensure the safe operation of bridges, encompassing real-time monitoring data analysis, load-bearing capacity evaluation, and other critical aspects. Utilizing modern sensors and data processing technologies for health monitoring can enhance the overall performance and safety of bridges.

The introduction of an improved clustering-silhouette coefficient (SC) algorithm for the performance evaluation of double-layer truss arch bridges addresses this gap. By integrating advanced computational techniques, such as the modified K-means clustering algorithm, SC evaluation, and cloud model theory, this comprehensive algorithm offers a dynamic performance assessment tool for bridge health monitoring. It allows for the precise detection of stress anomalies and the assessment of the overall performance of bridge components, which is crucial for timely maintenance and the preservation of structural integrity.

The significance of this study lies in its potential to revolutionize bridge health monitoring practices. By providing a more accurate, data-driven approach to performance evaluation, it supports informed decision-making in bridge maintenance and management. The proposed method not only enhances the predictive capabilities of bridge health assessments but also contributes to the body of knowledge in structural engineering and computational intelligence, paving the way for smarter and safer infrastructure management.

In essence, the integration of an improved clustering-silhouette coefficient within the cloud model framework for bridge health monitoring is a monumental step towards proactive, intelligent systems that ensure the enduring safety and reliability of our bridge networks.

1.2. Literature Review

1.2.1. Current Study's Bridge Performance Assessment in Structural Health Monitoring

Current research methods of bridge performance assessment in structural health monitoring (SHM) have been evolving to incorporate advanced technologies and techniques, covering various aspects from visual evaluations to complex dynamic analyses. Researchers have proposed multiple methods to enhance the accuracy and efficiency of bridge assessments. For instance, in the field of Sustainable Building Design (SBD), Gharehbaghi et al. [1] proposed a holistic sustainable framework for rating the sustainable performance of buildings. This framework includes a multi-dimensional assessment method and compares case studies from Australia, the United Kingdom, and the USA. The study highlights the importance of incorporating sustainable practices in building design and assessment. Nondestructive Evaluation (NDE) tools and SHM techniques [2,3] play a crucial role in assessing the condition of bridges. The Federal Highway Administration (FHWA) also provides resources such as the Robotic Assisted Bridge Inspection Tool (RABIT™) and NDE protocols for bridge assessment [2]. Additionally, vibration-based SHM methods are being explored for accurate damage quantification in steel bridges [4]. The development of deep learning-based approaches for bridge condition assessment is gaining traction as an alternative to traditional inspection methods [5]. These approaches leverage SHM systems to improve the accuracy and efficiency of bridge performance assessment. Overall, the current research landscape in bridge performance assessment in SHM is focused on integrating advanced technologies, such as NDE tools, and deep learning methods to enhance the accuracy and reliability of bridge condition assessment [6]. The diversification and refinement of bridge assessment technologies have significantly improved the safety and durability of bridge structures. These studies provide a rich theoretical and practical foundation for bridge assessments. Future research in this field, such as the integration of clustering methods and cloud models, aims to further improve the performance assessment of bridges and ensure their structural safety and resilience, offering more reliable data support for bridge maintenance and management.

1.2.2. The State of the Art of Clustering, Quality Assessment, and Cloud Algorithms

Predictive maintenance is a common goal for bridge performance assessment in bridge health monitoring. By using clustering and quality assessment algorithms, bridge health monitoring systems can predict when maintenance is needed, thus preventing failures and extending the lifespan of the bridge. In bridge health monitoring, sensors collect data on structural integrity, while clustering and quality assessment algorithms process these data to identify patterns and anomalies, which largely rely on the machine learning technologies.

Machine learning techniques [7] can be categorized into three types: supervised learning, unsupervised learning, and semi-supervised learning. In supervised learning, algorithms generate functions that map inputs to outputs, with applications in classification and regression. Unsupervised learning deals with unlabeled input data and unknown outcomes, finding applications in clustering, dimensionality reduction, and association rule learning. Semi-supervised learning involves a mix of labeled and unlabeled instances.

As a machine learning method, clustering analysis algorithms are numerous and widely applied. K-means clustering [8] is an algorithm that requires a preset number of clusters. Although efficient, it is sensitive to noise and outliers. Gaussian Mixture Models (GMM) [9,10] are based on probabilistic models and can handle complex data distributions. Density-Based Spatial Clustering of Applications with Noise (DBSCAN) [11] can identify noise and outliers. Hierarchical Clustering (HC) [12] constructs a hierarchical tree structure and is suitable for small dataset analysis. Density Peak Clustering (DPC) [13,14] forms clusters by identifying density peaks. Kohonen clustering [15] obtained classification patterns in normal operating conditions and is straightforward for outlier detection. Spectral Clustering (SC) [16,17] uses graph theory and matrix operations, making it suitable for data with complex shapes. Self-Organizing Maps (SOM) [18] are a neural network-based clustering algorithm. By constructing a low-dimensional grid and adjusting weight vectors, it clusters input data samples into different grid units, suitable for clustering analysis, feature extraction, and anomaly detection. Affinity Propagation (AP) [19] achieves clustering through message passing without needing a preset number of clusters. Mean Shift Clustering (MSC) [20] is based on density gradients and is suitable for clustering non-convex shapes.

These clustering algorithms demonstrate their advantages in handling different types of data and meeting specific application needs, but they also have certain limitations. These limitations include sensitivity to parameters and noise, an inability to automatically determine the optimal number of clusters, poor performance with high-dimensional data, dependence on measurement scales, subjectivity in result interpretation, the impact of algorithm choice on results, efficiency issues in handling large-scale data, assumptions about data distribution, and the stability of clustering results. These drawbacks limit the performance and universality of clustering algorithms in different application scenarios. Researchers often use a combination of these machine learning methods or integrate other methods for targeted analysis. Specifically, for the application of machine learning methods in the performance evaluation of cables/suspenders, Shunlong Li et al. [21] proposed a state evaluation method based on machine learning. They used the GMM and Bayesian Information Criterion (BIC) to analyze the tension ratio of stay cables, thereby assessing their safety status, and ensured the evaluation was independent of external factors through preprocessing.

To evaluate the quality of these clustering methods, several approaches of Cluster Quality Evaluation Methods can be used as follows. The Elbow Method (EM) [22] seeks for the optimal number of clusters by calculating the sum of squared errors (SSE) for different numbers of clusters. The silhouette coefficient method [23,24] evaluates the cohesion within clusters and the separation between clusters. The higher the value is, the better the clustering quality is. The Davies–Bouldin index (DBI) [25] measures the separation between clusters and the compactness within clusters. Lower values indicate better clustering. The Rand Index (RI) [26] and Adjusted Rand Index (ARI) [27,28] evaluate the quality by comparing the clustering results with the true labels. A Principal Component Analysis (PCA) [29] can be used for the dimensionality reduction of multiple parameters. Cross-Validation [30,31] assesses the stability and generalization ability of clustering by splitting the dataset.

The cloud model (CM) theory was originally introduced by Prof. Deyi Li in 1995 [32], grounded in the principles of probability and fuzzy set theory. Since its inception, the theory has been continuously refined and expanded through subsequent studies. In the context of the normal cloud model, a decision-making method has been applied in the evaluation of distribution network planning, showcasing the versatility of cloud models in different domains [33]. Furthermore, the cloud model method has also been introduced in the evaluation of rock slope stability, addressing the issue of fixed weights in traditional fuzzy analysis methods [34]. Additionally, the cloud model is introduced to construct an assessment model to account for the randomness and fuzziness [35]. Also, the cloud model can be combined with the variable weight theory to determine the indicator's weight [36].

1.3. Research Gaps and Objectives

Despite the extensive application of clustering analysis, silhouette coefficients, and structural performance assessment techniques within their individual domains, their comprehensive integration into performance evaluations of bridge component systems via SHM, particularly within truss arch composite frameworks, remains rare in the current applications. Cluster analysis may suffer from dependency on initial conditions, leading to local rather than global optima, while the silhouette coefficient has limited interpretability for high-dimensional data, primarily being applicable to two-dimensional contexts. Additionally, its application of the silhouette coefficient may not always provide clear boundaries for cluster evaluation. To simplify calculations and enhance the clarity of the cluster quality assessment, an improved silhouette coefficient is proposed. This modification is necessary to provide a more intuitive and quantitative indicator for evaluating the relative position of each data point within its assigned cluster and the overall quality of the clustering.

From the above, this paper will propose an innovative method for assessing the performance of various components in double-layer truss arch bridges, including arch ribs, trusses, and hangers. The approach integrates an improved K-means clustering algorithm, a silhouette coefficient assessment, and the cloud model theory. It analyzes field health monitoring data, applies nonlinear structural analysis, and utilizes the enhanced K-means for data clustering. The silhouette coefficient and cloud model theory assess the clustering quality, ensuring the accurate detection of stress anomalies across components and a holistic evaluation of their operational performance. A case study validates the method's effectiveness for comprehensive bridge health assessment and its utility in bridge maintenance strategy formulation.

2. Methodology

2.1. Framework Construction for Performance Assessment of Suspender Cable System

The research framework in Figure 1 comprehensively addresses the assessment and monitoring of bridge performance. In this study, we defined the research scope, collected monitoring data, set evaluation thresholds, analyzed data preliminarily, identified key features with PCA, grouped data through clustering analysis, conducted assessment with silhouette analysis, and predicted structural safety using the cloud model considering data uncertainty.

The methodology section meticulously constructs a framework for evaluating the performance of the suspender cable system of double-layer truss arch bridge (see Figure 1a), introduces a refined K-means clustering approach (see Figure 1e), and details the enhancement techniques that bolster the silhouette coefficient (SC) indicator (see Figure 1f). It further elaborates on the development of a cloud model-based framework for assessing clustering quality and the strategies for setting assessment levels and executing multi-parameter indicator cloud clustering.

Transitioning into the application phase, the case study (see Figure 1a) of a double-layer truss arch bridge is presented, offering a detailed account of the bridge's characteristics, the finite element method-driven stress threshold determination (see Figure 1b), and the intricacies of data acquisition (see Figure 1c) for field monitoring. This practical application is pivotal in validating the theoretical constructs developed earlier in the paper.

The results and discussion section delves into the analysis of key components identified through characteristic analysis and the PCA method (see Figure 1d), followed by an eigenvalue analysis corresponding to each cluster level post-modification. It scrutinizes the impact of the improved clustering algorithm on the case bridge components and assesses the effects on cluster quality and performance metrics (see Figure 1e). The section culminates in a cloud model-based cluster quality assessment (see Figure 1f) of the suspender cables, evaluating the impact of various single-parameter cloud model (see Figure 1g) parameters on clustering quality.

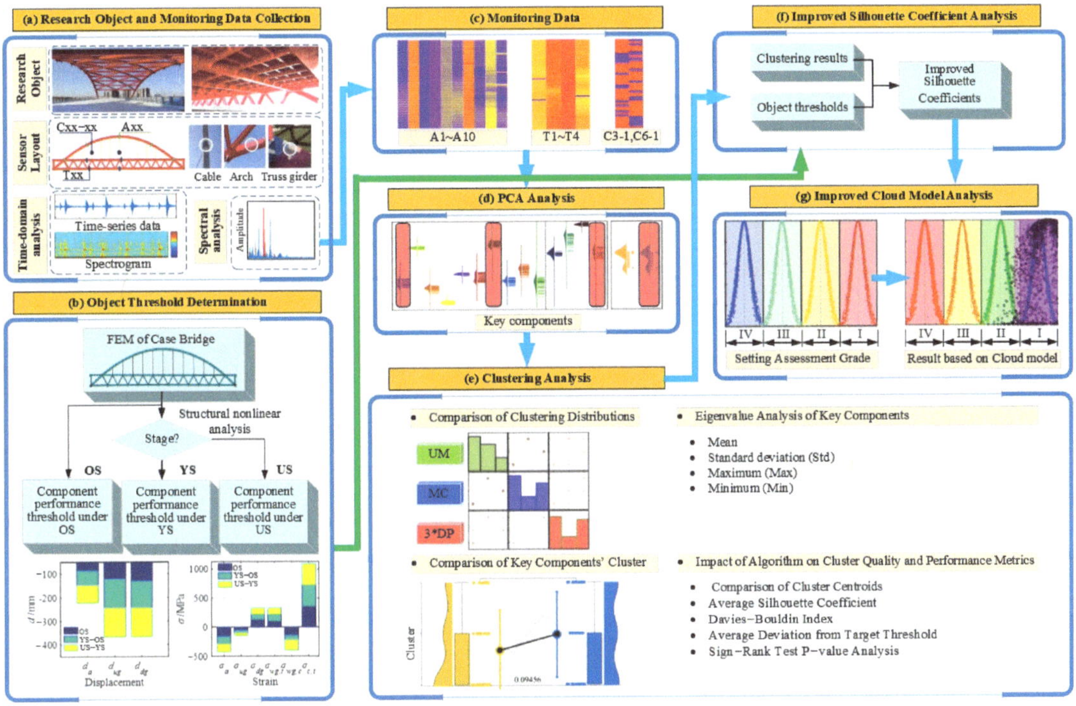

Figure 1. Overall research framework.

2.2. *Revised K-Means Clustering Method*

2.2.1. Description

The K-means clustering method is a process that begins with K initial cluster centers and refines them by reassigning data points based on Euclidean distance to the nearest center and then updating the center as the mean of its assigned points. This cycle repeats until the centers no longer change significantly, resulting in an optimized clustering where the sum of distances within each cluster is minimized, and an indicator function R efficiently classifies each data point into its respective cluster.

2.2.2. Enhancement Techniques

The enhanced K-means clustering algorithm is an iterative process designed to partition data into K distinct clusters. It begins by randomly selecting K clusters as initial centroids, $K = 3$ and $\mu_{k=}$ $[\mu_1, \mu_2, \ldots, \mu_K]$. After that, the raw data matrix $\sigma_p(t) = [\sigma_{p1}(t), \sigma_{p2}(t), \ldots, \sigma_{pn}(t)]$ was input and sorted into an ascending order matrix $x_p = [x_{p1}, x_{p2}, \ldots, x_{pn}]$. The algorithm then enters a loop, where each time-serial data point is assigned to the nearest centroid by using the Squared Euclidean Distance method [37] and determined by the expression $R_p(t) = \mathrm{argmin}\,dist(\sigma_{p,i}(t), \mu_K)$, and the original centroids are recalculated as the mean of their assigned data points. This process repeats until the centroids' positions change insignificantly, indicating convergence, or the algorithm reaches a predefined maximum number of iterations. The end result is a set of assigned labels and centroids that minimize the variance within each cluster, effectively organizing the data into distinct groups. Details can be seen in Appendix A.

2.3. *Enhanced Silhouette Coefficient Indicator*

2.3.1. Modifications to the Silhouette Coefficient

Assessing the goodness of clustering results was usually performed using some commonly used clustering metrics such as silhouette coefficient (SC) [23,38], Davies–Bouldin index (DBI) [39], and other indicators. The SC metric can be used to measure the similarity of each sample to other samples within its cluster and the degree of difference from other clusters. The application of these metrics provided an objective evaluation framework for clustering results, ensuring the rigor of the research and the reliability of the conclusions.

For the performance parameter stress state $\sigma_{pi}(t)$ of the i-th component at time t and the corresponding target threshold $\sigma_{pi,Obj}(t)$, the average SC $\bar{s}_i$ can be defined as follows:

$$\bar{s}_i = \sum_{t=t_0}^{t_{Total}} \left(\sigma_{pi,Obj}(t) - \sigma_{pi}(t)\right) / \max\left\{\sigma_{pi}(t),\ \sigma_{pi,Obj}(t)\ \right\} \quad (1)$$

where the value range from time point t_0 to t_{Total} is between $[-1, 1]$. The closer this metric is to 1, the better the clustering effect; the closer it is to -1, the worse the clustering effect; and a value close to 0 indicates overlapping samples. The SC provides a method for measuring clustering quality by considering the similarity and dissimilarity between samples.

To overcome these limitations, this paper proposed an improved strategy: by sorting the original data, the impact of randomness during system initialization and iteration can be effectively reduced, ensuring the consistency and accuracy of clustering results. This method improved the robustness of the K-means algorithm by optimizing the initialization process of cluster centers, making it more stable and effective when handling real data.

2.3.2. Construction of a Cloud Model-Based Clustering Quality Assessment Framework

To appraise the efficacy of the K-means clustering, the SC is a measure used to assess the quality of clustering, reflecting the cohesion and separation of clusters, which is also a standard metric for gauging clustering algorithm performance [23,38]. This indicator quantifies the affinity of each data point to its cluster peers and the divergence from points in alternate clusters. It is well suited for monitoring the variability in the mechanical behavior of structural elements—like arch ribs, trusses, and hangers—throughout the testing time serial t within the interval $[t_1, t_n]$. The SC effectively evaluates the clustering quality by measuring the alignment of component states with predefined thresholds.

The improved silhouette coefficient is calculated by considering both the cohesion within a cluster and the separation from neighboring clusters. The cohesion is measured by the average distance between data points within the same cluster, represented by $\lambda_i(t,k) = \sum_{t=t_0}^{t_n} dist(\mu_i(t_j,k), \mu_i(t_p,k))/(n-1)$, while separation is measured by the average distance to the nearest neighboring cluster center, denoted by $\beta_i(t,k) = \min_{j\neq k}\left(\sum_{l=1,l\neq k}^{K} dist(C_i(t,k), C_i(t,l))/(K-1)\right)$. The SC indicator $S_i(t,k) = \left(\mu_{i,Obj}(k) - \lambda_i(t,k)\right)/\max\left(\mu_{i,Obj}(k), \lambda_i(t,k)\right)$ is then computed using these values with a range of $[-1, 1]$, where values close to 1 indicate good clustering quality.

To simplify and make the silhouette coefficient more accessible, the separation component is set to a fixed value, ensuring clear cluster boundaries and good clustering effects. The modified SC $S'_i(t, k)$ is then adjusted to account for this fixed separation threshold, providing a clear assessment of the suitability of the stress dataset of the i-th hanger in cluster k at time t.

$$S'_i(t,k) = (S_i(t,k) - \min(S_i(t,k)))/(\max(S_i(t,k) - \min(S_i(t,k)) \quad (2)$$

Finally, to align with standard cloud model parameter intervals and reflect clustering effects more intuitively, the SC is normalized to a range of [0, 1] through a normalization process. The modified SC, $S'_i(t, k)$, provides a quantitative measure where values closer to

1 indicate better clustering effects, and values closer to 0 indicate poorer clustering effects. This normalized coefficient offers a clear and direct assessment of the clustering quality for the stress dataset of each hanger over the time period from t_1 to t_n.

2.4. Methods for Setting Assessment Levels and Multi-Parameter Indicator Cloud Clustering

2.4.1. Setting Assessment Grading Standards

In accordance with the activation mechanism of the positive cloud generator [40,41], the SC cluster quality assessment intervals and their respective standard cloud model parameters were established. Leveraging current experience, the cluster quality is categorized into four evaluation grades: Grade I, Grade II, Grade III, and Grade IV, corresponding to four distinct percentile ranges of the adjusted silhouette coefficient $S'_i(t, k)$, namely (75, 100], (50, 75], (25, 50], and (0, 25] (see in Figure 2).

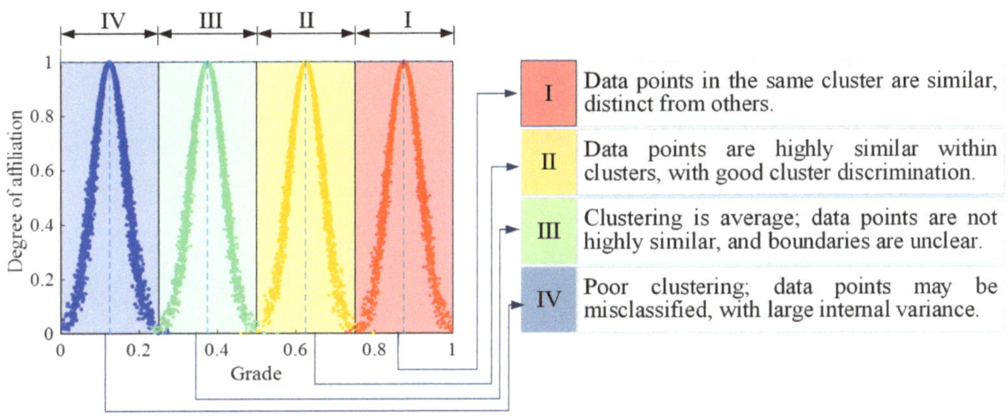

Figure 2. Score ranges and qualitative description for each level of clustering effect.

For the SC $S_i'(t, k)$ across these four percentile ranges, the standard cloud model's characteristic parameters are determined using a normal distribution, including the expected value EX_k, entropy EN_k, and hyper-entropy HE_k. The expression can be determined by EX_k= $(\overline{\gamma}_{k,max} + \overline{\gamma}_{k,min})/2$, EN_k= $(\overline{\gamma}_{k\cdot max} - \overline{\gamma}_{k,min})/6$, and $HE_k = 1/250$, with k ranging from 1 to 4. Here, $\overline{\gamma}_{k,max}$ and $\overline{\gamma}_{k,min}$ denote the maximum and minimum scores within each level's range. The EX value indicates the midpoint of the SC for each range, signifying the central tendency of the coefficient. Accordingly, the EX value for the standard cloud model characteristic parameters corresponding to the four SC intervals are 0.875, 0.625, 0.375, and 0.125. The entropy value EN, which mirrors the fluctuation within the range, is uniformly set at 1/24. The hyper-entropy value HE, reflecting the uncertainty of the entropy, is uniformly set at 0.004.

2.4.2. Cloud Clustering Quality Assessment of Single-Parameter Indicator

Based on the trigger mechanism of the reverse cloud generator [41], the cloud clustering quality assessment corresponding to the single-parameter indicator monitoring data silhouette coefficient $Si'(t,k)$ can be calculated as follows:

$$EX_i = \frac{1}{n}\sum_{j=1}^{n}\overline{\gamma}_{ij} \tag{3a}$$

$$EN_i = \sqrt{\frac{\pi}{2}}\cdot\frac{1}{n}\left|\overline{\gamma}_{ij} - EX_i\right| \tag{3b}$$

$$HE_i = \sqrt{\Lambda_i^2 - EN_i^2} \tag{3c}$$

where $\overline{\gamma}_{ij}$ represents the silhouette coefficient, EX_i is the expectation of the parameter indicator cloud, EN_i is the entropy of the indicator cloud, HE_{ij} is the hyper-entropy of the indicator cloud, and Λ_i is the standard deviation of the score for the ith indicator and $\Lambda_i \geq EN_i$.

To assess the overall cloud clustering quality, the average silhouette coefficient provided a comprehensive measurement of the clustering quality for the entire system at a given time. In cases where the reverse cloud generator operates, triggering based on the clustering quality, the mechanism would adjust the parameters to optimize the silhouette coefficient, thereby enhancing the clarity and separation of the clusters.

2.4.3. Comprehensive Cloud Clustering Quality Assessment of Multi-Parameter Indicator System

Based on the calculated parameters of the indicator clouds and combined with the corresponding indicator weights, the comprehensive cloud was obtained, with the specific formula being

$$EX_C = \sum_{i=1}^{m} (EX_{C,i} \cdot \omega_i) / \sum_{i=1}^{m} \omega_i \tag{4}$$

$$EN_C = \sum_{i=1}^{m} (EN_{C,i} \cdot \omega_i^2) / \sum_{i=1}^{m} \omega_i^2 \tag{5}$$

$$HE_C = \sum_{i=1}^{m} (HE_{C,i} \cdot \omega_i^2) / \sum_{i=1}^{m} \omega_i^2 \tag{6}$$

where ω_i is the weight assigned to the single-parameter indicators. Once the comprehensive assessment cloud was calculated, it was compared with the standard assessment cloud maps to identify the standard cloud map that bore the closest resemblance to the comprehensive assessment cloud. This comparison was then used to determine the clustering quality assessment level of the existing bridge.

3. Determination Method of Component Stress State and Structural Load-Bearing Threshold

3.1. Determining Method of Stress State Based on Field Monitoring

The field monitoring methodology for arch ribs and truss girders can be ascertained through the use of strain gauges. The readings from these gauges, when multiplied by the elastic modulus of the steel materials, provide a direct measurement of stress within the structures. This approach eliminates the need for complex calculations, offering a straightforward and reliable method to assess the structural integrity and performance under various load conditions. By continuously monitoring the strain data, engineers can identify potential issues early, ensuring timely maintenance and a high level of safety for the bridge.

For the cable system of the ith cable with unit length mass ρ_i, cable length l_i, material modulus of elasticity E_i, and section moment of inertia I_i, the time-domain signal measured by sensors is $x_i(t)$. By applying the fast Fourier transform (FFT), this signal is converted into the frequency-domain signal $F_i(\omega)$ at any time t. The fundamental frequency of the ith cable, representing the cable tension vibration frequency, can be denoted as $f_{i1}(t)$.

Assuming the cable is simply supported at both ends and has a small vertical span ratio, the cable vibrates only in the vertical plane. Using classical beam theory and neglecting bending or shear effects on the cross-section, the tension force $T_{C,i}$ of the ith cable can be expressed as $T_{C,i}(t) = 4\rho_i l_i^2 f_{1,i}^2(t) - \pi^2 E_i I_i / l_i^2$, referred to from Gui et al. [42] and Lei et al. [43], and can be calculated with its boundary conditions requiring both ends of the cable to have zero displacement and curvature. Since the tension force and frequency change inconsistently under the same temperature conditions due to the initial horizontal tension, the effect of temperature is disregarded [44], and the elongation rate change of the cable before and after loading is minimal.

When the cable's bending stiffness EI is low, the term $\pi^2 EI/l^2$ can be omitted from the formula. Accordingly, the stress $\sigma_{pi}(t)$ of the ith cable can be simplified to $\sigma_{pi}(t) = T_{C,i}/A_{C,i}$, where $A_{C,i}(t)$ is the cross-sectional area of the cable, assumed to be constant regardless of changes due to external forces or corrosion. The cable stress $\sigma_{C,i}(t)$ should not exceed the cable material's ultimate tensile strength $f_{u,C}$. The stress state matrix $\sigma_C(t)$ for n cables is thus $\sigma_C(t) = [\sigma_{C,1}(t_m), \ldots, \sigma_{C,n}(t_m)]$. For newly constructed bridges, the degradation of cable material properties need not be considered.

3.2. Model-Driven Target Threshold Determination

3.2.1. Explanation of the Model-Driven Method

The model-driven method is a sophisticated approach used in the performance assessment of truss arch composite systems, where the complexity of the structural ultimate state arises from the different degrees of redundancy in the forces experienced by each component. This method aims to ensure the safety and reliability of the entire structure by considering the simultaneous attainment of the ultimate state by all critical components, such as arch ribs, main trusses, and hangers.

In an ideal scenario, the model-driven method envisions that all key structural elements will reach their ultimate state concurrently. This simultaneous occurrence is crucial for maintaining the structural integrity and for preventing the failure of any single element that could compromise the entire system.

However, in practical situations, certain components, particularly hangers, may reach their ultimate state earlier than others due to material yield or the limitations of ultimate strength. This premature reaching of the ultimate state by some elements can lead to the entire bridge structure attaining its limit load-bearing capacity, while other components might still be operating well below their own ultimate states.

The model-driven method addresses this complexity by employing advanced modeling techniques and nonlinear analysis to predict and evaluate the behavior of each component under extreme conditions. It uses models such as the Von Mises bilinear model to simulate material nonlinearity and to determine the ultimate load-bearing capacity ($R_{u,max}$) of the bridge. This capacity is then used to ascertain the individual capacities of the main truss, arch ribs, and hangers, ensuring that the design accounts for the potential variability in the performance of different components.

By iteratively refining the model and adjusting parameters, the model-driven method converges towards a solution that closely represents the actual structural behavior. This approach is essential for the accurate assessment of the bridge's load-bearing capacity and for ensuring that the structure can safely accommodate the design loads without surpassing the critical thresholds of any of its components.

3.2.2. Determining the Target Threshold for the Suspender Cable System

To determine the bridge's capacity under extreme conditions, it was imperative to integrate considerations of geometric and material nonlinearities. Establishing the target threshold for the hanger system was a vital approach guided by model-driven methods. By employing material nonlinear models such as the Von Mises bilinear model [45], the bridge's ultimate load-bearing capacity, denoted as $R_{u,\max}$, was derived from structural nonlinear analysis, as illustrated in Equation (7a), which is placed in Table 1. This capacity was then leveraged to ascertain the individual load-bearing capacities of the main truss, arch ribs, and hanger system, as shown in Equation (7b). These steps constituted an iterative refinement process, where adjustments and corrections to the model were made to closely simulate the actual structural behavior.

Table 1. Calculation process of threshold determination of ultimate bearing capacity of components.

NO.	Step	Details Description
1	Determine structural model	Construct an appropriate stability calculation model based on the actual conditions of the bridge.
2	Define loading conditions	Identify the load conditions that the bridge must withstand, including dead weight, live load, and temperature effects.
3	Define nonlinear models	Consider both material nonlinearity and geometric nonlinearity of the structure to predict potential states of nonlinear failure or collapse.
4	Conduct nonlinear stability analysis	Utilize numerical analysis methods, such as the finite element method, to account for material nonlinearity and geometric nonlinearity. Gradually increase the load to analyze the structural response and assess stability.
5	Assess limit states	Determine if the bridge is at a limit state based on the analysis results. Evaluate the limit state of the structural system by analyzing parameters such as displacement, stress, and strain. $R_{u,\max}(\sigma_j, d_j) = \mathrm{argmin} R(\sigma_{\max,j}, d_{\max,j})$ (7a) $R_{u,\max}(\sigma_j, d_j) \to \sigma_{u,j}, d_{u,j}$ (7b)
6	Establish target limit thresholds for components	Establish the limit thresholds for each component's performance parameters based on the identified limit states of the structural system.

In the equations, $\sigma_j = (\sigma_T, \sigma_A, \sigma_C)$ represent the stresses in the main truss, arch ribs, and hangers, respectively, while $d_j = (d_T, d_A, d_C)$ denote the corresponding displacements. The term "max" signifies the maximum value of stress or displacement, and "u" indicates the ultimate state. Equation (7a) selects the minimum of the maximum strains among the main truss, arch ribs, and hangers to define the ultimate state of the bridge structure system, ensuring equilibrium and stability when each key component reaches its ultimate capacity. Equation (7b) calculates and determines the theoretical ultimate load-bearing capacities for each main truss, arch rib, and hanger when the bridge structure system is deemed to have reached its theoretical ultimate state. This process encompassed structural analysis and simulation, utilizing sophisticated numerical methods to predict the actual behavior and load-bearing capacity of each component under extreme loading conditions.

4. Case Study: Application in Double-Layer Truss Arch Bridges

4.1. Description of the Selected Bridge

To implement the algorithmic model introduced in this manuscript and to scrutinize the viability and efficacy of the proposed safety monitoring approaches, performance assessment metrics, and maintenance strategies, a fully constructed double-layer steel truss arch bridge (see Figure 3a,b) in Zhongshan City, Guangdong Province, has been selected for study. The bridge's primary truss features an all-welded integral node plate truss configuration, with a span of 153 m. Two primary trusses are positioned transversely, separated by 37.3 m, and the central lines of the upper and lower trusses exhibit a height difference of approximately 10 m (see Figure 3b), with a standard inter-segment distance of 10.6 m. The arch ribs are designed with a box-shaped cross-section reinforced by ribs, fully welded, and shaped in a parabolic form, with a central span of 150 m and a rise of 37.5 m, achieving a rise-to-span ratio of 1/4. The bridge deck is composed of an orthotropic steel deck, ultra-high-performance concrete (UHPC), and a SMA13 asphalt concrete surfacing layer. The bridge is equipped with 22 suspender cables, specified as 2 × PES7-91, with an ultimate tensile strength of 1770 MPa and an elastic modulus exceeding 1.9×10^5 MPa. The suspender cables are pin-jointed at both ends, encapsulated with PE protection and fitted with magnetic flux sensors that feed data into a data acquisition system. During the bridge's completion phase, all suspender cable vibration frequencies were tested. In the operational phase, magnetic flux sensors, data acquisition systems, and solar power systems were installed at the mid-span and quarter points to monitor and record the magnetic flux variations of the suspender cables in real time, which are then translated into suspender cable sectional strains using theoretical formulas.

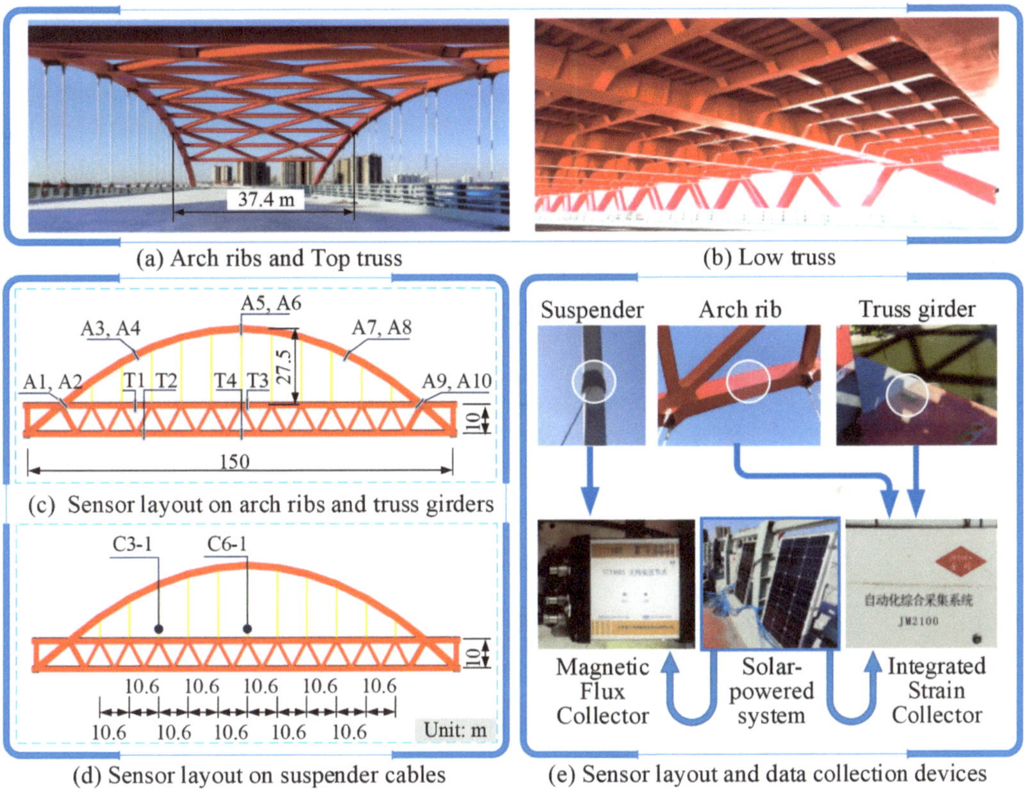

Figure 3. The tested example of the double-layer truss tied-arch bridge.

In Figure 3c–e, strain sensors and magnetic flux sensors are strategically placed on the bridge's main truss, arch ribs, and suspender cables, with four strain sensors on the main truss cross-section, ten on the arch rib cross-section, and two magnetic flux sensors on the suspender cable cross-section. The specific arrangement in Table 2 is as follows: strain sensors are positioned at $L/4$ and $L/2$ along the main truss, with two at the bottom of the upper and lower cross-sections at the mid-span and one at the quarter point; strain sensors are also placed at the arch foot and $L/4$ and $L/2$ on the arch ribs, with two at each of the quarter points and the apex of the arch ribs. This sensor arrangement is designed for the real-time monitoring and documentation of the bridge's data, providing a foundation for the assessment and maintenance of the bridge's structural integrity.

Table 2. Sensor placement, quantity, and code installed on the test bridge.

Member	Sensor Types	Test Point Code	Locations
Arch rib	Strain sensors	A1, A2, A3, A4, A5, A6, A7, A8, A9, A10	See Figure 3c
Truss girder	Strain sensors	T1, T2, T3, T4	See Figure 3c
Suspender cable	Magnetic Flux sensors	C3-1, C6-1	See Figure 3d

4.2. FEM-Driven Stress Objected Thresholds for Bridge Components

4.2.1. Description of Finite Element Model

In this case study, the main truss and steel bridge deck are constructed using Q345qD material, while the arch ribs are fabricated predominantly from Q420qD and Q345qD. The hangers are characterized by a nominal yield strength of 1670 MPa and an ultimate tensile strength of 1770 MPa (see Table 3). To capture the nonlinear behavior of these materials, a Von Mises bilinear model is applied across all three. For the modeling of the arch ribs and the main truss, beam elements are utilized to represent their cross-sections, whereas the hangers are represented by tension-only truss elements, which more accurately reflect their load-bearing characteristics and nonlinear responses.

Table 3. Material properties.

Material	f_y/MPa	f_u/MPa	Location	Reference
Q420qD	420	540	From arch foot to 1/4*L* and 3/4*L*	[46]
Q345qD	345	490	From top to 1/4*L* and 3/4*L*, Truss girder	[46–48]
C1670	1670	1770	Suspender cables	Design files

This study delves into the ultimate load-bearing capacity of bridges in the elastoplastic phase and evaluates the mechanical performance of the cable/stay system under extreme conditions through the application of a structural nonlinear analysis. The methodology includes steps such as an elastic buckling analysis, nonlinear data integration, parameter refinement, detailed computational simulations, and the derivation of load–displacement curves. By imposing a unit load on the bridge surface, the study determines the elastic buckling eigenvalues and eigenvectors and subsequently accounts for material and geometric nonlinearities to ascertain the critical load of a double-layer truss arch bridge. The bridge is deemed to have reached its ultimate load-bearing capacity when the load–displacement or load–strain curves exhibit a stable plateau.

Employing MIDAS CIVIL software for elastoplastic buckling analysis, the study comprehensively addresses the geometrric nonlinearity due to significant deformations and factors such as the material's yield strength. An incremental arc-length method is applied to ensure the precision of the computational outcomes. This approach offers a meticulous framework for comprehending the behavior of bridges under diverse loading conditions and corroborates the structural response via finite element computations, as depicted in Figure 4.

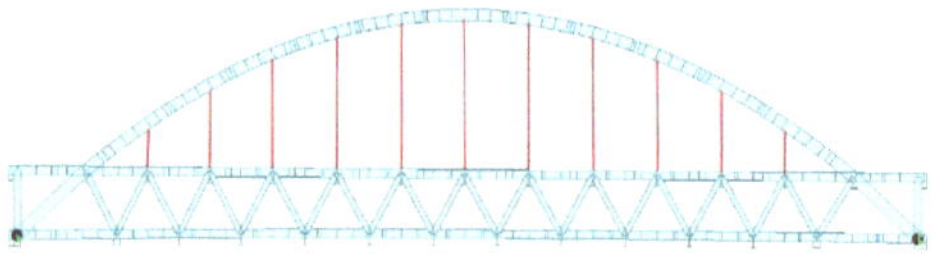

(**a**)

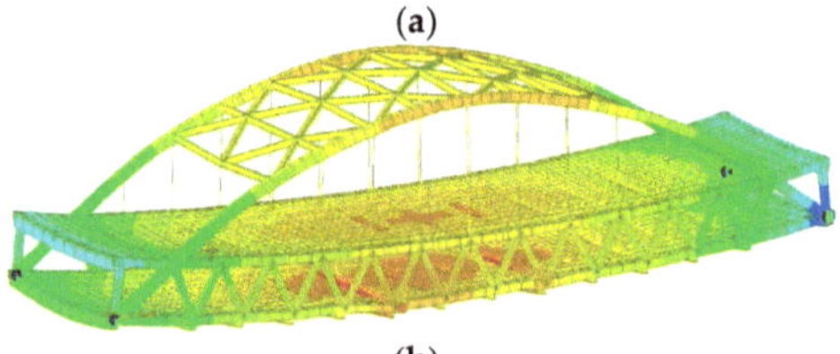

(**b**)

Figure 4. The ultimate stress state after the nonlinear analysis of the tested bridge structures. (**a**) The finite element model. (**b**) The ultimate stress state.

4.2.2. Verification of the Proposed Method's Effectiveness and Practicality

Figure 4 presents the finite element analysis results, highlighting that the transverse rib section of the upper main truss in the bridge is the first to exhibit local instability as it reaches the material's yield strength and, subsequently, its ultimate tensile and compressive strengths. These conditions are critical for assessing the nonlinear ultimate bearing capacity of the bridge's materials. Stress and displacement levels of the upper structural components of the double-layer truss arch bridge are derived from these states.

Figure 5 details the displacement and stress conditions for the arch ribs, chords, web members, and suspender cables of the bridge in three states: as built (OS), material yield (YS), and ultimate bearing (US). The ultimate state calculations indicate that while the transverse rib section of the upper main truss reaches yield and ultimate strength first, the arch rib section experiences the maximum compressive stress of −425 MPa at the bridge's ultimate bearing capacity. The web member section shows tensile and compressive strengths of 340 MPa and −397 MPa, respectively, exceeding the material's yield strength.

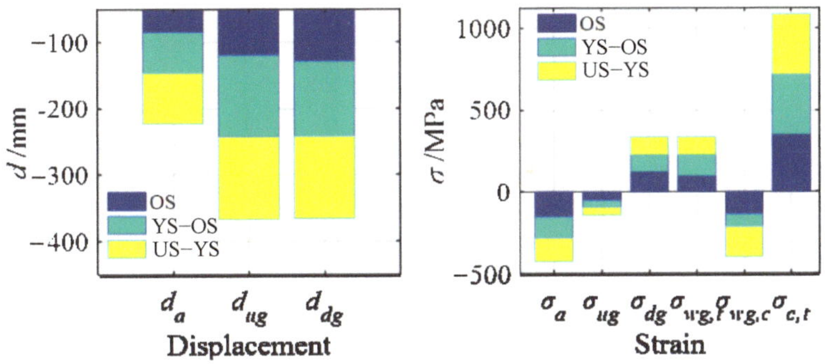

Figure 5. Maximum displacements and stresses of bridge members in OS, YS, and US stage. Note: d_a, d_{ug}, and d_{dg} represents the displacement of arch ribs, upper chord, and low chord, respectively; σ_a, σ_{ug}, σ_{dg}, $\sigma_{wg,t}$, $\sigma_{wg,c}$, and $\sigma_{c,t}$ represents the strain on arch ribs, upper chord, low chord, web member, and suspender cable, respectively.

The maximum displacements for the arch ribs, upper chords, and lower chords are recorded as −222.29 mm, −366.78 mm, and −365.52 mm, respectively. In the ultimate state, the maximum deflection-to-span ratios for these components are 1/410, nearing the specification limit of 1/500 for steel truss bridges, validating the reasonableness of the bridge's ultimate state determination.

Comparing displacement states across OS, YS, and US reveals the greatest vertical displacements in truss members, with the upper and lower chords showing similar magnitudes. The stress analysis across these states shows the highest tensile stress and largest variation in the suspender cables. The arch ribs exhibit the highest absolute stress values, increasing from 151 MPa in OS to 281 MPa in YS and peaking at 425 MPa in US. A consistent stress increase is observed in the chords and web members, reflecting the growing internal forces as the bridge approaches material limits.

For the suspender cables C3-1 and C6-1, the maximum stress values are 351.2 MPa, 717.4 MPa, and 1087 MPa in the OS, YS, and US stage, respectively. These values are crucial for establishing threshold ranges for each cable under different states. Exceeding these thresholds may indicate cable failure or the bridge reaching its ultimate capacity, which is vital for structural health and safety assessments, guiding maintenance and reinforcement measures.

4.3. Data Acquisition to Bridge Field Monitoring

4.3.1. Data Acquisition and Spectral Analysis of Suspender Cables

During the time interval from 00:14:19 to 23:12:43 on 23 July 2023, actual vibration time-domain data were collected for two suspension cables, designated as C3-1 and C6-1, situated at the mid-span and quarter points of the arch ribs. A Fourier transform approach was applied to perform a spectral analysis, identifying the maximum peak value of the vibration frequency at each time point as the representative frequency for the respective suspension cables.

Figure 6a–d display the measured vibration frequency time-domain data and corresponding spectrograms for the suspension cables C3-1 and C6-1 at four specific time points.

Figure 6e illustrates the outcomes of the spectral analysis, indicating that the vibration frequencies for both suspension cables fluctuate within the range of 5.344 to 10.562 Hz.

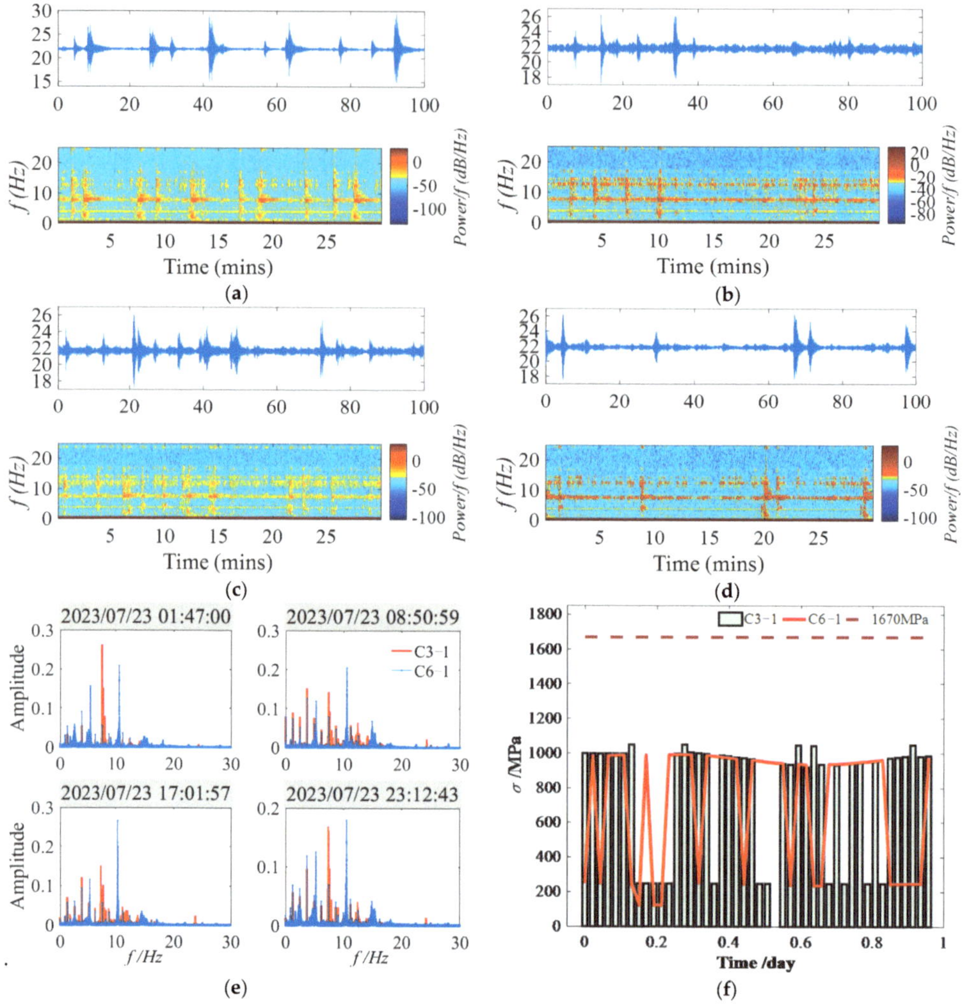

Figure 6. Partially measured vibration frequencies and measured stresses of suspender cable C3-1 and C6-1. (**a**) 23 July 2023 01:47:00 (**b**) 23 July 2023 08:50:59. (**c**) 23 July 2023 17:01:57. (**d**) 23 July 2023 23:12:43. (**e**) Spectrograms at 4 time points. (**f**) Measured stresses.

Through a spectral analysis, the actual cable tensions and the associated stress variations of the suspension cables were ascertained. Figure 6f demonstrates how the measured vibration frequencies of the C3-1 and C6-1 suspension cables were converted into stress variations. The mid-span suspension cable C6-1 of the arch rib has a corrected length of 22.515 m, accounting for a 2.5 m deduction for the anchor head, while the quarter-point suspension cable C3-1 has a corrected length of 15.875 m with the same deduction. During the testing interval, the recorded maximum stress values for C3-1 and C6-1 suspension cables were 1049.94 MPa and 992.26 MPa, respectively, and the minimum values were 247.57 MPa and 123.07 MPa, respectively, indicating a notable disparity between the extremes.

Upon comparing the maximum measured cable tensions of C3-1 and C6-1 with their theoretical load-bearing capacities of 1085 MPa and 1083 MPa, respectively, it is observed that the measured tensions approach the theoretical limits. Conversely, the minimum measured tensions significantly deviate from the theoretical capacities and are substantially below the material's ultimate tensile strength of 1770 MPa.

4.3.2. Monitored Data of Bridge Components

The data selected for the research time period in this paper spans from 00:00:00 on 23 July 2023, to 23:59:59 on the same day. The sensors on the bridge primarily consist of strain sensors and fluxgate sensors, with specific installation locations and quantities detailed in Figure 3 and Table 2. The sensors are powered and transmitted through a data acquisition system and solar power supply system installed on the outer side of the bridge deck at the mid-span.

The strain data collected from the arch ribs and main trusses are all obtained with vibrating wire strain gauges. Sensors are installed at different positions on the arch ribs, main trusses, and cables, with 10, 4, and 2 sensor measurement points, respectively. The monitored strain data are plotted in Figure 7.

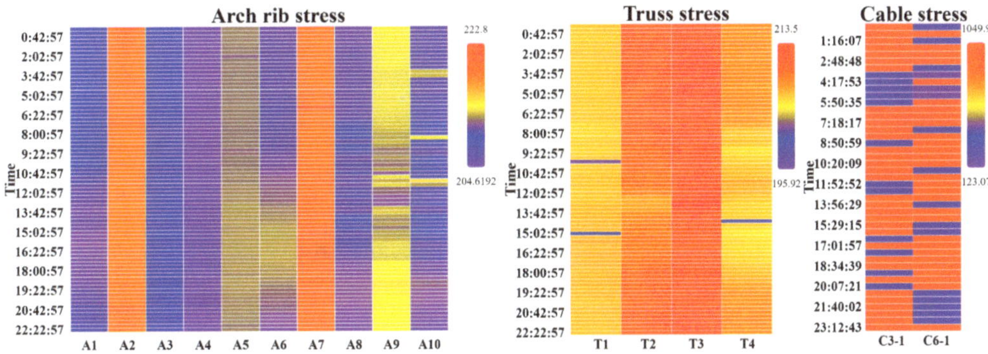

Figure 7. The monitored strain sensor of the case bridge.

5. Results and Discussion

5.1. Selection of Key Components Based on Characteristic Analysis and PCA Method

5.1.1. Characteristic of Monitoring Parameters Based on Cloud Rain Diagram

For the 10 arch rib strain measurement points (A1 to A10), 4 main truss section strain measurement points (T1 to T4), and 2 cable force measurement points (C3-1 and C6-1) on the double-layer truss arch bridge, the cloud rain diagram in Figure 8 offers a comprehensive view of the monitoring data variations in Figure 7 across the double-layer truss arch bridge, identifying several key observations:

The strain measurements at points A1 through A9 are notably consistent with minimal outliers, yet point A10 stands out with a higher frequency of extreme values within its dense data region, hinting at potential irregularities.

While T1 and T4 show some outliers within their dense data regions, possibly pointing to specific issues or variability, the other two measurement points on the truss section maintain a stress variation within a narrow 10 MPa range, indicating a general stability.

The monitoring data for C3-1 and C6-1 are centered in two separate dense regions, revealing a bimodal distribution. This pattern suggests the influence of different factors, such as loading conditions, material properties, or structural configurations, on these cables.

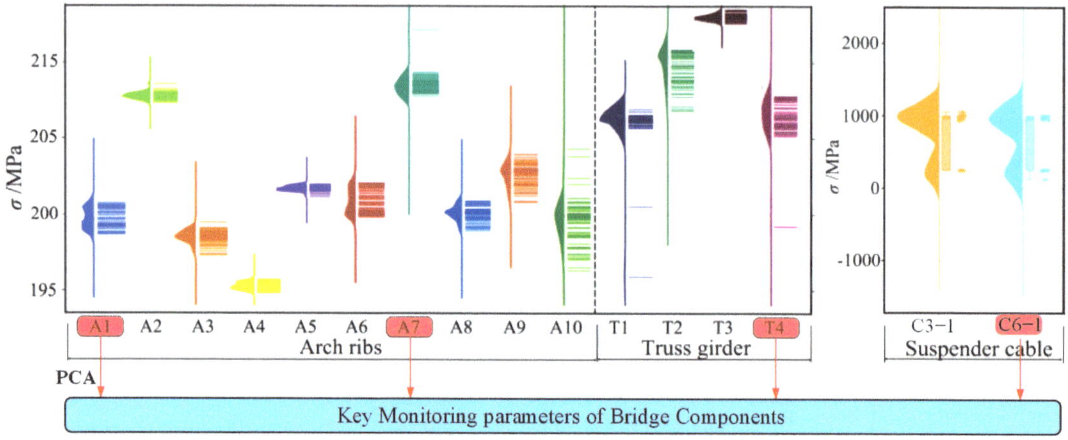

Figure 8. Cloud rain diagram and key monitoring parameters selection of components using PCA method.

The cable stress demonstrates greater sensitivity to changes under identical loads compared to the arch rib and truss section stresses, indicating a more dynamic response to external forces.

Despite the presence of outliers and the bimodal distribution in the suspender cables, the overall structural performance is stable, as indicated by the majority of the monitoring data. However, these exceptions necessitate a more in-depth examination to uphold the bridge's safety and reliability.

In essence, the monitoring data from the majority of points signals a reliable structural performance, yet the identified outliers and the distinct bimodal distribution in the suspender cables demand closer scrutiny to ensure the structural integrity of the bridge. The heightened sensitivity of the cable stress to load variations emphasizes the need for vigilant monitoring and analysis, which should be considered in the selection of key bridge components.

5.1.2. Selection of Key Components

The PCA method was applied to identify key measurement points on the arch ribs and main trusses as well as between the arch ribs and the cables. The results are illustrated in Figure 8.

Notably, the strain measurement point on the upstream side of the left arch foot section (A1), the strain measurement point on the upstream side of the left arch rib at the 3/4 span (A7), the strain measurement point on the upstream side of the main beam at mid-span (T4), and the position of cable C6-1 are all identified as critical measurement points.

The arch foot and mid-span areas are critical regions where stress concentrations are likely to occur due to the bending moments and shear forces. The 3/4*L* span location on the arch rib is a zone where the load distribution changes, affecting the overall stress pattern in the bridge. The cable force measurement points, such as C6-1, are crucial for understanding the tension forces in the cables, which are essential for the stability and safety of the bridge. Therefore, the application of a PCA to these measurement points ensures that the most representative and informative data are strategically rational for analysis, which is essential for the accurate assessment of the structural integrity and performance of the double-layer truss arch bridge.

5.2. Eigenvalue Analysis of Key Components after Modified Cluster

In this part, the K-means clustering analysis was applied to the stress data of key components during a specific testing period. The number of clusters was set to 3, with a maximum of 100 iterations, and the algorithm was terminated when the centroid movement was less than 1×10^{-4}. The algorithm ultimately returned the cluster labels for each data point and the coordinates of the cluster centers.

Table 4 provides statistical eigenvalue analysis results for the truss girder, arch ribs, and suspender cables, segmented by cluster level after a modified cluster. The mean, standard deviation (Std), maximum (Max), and minimum (Min) values are reported for each level, reflecting the distribution and variability of eigenvalues. The eigenvalues for the suspender cables show a substantial range, particularly at cluster level 2, where the standard deviation is markedly higher than in other components. The maximum eigenvalue for the suspender cables peaks at 992.26, while the minimum is notably lower at 123.07, suggesting a wide spectrum of dynamic responses within the structural system.

Table 4. Eigenvalue analysis results of key components corresponding to each cluster level/MPa.

Eigenvalue	Cluster Level	Arch Ribs		Truss Girder	Suspender Cable
		A1	A7	T4	C6
Mean (Std)	1	209.089 (1.051)	218.641 (1.008)	209.184 (0.000)	222.671 (49.631)
	2	209.826 (1.014)	219.367 (1.079)	216.255 (2.130)	947.944 (9.063)
	3	210.538 (0.900)	222.764 (0.000)	217.965 (1.092)	984.763 (7.755)
Min/Max	1	208.612/209.374	218.194/219.000	209.184/209.184	123.070/251.560
	2	209.471/210.123	219.035/219.847	215.465/217.041	935.160/961.710
	3	210.207/210.837	222.764/222.764	217.153/218.199	966.850/992.260

Additionally, an enhanced cluster analysis of on-site measured data has identified that the stress conditions of the main truss, arch ribs, and suspender cable cross-sections in the first cluster are quite consistent, with values predominantly around 210 MPa. Similar stress conditions are observed for the main truss and arch ribs in the second and third clusters, also around 210 MPa. However, there is a notable disparity in stress conditions when comparing these to the suspender cable sections, which can be largely attributed to the significant cable stretching during the construction phase. The variance analysis has shown that while the maximum variance for the main truss and arch ribs across the three clusters is 2.13 MPa, the variance for the suspender cables is considerably higher, peaking at 49.631 MPa in the first cluster, as opposed to 9.063 MPa for the second and 7.755 MPa for the third.

5.3. Assessment of Cluster Quality Using a Cloud Model for Suspender Cables

Using the improved clustering algorithm, cloud model formulas, and various clustering target thresholds, the single-parameter cloud model for the hangers was determined. A Principal Component Analysis (PCA) was employed to ascertain the weight coefficients of each parameter, ultimately assessing the quality of the clustering effect of the hanger system, as shown in Table 5.

Table 5. Single parameter and system parameter of suspender cable cloud model.

Suspender Cable	Single Parameter			System Parameter		
	EX	*EN*	*HE*	EX_C	EN_C	HE_C
C3-1	0.936	0.118	0.108	0.938	0.11	0.098
C6-1	0.942	0.095	0.08			

From the perspective of the expectation (EX) parameter, the value for the hanger C6-1 is higher than that for C3-1, indicating that C6-1 outperforms C3-1 in this parameter. In terms of entropy (EN), C6-1 has a lower value (0.0945) compared to C3-1 (0.1183), suggesting that C6-1 has less uncertainty or energy consumption in this parameter. Regarding hyper-entropy (HE), C6-1 also exhibits a lower value than C3-1, indicating less uncertainty in this parameter for C6-1. Consequently, when considering the single-parameter cloud model, C6-1 demonstrates superior performance across the expectation (EX), entropy (EN), and hyper-entropy (HE) parameters compared to C3-1, signifying that C6-1 has better overall performance.

The clustering quality effect assessment of the hanger system, calculated using the single-parameter and system cloud model parameters in Table 5 for different clustering target values, is presented in Figure 9. When the clustering target threshold is set to 1, the sample cloud center point obtained by normal sampling is 0.938, with a 96% confidence interval ranging from 0.701 to 1.175. The vast majority of the hanger system sample points fall within the I-Grade—excellent category. Therefore, the clustering quality achieved by the improved K-means clustering–silhouette coefficient–cloud model is effective.

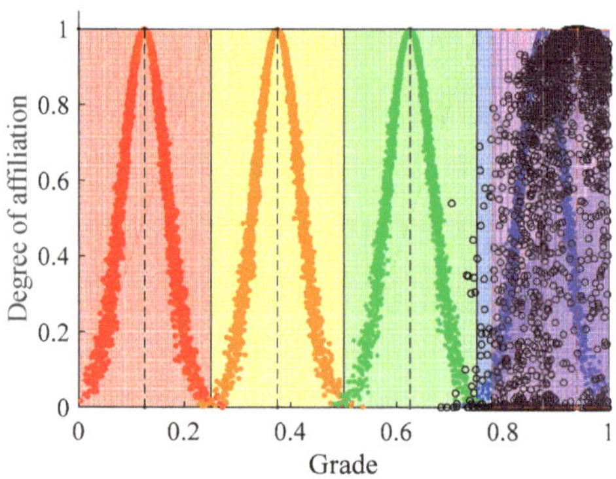

Figure 9. Clustering quality evaluation results.

6. Discussion

6.1. Effects of Clustering Algorithm Improvement on Case Bridge Components

6.1.1. Comparison of Clustering Distributions of All Components among UM, MC, and 3*DP Method

Figure 10 presents a comparative analysis of clustering distribution results obtained using the UM and MC methods as well as the dimensionless data derived from tripling the original monitoring data, referred to as three times the original dimensional data (3*DP).

The clustering results from both the UM and MC methods are categorized into three groups, numbered 1 to 3. In contrast, the 3*DP data, which represent the original monitoring data scaled by a factor of three, exhibit a broader range of categories, specifically from 0 to 3.

Among the ten measurement points on the arch rib section, points A1, A2, A4, A5, A6, A7, A8, and A9 show similar distribution patterns across the three clustering groups in their bar charts.

For the four measurement points on the truss girder section, the distribution patterns of the MC and 3*DP methods in the bar charts of the clustering results are consistently similar across all categories.

Regarding the two measurement points on the cable section, for the C3-1 cable, the distribution patterns of the MC and 3DP methods in the bar charts of the clustering results are alike, while there is a noticeable difference when compared to the UM method. For the C6-1 cable, the clustering results from the UM, MC, and 3DP methods all exhibit certain differences, indicating a more complex response to the clustering analysis.

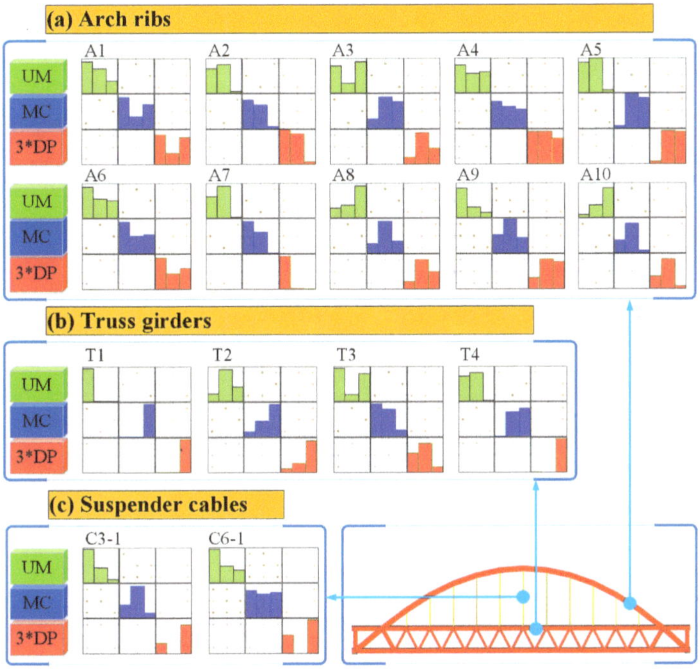

Figure 10. The cluster analysis results of bridge components' stresses during the experimental period.

After the cluster analysis, the cluster centers for the stress of suspender cable C3-1 are identified as 247.57 MPa, 957.73 MPa, and 1011.68 MPa. Correspondingly, the cluster centers for the stress of suspender cable C6-1 are 949.20 MPa, 222.67 MPa, and 986.04 MPa. The largest cluster center is located in the third category for both, indicating that this category corresponds to the most unfavorable stress on the cable structural system. Therefore, it is necessary to analyze separately the stress of the suspender cables with the largest cluster in the third category to assess the performance changes of the cable structural system. Of course, analyzing the cluster analysis results of the three types of cable stresses on the one hand can reflect the relative distances between the three cluster centers and on the other hand reflect the average change in the performance of the cables during the test period. Thus, it is essential to combine both situations to evaluate the performance of the cables.

6.1.2. Comparison of Key Components' Cluster between UM and MC Method

The cloud rain diagram in Figure 11 offers critical insights into the clustering outcomes of measurement points, delineating the differences between the UM and MC approaches.

An analysis of the semi-violin plot reveals a key distinction. While measurement point A7 maintains similar clustering patterns in both UM and MC states, the other points show notable variability. This suggests that the MC adjustments have influenced different measurement points in unique ways, indicating a need for a tailored strategy for each point.

Examining the box plots for the four key parameter measurement points, we observe a consistent alignment of the interquartile range and lower quartile with clustering points 2 and 1, irrespective of the clustering state. The uniformity of the whisker ranges, with minimums at 1 and maximums at 3, underscores a stable distribution pattern, indicative of reliable clustering across the board.

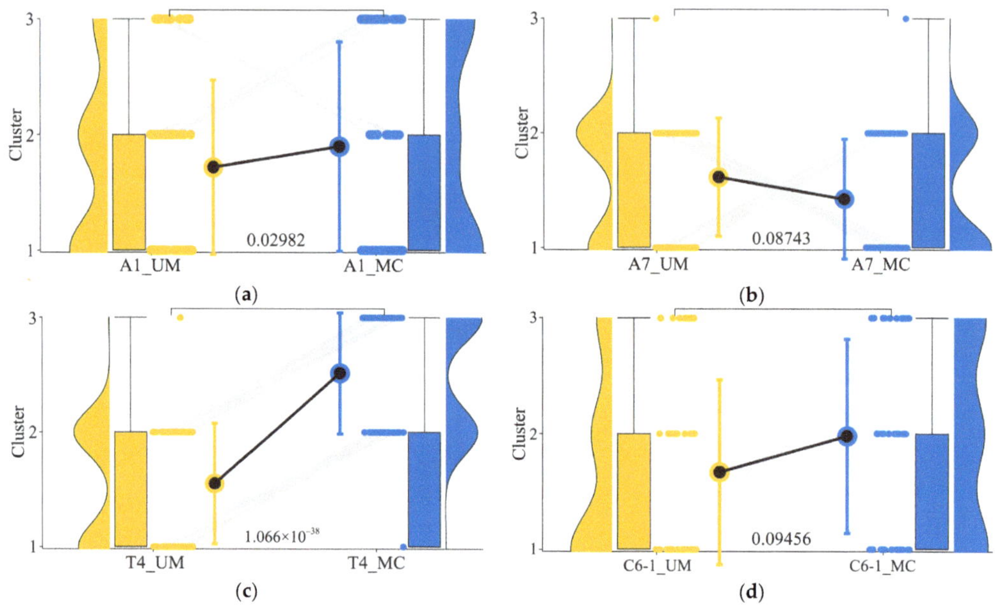

Figure 11. Cluster comparison of key components between UM and MC. (**a**) A1, (**b**) A7, (**c**) T4, (**d**) C6-1.

The jitter scatter plot provides further detail. Points A1 and C6-1 exhibit a balanced distribution across all three clustering points, suggesting a strong clustering characteristic. In contrast, A7 and T4 display an uneven distribution, hinting at sensitivity to fluctuations or a deviation from the overall trend, particularly in light of the MC adjustments.

Looking at the distribution of mean points, we see that for A1, T4, and C6-1, the MC state has shifted to higher values compared to the UM state. This shift indicates a more pronounced allocation of results to the three clusters following the improvements. However, for A7, the mean point in the MC state is lower than in the UM state, suggesting a potential decrease in its representativeness post-enhancement.

These observations highlight the varying effectiveness of the MC measures against the UM state and underscore the need for a nuanced strategy to ensure consistent clustering efficacy. It also points to a dual requirement: reinforcing robust points to maintain stability in the MC state and calibrating more sensitive points to ensure ongoing precision and reliability in the monitoring system.

6.2. Impact of Algorithm on Cluster Quality and Performance Metrics

Table 6 presents a comparative analysis of the clustering evaluation metrics for C3-1 and C6-1 hangers corresponding to the targeted stress thresholds, with the relative error $R_{\text{MC-UM}}$ = (MC-UM)/UM × 100%, including the average silhouette coefficient (ASC), Davies–Bouldin index (DBI), average deviation (AD), and p-value by the Sign-Rank test from target values. A lower DBI indicates a better clustering performance, with higher separation between clusters and tighter cohesion within clusters [39].

Table 6. Comparison of performance evaluation results of suspender cable by using different evaluation indicators.

Parameter		C3-1			C6-1		
		UM	MC	$R_{MC\text{-}UM}$ (%)	UM	MC	$R_{MC\text{-}UM}$ (%)
Cluster centroids	Cluster 1	247.566	247.566	0	222.671	222.671	0
	Cluster 2	974.479	957.732	−1.719	947.944	949.205	0.133
	Cluster 3	1047.374	1011.682	−3.408	984.764	986.044	−0.165
ASC of single-parameter clustering		0.816	0.734	−10.049	0.905	0.905	0
DBI of single-parameter clustering		0.221	0.498	125.339	0.273	0.266	−2.564
AD (%) from target threshold for Cluster 3		40.826	42.843	4.940	44.364	44.291	0.130
p-value ($\times10^{-5}$) of Sign-Rank test for Cluster 3		6250	28.678	−99.541	6.1035	12.207	100

(1) Comparison of Cluster Centroids

For Cluster 1, both C3-1 and C6-1 showed no difference in centroids calculated by the UM and MC methods. In Cluster 2, MC resulted in a slightly lower centroid for C3-1 by 1.719% and a slightly higher centroid for C6-1 by 0.133% compared to UM. In Cluster 3, MC lowered the centroid for C3-1 by 3.408% and for C6-1 by 0.165%. Overall, the difference in centroids calculated by UM and MC for C3-1 and C6-1 ranged from 0% to 3.5%.

(2) Average Silhouette Coefficient

For C3-1, MC showed a lower silhouette coefficient than UM, with a difference of −10.049%, suggesting better separation with UM. For C6-1, the silhouette coefficient remained stable at 0.905 with both methods.

Across all conditions, C6-1 had a higher ASC with MC (0.905) than C3-1 (0.816 and 0.734), indicating better clustering quality with higher coefficients. However, for C3-1, the silhouette coefficient was lower with MC (0.734) than UM (0.816), suggesting slightly better separation with MC.

(3) Davies–Bouldin Index

C3-1′s DBI was 125.339% higher with MC than UM, indicating a potential decrease in clustering separation. C6-1′s index was only 2.564% higher with MC, showing minimal change. Generally, a lower index indicates better clustering. C3-1 had a lower index with UM (0.221) than MC (0.498), while C6-1′s index was slightly lower with UM (0.273) than MC (0.266), suggesting better separation with UM for C3-1.

(4) Average Deviation from Target Threshold

Using the same material limit stress target threshold of 1770 MPa for Cluster 3, the average deviation for C3-1 and C6-1 with UM was 40.826% and 42.843%, respectively, compared to 44.364% and 44.291% with MC. The difference in deviation for C3-1 between MC and UM was two percentage points, with a difference rate of 4.94%. For C6-1, the deviation was almost identical between the two methods, with a difference rate of 0.13%, indicating stable clustering results.

(5) Sign-Rank Test p-value Analysis

In Cluster 3, the p-values for C3-1 and C6-1 showed significant differences between UM and MC. For C3-1, MC's p-value was significantly lower than UM's, with a difference rate of −99.541%, indicating superior statistical significance with MC. For C6-1, MC's p-value was double that of UM, suggesting a potential decrease in statistical significance with MC.

The Sign-Rank test results indicate that MC provided enhanced statistical significance for C3-1 in Cluster 3, while for C6-1, the significance was reduced. Since a lower p-value typically indicates higher statistical significance, MC may be more advantageous for C3-1.

In summary, the distribution, characteristics, and noise level of the data significantly impact the choice of algorithm and clustering results. The modified K-means (MC) algorithm showed different clustering effects compared to the unmodified K-means (UM) when processing data for C3-1 and C6-1. Statistical significance was higher for MC in Cluster 3 for C3-1, as evidenced by a lower p-value. In terms of centroid values, MC significantly affected the centroids for C3-1 in Cluster 2 and 3, usually yielding lower values than UM. In terms of clustering quality metrics, UM generally performed better in deviation and the Davies–Bouldin index, suggesting higher separation. For silhouette coefficient stability, C6-1 showed stable clustering effects with both methods, with a high silhouette coefficient. The choice of algorithm should be based on specific clustering objectives and data characteristics, considering multiple indicators, including statistical significance, centroid values, deviation, Davies–Bouldin index, and silhouette coefficient.

6.3. Influence of Single-Parameter Cloud Model Parameters on Clustering Quality

Figure 12 illustrates the predictive reliability and clustering quality of the clustering results for the hanger components C3-1 and C6-1 based on their single-parameter *EX*, *EN*, and *HE* values.

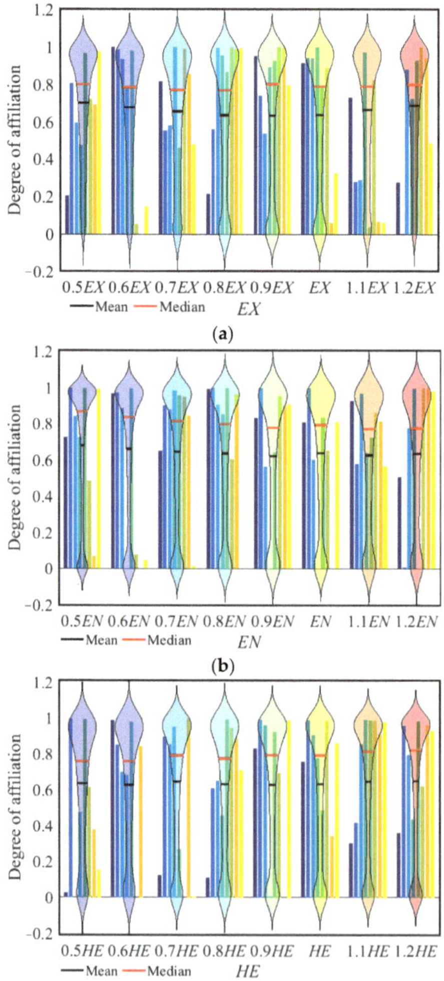

Figure 12. Effects of different parameters on clustering quality for single-parameter indicator clouds. (**a**) Different expectation values. (**b**) Different entropy values. (**c**) Different super entropy values.

(a) When analyzing the 96% confidence interval [$CI_{\text{lower},\ EX}$, $CI_{\text{upper},\ EX}$] corresponding to the structural system's expectation value, the width of the confidence interval, $CI_{HE} = CI_{\text{upper},\ EN} - CI_{\text{lower},\ EN}$, exhibits a trend of decreasing initially and then increasing with the increase in the single-parameter *EX* value. At an optimal *EX* value, the predictive reliability of the structural system is at its peak. However, as the *EX* value continues to rise or fall, the confidence interval widens, indicating increased uncertainty and decreased predictive reliability, reflecting a decline in the model's predictive capability under extreme conditions.

(b) As the single-parameter *EN* value increases, both the single-parameter cloud model and the system cloud model's entropy values also increase, while the *HE* value remains constant. Under the 96% confidence interval, the width of the confidence interval, CI_{HE}, gradually widens, signifying that the uncertainty in clustering quality rises with an increase in *EN*. Specifically, as the *EN* value doubles from 0.5 to 1.3 times, the CI_{HE} value increases from 0.224 to 0.59, indicating that predictions are more certain at lower entropy values, and uncertainty grows as the entropy value increases.

(c) The clustering quality of the single-parameter indicator cloud shows an increasing trend with the increase in *HE*. The hyper-entropy of the system cloud model increases with the hyper-entropy of the single-parameter cloud model, indicating an increase in system complexity or uncertainty. Nonetheless, the system's EX_C and EN_C values remain stable, suggesting that the system's predictive ability and uncertainty are maintained at a relatively constant state during changes in hyper-entropy. The evaluation of the confidence interval shows that $CI_{\text{lower, HE}}$ and $CI_{\text{upper},EN}$ are relatively stable across the entire range of hyper-entropy values, indicating consistency and reliability in the measurement results. The width of the confidence interval, CI_{HE}, varies within a narrow range, emphasizing the precision of the clustering results. The analysis of the confidence interval width, CI_{HE}, reveals that it remains essentially unchanged with the increase in the single-parameter hyper-entropy value, indicating that the uncertainty of the predictive results is maintained at a relatively stable level across the entire range of hyper-entropy values.

In summary, the research findings underscore the significant impact of expectation, entropy, and hyper-entropy values on the accuracy and uncertainty of predictions within structural system forecasting and the clustering analysis. By meticulously adjusting these parameters, it is possible to enhance predictive reliability while gaining a better understanding and management of the uncertainties inherent in the forecasting process.

7. Conclusions and Future Work

7.1. Summary of Findings

The research employs an enhanced clustering-silhouette coefficient–cloud model algorithm for the performance assessment of bridge components, specifically applied to operational maintenance of double-layer truss arch bridges, to guide maintenance decisions and ensure structural safety. The findings conclude that this method effectively supports the evaluation and preservation of critical bridge infrastructure.

(1) The study introduces an advanced clustering-silhouette coefficient–cloud model algorithm for assessing performance changes in bridge health monitoring. This integrated approach, capable of processing large datasets for pattern recognition and addressing uncertainty and fuzziness through expectation, entropy, and hyper-entropy parameters, offers a comprehensive data analysis tool for evaluating bridge component performance.

(2) The research concludes that the enhanced algorithm effectively assesses bridge health, emphasizing the need for the detailed scrutiny of suspender cable stress to safeguard structural integrity.

(3) PCA application ensures the strategic selection of data for precise structural evaluation, and the eigenvalue analysis points to notable stress variance, particularly within cluster levels, revealing the dynamic behavior of bridge components.

(4) The superior clustering quality, indicated by a high sample cloud center point and a 96% confidence interval in the excellent category, attests to the method's efficacy. The study

underscores the pivotal role of parameter tuning in enhancing predictive accuracy and managing uncertainties, advocating for algorithm selection tailored to specific clustering goals and data attributes for effective bridge maintenance and safety.

7.2. Contributions and Innovations

This study introduces an innovative approach to bridge health monitoring through the development and application of an enhanced clustering-silhouette coefficient–cloud model algorithm. The key contributions and innovations of this study are as follows:

(1) The creation of a novel algorithm that integrates a clustering analysis with silhouette coefficient optimization and cloud model theory, providing a sophisticated tool for handling large datasets and identifying patterns within bridge performance data.

(2) A comprehensive method for evaluating the performance of bridge components, particularly focusing on the dynamic responses of suspender cables, which are critical to the structural integrity of the bridge.

(3) The incorporation of expectation, entropy, and hyper-entropy parameters to address uncertainty and fuzziness in the monitoring data, leading to more reliable predictions and assessments, which demonstrates of the effectiveness of the improved K-means clustering method through high-quality clustering metrics, including a 96% confidence interval in the excellent category.

(4) Offering insights into the selection of algorithms based on specific clustering objectives and data characteristics, taking into account multiple indicators for optimal structural health monitoring, taken a double-layer truss arch bridge as the example.

These contributions and innovations represent a significant advancement in the field of structural health monitoring, providing a robust, reliable, and practical methodology for assessing and ensuring the safety and performance of critical bridge infrastructures.

7.3. Limitations and Recommendations for Future Research

It is important to note that the specific analysis approach depends on the selection of appropriate cloud model parameters and their sensitivity to the clustering quality assessment. This study focused on the development of an enhanced clustering algorithm and its application to bridge health monitoring, identifying the significant impact of cloud model parameters on the assessment process. Hence, future research shall focus on the following:

(1) Conducting a thorough sensitivity analysis of the cloud model parameters to determine their individual and collective influence on the clustering quality assessment, ensuring that the algorithm's response to parameter variations is well understood.

(2) Extending the research to different types of bridge structures and materials to evaluate the adaptability and effectiveness of the cloud model-based clustering assessment across a broader range of engineering applications.

(3) Investigating the potential of integrating the cloud model-based clustering method with other SHM techniques, such as machine learning and artificial intelligence, to enhance the predictive capabilities and diagnostic accuracy of the system.

Author Contributions: Conceptualization, C.G.; methodology, C.G.; validation, C.G.; formal analysis, C.G.; resources, D.H.; data curation, L.G., Y.Z., L.W., X.X. and Y.X.; writing—original draft preparation, C.G.; writing—review and editing, C.G.; supervision, D.H.; project administration, D.H. All authors have read and agreed to the published version of the manuscript.

Funding: This study was funded by the China Earthquake Administration Earthquake Science and Technology Spark Plan Project: Research on the Mechanism of Earthquake-Induced Instability and Failure of Cantilevered Flange Steel Box-Shaped Pressed and Bent Components (XH24063A); the Science and Technology Project Funded by PowerChina Corporation Limited (DJ-ZDXM-2022-35); the Langfang City Science and Technology Plan Project: Research on Collaborative Early Warning Assessment of Multi-Parameter Performance Indicators Based on Bridge Structural Health Monitoring (2022011066); Central Universities Basic Scientific Research Special Project: Research on Early Warning Technology for Highway Bridges Based on Health Monitoring (ZY20230202).

Data Availability Statement: Data are contained within the article.

Acknowledgments: The authors would like to express their gratitude to the individuals who have contributed significantly to the research presented in this paper. Dayong Han, serving as the project manager, has supplied essential research resources and access to the field site for bridge health monitoring. Chengzhong Gui in the Institute of Disaster Prevention, CEA, has provided invaluable guidance on the conceptual framework and research work of this study. His expertise and insights have been instrumental in shaping the overall approach and methodology of our research. Liang Gao, Yingai Zhao, Liang Wang, Xianglong Xu, and Yijun Xu have been integral to the execution of the field monitoring program, providing the analysis datasets. Additionally, Zhi Duan and other members in China Highway Engineering Consulting Group Co., Ltd., have been instrumental in preparing for the sensor installation and data collection vital to this research.

Conflicts of Interest: Chengzhong Gui has received research grants from Company Power China Road Bridge Group Co., LTD. Authors Dayong Han, Liang Gao, Yingai Zhao Liang Wang, Xianglong Xu, Yijun Xu were employed by the company Power China Road Bridge Group Co., LTD. The remaining authors declare that the research was conducted in the absence of any commercial or financial relationships.

Appendix A. The Calculation Steps of the Revised K-Means Clustering Method

NO.	Main Step and Equations	Descriptions
1	Input raw parameter data	$\sigma_p(t) = [\sigma_{p1}(t), \sigma_{p2}(t), \ldots, \sigma_{pn}(t)]$
2	Initialize random seed for reproducibility • rng(0);	-
3	Initialize centroids • K = 3 • centroids = X(randperm(N, K), :)	 Set the number of clusters Select K points randomly
4	Sort raw data in ascend order	$x_p = [x_{p1}, x_{p2}, \ldots, x_{pn}]$
4	Set iteration parameters • max_iters = 100; • tolerance = 1×10^{-4};	 Set the maximum number of iterations Set the tolerance for stopping the iteration
5	Main iteration loop • for iter = 1:max_iters	 Iterate until convergence or limit.
6	Assign samples to the nearest centroid • dist(i,j) = norm(X(i,:)—centroids(j,:)); • [~, assignments(i)] = min(distances(i,:));	 Calculate each sample's distance to every centroid. Assign each sample to the closest centroid.
7	Update centroids • new_centroids(j,:) = mean(cluster_samples);	 Recalculate centroids by mean for each cluster.
8	Check convergence • if norm(new_centroids—centroids) < tolerance • break;	 Terminate if change is minor.
9	Update and continue • idx • centroids = new_centroids;	 Update centroids and iterate.
10	End of the algorithm -	 Terminate at convergence or max iterations.

References

1. Gharehbaghi, K.; Georgy, M.; Robson, K.M.; Wilkinson, S.; Farnes, K. Versatility in sustainable building design (SBD) practices: An empirical study. *Int. J. Build. Pathol. Adapt.* **2021**. ahead-of-print. [CrossRef]
2. Administration, F.H. *Long-Term Bridge Performance (LTBP) Program Protocols, Version 1 NO. FHWA-HRT-16-007*; Research, Development, and Technology Turner-Fairbank Highway Research Center: McLean, VA, USA, 2016.
3. Liu, Z. Smart Sensors for Structural Health Monitoring and Nondestructive Evaluation. *Sensors* **2024**, *24*, 603. [CrossRef] [PubMed]
4. Azhar, A.S.; Kudus, S.A.; Jamadin, A.; Mustaffa, N.K.; Sugiura, K. Recent vibration-based structural health monitoring on steel bridges: Systematic literature review. *Ain Shams Eng. J.* **2024**, *15*, 102501. [CrossRef]
5. Wang, Y.; Cai, C.; Han, B.; Xie, H.; Bao, F.; Wu, H. A deep learning-based approach for assessment of bridge condition through fusion of multi-type inspection data. *Eng. Appl. Artif. Intell.* **2024**, *128*, 107468. [CrossRef]
6. Jamali, S.; Chan, T.H.; Nguyen, A.; Thambiratnam, D.P. Reliability-based load-carrying capacity assessment of bridges using structural health monitoring and nonlinear analysis. *Struct. Health Monit.* **2019**, *18*, 20–34. [CrossRef]
7. Hariri-Ardebili, M.A. Risk, Reliability, Resilience (R3) and beyond in dam engineering: A state-of-the-art review. *Int. J. Disaster Risk Reduct.* **2018**, *31*, 806–831. [CrossRef]
8. Sinaga, K.P.; Yang, M.-S. Unsupervised K-Means Clustering Algorithm. *IEEE Access* **2020**, *8*, 80716–80727. [CrossRef]
9. Zhou, J.; Qin, Y.; Luo, J.; Zhu, T. Remaining Useful Life Prediction by Distribution Contact Ratio Health Indicator and Consolidated Memory GRU. *IEEE Trans. Ind. Inform.* **2023**, *19*, 8472–8483. [CrossRef]
10. Dou, H.; Liu, Y.; Chen, S.; Zhao, H.; Bilal, H. A hybrid CEEMD-GMM scheme for enhancing the detection of traffic flow on highways. *Soft Comput.* **2023**, *27*, 16373–16388. [CrossRef]
11. Zhang, M.; Kujala, P.; Musharraf, M.; Zhang, J.; Hirdaris, S. A machine learning method for the prediction of ship motion trajectories in real operational conditions. *Ocean. Eng.* **2023**, *283*, 114905. [CrossRef]
12. Shi, Y.P.; Hu, J.P.; Wu, Y.Z.; Ghosh, B.K. Intermittent output tracking control of heterogeneous multi-agent systems over wide-area clustered communication networks. *Nonlinear Anal. -Hybrid Syst.* **2023**, *50*, 101387. [CrossRef]
13. Rodriguez, A.; Laio, A. Clustering by fast search and find of density peaks. *Science* **2014**, *344*, 1492–1496. [CrossRef] [PubMed]
14. Liu, R.; Wang, H.; Yu, X. Shared-nearest-neighbor-based clustering by fast search and find of density peaks. *Inf. Sci.* **2018**, *450*, 200–226. [CrossRef]
15. Guo, A.; Jiang, A.; Lin, J.; Li, X. Data mining algorithms for bridge health monitoring: Kohonen clustering and LSTM prediction approaches. *J. Supercomput.* **2020**, *76*, 932–947. [CrossRef]
16. Tang, C.; Li, Z.L.; Wang, J.; Liu, X.W.; Zhang, W. Unified One-Step Multi-View Spectral Clustering. *IEEE Trans. Knowl. Data Eng.* **2023**, *35*, 6449–6460. [CrossRef]
17. Huang, D.; Wang, C.-D.; Wu, J.-S.; Lai, J.-H.; Kwoh, C.-K. Ultra-Scalable Spectral Clustering and Ensemble Clustering. *IEEE Trans. Knowl. Data Eng.* **2020**, *32*, 1212–1226. [CrossRef]
18. Horton, D.E.; NJohnson, C.; Singh, D.; Swain, D.L.; Rajaratnam, B. Contribution of changes in atmospheric circulation patterns to extreme temperature trends. *Nature* **2015**, *522*, 465–469. [CrossRef]
19. Wei, Z.; He, D.; Jin, Z.; Liu, B.; Shan, S.; Chen, Y.; Miao, J. Density-Based Affinity Propagation Tensor Clustering for Intelligent Fault Diagnosis of Train Bogie Bearing. *IEEE Trans. Intell. Transp. Syst.* **2023**, *24*, 6053–6064. [CrossRef]
20. Zhu, T.; Wang, X.; Zhang, J.; Yu, S. Mean-shift clustering approach to the tracklets association with angular measurements of resident space objects. *Astron. Comput.* **2022**, *40*, 100588. [CrossRef]
21. Li, S.; Wei, S.; Bao, Y.; Li, H. Condition assessment of cables by pattern recognition of vehicle-induced cable tension ratio. *Eng. Struct.* **2018**, *155*, 1–15. [CrossRef]
22. Shi, C.; Wei, B.; Wei, S.; Wang, W.; Liu, H.; Liu, J. A quantitative discriminant method of elbow point for the optimal number of clusters in clustering algorithm. *EURASIP J. Wirel. Commun. Netw.* **2021**, *2021*, 31. [CrossRef]
23. Song, W.; Wang, Y.; Pan, Z. A novel cell partition method by introducing Silhouette Coefficient for fast approximate nearest neighbor search. *Inf. Sci.* **2023**, *642*, 119216. [CrossRef]
24. Lin, J.-Q.; Chen, M.-S.; Wang, C.-D.; Zhang, H. A Tensor Approach for Uncoupled Multiview Clustering. *IEEE Trans. Cybern.* **2024**, *54*, 1236–1249. [CrossRef] [PubMed]
25. Rojas Thomas, J.C.; Peñas, M.S.; Mora, M. New Version of Davies-Bouldin Index for Clustering Validation Based on Cylindrical Distance. In Proceedings of the 2013 32nd International Conference of the Chilean Computer Science Society (SCCC), Cautin, Chile, 11–15 November 2013; pp. 49–53.
26. Tsai, W.-H.; Wang, H.-M. Evolutionary minimization of the Rand index for speaker clustering. *Comput. Speech Lang.* **2009**, *23*, 165–175. [CrossRef]
27. Zhang, S.; Wong, H.-S.; Shen, Y. Generalized Adjusted Rand Indices for cluster ensembles. *Pattern Recognit.* **2012**, *45*, 2214–2226. [CrossRef]
28. de Vargas, R.R.; Bedregal, B.R.C. A way to obtain the quality of a partition by Adjusted Rand Index. In Proceedings of the 2013 2nd Workshop-School on Theoretical Computer Science (WEIT), Rio Grande, Brazil, 15–17 October 2013; pp. 67–71.
29. Dong, Y.N.; Qin, S.J. A novel dynamic PCA algorithm for dynamic data modeling and process monitoring. *J. Process Control.* **2018**, *67*, 1–11. [CrossRef]
30. Bates, S.; Hastie, T.; Tibshirani, R. Cross-Validation: What Does It Estimate and How Well Does It Do It? *arXiv* **2023**, arXiv:2104.00673. [CrossRef]

31. Yates, L.A.; Aandahl, Z.; Richards, S.A.; Brook, B.W. Cross validation for model selection: A review with examples from ecology. *Ecol. Monogr.* **2023**, *93*, e1557. [CrossRef]
32. Qin, K.; Xu, K.; Liu, F.; Li, D. Image segmentation based on histogram analysis utilizing the cloud model. *Comput. Math. Appl.* **2011**, *62*, 2824–2833. [CrossRef]
33. Song, W.; Zhu, J. A multistage risk decision making method for normal cloud model considering behavior characteristics. *Appl. Soft Comput.* **2019**, *78*, 393–406. [CrossRef]
34. Chen, Z.-Y.; Dai, Z.-H. Application of group decision-making AHP of confidence index and cloud model for rock slope stability evaluation. *Comput. Geosci.* **2021**, *155*, 104836. [CrossRef]
35. Zhang, Y.; Shang, K. Cloud model assessment of urban flood resilience based on PSR model and game theory. *Int. J. Disaster Risk Reduct.* **2023**, *97*, 104050. [CrossRef]
36. Chang, Z.; Mei, H.; Yan, C.; Shi, Y.; Zhu, X.; Lu, Z.; Jia, Z. Health status evaluation of highway tunnel inverted arch based on variable weight and extension cloud model. *Eng. Fail. Anal.* **2024**, *157*, 107939. [CrossRef]
37. van Ooyen, A. 2—Theoretical Aspects of Pattern Analysis. In *New Approaches for the Generation and Analysis of Microbial Typing Data*; Dijkshoorn, L., Towner, K.J., Struelens, M., Eds.; Elsevier Science B.V.: Amsterdam, The Netherlands, 2001; pp. 31–45.
38. Kaufman, L.; Rousseeuw, P.J. *Finding Groups in Data: An Introduction to Cluster Analysis*; John Wiley & Sons Inc.: Hoboken, NJ, USA, 1990.
39. Ros, F.; Riad, R.; Guillaume, S. PDBI: A partitioning Davies-Bouldin index for clustering evaluation. *Neurocomputing* **2023**, *528*, 178–199. [CrossRef]
40. Guo, Z.; Tian, Y.; Guo, X.; He, Z. Research on Measurement and Application of China's Regional Logistics Development Level under Low Carbon Environment. *Processes* **2021**, *9*, 2273. [CrossRef]
41. Zhang, H.; Sun, Y.; Zhang, W.; Song, Z.; Ding, Z.; Zhang, X. Comprehensive evaluation of the eco-environmental vulnerability in the Yellow River Delta wetland. *Ecol. Indic.* **2021**, *125*, 107514. [CrossRef]
42. Gui, C.Z.; Song, G.H.; Ke, Z.T.; Qi, F.L. Investigation on Rating Method of Cable Force Increment Adjustment Factor. *Adv. Eng. Sci.* **2019**, *51*, 75–83.
43. Lei, J.; Gui, C.; Lin, W.; Song, G.; Jia, S. A Maintenance Decision-Making Process for Steel Cable Structure System. *Arab. J. Sci. Eng.* **2023**, *48*, 13573–13590. [CrossRef]
44. Zhou, J.; Li, X.; Xia, R.; Yang, J.; Zhang, H. Health Monitoring and Evaluation of Long-Span Bridges Based on Sensing and Data Analysis: A Survey. *Sensors* **2017**, *17*, 603. [CrossRef]
45. Gui, C.; Lei, J.; Yoda, T.; Lin, W.; Zhang, Y. Effects of inclination angle on buckling of continuous composite bridges with inclined parabolic arch ribs. *Int. J. Steel Struct.* **2016**, *16*, 361–372. [CrossRef]
46. *GB/T714-2015*; Structural Steel for Bridge. General Administration of Quality Supervision, Inspection and Quarantine of the P.R.C. Standardization Administration of the P.R.C.: Beijing, China, 2015.
47. Qiang, B.; Li, Y.; Yao, C.; Wang, X.; Gu, Y. Through-thickness distribution of residual stresses in Q345qD butt-welded steel plates. *J. Mater. Process. Technol.* **2018**, *251*, 54–64. [CrossRef]
48. Yuan, Z.; Ji, B.; Fu, Z.; Ge, H. Local stress variation in welded joints by ICR treatment. *J. Constr. Steel Res.* **2016**, *120*, 45–51.

infrastructures

MDPI

Article

Evaluation of the Szapáry Long-Span Box Girder Bridge Using Static and Dynamic Load Tests

Marame Brinissat [1], Richard Paul Ray [1,*] and Rajmund Kuti [2]

1 Department of Structural and Geotechnical Engineering, Széchenyi István University, 9026 Győr, Hungary; brinissat.marame@hallgato.sze.hu

2 Department of Machine Design, Faculty of Mechanical Engineering, Informatics and Electrical Engineering, Széchenyi István University, 9026 Győr, Hungary; kuti.rajmund@sze.hu

* Correspondence: ray@sze.hu

Abstract: This paper presents the results of a recent field test carried out before the opening phases of the Szapáry motorway bridge across the Tisza River in central Hungary. The evaluation test was based on static and dynamic load tests that provided information on deflection, stresses, and dynamic mode shapes along the bridge. The structure has two large continuous independent steel box girders that cover spans across the floodplain and river. Various configurations of truck loading applied up to 6400 kN of loading on the deck. During the static tests, string potentiometers recorded deflections at mid-span. Additionally, strain gauges enabled strain/stress measurements at the mid-point of the longest span and directly above one support. Dynamic loadings showed variation in deflection response due to vehicle speed, and ambient vibration testing led to determining vibration modes and frequencies. A three-dimensional finite-element model produced similar deflections, stresses, and modal behavior. Measured and modeled deflections and stresses indicated that the bridge performed within design margins. The testing and analysis results will be part of a future program assessing conditioned-based maintenance.

Keywords: structural health monitoring; box-girder bridge; load testing; deflection; dynamic loads

check for updates

Citation: Brinissat, M.; Ray, R.P.; Kuti, R. Evaluation of the Szapáry Long-Span Box Girder Bridge Using Static and Dynamic Load Tests. *Infrastructures* **2023**, *8*, 91. https://doi.org/10.3390/infrastructures8050091

Academic Editors: Armin Mehrabi and Seyed Saman Khedmatgozar Dolati

Received: 6 April 2023
Revised: 25 April 2023
Accepted: 27 April 2023
Published: 9 May 2023

1. Introduction

Bridges are an important and costly part of any transportation infrastructure. Their design and performance often dictate the capacity of an entire highway or railway segment. During the design phase, conflicting goals may arise based on economy, capacity, durability, aesthetics, and road performance. Many design alternatives are available to the bridge engineer; however, this article focuses on a steel box girder design with multiple long spans and modest clearance. The Szapáry bridge crosses the Tisza River and floodway north of Szolnok, Hungary, as part of the new M4 motorway connecting Budapest to central Eastern Hungary and Romania. The design is especially suited to the Tisza, where there are broad floodways and significant changes in river level. The long spans reduce the number of piers in the floodway, limiting the environmental footprint and flow obstruction and minimizing costly foundation works. In the case of steel bridges, box girders were more suitable for constructing long bridges with larger spans and wider decks.

The serviceability of long-span bridges will decrease over time; the primary source of this decreasing function is the cyclic load on the girders caused by the vehicle loadings [1]. One parameter to determine bridge serviceability is the vertical deflection of the spans. Many factors contribute to vertical deflection, including the stiffness of the main box beam, vertical stiffness of the foundation systems, presence of expansion joints, and degree of composite action with the concrete bridge deck [2–4]. The ideal testing system would include locations throughout the bridge and evaluate each component independently, i.e., the relative stiffness of the box beam and foundation system. That would require a testing program that is quite complex, time-consuming, and costly. The next best option is to evaluate measurements by modeling the entire bridge [5]. Deflection measurement under static and moving loads is essential for bridge design, function, and structural health [6,7]. The Szapáry bridge was load-tested to demonstrate bridge safety to the public and assess its performance under static and moving loads.

When bridge deflections exceed specified design limits, they may experience damage or collapse, posing a severe threat to the public and likely causing significant economic losses. Thus, the deflections should be within design limits (L/400) given by the Hungarian annex [8] and in accordance with the AASHTO code (L/800) [9]. The project required a proof load test, which involved applying loads that were greater than those used in a diagnostic test [10]; several bridges have collapsed during such load tests including the steel framework bridge over the Morava River near Ljubitschewo in Serbia, the road bridge near Salez in Switzerland, and a suspension bridge in Maurin in France [11]. For this reason, testing bridges is a critical task for structural engineers. Careful monitoring enables the early detection of defective fractures, misalignments, and assembly errors, reducing maintenance and repair costs and lowering the risk of bridge collapses or other serious accidents [12–14].

Engineers often test bridges before they open to traffic. Fang et al. [15] conducted 40 different static field loading tests to examine the behavior of the Kao-Ping-His bridge, the longest bridge in Taiwan. Wei-Xin et al. [16] implemented field static load tests and numerical analyses on the Qingzhou cable-stayed bridge in Fuzhou, China. The deflection results of both tests revealed that the bridge had linear characteristics, which indicates that the bridge has an adequate load-carrying capacity. The analytical results agreed well with the test results.

An early study by Hassan et al. [17] evaluated the behavior of the North Halawa Valley viaduct on the island of Oahu, HI, USA, under numerous static truck-loading combinations. They reported that load testing is a crucial starting point for structural monitoring and bridge repair. In addition, the load test results verified the structure's suitability in the short term while predicting its potential long-term performance.

Galati et al. [18] conducted a series of static, dynamic field load tests of the A6102 Lexington Bridge in Hardin, MO, USA. They compared their measurements with AASHTO LRFD bridge design calculations and their finite element model (FEM). Their results revealed that the bridge would be safe for service, as the field tests produced lower deflection values than the results of the AASHTO analysis. Similar findings were found by Wang et al. [19] after comparing the complex Pingnan Xiangsizhou bridge deflection results produced by the non-destructive load tests and numerical simulation, followed by an evaluation of the bridge's fundamental characteristics in order to ensure that the structure was up to standard.

The Szczercowska bridge required testing before the structure reinforcement to better assess the benefits of strengthening the bridge structure. The tests and numerical modeling results showed that the bridge has sufficient strength reliability, as the field tests always underestimate the results of the finite element modeling analysis [19]. According to Zheng et al.'s [20] analysis, the experimental deflection values were higher than the theoretical results as the material exhibited less rigid behavior at the time of the test; thus, the bridge structural stiffness was different from what theoretical calculations predicted, and there was insufficient safety reserve.

Typically, engineers and designers perform several checks and tests throughout the construction phase of complex structures like the steel box bridge in this study. The final testing phase may require more exhaustive static and dynamic tests [21,22]. According to Sun et al. [23], the dynamic vehicle loading test evaluates the load-carrying capacity of a bridge better than the analysis. The moving trucks used to load the bridge are typically modeled numerically as additional masses. They are, in fact, complex dynamic systems that may have significant effects when interacting with the bridge.

More recently, researchers have investigated how moving vehicles interact dynamically with bridges, focusing primarily on how the characteristics of the road and moving vehicles affect the bridge dynamics [24–26]. The stress and deflection values increase with increasing vehicle load and velocity, making these two factors the most important in determining girder deflections [27]. According to the literature review, trucks weighing more than 20 tons and traveling at speeds ranging from 10 to 80 km/h are commonly used in field testing [28,29].

Numerical models rely on full-scale tests under traffic [30] for validation. Several model validation studies describe the process of direct measurements under static and moving loads, such as strains, displacements, and accelerations [15,31]; or derived from acceleration records, such as natural frequencies and mode shapes [22,28,32].

The load tests aim to understand better the bridge response to static and dynamic loadings and calibrate the initial FEM model. The measured data provides an initial database for future health monitoring of the new structure. Future condition evaluations will then compare their assessment to the initial conditions. In their studies, Sousa et al. [33] stated that vertical deflection is related to bridge performance under short-term and long-term observations.

This paper reports a recent investigation as the initial step in the structural health behavior of a continuous long-span box girder bridge with a main span that extends to 148 m in Hungary. The field load tests of the Szapáry bridge performed before its opening will be summarized. The geometry of the bridge and the truck's position will be considered throughout the comparison to accurately reflect the actual field conditions. As a primary goal, the study will follow the load testing process to validate our numerical model and provide a baseline for further numerical studies of what-if structural health situations. Our future goal is to determine the feasibility of structural health monitoring by measuring response to dynamic loads.

2. Testing the Szapáry Bridge on the Tisza River in Hungary

2.1. Description of the Szapáry Bridge

Hungary's new M4 Tisza motorway bridge is a continuous seven-span box-girder bridge. It measures 756 m overall, with span lengths of (96 + 120 + 148 + 120 + 96 + 96 + 80) m, respectively, as shown in Figure 1a. The 148 m span of the bridge over the Tisza River is the longest in Hungary for this type of design. The bridge consists of two separated trapezoidal single-box girders supporting two traffic directions (Figure 1b). The girder's top slab extends 13.815 m wide and cantilevers 3.309 m and 2.886 m along the outer and inner edges. The base flange is 5.50 m wide. The height of the box girder ranges from 4.60 m to 4.412 m on the right and left sides (Figure 1c).

(a)

(b) (c)

Figure 1. The bridge on the Tisza River: during construction: (**a**) aerial view; (**b**) lateral viaduct view; (**c**) steel box section view.

The trapezoidal deck consists of plate girders with 12–50 mm thicknesses. Longitudinal stiffeners and crossbeams support it in perpendicular directions. The main girders span the box's upper section to support the deck (see Figure 1c). The longitudinal stiffeners are U-shaped sections spaced 600 mm on the top, 1100 mm on the bottom, and 221mm and 229 mm on the sides. Transverse stiffeners are I-shaped sections placed at 4000 mm intervals. In addition, steel diaphragm sections stiffen the structure above each support. The orthotropic steel deck connects to the junction with trusses type HEA220, each 8000 mm, and assists in providing lateral stability to the girders. Welding these elements increases durability, eases inspection, and improves aesthetic quality. The following sections concisely describe the measurement results, their interpretation, and conclusions.

2.2. Experimental Program

In this study, load testing and monitoring of the bridge provided information on its response under static and moving loads. Specifically, the measurements provided vertical deflection of the bridge deck at mid-span and other critical locations. The test involved placing the trucks in various positions and measuring the resulting deflections at the mid-span locations along all the bridge's seven spans.

Testing started with static tests, wherein up to 16,400 kN (5-axle) trucks loaded the deck. The field test configuration appears in Figure 2, and tables presented below detailed results showing the testing order. The test was performed on the right track of the Tisza bridge.

Figure 2. Bridge Truck loading test in 2 × 2 configuration.

Following the static loading, one pair of trucks (total 800 kN), running side by side, drove the length of the bridge at constant speeds from 10 to 90 km/h. The moving load tests provided data on stiffness and vibration response. In addition, the testing program measured the bridge response to ambient excitation using acceleration-time records, and the bridge modal parameter results will appear in a forthcoming paper.

Measurement of Deflections and Stresses

The vertical deflection measurements occurred on each of the seven segments at mid-span, delivering 14 continuous readings. Deflection and strain gauge positions marked high and low sides appear in Figure 3a. High refers to the side of the deck near the bridge centerline where the elevation was slightly higher due to the slight inclination for deck drainage.

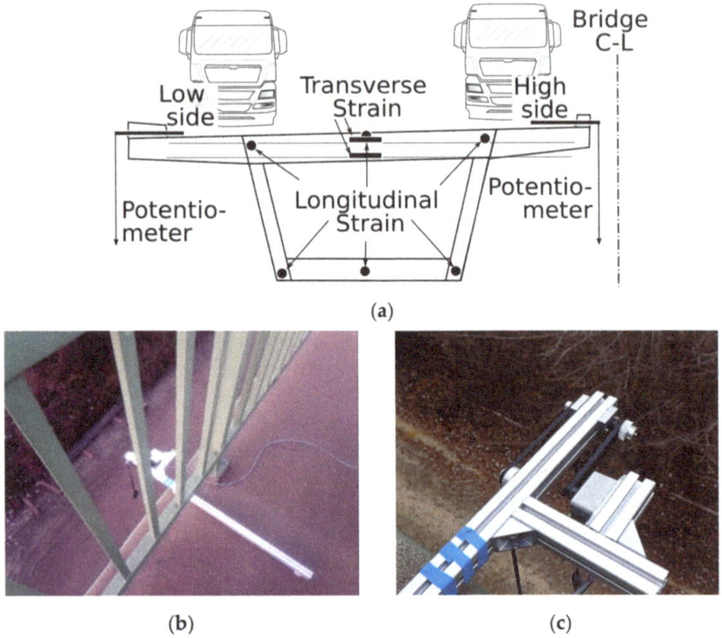

Figure 3. Measurement of deflections: (**a**) plan of the test measurement; (**b**,**c**) measurement instrument.

A rotary resistance meter fastened to the loaded bridge provided measurements with its reference cable anchored to the ground surface. The error of the device remained well below 0.5% full scale. The error translated to less than 1 mm for the range of measurements performed during the load tests. The longest loaded span used the adjacent unloaded span as a stable reference point since river traffic made anchors problematic. Additionally, the deck inclination was measured to produce a pair of deflections similar to the other spans. Optical leveling provided an additional deflection check along the loaded and unloaded bridges. The deflection of foundation piers was negligible.

Twelve longitudinal and two transverse locations measured strain continuously throughout testing. The gauges measured longitudinal stresses in the main girder across two sections: above the third support and in the middle of the longest span. Three-gauge locations provided readings at the top and three at the bottom plates, with one on the bridge's axis and two near the spines (Figure 3a). One pair of gauges measured transverse strain on the upper girder of the cross-section (Figure 3a). The testing was conducted during overcast weather conditions with a slight temperature variation of 2 °C, which had no significant impact on the measurement readings. The strain gauges primarily contributed to relative changes between loaded and unloaded conditions for static and dynamic tests.

2.3. Tests Interpretation through Numerical Modeling

In her investigations, Gouda [34] demonstrated that FEM modeling could predict the linear structural response of such bridges with high accuracy. A finite-element model of the Szapáry bridge superstructure was developed in Midas Civil [35] to provide an interpretation of the main results of the experimental tests. The analysis also offered insight into the effects of load placement on bridge deflection.

Concerning the bridge, the deck consisted of straight elastic beam elements with variable cross-sections representing steel stiffeners, cross beams, and diaphragms. Truss elements with uniform cross-sections modeled truss members. End restraint conditions duplicated field conditions. Plate elements having different thicknesses and located in the midplane of the slabs simulated the orthotropic deck slabs, as shown in Figure 4b. Elastic links represented the stability of the elements with rollers and pinned constraints providing boundary support. The constraints allowed horizontal translation in the transverse direction. Rigid links (Figure 4a) attached the trusses to the deck. The connections are between plate elements and trusses at the bottom and top of the deck. This procedure guaranteed rotational continuity.

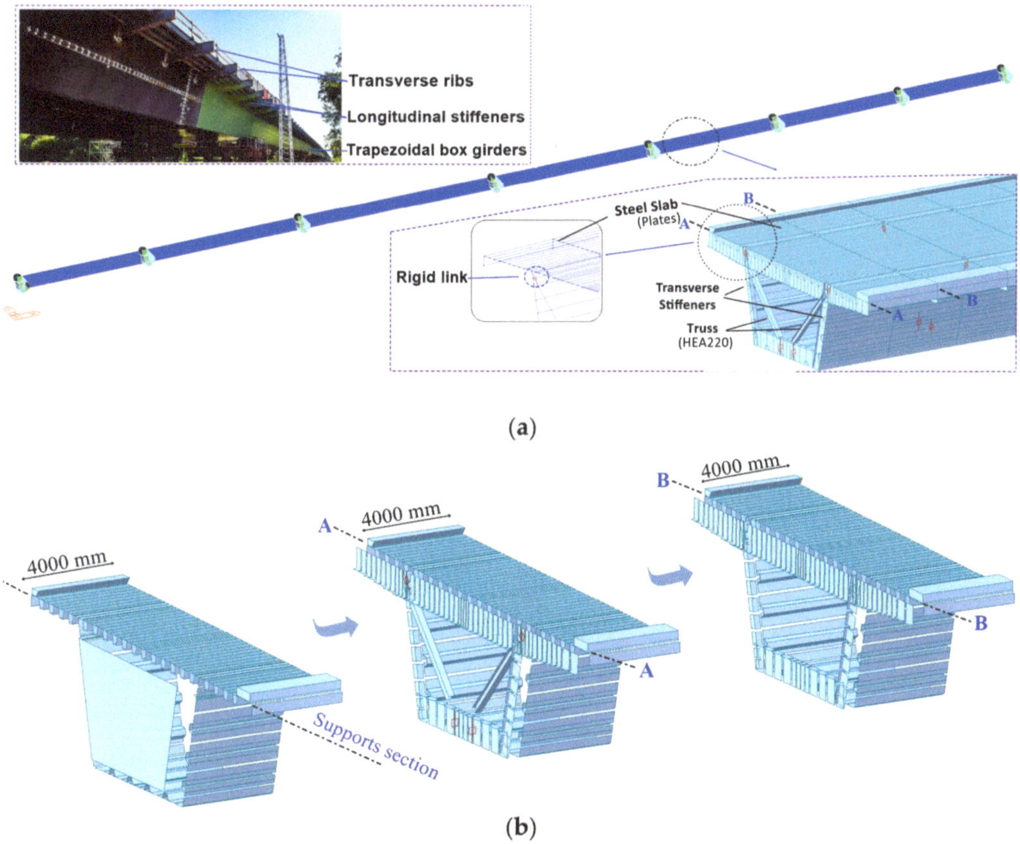

Figure 4. Finite element model of the Szapáry bridge: (**a**) Details of the FE model; (**b**) Arrangement of bridge cross-sections: along support, Sec A-A, Sec B-B, as marked in (**a**).

All components represented field conditions with identical geometries and materials. The model consisted of 16,329 plates, 21,336 beams, and 196 truss elements, connecting 23,759 nodes. A perspective view of a model section appears in Figure 4.

Material properties used for all elements are S355 grade steel (EN 1993-2) [36], with Young's modulus of 210,000 MPa (N/mm^2), Poisson's ratio of 0.3, and a weight density value of 76.98 kN/m^3. All beam member sections represented identical field members in the as-built structure based on measurements from the geometrical survey. Permanent loads (structural self-weight) act on the bridge everywhere, while variable loads represent the positions of traffic (loaded trucks for testing). Detailed truck data appears in Figure 5, and the loading patterns for the static and dynamic tests.

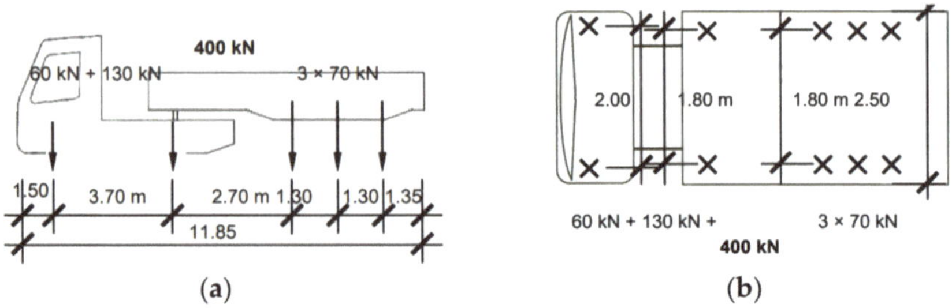

Figure 5. Design truck: (**a**) elevation and total applied axle loads; (**b**) plan view.

3. Field Tests and FEM Model: Results and Discussions

Comparing the FEM deflections due to static and moving load to measured deflections will validate the bridge model and define the bridge's global behavior. The proper modeling approach for this study provides results necessary in determining the existence of systematic errors. Therefore, the fundamental knowledge and information for the modeling should be adequately understood and applied correctly to ensure accurate and quality data [37].

3.1. Bridge Deflection under Static Loadings

As mentioned before, the longitudinal arrangements of the load combinations and deflection results appear in the following sections. Tables 1–4 summarize the static loading case arrangements where every colored rectangle represents a 15 m space for a 400 kN truck. The blued curve represents measured vertical deflections interpolated between the mid-span and each pier. The red curve represents deflections at every 4 m along the bridge model.

Table 1. Longitudinal arrangements of load configuration 01.

	Span 1 96 m 6 db	Span 2 120 m 8 db	Span 3 148 m 10 db	Span 4 120 m 8 db	Span 5 96 m 6 db	Span 6 96 m 6 db	Span 7 80 m 5 db
Case 1							
Case 2							
Case 3							
Case 4							

Table 2. Longitudinal arrangement of load configuration 02.

	Span 1 96 m 6 db	Span 2 120 m 8 db	Span 3 148 m 10 db	Span 4 120 m 8 db	Span 5 96 m 6 db	Span 6 96 m 6 db	Span 7 80 m 5 db
Case 5							
Case 6							
Case 7							
Case 8							
Case 9							
Case 10							

Table 3. Longitudinal arrangement of load configuration 03.

	Span 1 96 m 6 db	Span 2 120 m 8 db	Span 3 148 m 10 db	Span 4 120 m 8 db	Span 5 96 m 6 db	Span 6 96 m 6 db	Span 7 80 m 5 db
Case 11		▷	▷	▷			
Case 12	▷	▷		▷	▷		
Case 13					▷	▷	▷

Table 4. Longitudinal arrangement of load configuration 04.

	Span 1 96 m 6 db	Span 2 120 m 8 db	Span 3 148 m 10 db	Span 4 120 m 8 db	Span 5 96 m 6 db	Span 6 96 m 6 db	Span 7 80 m 5 db
Case 14	▷	▷	▷				
Case 15		▷	▷	▷			
Case 16			▷	▷	▷		
Case 17				▷	▷	▷	▷

(a) Load Configuration 01 (LC1)

The LC1 load cases followed the sequence shown in Table 1. The deck deflection measurements were taken during load cases 1, 2, 3 and 4. Figure 6 shows the vertical deflection predictions and calculations graphs for two loaded lanes at span 3 (Case 1,2), spans 2 and 4 (Case 3) and spans 5 and 7 (Case 5). Note the different vertical scales for deflection. As expected from Cases 1 and 2, the girder deflections increase as the truckload number increases in the center of the span. Therefore, the backspan upward deflections (the same graph, spans 2 and 4) also increase. From load cases 04 and 05, a critical upward deflection between the loaded spans appears, equal to 75% of the maximum deflection for case 03 and 63% for case 04. Note that upward deflection means moment reversal that may harm the plate deck. An example of the Case 2 field test and analytical results appears in Figure 7.

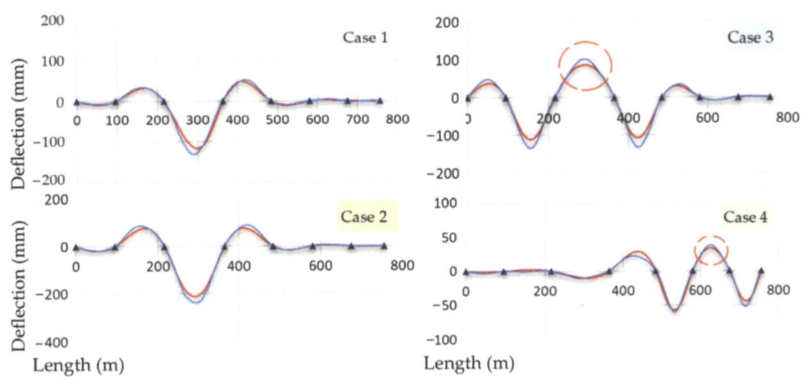

Figure 6. Comparison LC 1 of experimental tests (blue) with numerical analyses (red).

(**a**)

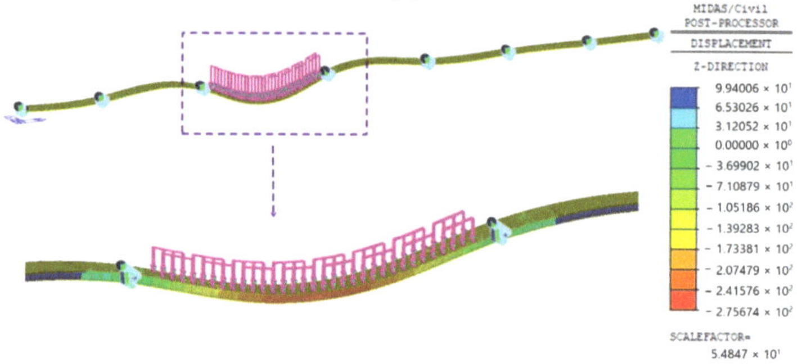

(**b**)

Figure 7. Load case 02 with 16 trucks: (**a**) load test; (**b**) vertical deflection result by FEM.

(b) Load Configuration 02 (LC2)

LC2 creates strong negative bending at the supports. Eight trucks apply load on one side (typically) and eight on the other. The loadings progress along the bridge deck at each support location, as shown in Table 2. The support sections are critical locations that require test loading. Figure 8 reveals that deflections occur only locally. Note the changing vertical scales again. For every case, except 07 and 08, spans adjacent to the load deflect slightly, and those farther away very little. Case 07 produces a lower deflection of span 4 than span 3 due to the different span lengths. The figure shows noticeable upward deflections in span 2 for this loading. Case 08 yields a more significant deflection in span 4 and less deflection in span 5 compared with the symmetrical loadings in LC-2. Loading on span 4 consists of 10 trucks, while on span 5, 6 trucks. While the 10-truck load dominates the deflection shape, the 6-truck loading attenuates deflection in span 5. Another effect from Case 08 is significant upward deflection in span 3, the longest span.

(c) Load Configuration 03 (LC3)

Load configuration 03 creates torsional deflections by alternating lanes in successive spans. The arrangement of trucks forces a downward twist on the loaded edge and an upward twist on neighboring spans in the same lane. The twist increases further due to the loads applied on neighboring spans but in the other lane, as shown in Table 3. Figure 9 shows the results of a typical torsional truck loading. There was a significant difference between the measured and analytical results in the loaded area because the number of loaded spans has increased compared to the other load cases. Differences in torsional stiffness, loading location, and measurement may account for the variance between measured and calculated deflections. Deflection estimates appear slightly overestimated in case 12 (red circle). In the backspan of case 13 (red circle), which represents 54 % of the maximum deflection, these deflections are also due to the torsional effect. An example of this effect appears in Figure 10.

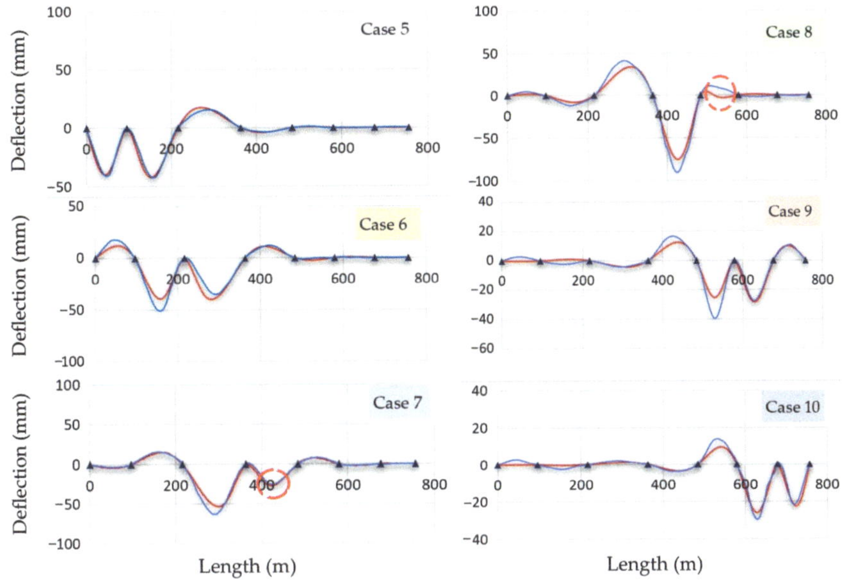

Figure 8. Comparison LC 2 of experimental tests (blue) with numerical analyses (red).

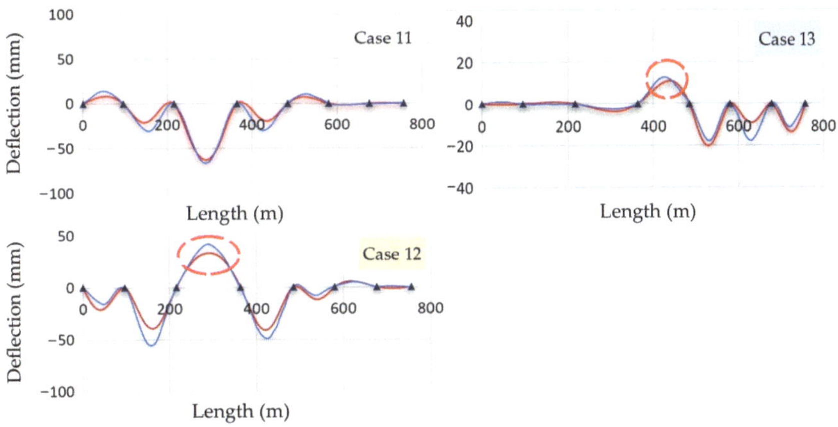

Figure 9. Comparison LC 2 of experimental tests (blue graphs) with numerical analyses (red graphs).

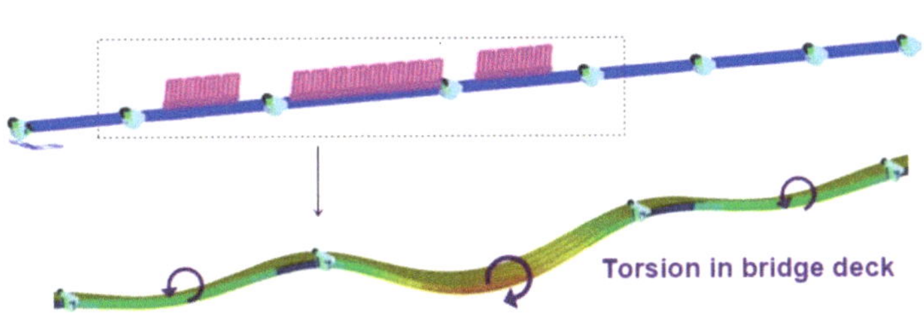

Figure 10. Load case 11, deflections due to torsional loading.

(d) Load Configuration 04 (LC4)

The one-side loadings arrangement appears in Table 4. Figure 11 shows the resulting deck deflections caused by these loads, a significant vertical deflection appears only in the third span, and this is due to the number of trucks (8 trucks) compared to the other spans (4 trucks). For the other spans (4 trucks), a lower deflection is noticeable, representing 1/3 of the maximum deflection for load cases 14, 15 and 17 and 1/6 for case 16.

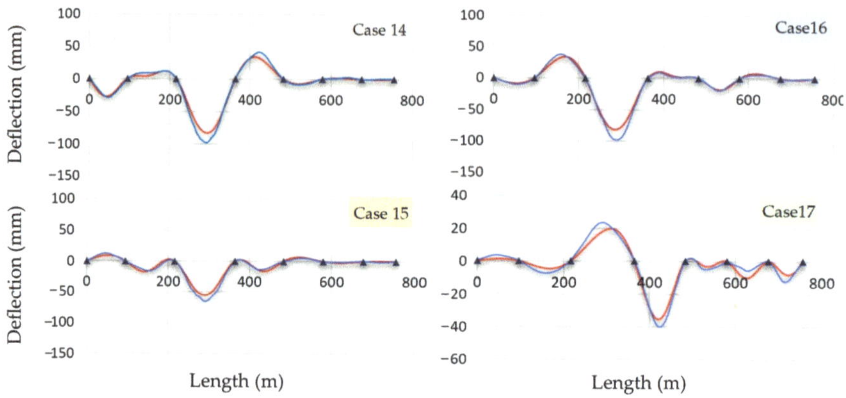

Figure 11. Comparison LC 1 of experimental tests (blue graphs) with numerical analyses (red graphs).

3.1.1. Discussion of Static Deflection Results

There were 17 load tests subject to the weight of various operational trucks. The measured and computed deflections along the bridge appear in graphs. Based on the comparison figures above, the computed deflections in the relevant places and situations are considerably not different from the real state. Observations confirmed that the supports did not settle during the load test, and a visual inspection of the bridges showed no other noticeable damage.

Static load configuration 3, an alternating lane loading case, was found to be a critical design loading position for the box girder bridge due to the possibility of a torsional effect. Furthermore, in static load tests 3 and 4, the large spacing between the loaded spans produces an adverse upward deflection. In general, most of the test cases revealed that the viaduct behaved elastically, with deformations proportional to the applied truck loads and with negligible residual displacements (significantly below an average of 15 mm of the elastic deformation).

For the approval of the constructed bridge, the Hungarian national authority annex [8] allows an error range of no more than 20% between the maximum remaining elastic deformation and maximum detected deformation for steel bridge structures. From Table 5, the results show an average error between 1–21%. These differences may occur because of the deck stiffness variation caused by the non-modeled secondary elements (guard rails and pavement layers); thus, the tested bridge meets the authority requirements.

Table 5. The maximum error between the maximum calculated and measured deflections.

Cases	1	2	3	4	5	6	7	8	9	10	11	12	13	14	15	16	17
MaxError (mm)	10.30	25.26	23.57	2.78	0.48	6.56	7.10	15.42	0.58	3.82	2.73	7.73	2.53	14.99	10.12	16.42	4.33
MaxError (%)	8.78	11.88	21.04	4.85	1.14	16.36	13.27	20.50	2.06	14.68	4.34	18.78	12.40	18.08	18.19	20.09	12.21

3.1.2. Deflection Limit Check

The deflection limit check applies to the center of the span. The critical value of maximum displacement is more significant than (L/400) according to the National Annex e-UT 07.01.12 [8], as presented in Equation (1). Bridge engineers commonly use this limit to evaluate the serviceability of the bridge. Here L is the span length: 96, 120, 148, 120, 96, 96 and 80 m, respectively. So, the calculated allowed deflection is, L/400 = 240, 300, 370, 300, 240, 240 and 200 mm for each span. The maximum displacements of the mid-spans under static vehicle load appear in Figure 12 and fall within the serviceability deflection limit. In conclusion, the bridge fulfills the standard's serviceability requirement.

$$DZ_{\text{ELS (MIDAS)}} \,\&\, DZ_{\text{ELS (Field test)}} < L/400, \quad (1)$$

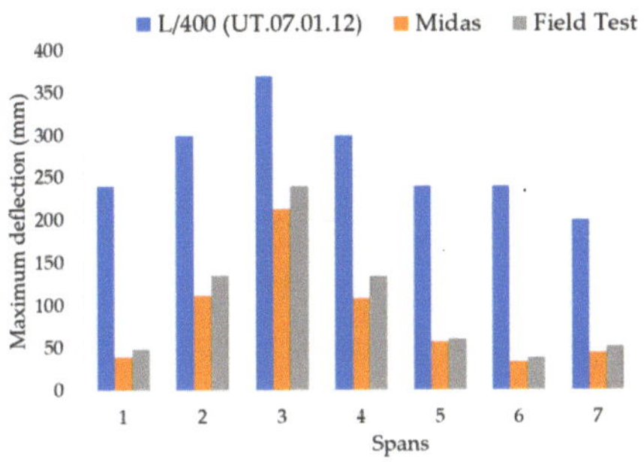

Figure 12. Maximum mid-span deflections of the bridge.

3.2. Bridge Deflection under Dynamic Loadings

A linear time history analysis predicted the moving load effect on the model; the results of the dynamic deflection responses at the midpoint of the first three spans appear in Figures 13 and 14. The applied loads in the model matched the test trucks' weight and velocity.

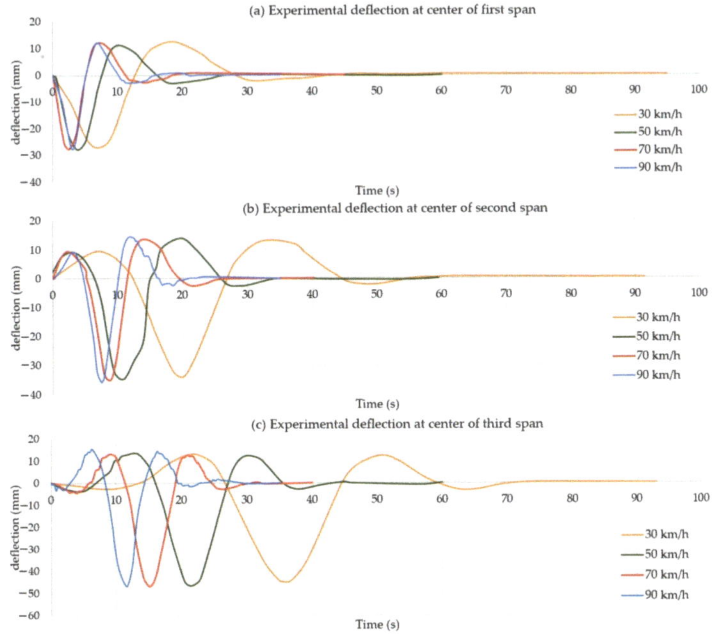

Figure 13. Recorded data of dynamic deflection under moving trucks.

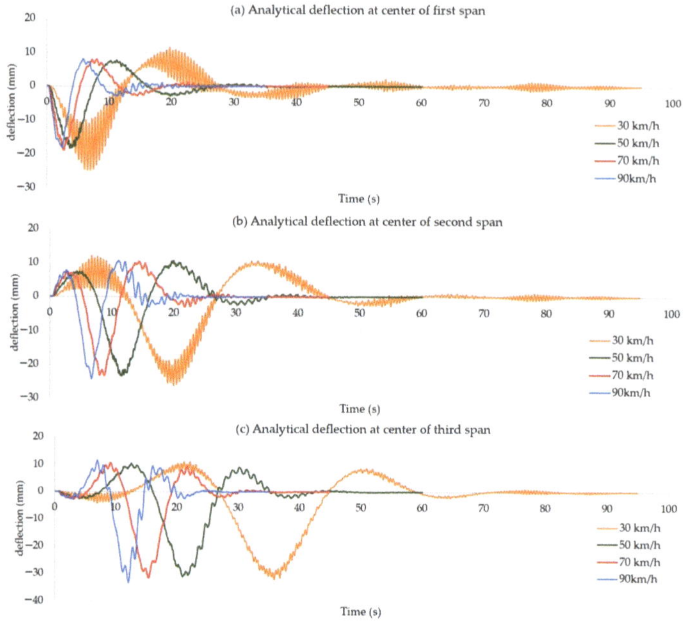

Figure 14. FEM results of dynamic deflection under moving trucks.

As can be seen above, the differences between the recorded and calculated dynamic deflections of the deck along the longitudinal axis under dynamic loadings for different speeds agree closely, with some residual deflections reaching an average of 12 mm at the longest span. These differences may be caused by the truck FE-Model simulated using constantly moving nodal loads without considering the truck's dynamic properties. In addition, there is no significant variation between the maximum amplitudes of deflection that occur at the midpoint of the seven spans.

When the trucks travel at a speed of 30 km/h, the dynamic deflection curves tend to fluctuate more across most of the bridge spans. However, the amplitude of the variability is relatively low. In contrast, at a speed of 90 km/h, the fluctuation of the deflection curve is less pronounced, but the amplitude of the variation is significantly higher. The cause of this behavior is the resonant frequencies of the dynamic system that consists of the moving trucks and the bridge.

Dynamic Amplification Factor

According to Eurocode 1 [38], the dynamic amplification factor φ (DAF), calculated by Equation (2), is the quotient between the dynamic and static response. It is included in traffic load models and depends on several parameters and actions. Eurocode limits φ to a maximum value of 1.7.

$$\varphi_{\,max} = \frac{DZ_{dyn\ max}}{DZ_{static\ max}},\ \varphi_{\,max(Field\ test,\ Midas)} < 1.7 \quad (2)$$

DAF was difficult to assess since the static loading configurations differed significantly from the dynamic loads (8–10 trucks vs. 2 trucks). The alternative here substitutes the 10 km/h test for static conditions. Figure 15 shows that the maximum DAF value is about 1.21 for the field test at 90 km/h on span 5. All test results and computations produced values much less than 1.7, so the bridge works within the appropriate limits with an elastic behavior.

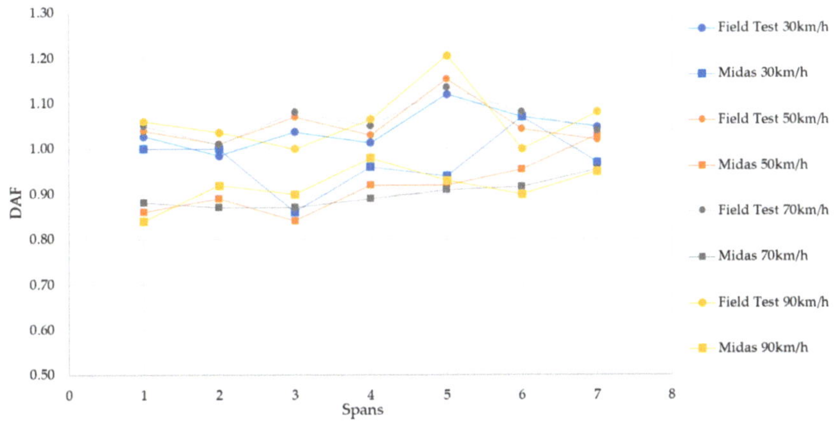

Figure 15. Dynamic amplification factor DAF as a function of truck speed.

4. Stresses in the Steel-Box Deck

During the static load tests, embedded strain gauges recorded stresses in the top and bottom bridge deck panels. A comparison of the longitudinal stresses to the finite element model is presented in Table 6 for the first 10 static loadings in the middle of the longest span and one of its supports, respectively (negative values indicate compression).

Table 6. Stress variations from static load cases.

Load Cases	Stress Variations (N/mm^2)							
	Middle Top Plate Midspan		Middle Bottom Plate Midspan		Middle Top Plate Near Support		Middle Bottom Plate Near Support	
	Measured	Analytical	Measured	Analytical	Measured	Analytical	Measured	Analytical
1	−25.6	−25.1	35.4	37.3	12.2	14	−14.8	−16.2
2	−46.4	−46.9	69.2	71.7	32.4	32.3	−33.5	−37.6
3	13.9	13.7	−20.9	−20.9	11.3	11.9	−9.5	−15.7
4	0.2	−1.4	2.5	2	2.7	1	−0.3	−1.2
5	2	3	−3.5	−4.6	4.3	6.1	−4.1	−7.9
6	−6.3	−10.4	9.6	13.3	26.6	29.1	−25.7	−29.4
7	−10.7	−11.8	15.9	17.4	6	6.5	−6.1	−7.5
8	5.1	4.6	−7.7	−6.7	−3.5	−3.3	3.9	3.9
9	−0.9	−0.6	0.8	0.9	0.2	0.4	−0.3	−0.5
10	0.2	0.2	−0.3	−0.2	−0.2	−0.1	0.3	0.1

Table 6 shows that the measured stresses are similar to those calculated. The magnitude of the deviations is generally small since the changes in stiffness due to pavement layers may cause variations. The trends are very similar and show an acceptable variation.

5. Conclusions

The Szapáry Bridge is one of Hungary's longest bridges constructed with steel box girders. A comprehensive FEM of the bridge was modeled and calibrated to the proof load test measurements to demonstrate the bridge's serviceability before opening to the traffic. Based on the results obtained from the calibrated model, the following conclusions were derived:

- The finite-element model sufficiently represented the bridge field behavior and accurately estimated the deck deflection. This model can serve as a baseline for the as-built conditions of the structure and a starting point for the long-term monitoring of the Szapáry bridge.

- On the basis of the deflection comparison, the method proposed in this paper can be used for reliability and safety assessment of the bridge's state under different loads. The information is more direct when compared to working exclusively with components and structural database. When combined with inspection and maintenance, it will provide valuable information on the bridge's condition over time.
- The structural load-carrying capacity fulfilled the requirements of the original design. It confirmed an appropriate response for serviceability since the experimental deflections align well with the computed values according to the design provisions.
- However, critical load arrangements could damage the long-term bridge serviceability. Thus, designers should consider the many possible load combinations when evaluating deflections.
- For the box girder bridge design, multiple truck loading arrangements must account for any possible torsional action and damage that may occur.

Author Contributions: Conceptualization, M.B. and R.P.R.; methodology, R.P.R.; software, M.B.; validation, R.P.R. and R.K.; resources, M.B. and R.P.R.; data curation, R.P.R.; investigation, R.K.; writing—original draft preparation, M.B.; writing—review and editing, R.P.R. and R.K.; supervision, R.P.R. and R.K. All authors have read and agreed to the published version of the manuscript.

Funding: This research received no external funding.

Institutional Review Board Statement: Not applicable.

Informed Consent Statement: Not applicable.

Data Availability Statement: The data presented in this study are available from the corresponding author upon reasonable request.

Acknowledgments: The authors thank Teiter Zoltan and Péter Hudacsek for providing the data and design information.

Conflicts of Interest: The authors declare no conflict of interest.

References

1. Hidayat, I.; Suangga, M.; Maulana, M.R. The effect of load position to the accuracy of deflection measured with LVDT sensor in I-girder bridge. *IOP Conf. Ser. Earth Environ. Sci.* **2018**, *109*, 012024. [CrossRef]
2. Minsili, L.S.; Tchemou, G.; Louis, M.A.O.; Josette, M.L.A. Parametric Analysis of Vertical Deflections of Bridges under Moving. *Sch. Bull.* **2017**, *3*, 685–694.
3. Birhane, F.N.; Kim, S.I.; Jang, S.Y. Long-term deflection of prestressed concrete bridge considering nonuniform shrinkage and crack propagation by equivalent load approach. *Appl. Sci.* **2020**, *10*, 7754. [CrossRef]
4. Ma, Y.; Chen, H.; Han, Q. Sensitivity Analysis of Factors Affecting down Deflection of Long-span Continuous Rigid Frame Bridge. *IOP Conf. Ser.* **2021**, *781*, 022048. [CrossRef]
5. Robertson, I.N. Prediction of vertical deflections for a long-span prestressed concrete bridge structure. *Eng. Struct.* **2005**, *27*, 1820–1827. [CrossRef]
6. Brownjohn, J.M.W.; Magalhaes, F.; Caetano, E.; Cunha, A. Ambient vibration re-testing and operational modal analysis of the Humber Bridge. *Eng. Struct.* **2010**, *32*, 2003–2018. [CrossRef]
7. Bardakis, V.G.; Fardis, M.N. A displacement-based seismic design procedure for concrete bridges having deck integral with the piers. *Bull. Earthq. Eng.* **2011**, *9*, 537–560. [CrossRef]
8. Ájpli, B.; Borzsák, P. *e-UT 07.01.12—Structural Calculation. Design Code of Highway Bridges 2*; Hungarian roads association: Budapest, Hungary, 2011. (In Hungarian)
9. Dong, C.; Bas, S.; Debees, M.; Alver, N.; Catbas, F.N. Bridge Load Testing for Identifying Live Load Distribution, Load Rating, Serviceability and Dynamic Response. *Front. Built Environ.* **2020**, *6*, 1–14. [CrossRef]
10. Lantsoght, E.O. *Load Testing of Bridges; Proof Load Testing and the Future of Load Testing*; CRC Press: London, UK, 2019; Volume 13. [CrossRef]
11. Schacht, G.; Krontal, M.; Marx, S.; Bolle, G.; Wismar, H. Loading Tests of Existing Concrete Structures-Historical Development and Present Practise. In Proceedings of the Fib Symposium, Prague, Czech Republic, 8–10 June 2011.
12. Jenkins, C.H.; Kjerengtroen, L.; Oestensen, H. Sensitivity of Parameter Changes in Structural Damage Detection. *Shock. Vib.* **1997**, *4*, 807239. [CrossRef]
13. Dolati, S.S.K.; Caluk, N.; Armin, B.M.; Dolati, S.S.K. Non-Destructive Testing Applications for Steel Bridges. *Appl. Sci.* **2021**, *11*, 9757. [CrossRef]

14. Zhu, Y.F.; Ren, W.X.; Wang, Y.F. Structural health monitoring on Yangluo Yangtze River Bridge: Implementation and demonstration. *Adv. Struct. Eng.* **2022**, *25*, 1431–1448. [CrossRef]
15. Fang, I.-K.; Chen, C.-R.; Chang, I.-S. Field Static Load Test on Kao-Ping-Hsi Cable-Stayed Bridge. *J. Bridg. Eng.* **2004**, *9*, 531–540. [CrossRef]
16. Ren, W.-X.; Lin, Y.-Q.; Peng, X.-L. Field Load Tests and Numerical Analysis of Qingzhou Cable-Stayed Bridge. *J. Bridg. Eng.* **2007**, *12*, 261–270. [CrossRef]
17. Hassan, M.; Burdet, O.; Favre, R. Analysis and Evaluation of Bridge Behavior Under Static Load Testing Leading to Better Design and Judgment Criteria. In Proceedings of the Fourth Bridge Engineering Conference, Transportation Research Board (TRB), San Francisco, CA, USA, 28–30 August 1995.
18. Galati, N.; Casadei, P.; Nanni, A. *In-Situ Load Testing of Bridge a6101 Lexington, MO*; Center for infrastructure engineering studies: Washington, DC, USA, 2005.
19. Wang, X.; Wang, L.; Wang, H.; Ning, Y.; Huang, K.; Wang, W. Performance evaluation of a long-span cable-stayed bridge using non-destructive field loading tests. *Appl. Sci.* **2022**, *12*, 2367. [CrossRef]
20. Zheng, X.; Du, H.; Zhang, R.; Ma, Q.; Gao, H. Static load test of RC T-beam bridge flexural strengthening with externally bonded steel plates. *IOP Conf. Ser. Earth Environ. Sci.* **2021**, *791*, 012018. [CrossRef]
21. Cao, W.J.; Koh, C.G.; Smith, I.F.C. Enhancing static-load-test identification of bridges using dynamic data. *Eng. Struct.* **2019**, *186*, 410–420. [CrossRef]
22. Bayraktar, A.; Türker, T.; Tadla, J.; Kurşun, A.; Erdiş, A. Static and dynamic field load testing of the long span Nissibi cable-stayed bridge. *Soil Dyn. Earthq. Eng.* **2017**, *94*, 136–157. [CrossRef]
23. Sun, Z.; Siringoringo, D.M.; Fujino, Y. Load-carrying capacity evaluation of girder bridge using moving vehicle. *Eng. Struct.* **2021**, *229*, 111645. [CrossRef]
24. Cantero, D.; Gonzáez, A.; Obrien, E.J. Comparison of bridge dynamic amplifications due to articulated 5-axle trucks and large cranes. *Balt. J. Road Bridg. Eng.* **2011**, *6*, 39–47. [CrossRef]
25. Lak, M.A.; Degrande, G.; Lombaert, G. The effect of road unevenness on the dynamic vehicle response and ground-borne vibrations due to road traffic. *Soil Dyn. Earthq. Eng.* **2011**, *31*, 1357–1377. [CrossRef]
26. Cantero, D.; Hester, D.; Brownjohn, J. Evolution of bridge frequencies and modes of vibration during truck passage. *Eng. Struct.* **2017**, *152*, 452–464. [CrossRef]
27. Moussa, A.I.; Shahawy, M.A. Dynamic and static tests of prestressed concrete girder bridges in Florida. *Struct. Res. Cent.* **1993**, *80*.
28. Hernandez, W.; Viviescas, A.; Riveros-Jerez, C.A. Vehicle Bump Testing Parameters Influencing Modal Identification of Long-Span Segmental Prestressed Concrete Bridges. *Sensors* **2022**, *22*, 1219. [CrossRef] [PubMed]
29. Gatti, M. Structural health monitoring of an operational bridge: A case study. *Eng. Struct.* **2019**, *195*, 200–209. [CrossRef]
30. Meixedo, A.; Ribeiro, D.; Santos, J.; Calçada, R.; Todd, M. Progressive numerical model validation of a bowstring-arch railway bridge based on a structural health monitoring system. *J. Civ. Struct. Health Monit.* **2021**, *11*, 421–449. [CrossRef]
31. De Angelis, A.; Pecce, M.R. Model assessment of a bridge by load and dynamic tests. *Eng. Struct.* **2023**, *275*, 115282. [CrossRef]
32. Qin, S.; Zhang, Y.; Zhou, Y.L.; Kang, J. Dynamic model updating for bridge structures using the kriging model and PSO algorithm ensemble with higher vibration modes. *Sensors* **2018**, *18*, 1879. [CrossRef] [PubMed]
33. Sousa, H.; Cavadas, F.; Henriques, A.; Bento, J.; Figueiras, J. Bridge deflection evaluation using strain and rotation measurements. *Smart Struct. Syst.* **2013**, *11*, 365–386. [CrossRef]
34. Gouda, L.P. Study on Parametric Behaviour of Single Cell Box Girder under Different Radius of Curvature. Master's Thesis, National Institute of Technology Rourkela, Rourkela, India, 2013.
35. MIDAS Civil. *Midas User Manual*; MIDAS Information Technology Co.: Seongnam, Republic of Korea, 2018.
36. *EN 1993-2*; Eurocode 3: Design of steel structures—Part 2: Steel bridges. CEN-European Committee for Standardization: Brussels, Belgium, 2006.
37. Gupta, P.K.; Singh, K.K.; Mishra, A. Parametric Study on Behaviour of Box-Girder Bridges Using Finite Element Method. *Asian J. Civ. Eng.* **2010**, *11*, 135–148.
38. *EN 1991-3*; Eurocode 1: Basis of design and actions on structures—Part 2: Traffic loads on bridges. CEN-European Committee for Standardization: Brussels, Belgium, 2003.

infrastructures

MDPI

Article

Comprehensive Empirical Modeling of Shear Strength Prediction in Reinforced Concrete Deep Beams

Eyad K. Sayhood [1], Nisreen S. Mohammed [1], Salam J. Hilo [1,*] and Salih S. Salih [2]

1 Civil Engineering Department, University of Technology—Iraq, Baghdad 10071, Iraq; 40075@uotechnology.edu.iq (E.K.S.); 40055@uotechnology.edu.iq (N.S.M.)
2 Information Technology Centre, University of Technology—Iraq, Baghdad 10071, Iraq; salih.s.salih@uotechnology.edu.iq
* Correspondence: eng.salamjh@gmail.com; Tel.: +964-7814-395-815

Abstract: This paper presents comprehensive empirical equations to predict the shear strength capacity of reinforced concrete deep beams, with a focus on improving the accuracy of existing codes. Analyzing 198 deep beams imported from 15 existing investigations, this study considers various parameters such as concrete compressive strength (f'_c), the shear span-to-effective depth ratio (a_v/d), and reinforcement ratios (p_s, p_v, and p_h). Introducing a novel predictive empirical equation, this study conducts a rigorous evaluation using statistical metrics and a linear regression analysis (MAE, RMSE, and R^2). The proposed model demonstrates a significant reduction in the coefficient of variation (CV) to 27.08%, compared to the existing codes' limitations. Comparative analyses highlight the accuracy of the empirical equation, revealing an improved convergence of data points and minimal sensitivity to variations in key parameters. The results proved that the proposed empirical equation enhanced the accuracy to predict the shear strength capacity of the reinforced concrete deep beams in various scenarios, making it a valuable tool for structural engineers. This research contributes to advancing the understanding of shear strength capacity in reinforced concrete deep beams, offering a reliable empirical equation with implications for refining design methodologies and enhancing safety with the efficiency of structural systems.

Keywords: reinforced concrete structures; shear strength capacity; structural analysis; empirical equations; concrete compressive strength; coefficient of variation

Citation: Sayhood, E.K.; Mohammed, N.S.; Hilo, S.J.; Salih, S.S. Comprehensive Empirical Modeling of Shear Strength Prediction in Reinforced Concrete Deep Beams. *Infrastructures* **2024**, *9*, 67. https://doi.org/10.3390/infrastructures9040067

Academic Editors: Armin Mehrabi and Seyed Saman Khedmatgozar Dolati

Received: 19 February 2024
Revised: 19 March 2024
Accepted: 22 March 2024
Published: 28 March 2024

1. Introduction

A deep beam, conventionally defined by a span-to-depth ratio (h/L) of ≤4 or with a shear span smaller than twice its depth, is primarily governed by shear strength rather than flexure, given sufficient longitudinal reinforcement utilization, as depicted in Figure 1b [1]. Additionally, deep beams with a span ratio (h/L) ≤ 2.5 are classified as such and have extensive applications in constructions like squat walls, foundation pile caps, and deep foundations, as illustrated in Figure 1c [2].

Numerous studies have explored the structural behavior of reinforced concrete (RC) deep beams, employing experimental, analytical, and numerical approaches. Eyad et al. (2018) [3] investigated a simply supported deep beam subjected to a uniform distributed load, providing a comprehensive analysis of cracking effects and ultimate shear strength.

Albidah (2023) [4] conducted tests on six metakaolin–fly ash-based geopolymer concrete beams, considering parameters such as the steel fiber content and shear reinforcement percentage. The study demonstrated significant enhancements in shear strength by 16.7% and 31.6% with the addition of steel fibers at rates of 0.35% and 0.70%, respectively. Eyad et al. [5] explored the impact of confining the strut region through the use of strut reinforcements, and Eyad et al. [6] proposed an empirical formula for the strut efficiency factor (B_s) in RC deep beams which was derived from a comprehensive analytical study based on the strut-and-tie model. Other researchers have numerically investigated the main parameters influencing the behavior and shear capacity of RC deep beams [7,8].

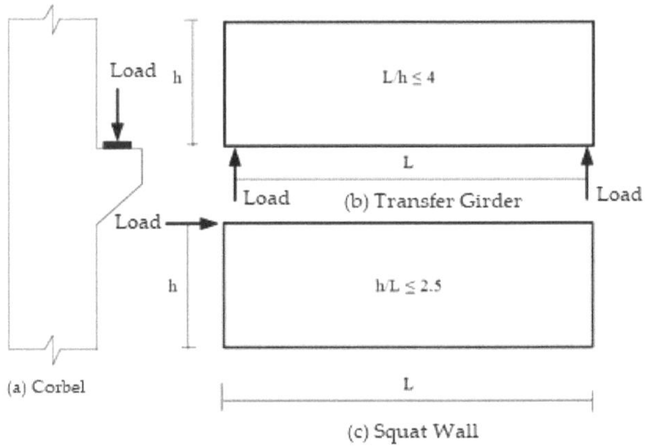

Figure 1. Application of deep beams in reinforced concrete structures.

Despite numerous experimental and numerical investigations into reinforced concrete deep beams which have considered factors such as the concrete compressive strength, the a_v/d ratio, and reinforcement directions in relation to shear capacity, the current ACI 318R-5 [1] and BS 8110 [9] codes still do not incorporate these factors comprehensively. The formulas provided in these codes are primarily limited to the concrete compressive strength, web width (b_w), and depth (d) factors only, as presented in Equations (1)–(3), respectively. Due to this fact, the current study found that the predictive accuracy of code ACI 318R-5 [1] and code BS 8110 [9] is restricted, with coefficients of variation (CVs) for shear capacity prediction. This study offers a thorough examination of deep beams, taking into account pivotal factors that influence the concrete compressive strength (f'_c), the shear span-to-depth ratio (a_v/d), the web width (b_w), the ratios of longitudinal (P_s), vertical (P_v), and horizontal (P_h) reinforcements, the depth (d), the yield strength of vertical stirrups (f_{yv}), and the concrete area (web width × depth ($b_w \times d$)). The evaluation was based on a large dataset that includes 198 deep beams imported from 15 different studies [10–25]. These studies were chosen for their detailed information on test conditions and material properties, thus forming a robust database conducive to scrutinizing code provisions and affirming the proposed predictive model.

This paper introduces a novel model for predicting the shear strength of reinforced concrete deep beams generated from an analysis of 198 experimental simply supported RC deep beams subjected to concentrated and uniform loads. The proposed model demonstrates a remarkable improvement in accuracy, outperforming the predictions of both ACI 318R-5 and BS 8110 by a remarkable percentage. This significant enhancement can be credited to the model's comprehensive consideration of 10 factors influencing the shear strength of reinforced concrete deep beams, unlike the restrictive focus on concrete compressive strength only found in the ACI and BS standards.

2. Methodology

To achieve the research objective, twelve empirical equations were developed to theoretically predict the shear strength of reinforced concrete deep beams. A total of 198 experimental simply supported RC deep beams subjected to concentrated and uniform loads wer eused to establish the empirical factors. The effects of several parameters (Table 1) were considered, such as the concrete compressive strength (f′c), the shear span-to-depth ratio (a_v/d), the web width (b_w), the ratios of longitudinal (P_s), vertical (P_v), and horizontal (P_h) reinforcements, the depth (d), the yield strength of vertical stirrups (f_{yv}), and the concrete area (web width × depth (b_w × d)). The results obtained from the experimental tests were used to verify the developed empirical formula.

To develop the empirical equations, various input data for the selected parameters as well as experimental results for the simply supported RC deep beams were exported manually into Microsoft Excel software version 2007. This study wrote a program using "Microsoft Visual Basic" (MVB) in Microsoft Excel software to calculate the relevant results/values required to establish twelve empirical equations. In addition, MVB was used to compute the relevant results based on code methods and formulas for a comparison with the developed empirical formula.

Table 1. Data ranges for the selected variables.

No.	Variable	Unit	Range
1	f'_c	MPa	16.08–47.6
2	a_v/d	-----	0.19–2.5
3	b_w	mm	76–914.4
4	P_s	-----	0.176–3.1%
5	P_v	-----	0.13–2.45%
6	P_h	-----	0–1%
7	d	mm	215.9–1752
8	f_{yv}	MPa	230–590
9	$b_w \times d$	mm^2	16,416–939,667.7
10	Vu	kN	77.8–6294

To improve readability and clarity, this methodology of this study is organized into three separate phases: existing experimental investigations, code methods for determining RC deep beam shear strength, and an examination of the statistical properties of the dataset used.

2.1. Existing Experimental Investigations

The dataset of the current study, generated from 15 references in the literature [10–25], consists of 198 simply supported RC deep beams subjected to shear testing. The main variables imported from these tests include the concrete compressive strength (f'_c), the shear span-to-depth ratio (a_v/d), web width (b_w), the ratios of longitudinal (P_s), vertical (P_v), and horizontal (P_h) reinforcements, depth (d), the yield strength of vertical stirrups (f_{yv}), and the concrete area (web width × depth (b_w × d)),as presented in Table 1. This study found that the resulting range of ultimate shear forces (V_u) varied from 77.8 kN to 6294 kN, providing a comprehensive dataset for assessing the shear behavior of RC deep beams. This study evaluated the literature to identify all of the important variables that are considered influential shear failure. By gathering such a broad dataset, this study allows for a thorough examination of how various variables affect the shear capacity of deep beams. This approach provides a thorough understanding of the relationship between these variables and final shear strength, boosting the potential to create a new empirical equation for predicting and interpreting accurate shear failure. Figure 2 shows the typical design, geometry, and failure pattern of a shear diagonal fracture in a representative deep beam, together with all of the variables investigated in this work.

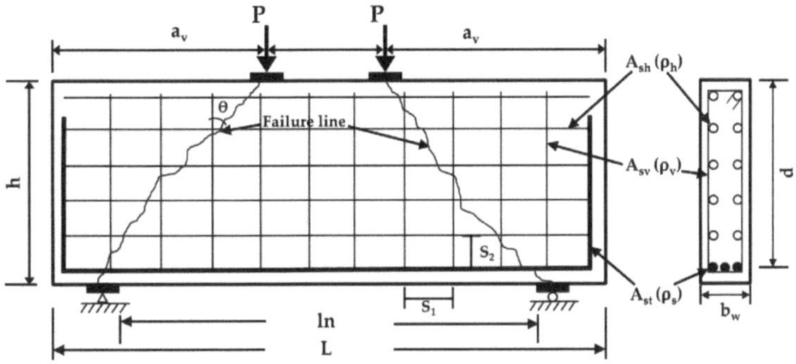

Figure 2. The geometry and failure pattern of simply supported RC deep beam with.

2.2. Code Procedures for Calculating the Resistance of RC Beams

There are many design codes provide unique methods to calculate the shear strength of RC deep beams. This study suggests choosing the most popular two standards, ACI 318 R 1515 [1]] and BS 8110 [9], and explain their formulas in details in term of computing the nominal shear strength. The ACI and BS design codes are available to guide the structural designers to design RC deep beams and calculate the ultimate shear strength. This study found that, there are nine variables effect on calculating the nominal shear strength of RC deep beams such as: shear span-to-depth ratio (a_v/d), the ratios of longitudinal (P_s), vertical (P_v), and horizontal (P_h) reinforcements, the yield strength of vertical stirrups (f_{yv}), concrete area (web width × depth ($b_w \times d$)), compressive strength of concrete (f'_c), web width (b_w), and depth (d). On the other hand, the ACI 318 R 15 and BS 8110 codes equations used only three variables such as: concrete compressive strength (f'_c), web width (b_w), and depth (d) only, which it is considered as very limited variables compared to the nine mentioned variables. The details of the ACI 318 R 15 and BS 8110 codes equations are presented bellow.

- ACI 318 R-15 [1]

The ACI code is a very popular calculation method for calculating the nominal shear strength in RC deep beams. The formula of ACI code that involve in calculating the nominal shear stress of RC deep beams was limited to the concrete compressive strength (f'_c), web width (b_w), and depth (d) only as presented in Equation (1).

$$V_{nBS} = \frac{5}{6}\sqrt{f'_c} b_w d \tag{1}$$

- BS 8110 [9]

The BS code is a very popular standardization for calculating the nominal shear strength in RC deep beams. The equations of BS code that involve in predicting the nominal shear stress of RC deep beams were restricted to the compressive strength of concrete (f'_c), width of web (b_w), and depth (d) only, as presented in Equations (2) and (3).

$$v_{nBS} \leq \begin{cases} 0.8\sqrt{f'_c} \\ \text{or} \\ 5\ \text{N/mm}^2 \end{cases} \tag{2}$$

$$V_{nBS} = v_{nBS} b_w d \tag{3}$$

2.3. Statistical Properties of the Dataset

This study suggests presenting a very simple equation (Equation (4)) to calculate the nominal shear strength (V_n) of RC deep beams with more accurate value compare to the

ACI 318 R 15 and BS 8110 codes. Simply, the suggested equation can be divided into two terms, the shear strength of concrete (V_c) and the strength of shear reinforcement (V_s).

$$V_n = V_c + V_s \tag{4}$$

The current study suggests presenting the details of the shear strength of concrete (V_c) and the strength of shear reinforcement (V_s) terms, in separate sections to provide a clear understanding of how this investigation developed new empirical equation covering the effect of the mentioned variables on the nominal shear strength of RC deep beams.

- Concrete Shear Strength Term (Vc)

To cover the effect of the concrete compressive strength (f'_c), longitudinal reinforcing percentage (p_s), modulus of elasticity of steel (E_s), modulus of elasticity of concrete (E_c), and shear span-to-effective depth ratio (a_v/d), this study suggests formulating the concrete shear strength (V_c) term in eight different equations as presented in Equations (5)–(8) and Equations (11)–(14), each equation customized to specific variables. To present these equations in details it was suggested to divide it into three different stages.

Stage one, comprises four equations as presented in Equations (5)–(8) which are related to the compressive strength of concrete (f'_c) and the percentage of longitudinal reinforcing (p_s) variables. The value of the uncracked compression zone depth (K) can be calculated based on multiplying the longitudinal reinforcing percentage (p_s) by the ratio of the modulus of elasticity of the longitudinal reinforcement (E_s) to the modulus of elasticity of concrete (E_c), as presented in Equations (9) and (10).

$$A \times {f'_c}^{B} \times p_s^C \tag{5}$$

$$A \times \left({f'_c}^{B} + p_s^C\right) \tag{6}$$

$$A \times \left({f'_c}^{B} + C \times p_s^C\right) \tag{7}$$

$$A \times {f'_c}^{B} \times K^C \tag{8}$$

The value of the uncracked compression zone depth (K) is introduced as an essential parameter in this term; it can be calculated based on multiplying the longitudinal reinforcing percentage (p_s) by the ratio of the modulus of elasticity of the longitudinal reinforcement (E_s) to the modulus of elasticity of concrete (E_c), as presented in Equations (9) and (10).

$$K = \sqrt{p_s^C n^2 + 2p - {p_s^C}_s^C\, n\, n} \tag{9}$$

$$n = \frac{E_s}{E_c} \tag{10}$$

Derived from deep beam geometry, the second stage comprises three equations (Equations (11)–(13)) which are related to the shear span-to-effective depth ratio (a_v/d).

$$\left(a_v/d\right)^F \tag{11}$$

$$\frac{D}{E + (a_v/d)^F} \tag{12}$$

$$D + \frac{E}{(a_v/d)^F} \tag{13}$$

Accounting for the size effect of diagonal shear strength in deep beams, the third stage comprises one equation (Equation (14)) which is related to the effective depth (d) variable.

$$\left(\frac{G}{d}\right)^H \tag{14}$$

- Shear Reinforcement Strength Term (V_s)

The shear reinforcement strength (V_s) term is expressed in four equations (Equation (15) to Equation (17)), each customized for specific conditions. These equations will be segmented into two different stages, with each stage reflecting the influence of distinct variables.

The first stage comprises two equations (Equations (15) and (16)) related to the transverse vertical shear reinforcement (p_v), which can be presented as follows:

$$k_v \times p_v \times f_{yv} \tag{15}$$

$$k_v = \left(\frac{1+\left(\frac{a_v}{d}\right)}{6}\right) \tag{16}$$

The second stage comprises two equations as well (Equations (17) and (18)), but these equations are related to the transverse horizontal shear reinforcement (p_h), which can be represented as follows:

$$k_h \times p_v \times f_{yv} \tag{17}$$

$$k_h = \left(\frac{5-\left(\frac{a_v}{d}\right)}{6}\right) \tag{18}$$

The coefficients k_v and k_h are determined based on the a_v/d ratio instead of the clear span-to-effective depth ratio (l_n/d), reflecting the influence of shear reinforcement on the deep beam. The sum of k_v and k_h is constrained to unity ($k_v + k_h = 1$), emphasizing the proportionality of vertical and horizontal shear reinforcements. Figure 3 shows the variation of the coefficients (k_v and k_h) with respect to (a_v/d).

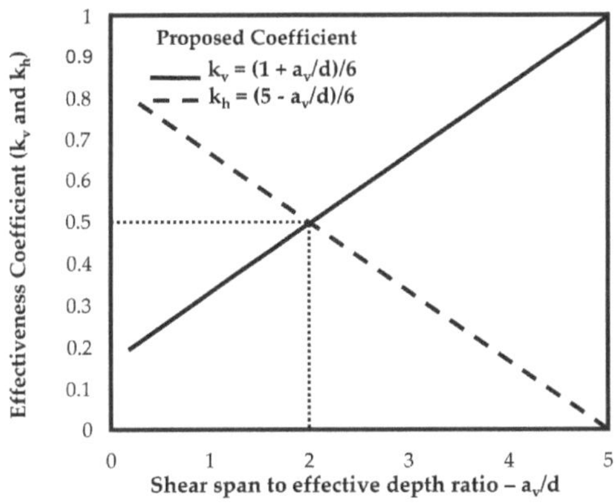

Figure 3. Coefficient of effectiveness for vertical and horizontal transverse shear reinforcements.

When the a_v/d ratio is low, the angle (θ) between the vertical reinforcement and the failure line (diagonal shear crack) is minimal. In such cases, a horizontal reinforcement proves more effective in resisting tension stresses (those that cannot be borne by the concrete) compared to a vertical reinforcement, ($k_v > k_h$); as the a_v/d ratio increases, the significance of the vertical reinforcement in resisting tension stresses becomes more pronounced. When (a_v/d) = 2, both types of reinforcement exhibit equal effectiveness, with $k_v = k_h = 0.5$.

3. The Proposed Empirical Equations

This study used 12 different combinations to develop empirical equations for estimating the nominal shear strength (V_n) of RC deep beams, as presented in Table 2. Then, a collinear regression analysis served as a cornerstone for determining the coefficients A–H in the proposed empirical equations. This intricate process involved the utilization of Microsoft Excel software, which replaced the test results for the nominal shear strength values of the selected 198 deep beams (V_n) in these calculations. The main objective of this research was to increase the accuracy of estimating the capacity of the nominal shear strength in RC deep beams through the proposed empirical equations. So, the presented study adopts and calculates the error values to assess the efficacy and precision of each suggested term in each developed equation. Three key statistical metrics were adapted to verify the developed empirical equations such as, the mean absolute error (MAE), the root mean square error (RMSE), and the coefficient of multiple determinations (R^2).

Table 2. Empirical equations for estimating the nominal shear strength (V_n) of an RC deep beam.

Proposal No.	Combination of Equations	Proposed Empirical Equations
1	5 × 11 × 14+ (15+17)	$A \times {f'_c}^{B} \times {p_s}^{C} \times \left(\frac{a_v}{d}\right)^F \times \left(\frac{G}{d}\right)^H \times (b_w d) + (k_v p_v + k_h p_h) \times f_{yv} \times (b_w d)$
2	5 × 12 × 14 + (15 + 17)	$A \times {f'_c}^{B} \times {p_s}^{C} \times \left(\frac{D}{E+(a_v/d)^F}\right) \times \left(\frac{G}{d}\right)^H \times (b_w d) + (k_v p_v + k_h p_h) \times f_{yv} \times (b_w d)$
3	5 × 13 × 14 + (15 + 17)	$A \times {f'_c}^{B} \times {p_s}^{C} \times \left(D + E/(a_v/d)^F\right) \times$ $\left(\frac{G}{d}\right)^H \times (b_w d) + (k_v p_v + k_h p_h) \times f_{yv} \times (b_w d)$
4	6 × 11 × 14 + (15 + 17)	$A \times \left({f'_c}^{B} + {p_s}^{C}\right) \times \left(\frac{a_v}{d}\right)^F \times \left(\frac{G}{d}\right)^H \times (b_w d) + (k_v p_v + k_h p_h) \times f_{yv} \times (b_w d)$
5	6 × 12 × 14 + (15 + 17)	$A \times \left({f'_c}^{B} + {p_s}^{C}\right) \times \left(\frac{D}{E+(a_v/d)^F}\right) \times$ $\left(\frac{G}{d}\right)^H \times (b_w d) + (k_v p_v + k_h p_h) \times f_{yv} \times (b_w d)$
6	6 × 13 × 14 + (15 + 17)	$A \times \left({f'_c}^{B} + {p_s}^{C}\right) \times \left(D + \frac{E}{(a_v/d)^F}\right) \times$ $\left(\frac{G}{d}\right)^H \times (b_w d) + (k_v p_v + k_h p_h) \times f_{yv} \times (b_w d)$
7	7 ×11 × 14 + (15 + 17)	$A \times \left({f'_c}^{B} + C \times p_s\right) \times \left(\frac{a_v}{d}\right)^F \times \left(\frac{G}{d}\right)^H \times (b_w d) + (k_v p_v + k_h p_h) \times f_{yv} \times (b_w d)$
8	7 × 12 × 14 + (15 + 17)	$A \times \left({f'_c}^{B} + C \times p_s\right) \times \left(\frac{D}{E+ a_v/d)^F}\right) \times$ $\left(\frac{G}{d}\right)^H \times (b_w\, d) + (k_v p_v + k_h p_h) \times f_{yv} \times (b_w\, d)$
9	7 × 13 × 14 + (15 + 17)	$A \times \left({f'_c}^{B} + C \times p_s\right) \times \left(D + \frac{E}{(a_v/d)^F}\right) \times$ $\left(\frac{G}{d}\right)^H \times (b_w\, d) + (k_v p_v + k_h p_h) \times f_{yv} \times (b_w\, d)$
10	8 × 11 × 14 + (15 + 17)	$A \times {f'_c}^{B} \times K^C \times \left(\frac{a_v}{d}\right)^F \times \left(\frac{G}{d}\right)^H \times (b_w\, d) + (k_v p_v + k_h p_h) \times f_{yv} \times (b_w\, d)$
11	8 × 12 × 14 + (15 + 17)	$A \times {f'_c}^{B} \times K^C \times \left(\frac{D}{E+(a_v/d)^F}\right) \times \left(\frac{G}{d}\right)^H \times (b_w\, d) + (k_v p_v + k_h p_h) \times f_{yv} \times (b_w\, d)$
12	8 × 13 × 14 + (15 + 17)	$A \times {f'_c}^{B} \times K^C \times \left(D + \frac{E}{(a_v/d)^F}\right) \times \left(\frac{G}{d}\right)^H \times (b_w\, d) + (k_v p_v + k_h p) \times f_{yv} \times (b_w\, d)$

The MAE is a collinear score that calculates the likelihood of each variable; it represents the average absolute discrepancy between expected and observed values in the verification model. The RMSE is the root mean square error, which is a squared and averaged calculation of the difference between expected and imported values. The MAE and RMSE work in parallel to identify the variations of errors in the predicted values, the MAE always being equal to or greater than the RMSE. R^2 determines the amount of variability that a regression model can count for. If the value of $R^2 = 1$ suggests that the regression model describes the data accurately, meanwhile if $R^2 = 0$ indicates that the regression model describes the data inaccurately. To determine the optimum empirical equations, the recommended models were chosen based on having the lowest MAE and RMSE values and the highest (R^2) value, aligning with the desired precision and reliability. These coefficients were calculated using Equations (19)–(21), respectively.

$$\mathrm{MAE} = \frac{1}{N}\sum_{i=1}^{N} |x_i - y_i| \tag{19}$$

$$\mathrm{RMSE} = \sqrt{\frac{\sum_{i=1}^{N}(x_i - y_i)^2}{N}} \quad (20)$$

$$R^2 = 1 - \frac{\mathrm{SSE}}{\mathrm{SST}} = 1 - \frac{\sum_{i=1}^{N}(x_i - y_i)^2}{\sum_{i=1}^{N}\left(x_i - \bar{x}\right)^2} \quad (21)$$

Figures 4–8, show the relationship between the effect of the main parameters on the shear resistance prediction and the value of V_{exp}/V_n.

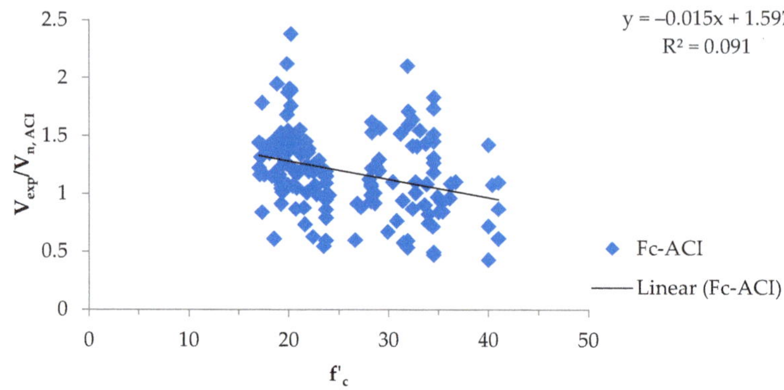

(**a**)

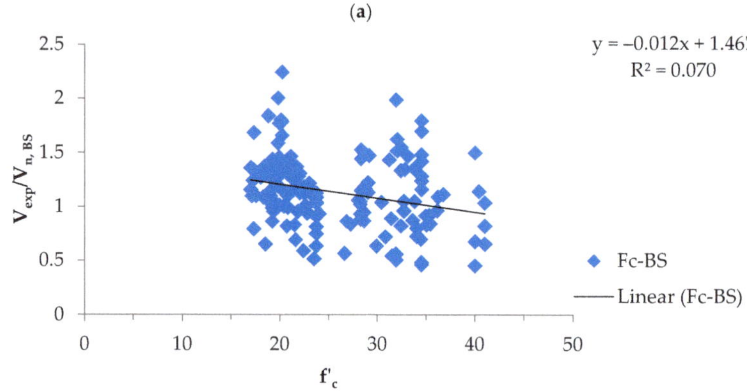

(**b**)

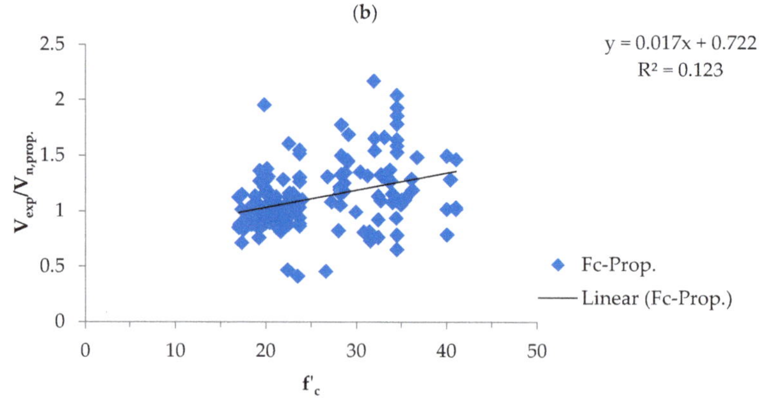

(**c**)

Figure 4. Effect of $f_c^{'}$ (MPa) on V_{exp}/V_n ratio for the ACI, BS, and proposed methods. (**a**) Effect of (f'_c) on $V_{exp}/V_{n,ACI}$. ratio. (**b**) Effect of (f'_c) on $V_{exp}/V_{n,BS}$. ratio. (**c**) Effect of (f′c) on $V_{exp}/V_{n,prop}$. ratio.

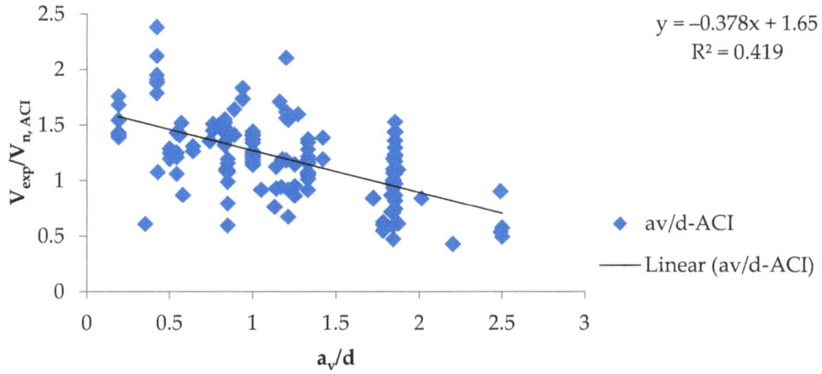

(a)

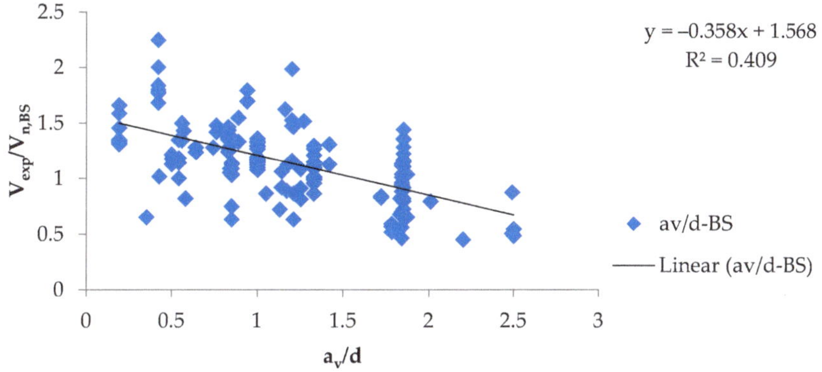

(b)

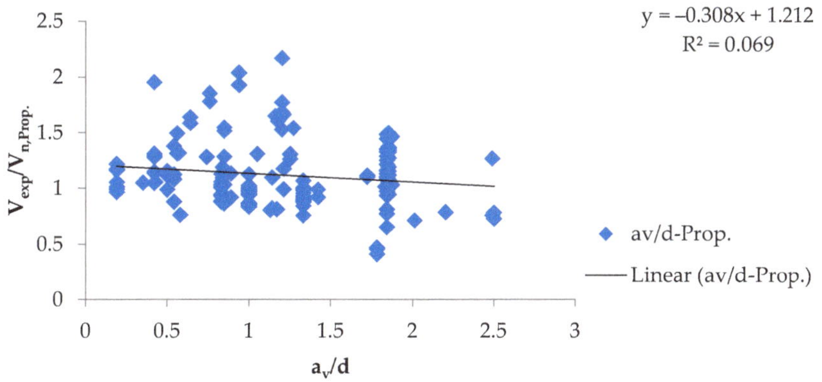

(c)

Figure 5. Effect of a_v/d (MPa) on V_{exp}/V_n ratio for the ACI, BS, and proposed methods. (**a**) Effect of a_v/d on $V_{exp}/V_{n,ACI}$. ratio. (**b**) Effect of a_v/d on $V_{exp}/V_{n,BS}$. ratio. (**c**) Effect of a_v/d on $V_{exp}/V_{n,prop}$. ratio.

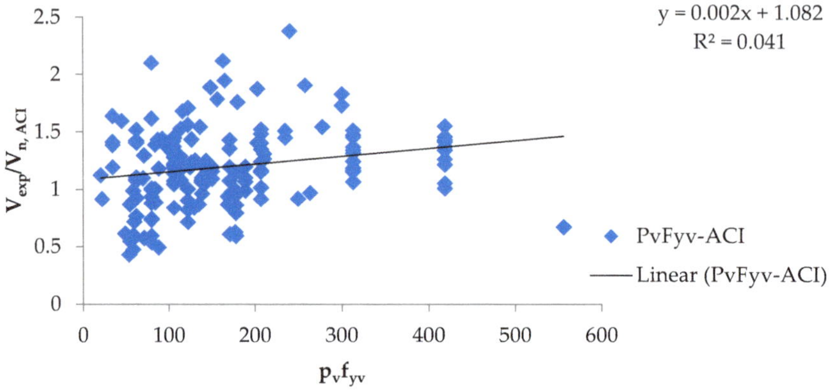

(a)

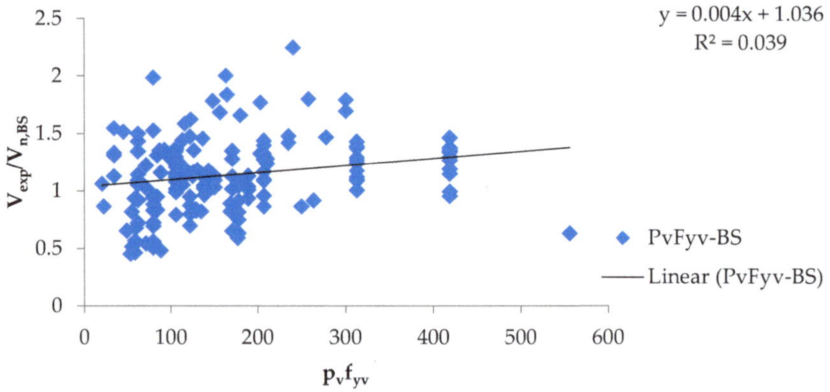

(b)

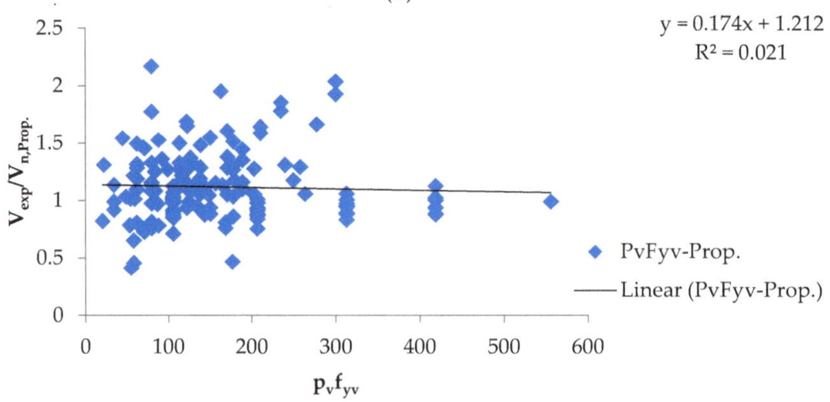

(c)

Figure 6. Effect of $p_v f_{yv}$ (MPa) on V_{exp}/V_n ratio for the ACI, BS, and proposed methods. (**a**) Effect of $p_v f_{yv}$ on $V_{exp}/V_{n,ACI}$. ratio. (**b**) Effect of $p_v f_{yv}$ on $V_{exp}/V_{n,BS}$. ratio. (**c**) Effect of $p_v f_{yv}$ on $V_{exp}/V_{n,prop}$. ratio

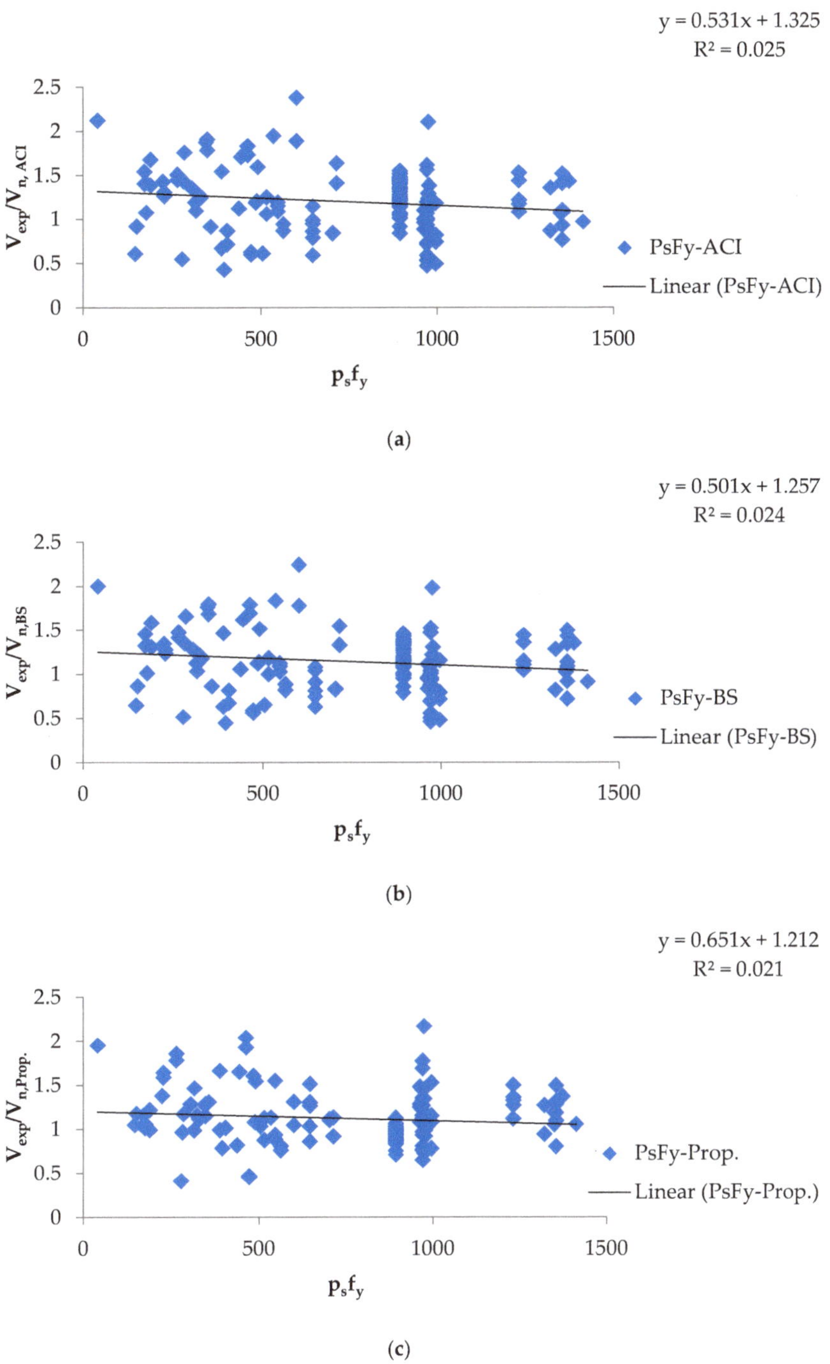

Figure 7. Effect of p_sf_y (MPa) on V_{exp}/V_n ratio for the ACI, BS, and proposed methods. (**a**) Effect of p_sf_y on $V_{exp}/V_{n,ACI}$. ratio. (**b**) Effect of p_sf_y on $V_{exp}/V_{n,BS}$. ratio. (**c**) Effect of p_sf_y on $V_{exp}/V_{n,prop}$. ratio.

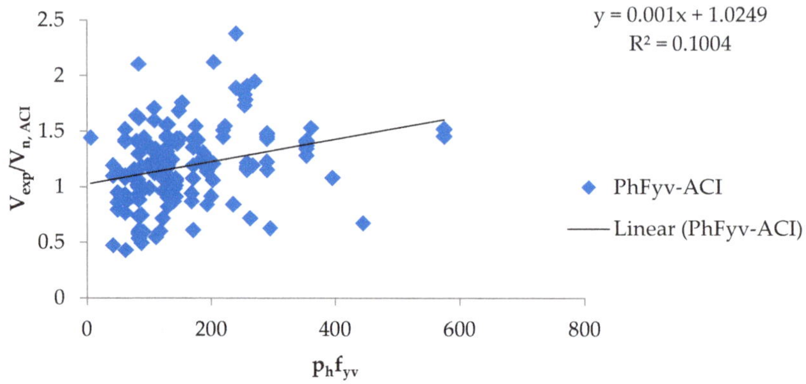

(a)

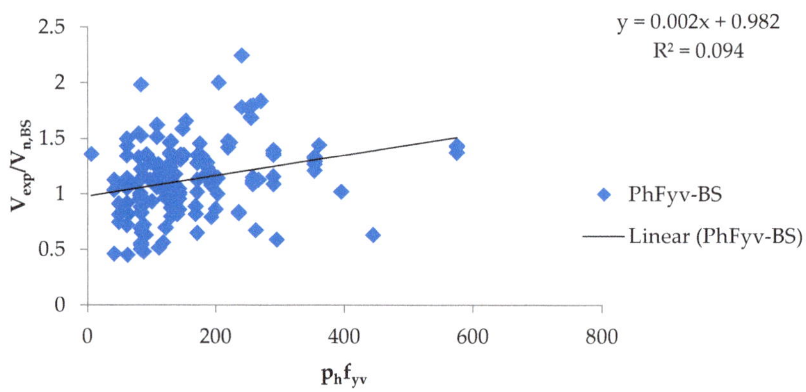

(b)

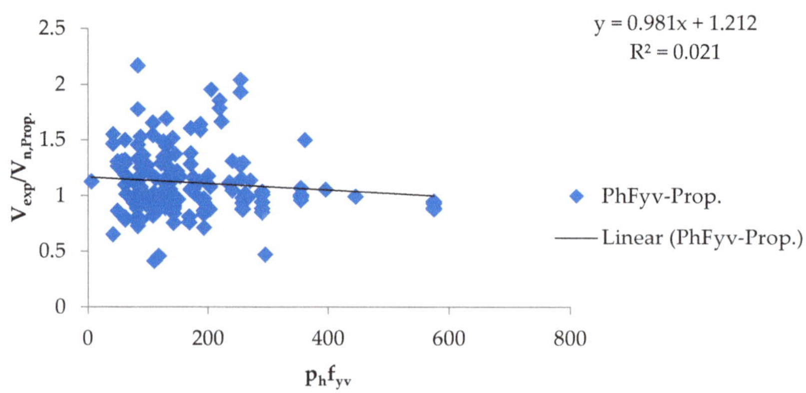

(c)

Figure 8. Effect of $p_h f_{yv}$ (MPa) on V_{exp}/V_n ratio for the ACI, BS, and proposed methods. (**a**) Effect of $p_h f_{yv}$ on $V_{exp}/V_{n,ACI}$. ratio. (**b**) Effect of $p_h f_{yv}$ on $V_{exp}/V_{n,BS}$. ratio. (**c**) Effect of $p_h f_{yv}$ on $V_{exp}/V_{n,prop}$. ratio..

Despite variations in f'_c between 16.08 and 47.6 MPa, as presented in Figure 4, the proposed method (proposal 7) exhibits minimal change, contrasting with other methods (ACI and BS) that yield significantly uneconomic strength predictions with increasing f'_c values.

Figure 5 shows little change in prediction for the proposed method, with a_v/d ranging from 0.19 to 2.5. Conversely, other methods experience a decline in the ratio of V_{exp}/V_n with the highest a_v/d values.

Figures 6–8 highlight the substantial influence of (p_{vfyv}), (p_{sfy}), and (p_{hfyv}) on the proposed method, varying between 0.053 and 10.29 MPa, from 0.405 to 14.11 MPa, and from 0 to 5.56 MPa, respectively. This influence surpasses that of other methods.

Following a meticulous regression analysis, the proposed Equation (22) is selected as the forecast model for the nominal shear strength of deep beams ($V_{n,\,Prop}$) (proposal 7 in Table 2). This decision is grounded in its exceptional performance, boasting the lowest MAE and highest RMSE values, coupled with an (R^2) value that closely approaches unity. The study suggests presenting the redemption factors that adopted in the proposed equation in a tabular form (Table 3) for clear understanding.

$$V_{n,Prop.} = 0.004\,({f'_c}^{0.17} + 0.65\,p_s)\left(\frac{a_v}{d}\right)^{-0.3}\left(\frac{1}{d}\right)^{0.17})\,b_w d + \left((\frac{1+(\frac{a_v}{d})}{6})p_v + (\frac{5-(\frac{a_v}{d})}{6})p_h \right) \times 10^{-6} f_{yv}\,b_w d \quad (22)$$

Table 3. Values of the coefficients (A-C and F-H) used in the selected empirical equation.

Coefficients	A	B	C	F	G	H
Values	0.004	0.17	0.65	−0.3	1	0.17

4. Evaluation of the Developed Empirical Equation

As a result, this study assesses the performance of predictions of the nominal shear strength of RC deep beams by comparing the proposed method with existing approaches through a comprehensive evaluation of existing experimental results.

In the comparison shown in Figure 9, the proposed empirical equation demonstrates a strong correlation between the experimental and theoretical results. The data points of the proposed equation are more convergent compared to other methods, indicating its superior predictive accuracy.

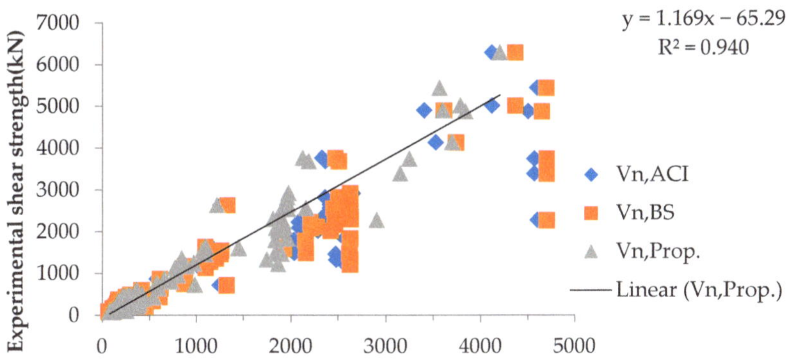

Figure 9. Comparison between experimental V_{exp} and predicted V_n shear strength values for existing and proposed equations.

Mean value (Mean), Equation (23): represents the average of the ratios of the experimental (V_{exp}) to the predicted shear strength values (V_n) for all deep beams, where (N) is the total number of deep beams, equal to 198 in this study.

$$\text{Mean} = \sum_{i=1}^{N} (V_{exp}/V_n)_i / N \tag{23}$$

Standard deviation (SD), Equation (24): measures the dispersion of the values of (V_{exp}/V_n), with Avg. representing the average of the V_{exp}/V_n values.

$$\text{S.D.} = \sqrt{\frac{\sum_{i=1}^{N} ((V_{exp}/V_n)_i - \text{Avg.})^2}{N}} \tag{24}$$

Coefficient of variation (CV %), Equation (25): indicates the relative variability in the values of V_{exp}/V_n, with Avg. representing the average of the V_{exp}/V_n values.

$$\text{CV}(\%) = \frac{\text{S.D.}}{\text{Avg.}} \times 100 \tag{25}$$

Maximum value (Max.): represents the maximum shear strength ratio.
Minimum value (Min.): represents the minimum shear strength ratio.
Range value (Range), Equation (26): indicates the spread between the maximum and minimum values.

$$\text{Range} = \text{Max./Min.} \tag{26}$$

The detailed comparison involved examining the ratio of the shear resistance of a tested beam (V_{exp}) to the calculated nominal shear resistance based on different methods of prediction (V_n), denoted as V_{exp}/V_n. This evaluation is detailed in Appendix A, and Table 4 presents the outcomes for all 198 tested beams using different prediction methods. The last column in Table 3 illustrates the results of the proposed method (Equation (7)). Notably, the coefficient of variation (CV %) values range between 29.03% and 29.53% for the ACI and BS methods. However, by incorporating the effects of the vertical and horizontal reinforcement ratios, the proposed method significantly improves the CV% to a value of 27.08%.

Table 4. Comparative analysis of shear strength ratios.

Details	ACI Method [1]	BS 8110 Method [9]	Proposed Method
Equation	(1)	(3)	(7)
Mean	1.15	1.10	1.15
Standard deviation	0.34	0.32	0.31
CV%	29.53	29.03	27.08
Max. ratio	2.49	2.25	2.23
Min. ratio	0.43	0.45	0.44
Range (max/min)	5.55	4.97	5.11
Number of tested beams for which $V_{exp} < V_n$	70	80	65

5. Discussion

The findings of this study underscore the significance of considering multiple factors such as the concrete compressive strength (f'_c), the shear span-to-depth ratio (a_v/d), the web width (b_w), the ratios of longitudinal (P_s), vertical (P_v), and horizontal (P_h) reinforcements, the depth (d), the yield strength of vertical stirrups (f_{yv}), and the concrete area (web width × depth ($b_w \times d$)). By analyzing a comprehensive dataset consisting of 198 experimental simply supported RC deep beams, this study proposes a novel empirical equation for predicting the nominal shear strength of RC deep beams which out performs existing codes such as ACI and BS codes.

Twelve different sets of empirical equations were developed to create a new empirical equation to estimate the nominal shear strength (V_n) of RC deep beams. Hence, a collinear regression analysis served as the cornerstone for determining the coefficients A–H in the proposed empirical equations.

The coefficient of multiple determinations (R^2) was used to make a comprehensive comparison between the results of using the proposed variables. The R^2 results showed a difference in the range from 0.001 to 0.4, which is less than 0.5, but this study specified the coefficients (A–H) and applied them to the proposed components of the equation. Then, the coefficient (R^2) was calculated for the results of the proposed empirical equation, and it was equal to 0.94, which is considered a truly acceptable value, especially when compared to the values of R^2 for the ACI and BS methods, which were equal to 0.8 and 0.7, respectively.

To evaluate the ratio of the calculated nominal shear strength to the literature (experimental) shear strength (V_{exp}/V_n) between the presented empirical equation and the ACI and BS codes' equations, a detailed comparison was provided and its results proved that the proposed empirical equation outperformed the equations of the ACI and BS codes in all aspects, such as the mean, standard deviation, CV%, Max. ratio, Min. ratio, and Range(max/min).

The findings of this study bring positive improvements for code development and structural design. By using the proposed empirical equation, structural designers can improve the effectiveness of structural designs by obtaining a higher degree of accuracy in predictions of the nominal shear strength of RC deep beams. As presented earlier, code committees may think about including the examined factors that effect shear behavior into existing standards, like ACI 318R-15 and BS 8110.

6. Conclusions

This study investigated the effect of various parameters on the shear strength capacity of RC deep beams, including the concrete compressive strength (f'_c), the shear span-to-depth ratio (a_v/d), the web width (b_w), the ratios of longitudinal (P_s), vertical (P_v), and horizontal (P_h) reinforcements, the depth (d), the vertical stirrup yield strength (f_{yv}), and the concrete area (web width × depth ($b_w \times d$)). This inquiry went beyond what can be achieved using current codes, such BS 8110 and ACI 318R-5, which have large coefficients of variation (CVs) when it comes to forecasting shear capacity.

Through a comprehensive evaluation of 198 deep beams, using data imported from an extensive dataset comprising around 15 investigations, this research proposes a novel predictive empirical equation for shear strength. The proposed equation, Equation (22), takes into account all the above-mentioned key parameters and was rigorously assessed through a collinear regression analysis and statistical metrics (the MAE, RMSE, and R^2).

The results proved that the proposed model significantly enhanced the prediction accuracy of calculating the nominal shear strength compared to the ACI and BS codes, achieving a CV equal to 27.08%, SD equal to 31.10%, Max. ratio equal to 2.23%, Min. ratio equal to 0.44%, and the range was equal to 5.11. The reinforcement ratios in both directions (vertical and horizontal) were considered in the proposed empirical equation and lead to improve the accuracy of calculating the nominal shear strength of the RC deep beams. The suggested model outperforms the ACI and BS codes because it is take in to account the effect of wide range of factors and not limited to f'_c and a_v/d.

The results of this study bring significant improvements for ACI and BS codes development and structural design. Furthermore, by using the presented empirical equation, structural designers can improve the effectiveness of structural designs by obtaining a higher degree of accuracy in predictions of the nominal shear strength of RC deep beams. On the other hand, the study proves that the suggested model was with a higher accuracy to predict the shear strength, which makes it a useful tool for civil and structural engineers.

Author Contributions: Conceptualization: E.K.S., N.S.M. and S.J.H.; data curation: E.K.S., N.S.M., S.J.H. and S.S.S.; formal analysis: E.K.S. and S.J.H.; funding acquisition: E.K.S., N.S.M., S.J.H. and S.S.S.; investigation: E.K.S., N.S.M., S.J.H. and S.S.S.; methodology: E.K.S., N.S.M., S.J.H. and S.S.S.; project administration: E.K.S. and N.S.M.; validation: E.K.S., N.S.M., S.J.H. and S.S.S.; resources: E.K.S. and S.J.H.; software: N.S.M., S.J.H. and S.S.S.; supervision: E.K.S. and N.S.M.; visualization: E.K.S., N.S.M., S.J.H. and S.S.S.; writing—original draft: E.K.S., N.S.M., S.J.H. and S.S.S.; writing—review and editing: E.K.S., N.S.M., S.J.H. and S.S.S. All authors have read and agreed to the published version of the manuscript.

Funding: This research received no external funding.

Data Availability Statement: The data are presented in the article.

Acknowledgments: The authors greatly acknowledge their institutes and universities for their support in preparing and completing this research.

Conflicts of Interest: The authors declare no conflicts of interest.

Appendix A

Table A1. Comparison of the V_{exp}/V_n ratios for all 198 beams with codes.

Ref. No.	Beam No.	$V_{Exp.}$(kN)	V_{nACI}(kN)	$V_{Exp.}/V_{nACI}$	V_{nBS}(kN)	$V_{Exp.}/V_{nBS}$	$V_{n,PROP.}$(kN)	$V_{Exp.}/V_{n,PROP.}$
	1	2530	2627.04	0.96306	2623.4	0.96440	1964	1.28819
	2	2922	2648.78	1.10315	2623.4	1.11382	1971	1.48250
	3	2019	2280.33	0.88540	2417.48	0.83517	1865	1.08257
	4	2348	2338.28	1.00416	2478.91	0.94719	1877	1.25093
	5	2224	2078.59	1.06996	2203.6	1.00926	1916	1.16075
	6	2121	2142	0.99020	2270.82	0.93402	1925	1.10182
	7	2824	2354.58	1.19936	2496.19	1.13132	1956	1.44376
	8	2655	2410.74	1.10132	2555.73	1.03884	1964	1.35183
	9	1783	2027.37	0.87946	2149.3	0.82957	1833	0.97272
	10	1490	2032.08	0.73324	2154.29	0.69164	1834	0.81243
	11	1463	2467.56	0.59289	2615.97	0.55926	1897	0.77122
	12	2522	2499.51	1.00900	2623.4	0.96135	1901	1.32667
	13	2170	2321.87	0.93459	2461.51	0.88157	1877	1.15610
	14	2295	2321.87	0.98843	2461.51	0.93235	1888	1.21557
	15	1832	2564.44	0.71439	2623.4	0.69833	1959	0.93517
	16	1214	2569.65	0.47244	2623.4	0.46276	1862	0.65199
	17	2095	2085.47	1.00457	2210.9	0.94758	1900	1.10263
[19]	18	2081	2085.47	0.99786	2210.9	0.94125	1842	1.12975
	19	3763	2326.49	1.61746	2466.41	1.52570	2121	1.77416
	20	3687	2360.82	1.56175	2502.8	1.47315	2184	1.68819
	21	1325	2474.46	0.53547	2623.28	0.50509	1744	0.75975
	22	2295	2539.12	0.90386	2627.85	0.87334	1810	1.26796
	23	3393	4563.6	0.74349	4696.6	0.72244	3148	1.07783
	24	3745	4563.6	0.82062	4696.6	0.79739	3244	1.15444
	25	2268	4599.41	0.49311	4698.34	0.48272	2902	0.78153
	26	5440	4594.16	1.18411	4692.98	1.15918	3563	1.52680
	27	1463	1183.85	1.23580	1255.05	1.16569	1088	1.34467
	28	1543	1190.44	1.29616	1262.04	1.22262	1061	1.45429
	29	716	1244.22	0.57546	1319.05	0.54281	983	0.72838
	30	2633	1252.06	2.10293	1327.37	1.98362	1214	2.16886
	31	5017	4116.49	1.21876	4364.07	1.14961	3781	1.32690
	32	4136	3521.41	1.17453	3733.19	1.10790	3694	1.11965
	33	6294	4116.49	1.52897	4364.07	1.44223	4200	1.49857
	34	4901	3401.89	1.44067	3606.48	1.35894	3598	1.36215
	35	4875	4500.61	1.08319	4646.14	1.04926	3833	1.27185

Table A1. *Cont.*

Ref. No.	Beam No.	$V_{Exp.}$(kN)	V_{nACI}(kN)	$V_{Exp.}/V_{nACI}$	V_{nBS}(kN)	$V_{Exp.}/V_{nBS}$	$V_{n,PROP.}$(kN)	$V_{Exp.}/V_{n,PROP.}$
[10]	36	369.35	259.958	1.42081	275.592	1.34021	364.4	1.01358
	37	467.25	277.967	1.68095	294.685	1.58559	383.5	1.21838
	38	493.95	280.903	1.75844	297.797	1.65868	421	1.17328
	39	407.15	289.763	1.40511	307.19	1.32540	386	1.05479
	40	416.05	290.266	1.43334	307.723	1.35203	430	0.96756
	41	445	288.247	1.54381	305.583	1.45623	383	1.16188
	42	389.35	280.176	1.38966	297.026	1.31083	392.1	0.99299
	43	262.55	147.103	1.78480	155.95	1.68355	228.2	1.15053
	44	333.75	157.294	2.12182	166.754	2.00145	170.9	1.95290
	45	378.25	158.955	2.37960	168.515	2.24461	288.6	1.31064
	46	302.6	158.68	1.90698	168.223	1.79880	234.1	1.29261
	47	300.35	158.955	1.88953	168.515	1.78233	286.7	1.04761
	48	295.9	157.85	1.87456	167.343	1.76822	231.7	1.27708
	49	289.25	148.48	1.94807	157.41	1.83756	255	1.13431
[12]	50	449.7	355.34	1.26555	363.141	1.23836	283.4	1.58680
	51	465.2	355.34	1.30917	363.141	1.28105	283.4	1.64150
	52	434.1	299.234	1.45070	305.803	1.41954	243.6	1.78202
	53	452.1	299.234	1.51086	305.803	1.47840	243.6	1.85591
	54	443	241.934	1.83108	247.245	1.79175	217.3	2.03866
	55	419.1	241.934	1.73229	247.245	1.69508	217.3	1.92867
[22]	56	161	111.669	1.44176	118.385	1.35997	156.7	1.02744
	57	148	109.559	1.35087	116.148	1.27424	157.7	0.93849
	58	141	107.097	1.31656	113.538	1.24187	158.6	0.88903
	59	170.5	117.205	1.45472	124.254	1.37219	161.2	1.05769
	60	184	118.619	1.55118	125.753	1.46319	163.1	1.12814
	61	174.5	120.294	1.45061	127.528	1.36833	168.6	1.03499
	62	170.5	114.907	1.48381	121.817	1.39964	168.7	1.01067
	63	171.5	116.348	1.47403	123.346	1.39040	170.7	1.00469
	64	161.5	112.857	1.43101	119.645	1.34983	171.4	0.94224
	65	161	110.77	1.45346	117.432	1.37101	183.4	0.87786
	66	172.5	113.152	1.52450	119.957	1.43802	185.5	0.92992
	67	178.5	117.773	1.51563	124.856	1.42965	188.3	0.94796
	68	168	115.196	1.45838	122.125	1.37564	189.2	0.88795
	69	147	121.397	1.21090	128.698	1.14221	146.3	1.00478
	70	143.5	115.774	1.23948	122.737	1.16917	146.5	0.97952
	71	140	117.773	1.18873	124.856	1.12129	148.5	0.94276
	72	153	114.033	1.34172	120.891	1.26560	149.3	1.02478
	73	128.5	112.561	1.14160	119.331	1.07684	149.3	0.86068
	74	131	112.561	1.16381	119.331	1.09779	150.8	0.86870
	75	126	108.027	1.16638	114.524	1.10021	151.3	0.83278
	76	150	120.57	1.24409	127.822	1.17351	154.2	0.97276
	77	145	114.907	1.26189	121.817	1.19031	154.5	0.93851
	78	130.5	106.472	1.22567	112.876	1.15614	153.7	0.84906
	79	158.5	115.485	1.37247	122.431	1.29461	159.7	0.99249
	80	158	112.561	1.40368	119.331	1.32405	160.5	0.98442
	81	155	113.152	1.36984	119.957	1.29213	162.2	0.95561
	82	166	117.205	1.41632	124.254	1.33597	164.7	1.00789
	83	153.5	106.472	1.44169	112.876	1.35990	136.5	1.12454
	84	118.5	113.152	1.04726	119.957	0.98785	131.7	0.89977
	85	123	120.847	1.01782	128.115	0.96007	134.8	0.91246
	86	131	123.034	1.06475	130.434	1.00434	136.8	0.95760
	87	122	120.57	1.01186	127.822	0.95445	137.9	0.88470
	88	124	115.196	1.07643	122.125	1.01535	135.7	0.91378
	89	103.5	113.152	0.91470	119.957	0.86281	136.8	0.75658
	90	115	113.446	1.01370	120.269	0.95619	136.8	0.84064
	91	124.5	116.635	1.06743	123.649	1.00688	139.1	0.89504
	92	124	117.773	1.05287	124.856	0.99314	140.9	0.88006

Table A1. *Cont.*

Ref. No.	Beam No.	$V_{Exp.}$(kN)	V_{nACI}(kN)	$V_{Exp.}/V_{nACI}$	V_{nBS}(kN)	$V_{Exp.}/V_{nBS}$	$V_{n,PROP.}$(kN)	$V_{Exp.}/V_{n,PROP.}$
	93	140.5	118.337	1.18729	125.455	1.11992	143.3	0.98046
	94	124.5	106.785	1.16589	113.207	1.09976	142.3	0.87491
	95	127.5	110.468	1.15418	117.112	1.08870	144.7	0.88113
	96	137	112.561	1.21712	119.331	1.14807	146.8	0.93324
	97	146.5	114.325	1.28143	121.201	1.20874	148.3	0.98786
	98	128.5	111.37	1.15381	118.068	1.08836	145.3	0.88438
	99	152	113.152	1.34333	119.957	1.26712	151.1	1.00596
	100	152.5	111.07	1.37301	117.75	1.29512	152.3	1.00131
	101	159.5	118.9	1.34146	126.05	1.26537	149.3	1.06832
	102	87	103.551	0.84017	109.779	0.79250	122.1	0.71253
[26]	103	754	822.787	0.91640	872.272	0.86441	576.3	1.30835
	104	350.3	572.572	0.61180	536.524	0.65291	339.3	1.03242
	105	206	480.714	0.42853	456.045	0.45171	262.3	0.78536
[23]	106	874.2	575.944	1.51786	610.583	1.43175	663.1	1.31835
	107	650.9	589.867	1.10347	618.963	1.05160	594.9	1.09413
	108	437.4	572.598	0.76389	607.036	0.72055	542.9	0.80567
	109	1175	834.823	1.40748	874.713	1.34330	893.7	1.31476
	110	952.3	877.025	1.08583	874.713	1.08870	801.1	1.18874
	111	804.4	866.789	0.92802	875.326	0.91897	732.1	1.09876
	112	1636.3	1151.22	1.42136	1092.14	1.49825	1094	1.49570
	113	1244	1155.41	1.07667	1090.82	1.14043	969.2	1.28353
[24]	114	1615.5	1908.61	0.84643	1924.72	0.83934	1448	1.11568
	115	1592.9	1897.8	0.83934	1924.72	0.82760	1446	1.10159
[26]	116	2563.7	2410.19	1.06369	2447.87	1.04732	2155	1.18965
[17]	117	284.8	263.28	1.08174	279.114	1.02037	270.2	1.05403
	118	377.6	263.28	1.43421	279.114	1.35285	275.5	1.37060
	119	358.1	263.28	1.36015	279.114	1.28299	281.8	1.27076
	120	228.7	263.28	0.86866	279.114	0.81938	242	0.94504
	121	255.7	263.28	0.97121	279.114	0.91611	242	1.05661
	122	208.7	263.28	0.79269	279.114	0.74772	242	0.86240
[20]	123	276.9	466.875	0.59309	437.482	0.63294	267.9	1.03359
	124	455.8	461.146	0.98841	437.482	1.04187	301	1.51429
[13]	125	350.8	305.71	1.14749	324.096	1.08240	269.3	1.30264
	126	305.8	321.653	0.95071	335.448	0.91162	234	1.30684
	127	257.8	298.99	0.86224	316.972	0.81332	203.8	1.26497
	128	156.1	179.872	0.86784	190.69	0.81861	153.9	1.01429
	129	140.4	196.011	0.71629	207.799	0.67565	138.4	1.01445
	130	123.6	184.235	0.67088	195.316	0.63282	124.8	0.99038
[25]	131	606.7	392.608	1.54531	413.845	1.46601	364.9	1.66265
	132	351.8	383.198	0.91806	406.245	0.86598	298.9	1.17698
[21]	133	116.75	186.528	0.62591	197.746	0.59040	249.9	0.46719
	134	114.53	191.167	0.59911	202.664	0.56512	251.5	0.45539
	135	105.65	192.993	0.54743	204.601	0.51637	256.2	0.41237
	136	166.8	191.952	0.86897	203.497	0.81967	219.3	0.76060
	137	177.93	188.856	0.94215	200.214	0.88870	219.5	0.81062
	138	205.75	193.788	1.06173	205.442	1.00150	234.4	0.87777
[16]	139	239.2	197.718	1.20980	209.609	1.14117	221.2	1.08137
	140	208.1	166.239	1.25181	176.237	1.18080	184.9	1.12547
	141	172.5	133.883	1.28844	141.935	1.21535	149.1	1.15694
	142	127.16	101.956	1.24720	108.087	1.17646	114	1.11544
	143	77.8	65.0963	1.19515	69.0114	1.12735	78.68	0.98882
[26]	144	348	570.945	0.60952	535	0.65047	331.1	1.05104

Table A1. *Cont.*

Ref. No.	Beam No.	$V_{Exp.}$(kN)	V_{nACI}(kN)	$V_{Exp.}/V_{nACI}$	V_{nBS}(kN)	$V_{Exp.}/V_{nBS}$	$V_{n,PROP.}$(kN)	$V_{Exp.}/V_{n,PROP.}$
[17]	145	284.1	263.698	1.07737	279.557	1.01625	268.8	1.05692
	146	377	263.698	1.42967	279.557	1.34856	273.3	1.37944
	147	357.5	263.698	1.35572	279.557	1.27881	278.6	1.28320
[11]	148	1357	1137.29	1.19319	1205.69	1.12550	845.3	1.60535
	149	1134	1032.33	1.09849	1094.42	1.03617	774.2	1.46474
	150	1286	1077.28	1.19375	1142.07	1.12603	830.4	1.54865
[18]	151	251	216.932	1.15704	229.979	1.09140	267.8	0.93727
	152	237	216.932	1.09251	229.979	1.03053	267.8	0.88499
	153	456	266.817	1.70904	281.25	1.62133	276.2	1.65098
	154	426	266.817	1.59660	281.25	1.51467	276.2	1.54236
[16]	155	239	212.613	1.12411	225.4	1.06034	291.1	0.82102
	156	224	187.532	1.19446	198.81	1.12670	243.6	0.91954
	157	190	137.056	1.38629	145.299	1.30765	192.2	0.98855
	158	164	100.022	1.63964	106.037	1.54663	144.7	1.13338
	159	90	63.7259	1.41230	67.5585	1.33218	97.9	0.91931
	160	249	200.919	1.23931	213.003	1.16900	228.3	1.09067
	161	224	163.066	1.37368	172.873	1.29575	188	1.19149
	162	216	132.787	1.62667	140.774	1.53437	152.4	1.41732
	163	140	103.727	1.34970	109.966	1.27312	117.3	1.19352
	164	100	61.3316	1.63048	65.0202	1.53798	79.54	1.25723
[14]	165	222.5	327.225	0.67996	346.906	0.64138	356.2	0.62465
	166	209.1	320.505	0.65241	339.781	0.61540	355.1	0.58885
	167	222.5	319.144	0.69718	338.339	0.65762	354.9	0.62694
	168	244.7	328.553	0.74478	348.313	0.70253	356.4	0.68659
	169	278.8	319.144	0.87359	338.339	0.82403	372.9	0.74765
	170	256.6	332.504	0.77172	352.501	0.72794	375.2	0.68390
	171	284.8	321.184	0.88672	340.501	0.83641	373.3	0.76293
	172	268.1	318.462	0.84186	337.615	0.79410	372.8	0.71915
	173	241.5	327.225	0.73802	346.906	0.69615	374.3	0.64520
	174	301.1	317.778	0.94752	336.89	0.89376	372.7	0.80789
	175	322.2	338.343	0.95229	358.692	0.89826	376.1	0.85669
	176	334.9	329.215	1.01727	349.014	0.95956	374.6	0.89402
	177	379.3	428.076	0.88606	395.85	0.95819	389.8	0.97306
	178	277.7	333.81	0.83191	353.886	0.78472	330.7	0.83973
	179	311.1	338.343	0.91948	358.692	0.86732	331.5	0.93846
	180	245.9	323.21	0.76081	342.649	0.71764	328.7	0.74810
	181	285.9	355.286	0.80470	376.654	0.75905	334.4	0.85496
	182	290	320.505	0.90482	339.781	0.85349	349.9	0.82881
	183	301.1	329.875	0.91277	349.715	0.86099	351.6	0.85637
	184	323.7	323.883	0.99943	343.362	0.94274	350.5	0.92354
	185	288.2	342.816	0.84068	363.434	0.79299	353.9	0.81435
	186	309.3	326.56	0.94715	346.2	0.89341	390.8	0.79145
	187	423.8	443.556	0.95546	395.85	1.07061	409.8	1.03416
	188	434.9	441.096	0.98595	395.85	1.09865	409.4	1.06229
	189	428.6	455.18	0.94161	395.85	1.08273	411.5	1.04156
	190	301.1	337.699	0.89162	358.009	0.84104	335.4	0.89773
	191	356.7	337.054	1.05829	357.325	0.99825	335.3	1.06382
	192	256.6	326.56	0.78577	346.2	0.74119	333.2	0.77011
	193	290	323.21	0.89725	342.649	0.84635	341.8	0.84845
	194	312.2	335.76	0.92983	355.954	0.87708	344.3	0.90677
	195	334.4	328.553	1.01780	348.313	0.96006	342.9	0.97521
	196	334.9	326.56	1.02554	346.2	0.96736	342.5	0.97781
	197	394.9	350.351	1.12716	371.423	1.06321	423.9	0.93159
	198	312.2	317.092	0.98457	336.163	0.92872	378.3	0.82527

References

1. ACI Committee. *Building Code Requirements for Structural Concrete (ACI 318-05) and Commentary (ACI 318R-05)*; American Concrete Institute: Farmington Hills, MI, USA, 2005.
2. Almasabha, G.; Murad, Y.; Alghossoon, A.; Saleh, E.; Tarawneh, A. Sustainability of Using Steel Fibers in Reinforced Concrete Deep Beams without Stirrups. *Sustainability* **2023**, *15*, 4721. [CrossRef]
3. Sayhood, E.K.; Resheq, A.S.; Habeeb, A.J. Shear strength of concrete deep beam subjected to uniformly distributed load. *Eng. Technol. J.* **2018**, *36*, 125–135. [CrossRef]
4. Albidah, A.S. Shear behviour of metakaolin-fly ash based geopolymer concrete deep beams. *Eng. Struct.* **2023**, *275*, 115271. [CrossRef]
5. Sayhood, E.K.; Abdullah, K.N.; Kazem, S.J. Strut Confinement of Simply Supports Deep Beam Using Strut Reinforcement. *Eng. Technol. J.* **2020**, *38*, 605–613. [CrossRef]
6. Sayhood, E.; Mohammed, N.S.; Alnimer, N. Effect Factors of Strut Strength for Reinforcement Deep Beams. *ARPN J. Eng. Appl. Sci.* **2019**, *14*, 2843–2856.
7. Arabzadeh, A.; Hizaji, R. A simple approach to predict the shear capacity and failure mode of fix-ended reinforced concrete deep beams based on experimental study. *Int. J. Eng.* **2019**, *32*, 474–483.
8. Rahimi, S.B.; Jalali, A.; Mirhoseini, S.M.; Zeighami, E. Experimental Comparison of Different Types of FRP Wrapping in Repairing of RC Deep Beams with Circular Openings. *Int. J. Eng.* **2021**, *34*, 1961–1973.
9. *BS 8110: Part 1*; Structural Use of Concrete, Part 1, Code of Practice for Design and Construction. British Standards Institution: Milton Keynes, UK, 1985.
10. Rashid, M.A.; Kabir, A. Behaviour of reinforced concrete deep beam under uniform loading. *J. Civ. Eng. Inst. Eng. Bangladesh Vol. CE* **1996**, 24.
11. Aguilar, G.; Matamoros, A.B.; Parra-Montesinos, G.; Ramírez, J.A.; Wight, J.K. Experimental evaluation of design procedures for shear strength of deep reinfoced concrete beams. *ACI Struct. J.* **2002**, *99*, 539–548.
12. Ahmad, S.; Shah, A.; Zamani, K.N.; Salimullah, K. Design and evaluation of the shear strength of deep beams by strut and tie model (STM). *Iran. J. Sci. Technol. Trans. B Eng.* **2011**, *35*, 1–13.
13. Ashour, A. Shear capacity of reinforced concrete deep beams. *J. Struct. Eng.* **2000**, *126*, 1045–1052. [CrossRef]
14. Clark, A.P. Diagonal tension in reinforced concrete beams. *J. Proc.* **1951**, 145–156.
15. Birrcher, D.B. *Design of Reinforced Concrete Deep Beams for Strength and Serviceability*; The University of Texas at Austin: Austin, TX, USA, 2009.
16. Kong, F.-K.; Robins, P.J.; Cole, D.F. Web reinforcement effects on deep beams. *J. Proc.* **1970**, 1010–1018.
17. Oh, J.-K.; Shin, S.-W. Shear strength of reinforced high-strength concrete deep beams. *Struct. J.* **2001**, *98*, 164–173.
18. Quintero-Febres, C.G.; Parra-Montesinos, G.; Wight, J.K. Strength of struts in deep concrete members designed using strut-and-tie method. *ACI Mater. J.* **2006**, *103*, 577.
19. Tuchscherer, R.G. *Strut-and-Tie Modeling of Reinforced Concrete Deep Beams: Experiments and Design Provisions*; The University of Texas at Austin: Austin, TX, USA, 2008.
20. Lu, W.-Y.; Hsiao, H.-T.; Chen, C.-L.; Huang, S.-M.; Lin, M.-C. Tests of reinforced concrete deep beams. *Comput. Concr. Int. J.* **2015**, *15*, 357–372. [CrossRef]
21. Sazzad, M.M.; Ali, M.; Nizamud-Ooulah, S. Study of behaviour of reinforced concrete deep beam under two points loading and the effect of shear reinforcement. In Proceedings of the 27th Conference on Our World in Concrete and Structures, Singapore, 29–30 August 2002; pp. 29–30.
22. Smith, K.; Vantsiotis, A. Shear strength of deep beams. *J. Proc.* **1982**, *7*, 201–213.
23. Tan, K.; Lu, H. Shear behavior of large reinforced concrete deep beams and code comparisons. *Struct. J.* **1999**, *96*, 836–846.
24. Brown, M.D.; Sankovich, C.L.; Bayrak, O.; Jirsa, J.O.; Breen, J.E.; Wood, S.L. *Design for Shear in Reinforced Concrete Using Strut-and-Tiemodels*; No. FHWA/TX-06/0-4371-2; The National Academies of Sciences, Engineering and Medicine: Washington, DC, USA, 2006; Available online: https://trid.trb.org/View/787671 (accessed on 19 January 2024).
25. Yang, K.-H.; Chung, H.-S.; Ashour, A.F. Influence of inclined web reinforcement on reinforced concrete deep beams with web openings. *ACI Struct. J.* **2007**, *104*, 580–589.
26. Williams, C.S.; Deschenes, D.J. A Pragmatic Approach to Strut-And-Tie Modeling. In Proceedings of the PCI Convention and National Bridge Conference, Washington, DC, USA, 6–9 September 2014.

infrastructures

Article

Study on Deflection-Span Ratio of Cable-Stayed Suspension Cooperative System with Single-Tower Space Cable

Lin Xiao, Yaxi Huang and Xing Wei *

School of Civil Engineering, Southwest Jiaotong University, Chengdu 610032, China; xiaolin@home.swjtu.edu.cn (L.X.); yaxihuang1999@163.com (Y.H.)
* Correspondence: we_star@swjtu.edu.cn

Abstract: This study uses the wind–vehicle–bridge coupling vibration analysis method to investigate the bridge stiffness problem of a large-span cable-stayed-suspension cooperative system. On the basis of the particle-damping-spring vehicle model, the TMeasy surface contact tire model is introduced, and a set of universal wind–vehicle–bridge coupling analysis algorithm is built in the framework of the whole process iterative method. Based on the Latin supercube sampling principle, random traffic flow is generated and loaded onto bridge structures with different stiffness conditions to analyze the driving comfort and safety under each stiffness condition. Combining the specification requirements, engineering experience, and research results, the vertical stiffness limit applicable to the bridge of the highway cable-stayed-suspension collaborative system is proposed. Existing engineering experience shows that the vertical deflection-to-span ratio of a cable-stayed bridge under live load is distributed between 1/400 and 1/1600, and the vertical deflection span ratio under the action of lane load is recommended based on numerical analysis.

Keywords: cable-stayed-suspension cooperative system bridge; stiffness standards; wind-vehicle-bridge coupling vibration; tire model; driving comfort; driving safety

check for updates

Citation: Xiao, L.; Huang, Y.; Wei, X. Study on Deflection-Span Ratio of Cable-Stayed Suspension Cooperative System with Single-Tower Space Cable. *Infrastructures* **2023**, *8*, 62. https://doi.org/10.3390/infrastructures8030062

Academic Editors: Armin Mehrabi and Seyed Saman Khedmatgozar Dolati

Received: 27 February 2023
Revised: 20 March 2023
Accepted: 21 March 2023
Published: 22 March 2023

1. Introduction

The cable-stayed-suspension cooperative system bridge adds cable-stayed cables to the long-span suspension bridge to improve its wind resistance stability, which makes up for the lack of rigidity of the suspension system [1]. It also provides a new bridge form for mountainous and oceanic long-span bridges. At present, the long-span cable-stayed-suspension collaborative bridges in the world include the Third Bosphorus Bridge (main span 1408 m, two-way eight-lane high-speed + double-track railway). The Xihoumen Highway-Railway Bridge (main span 1488 m) and the Jingzhou Libu Yangtze River Highway-Railway Bridge (main span 1120 m) under construction in China also adopt the cable-stayed-suspension coordination system. The design scheme of a bridge across the Xunjiang River adopts a (638 + 638) m space cable single tower cable-stayed-suspension collaborative system scheme, the tower height is 231 m, and the spacing is 16 m, with a total of 19 m (Figure 1).

Both cable-stayed bridges and suspension bridges are classic cable-supported bridge systems, and the cable structure is prone to vibration. Bryja found that the cable car would exhibit unstable behavior under wind loads, such as violent swinging or sliding, when studying the interaction between aerial cable cars and cables [2]. The irregular shape of the contact line in the railway electrification system can lead to oscillation and resonance between the pantograph and the contact line, affecting the stable contact performance of the pantograph with the contact line [3]. Moreover, when the wind attack angle is at a specific value, the damping of the catenary becomes negative, and the running vibration reaches the maximum amplitude [4]. Both of the above structures are simple cable-supported structures, while bridge structures composed of multiple cables, such as cable-stayed and suspension bridges, are prone to vibration behavior under dynamic loads [5]. Under wind loads, both cable-stayed and suspension bridges are prone to wind-induced vibration and flutter, but suspension bridges are more susceptible to nonlinear response due to their lower structural stiffness. The cable-stayed-suspension combined system bridge, as a combination of cable-stayed and suspension bridges, has a more complex response under wind and moving loads. Theoretical studies show that the structural nonlinear effect of cable-stayed-suspension cooperative system bridges is obvious, and the stiffness problem is more complicated than that of suspension bridges or cable-stayed bridges. The limit value of bridge stiffness involves the comprehensive consideration of structural safety and economy. At present, few studies examine the stiffness of bridges with cable-stayed-suspension cooperative system, and no relevant regulations are in place. Bridge stiffness mainly refers to its vertical stiffness and transverse stiffness [6], which are often measured by indicators such as vertical and transverse deflection–span ratios. China's "Code for Design of Highway Suspension Bridges" (JTGT D65-05-2015) stipulates that "the maximum vertical deflection value of stiffened beams caused by the frequent occurrence of lane loads should not be greater than 1/250 of the span." The "Code for Design of Highway Cable-Stayed Bridges" requires that the vertical deflection-span ratio shall not exceed 1/400. Japan's "Road and Bridge Indication Book" specifies that the deflection–span ratio of cable bridges shall not exceed 1/350. Bridge stiffness affects the dynamic response of vehicles passing on the bridge. Reasonable bridge stiffness should take the comfort and safety of driving on the bridge into consideration [7]. Given the space of a cable-stayed-suspension cooperative system bridge in Figure 1, this study uses the wind–vehicle–bridge coupling analysis as a means to establish a tire mechanics model considering the tire deformation surface. This study also explores the vertical deflection–span ratio limit under the effect of satisfying driving comfort. The value provides a reference for the limit value of the bridge stiffness of the cable-stayed-suspension cooperative system. As the transverse stiffness of the space cable structure is greater than that of the parallel cable structure, the study mainly discusses the vertical stiffness of the bridge.

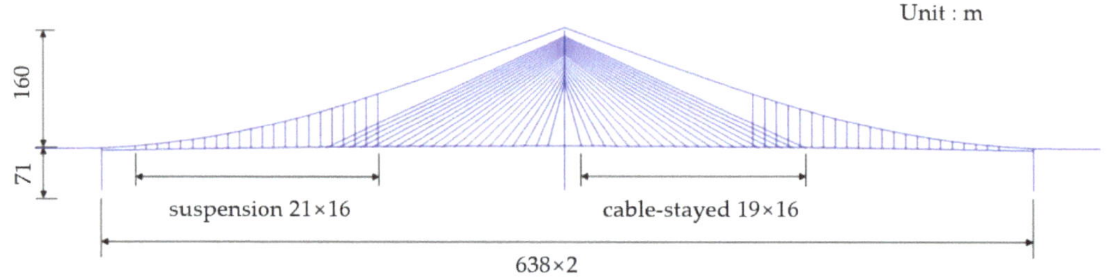

Figure 1. Schematic diagram of cable-stayed-suspension cooperative system with single tower of space cable.

2. Evaluation Method of Bridge Driving Comfort and Safety

Driving comfort and safety are directly related to vehicle dynamic response. Based on the coupled vibration analysis theory of wind–vehicle–bridge, the dynamic interaction between vehicle and bridge can be considered, and the response of bridge structure and vehicle can be accurately solved. It is an effective method for bridge stiffness analysis and evaluation, which is widely used in railway bridge engineering [8]. In particular, it is applied in the setting of safe wind speed and safe driving speed [9], the evaluation and optimization of wind barrier performance [10], the research on bridge structure vibration and vibration reduction measures [11], and the evaluation of track irregularity of high-speed railways [12]. However, the application research in highway bridges is relatively scarce. Basing on the design load of Chinese bridge code, investigation and sampling results, and equivalent theory, Deng Luji put forward a vehicle model suitable for wind–vehicle–bridge coupling calculation in China, including the values of geometric dimensions, mass, stiffness, and damping, and formed a systematic vehicle model library [13]. Polish scholars established high-precision vehicle models through the multi-body dynamics simulation software ADAMS and LS-DYNA. This kind of model has a strong application significance for the response of the car body, but its calculation efficiency is too slow to be suitable for the large-scale windmill-bridge coupling analysis [14].

2.1. Dynamic Model of Bridges and Vehicles

The bridge model is simplified as a beam model, and the bridge model is established in the finite element analysis software ANSYS (Figure 2). The beam element Beam4 is used for the main girder and tower, whereas Link10 is used for the simulation of the main cable, suspenders, and cables. The coupling function of degrees of freedom is used at the connection of tower and beam to constrain the longitudinal torsion, longitudinal translation, and lateral translation of the main beam. The end of the main girder constrains the longitudinal, transverse, and torsional degrees of freedom around the longitudinal axis. Both the rod element and the beam element are shown in blue, and the constraints are shown in green. In the vehicle–bridge coupling analysis, the vehicle is usually abstracted as a particle-damping-spring model, in which rigid bodies, such as car body, axle, and wheel, are connected with one another through dampers and elastic elements. The motion equation of the vehicle is constructed using the D'Alembert principle. The vehicle suspension system and wheel are abstracted as spring-damping elements, and the vehicle mass is assumed to be distributed in the center of mass of each rigid body. The car body has six degrees of freedom in space, namely, floating, swinging, stretching, nodding, shaking, and rolling. However, because the longitudinal vibration of the car along the driving direction has little influence on the vertical and lateral vibration of the bridge, the longitudinal freedom of the car body, that is, the degree of freedom of stretching, is generally ignored [15].

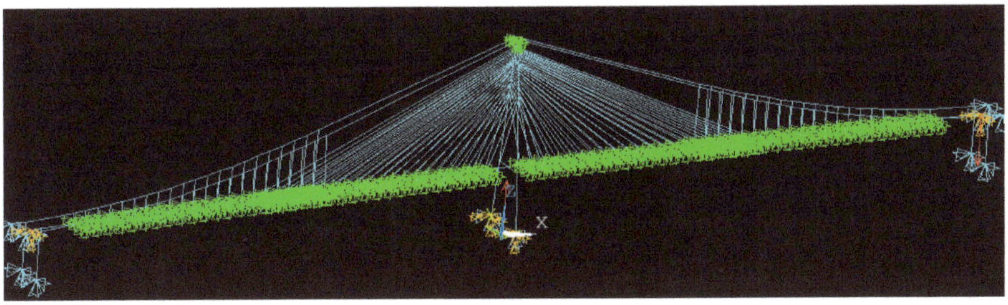

Figure 2. Finite element model diagram of bridge.

The difficulty in the analysis of vehicle–bridge coupling vibration is the contact model between the wheel and the bridge deck, and the key is the tire model. Wheel motion parameters include longitudinal slip rate s, slip angle α, radial deformation ρ, camber angle γ, wheel speed ω, and yaw angle β_t. According to different research purposes, tire dynamics establishes tire longitudinal sliding model, tire cornering model, and tire vertical vibration model. According to the applicable analysis state, it is subdivided into steady-state model, dynamic model, and tire model for the dynamic simulation [16]. TMeasy tire model is a tire model provided by Simpack software, which is suitable for dynamic simulation. The calculation of the sideslip force of the tire comes from the Lugre dynamic model [17], and the vertical stiffness nonlinear model is used. The vertical load is divided into two parts: static load F_z^{st} and dynamic load F_z^D. The static load is expressed as a nonlinear function related to vertical displacement, and the dynamic load is approximately expressed as a linear damping model, namely:

$$F_z = F_z^{st} + F_z^D = a_1 \Delta z + a_2 (\Delta z)^2 + d_T \Delta \dot{z} \tag{1}$$

where F_z is the vertical load of the tire; Δz is the vertical displacement of the tire; $\Delta \dot{z}$ is the derivative of vertical displacement; a_1 and a_2 are the effective load and the radial stiffness at twice the effective load, respectively; and d_T is the vertical damping of the tire.

$$F_y = c_y \Delta y + d_y \Delta \dot{y} \quad (2)$$

where F_y is the lateral force of the wheel, c_y is the lateral stiffness of the tire, d_y is the lateral damping of tires, and Δy is the lateral deformation of the tire.

Taking a two-axle vehicle as an example, the vehicle space dynamics model is established in Simpack, and its topology model and entity model diagram are shown in Figures 3 and 4, respectively. In Figure 3, the blue part represents the degree-of-freedom coupling, the red part represents the force element, and the different numbers represent the different force elements; number 5 in the figure represents the spring-damping force element, number 253 represents the TMeasy force element, and the green part is the constraint equation.

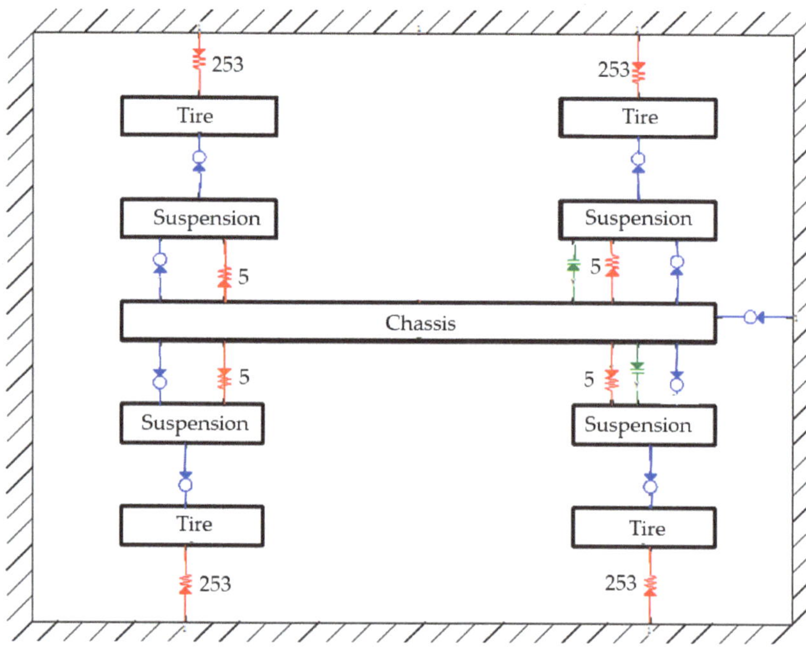

Figure 3. Topological model diagram of two-axle vehicle.

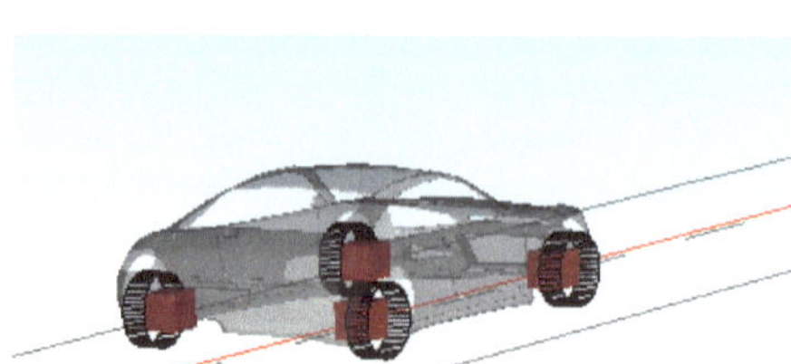

Figure 4. Solid model diagram of two-axle vehicle.

2.2. *Effect of Fluctuating Wind Field on Vehicle–Bridge System*

The wind-induced response of the bridge considers the resistance, lift, torque, and buffeting force of the stiffening beam caused by the average wind. The average wind-induced response is taken according to the Code for Wind-resistant Design of Highway Bridges, and the buffeting force is calculated by considering the unsteady and local spatial correlation of buffeting force on the basis of the Scanlan quasi-steady aerodynamic formula [18]. The existing aerodynamic admittance function is usually a function in frequency domain, which cannot be directly used to solve the buffeting force of bridges in time domain. Usually, the equivalent wind spectrum method and buffeting force spectrum method can be used for time domain analysis. The cross-section of bridge structure is very complex and often blunt, so obtaining the accurate expression of aerodynamic admittance function of specific cross-section is difficult at this stage [19]. In the absence of accurate aerodynamic admittance function, the simplified flat aerodynamic admittance function expression proposed by Liepmann is generally used for flat streamlined sections, whereas the aerodynamic admittance function can be safely set to one for relatively passive bridge structural sections without considering the influence of aerodynamic admittance. The aerodynamic effect of wind on bridge is shown in Figure 5 and the buffeting resistance, buffeting lift, and buffeting moment per unit length of main girder are expressed as follows:

$$F_{b,d}^{H} = \rho UB\{\frac{H}{BC_{H,d}(\alpha)\chi_{Hu}u(t)} + 1/2[H/B\frac{dC_{H,d(\alpha)}}{d\alpha}\chi_{Hw}w(t)]\} \tag{3}$$

$$F_{b,d}^{v} = \rho UB\{C_{V,d}(\alpha)\chi_{Vu}u(t) + +1/2[\frac{dC_{H,d(\alpha)}}{d\alpha} + H/BC_{H,d}(\alpha)]\chi_{Vw}w(t)\} \tag{4}$$

$$F_{b,d}^{M} = \rho UB^2[C_{M,d}(\alpha)\chi_{Vu}u(t) + +1/2\frac{dC_{M,d(\alpha)}}{d\alpha}\chi_{Mw}w(t)] \tag{5}$$

where $F_{b,d}^{H}$, $F_{b,d}^{v}$, and $F_{b,d}^{M}$ are buffeting resistance, buffeting lift, and buffeting moment of main girder caused by pulsating wind, respectively; u and w are horizontal and vertical fluctuating wind speed components; $d/d\alpha$ is the derivation of wind attack angle; χ_{Hu}, χ_{Hw} are aerodynamic admittance functions related to buffeting resistance; χ_{Vu},χ_{Vw} are aerodynamic admittance functions related to buffeting lift; and χ_{Mu}, χ_{Mw} are aerodynamic admittance functions related to buffeting torque.

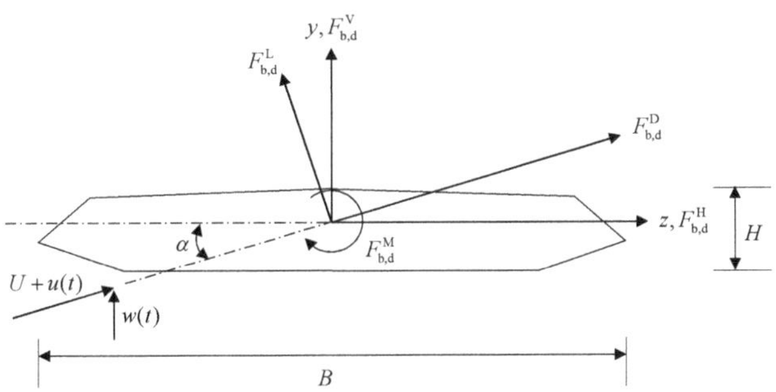

Figure 5. Aerodynamic effect of wind on bridge.

The wind load borne by vehicles is usually described by Baker's expression based on quasi-steady theory:

$$F_S = 0.5\rho U_r^2(t)C_S(\Psi)A_0 \tag{6}$$

$$F_L = 0.5\rho U_r^2(t)A_0C_L(\Psi) \tag{7}$$

$$M_R = 0.5\rho U_r^2(t)C_R(\Psi)A_0h_v \tag{8}$$

where F_S, F_L, and M_R are the lateral force, lifting force, and overturning moment acting on the center of mass of the car body, respectively; $C_S(\psi)$, $C_L(\psi)$, and $C_R(\psi)$ are the lateral force, lift, and overturning moment coefficients of the vehicle, respectively, which are obtained by CFD test; A_0 is the windward area of the vehicle; h_v is the distance from the center of mass of the car body to the road surface; U_r is the relative speed between the vehicle and the wind; and ψ is the corresponding deflection angle. Assume that the wind with the speed of U acts vertically on the longitudinal axis of the road, and the vehicle travels at the speed of U_v. The relative speed and its deflection angle are calculated according to the following formula, where $u(x,t)$ represents the turbulent wind acting on the vehicle at time t.

$$U_r = \sqrt{(U + u(x,t))^2 + U_y^2} \tag{9}$$

$$\psi = arctan((U + u(x,t))/U_y) \tag{10}$$

2.3. Vehicle–Bridge Interaction and Solution

According to the static and dynamic interaction and displacement coordination among vehicle subsystem, bridge subsystem and wind, the vibration equation of a vehicle–bridge–wind coupling system can be established as shown in Equations (11) and (12):

$$M_c\ddot{X}_c + C_c\dot{X}_c + K_cX_c = F_c^b + F_c^w \tag{11}$$

$$M_b\ddot{X}_b + C_b\dot{X}_b + K_bX_b = F_b^c + F_b^w \tag{12}$$

where M_c, C_c, and K_c are the mass, damping, and stiffness matrices of the vehicle, respectively; F_c^b is the force of the bridge on the vehicle;F_c^w is the force of the wind on the vehicle; M_b, C_b, and K_b are the mass, damping, and stiffness matrices of the bridge, respectively; F_b^c is the force of the vehicle on the bridge;F_b^w is the force of the wind on the bridge; X_c is the displacement of the vehicle center of mass; and X_b is the displacement at the contact point of the vehicle axle.

The irregularity of the bridge deck is regarded as a spatial random function $z = z(x,y)$, and the position and direction of the tire are respectively represented by the position vector r_{0C} and the unit vector e_{yR}. The geodetic coordinate system is taken as the reference system (Figure 6b), where r_{0C} is the vector formed by connecting the point C of the tire centroid with the origin of the geodetic coordinate system, indicating the displacement of the tire centroid relative to the geodetic coordinate system. The unit vector e_{yR} determines the direction of the tire plane.

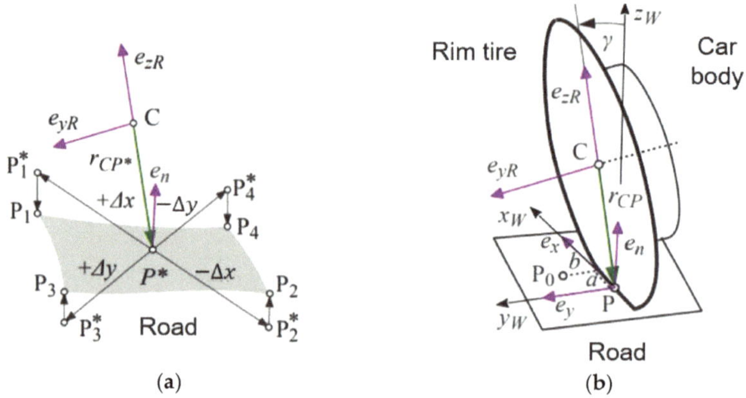

Figure 6. (**a**) Spatial relation diagram of P point; (**b**) tire coordinate system.

When the tire is on an uneven road surface, the position of the projection point P of the tire center of mass on the ground along the tire center plane cannot be directly calculated. To approximate the position of the contact point, the wheel–ground contact surface is assumed to be a plane, which is determined by the road conditions where the tire is located. Figure 6a shows a schematic diagram of this calculation method, taking the tire centroid C as the center point and the plane passing through point C and perpendicular to the tire center plane as the reference plane, a rectangle with a semi-longitudinal length of $\Delta x = 0.4 \times r_0$ and a semi-transverse length of $\Delta y = 0.4 \times b_0$ is selected, where r_0 and b_0 represent the unloaded state. The intersection of the four vertices of the rectangle with the road surface along the tangent direction line of the tire center plane is the road surface characteristic point. If the four feature points are not in the same plane, the elevation of the feature points is automatically adjusted to place them in the same plane. The normal vector e_n of the wheel pavement plane determines the Z-axis of the track at this moment, and the intersection of the tire center plane and the pavement determines the X-axis and Y-axis of the track at this moment. The midpoint of this intersection is the P-point, hereinafter referred to as the geometric contact point (Figure 7).

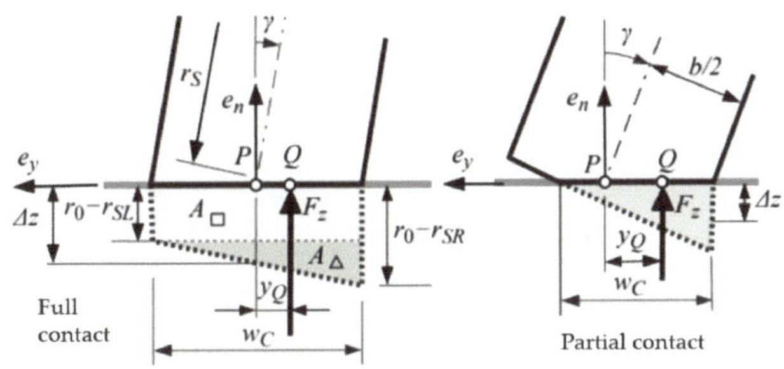

Figure 7. Schematic diagram of wheel–ground contact surface.

When solving, the bridge subsystem is assumed to be rigid, and the vehicle motion and vehicle force time history are obtained by solving the independent vehicle equation. Then, the wheel force is applied to the bridge, and the bridge deck motion state is obtained by solving the independent bridge equation. The superposition of the bridge deck motion time history and bridge deck irregularity is used as a new vehicle system excitation for the next iteration until the force between the vehicle and the bridge deck converges (Figure 8).

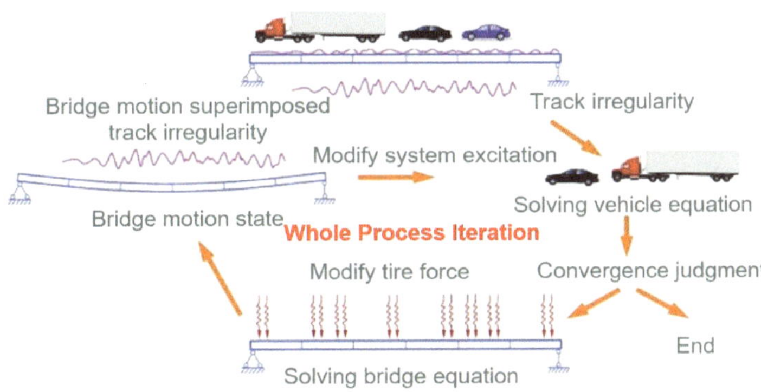

Figure 8. Schematic diagram of whole process iteration method.

2.4. *Driving Comfort and Safety*

Comfort and safety evaluation is based on the vehicle–bridge coupling vibration theory, which is an evaluation index that meets the human sensory and safety requirements. The most widely used comfort evaluation method is the 1/3 frequency doubling method [20] and the total weighted acceleration method (RMSV) proposed in ISO2631 standard. In recent years, the RMSV method has been widely used in research [21], which is more comprehensive than the frequency doubling method. The former obtains weighted acceleration root mean square value a_w by considering multi-directional vibration after weighting filtering the vehicle acceleration time domain signal. Its calculation formula is as follows:

$$a_w = \left[\int_{0.9}^{80} W^2(f) G_a(f) df \right]^{0.5} \tag{13}$$

where $W(f)$ is the frequency weighting function, and there are different weighting functions in ISO2631 standard for different vibration directions; $G_a(f)$ is the power spectral density function of the acceleration time domain signal.

Judging human comfort by a_w value (Table 1), this study uses 0.315 m/s^2 as the comfort limit for subsequent analysis.

Table 1. Weighted acceleration grading table.

Level	a_w (m/s^2)	Comments
1	<0.315	Comfortable
2	0.315–0.630	A little uncomfortable
3	0.500–1.000	Less comfortable
4	0.800–1.600	Uncomfortable
5	1.250–2.500	Very uncomfortable
6	>2.000	Extremely uncomfortable

The roll safety factor RSF and sideslip safety factor SSF are often used as evaluation indicators [22]. The evaluation criteria of vehicle roll accident in the Specification for Dynamic Performance Evaluation and Test Appraisal of Locomotive and Rolling Stock (GB/T 5599-2019) are as follows:

$$\text{RSF} = \min\left\{ \left| \frac{\sum_{i=1}^{k} (F_{Li} + F_{Ri})}{\sum_{i=1}^{k} (F_{Li} - F_{Ri})} \right| \right\} \geq 1.2 \tag{14}$$

where F_{Li} is the axle load of each wheel on the left, and F_{Ri} is the axle load of each wheel on the right.

The sideslip resistance of a vehicle is defined as:

$$F_{SR} = \mu_s (F_{vl} + F_{vr}) - (F_{hl} + F_{hr}) \tag{15}$$

where F_{vl} and F_{vr}, respectively, represent the vertical contact forces of the windward wheel and the leeward wheel of an axle, and F_{hl} and F_{hr}, respectively, represent the lateral contact forces of the windward wheel and the leeward wheel of an axle; μ_s is the lateral attach rate between the wheel and the road surface, which can be set to 0.7, 0.5, 0.15, and 0.07 according to the road conditions, representing the four basic road conditions of dry, wet, snow, and ice, respectively.

When the vehicle has the tendency of lateral sliding, the vehicle is considered to have already a sideslip safety accident. Therefore, the dimensionless SSF after considering the safety reserve factor can be defined as:

$$\text{SSF} = \frac{\overline{F_{SR}} - 1.645\sigma_{SR}}{0.2\mu_s G_a} \geq 1.0 \tag{16}$$

where G_a is the gravity of an axle of a vehicle, and generally the axle with light axle weight is taken; 0.2 is the safety reserve factor; $\overline{F_{SR}}$ is the average value of vehicle side-slip resistance, and σ_{SR} is the variance of vehicle side-slip resistance.

3. Simulation of Random Traffic Flow Based on Latin Hypercube Sampling

Sampling is a universal means to obtain random samples. For random variables with specific distribution, the Monte Carlo (MC) sampling method or Latin hypercube sampling method (LHS) can be used [23]. It is difficult for the random variables generated by the

MC method to approximate the actual probability distribution accurately when the sample size is small. In contrast, LHS sampling can effectively reduce the complexity of calculation by layering the probability and then sampling layer by layer, and its approximation effect is obviously better than that of the MC method when the sample size is small [24].

Taking one-dimensional LHS sampling as an example (Figure 9), the cumulative distribution curve is divided into several equal intervals through stratification. Only one sample is randomly selected in each stratification to form a group of one-dimensional random variables, which is suitable for single parameter sample sampling. LHS sampling is a memory-related sampling method, which takes into account the samples that have been used before, and it has a good approximation effect. It is suitable for the simulation of random traffic flow with a sample size of 10 digits and 100 digits.

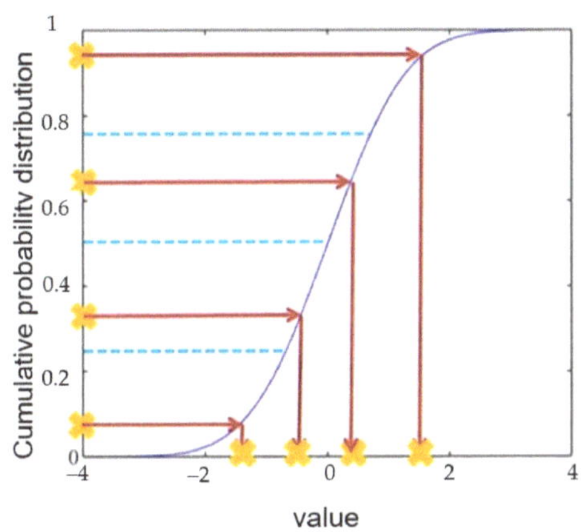

Figure 9. Schematic diagram of Vera Ding Chao cubic sampling.

Random traffic flow is mainly composed of four parts: the number of vehicles, the distribution of vehicle types, the distribution of axle load and wheelbase, and the distribution of vehicle speed and vehicle weight. Traffic volume forecast usually converts motor vehicles into the equivalent traffic volume of a standard vehicle according to their occupation of the road when driving on the road for statistical study. Generally, it is converted into the number of standard cars (pcu). According to the road traffic forecast report, the fleet with a total number of vehicles of 30 pcu is selected as the total number of sampling samples. LHS is used to stratify the distribution of vehicle type, axle load wheelbase, speed, and vehicle weight. Among them, the proportion of vehicle types conforms to the uniform distribution (Table 2), the distribution of axle load and wheelbase conforms to the lognormal distribution, and the distribution of vehicle speed and distance conforms to the normal distribution [25]. The axle load distribution of V1 car generated by Python language programming is shown in Figure 10, and the typical random fleet pattern in six lanes is shown in Figure 11.

Table 2. Vehicle distribution table.

Symbol	Category	Ratio	Conversion Coefficient
V1	Car	40.6%	1
V2	Bus	5.8%	2.0
V3	Buggy	13.8%	2.0
V4	Medium truck	10.9%	3.0
V5	Big truck	10.1%	4.0
V6	Trailers	10.2%	4.0
V7	Container truck	8.6%	4.0

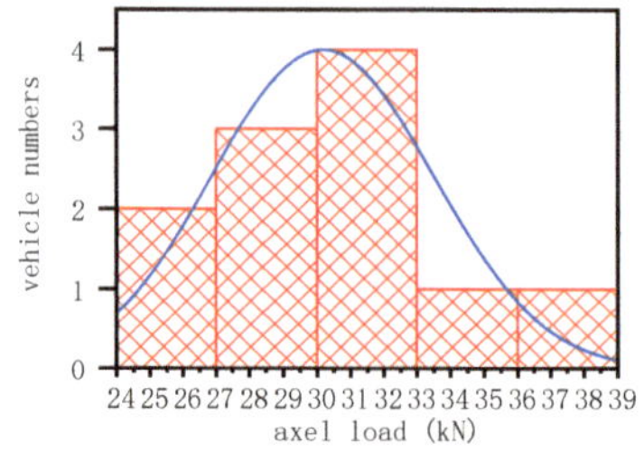

Figure 10. Axle load distribution diagram of V1 vehicle.

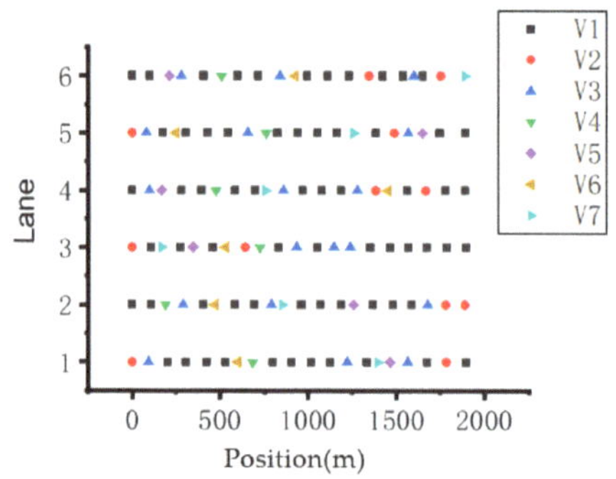

Figure 11. Random traffic flow simulation diagram.

4. Bridge Driving Safety and Comfort Analysis for Cable-Stayed-Suspension Cooperative System

4.1. Conditions

Aiming at the cable-stayed suspension bridge in Figure 1, an analysis model is established based on the aforementioned wind–vehicle–bridge coupling analysis method to study the influence of bridge stiffness change on driving safety and comfort.

The bridge is a single tower cable-stayed suspension cooperative system bridge with a main span of 2 × 638 m, and the main cable span is 2 × 730 m. Each side of the pylon is provided with 20 pairs of stay cables and 21 pairs of ordinary slings, in which the stay cables are arranged in a harp shape. The longitudinal anchorage spacing between the slings and stay cable beams is 16 m, and the crossing section spacing is 8 m, which is staggered. The length of the main girder of the single-span stay cable section is 327 m, and the length of the main girder of the sling section is 351 m. The cross section of the main girder is 26 m wide and 3 m high, and its standard section layout is shown in Figure 12.

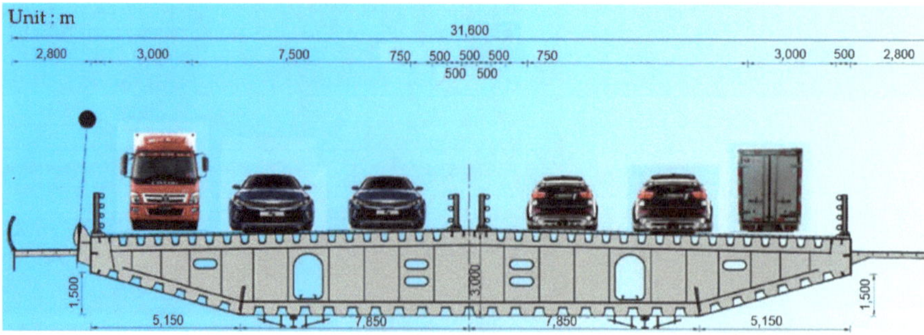

Figure 12. Layout of a standard bridge section.

To facilitate parameter analysis, based on the vertical static deflection span ratio of 1/330 under the load of two-way six lanes in the preliminary design scheme, the vertical deflection span ratio is adjusted by reducing the elastic modulus of the main girder, main cable, and stay cable, as shown in Table 3. Considering the driving situation in actual operation, two kinds of driving conditions, namely, ordinary driving conditions and extreme driving conditions, are compared under each torsion–span ratio. Ten kinds of analysis conditions are obtained. Among them, under normal driving conditions, random traffic loads are arranged on two-way six lanes, with 20 vehicles (30 pcu) in each lane, of which the proportion of heavy vehicles is 60%, and the maximum speed is 120 km/h. Under extreme driving conditions, lane loads (static loads) are arranged in five lanes as simulated traffic jams according to design specifications. Random traffic flow (20 vehicles, 120 km/h) is arranged in the sixth lane for loading to obtain dynamic response. In the analysis, the bridge deck wind speed is considered as the design wind speed of 25 m/s, and the pavement grade is Grade B.

Table 3. Stiffness parameters of bridge under extreme working conditions.

Conditions Number	Deflection–Span Ratio under Lane Load
1	1/330
2	1/300
3	1/280
4	1/250
5	1/200

Figure 13 shows the time history of vertical displacement in the left span under ordinary driving conditions (six-lane random traffic flow). Table 4 shows the static deflection and dynamic deflection under the same load. The influence of vehicle–bridge coupling vibration on deflection is analyzed by examining the vertical static displacement in the left span corresponding to vehicle load when the vertical displacement is maximum. The figure shows that the vertical dynamic deflection of the bridge increases gradually with the decrease of stiffness. When the bridge stiffness drops to 1/200, the vertical response value increases obviously because the whole bridge softens.

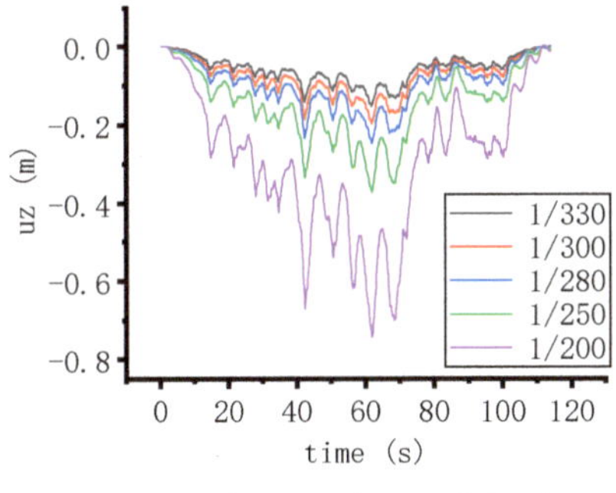

Figure 13. Time history of vertical displacement in the middle span of left girder under ordinary driving conditions.

Figure 14 shows the vertical dynamic displacement time history of the left span under extreme driving conditions (i.e., the static deflection caused by the load of five lanes is deducted). Table 5 shows the static deflection and dynamic deflection under the same load. From the head car to the bridge tower, the vertical displacement in the middle of the span first increases and then decreases to zero level. As the team drives to the right half span, the middle node of the left span is warped by the right load and presents a periodic vibration trend. The variation trend of the vertical response of the bridge under various stiffness conditions is roughly the same as that under ordinary driving conditions.

Table 4. Comparison table of static and dynamic displacement caused by vehicles under ordinary driving conditions.

Conditions Number	Maximum Displacement in Left Span (cm)	
	Vehicle-Induced Static Displacement	Coupled Vibration Displacement
1	13.79	15.18
2	17.94	19.63
3	22.87	24.67
4	34.54	37.14
5	67.28	74.27

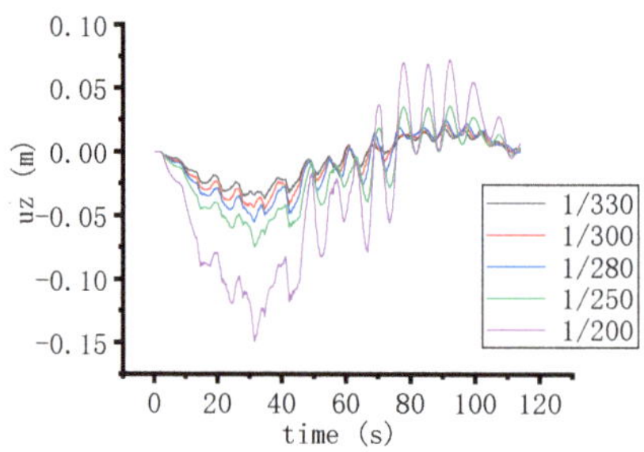

Figure 14. Time-history diagram of vertical displacement of left span.

Table 5. Comparison table of vehicle-induced static and dynamic displacement under extreme driving conditions.

Condition Number	Maximum Displacement in Left Span (cm)	
	Vehicle-Induced Static Displacement	Coupled Vibration Displacement
1	2.7	3.71
2	4.37	4.46
3	5.32	5.50
4	6.35	7.44
5	12.35	14.88

4.2. Vehicle Dynamic Response Analysis

Under normal driving conditions, the vertical acceleration time history of the center of mass of the vehicle in the most unfavorable driving state (the driving speed is 120 km/h) is shown in Figure 15. When the vehicle is driving in the pure suspension section, the acceleration change is not obvious. When the vehicle approaches the transition section of cable-stayed suspension, the ride comfort of the bridge deck worsens, and the vertical acceleration of the vehicle increases owing to the change of structural stiffness. When driving to the cable-stayed section, the acceleration changes tend to be gentle, the displacement at the bridge tower is very small due to the constraint, and the corner at the junction of the tower and beam is discontinuous under the action of vehicle load. Therefore, the vertical acceleration of the vehicle here increases obviously. The vertical acceleration trend of vehicles under each working condition is basically the same. Owing to the large response of the bridge under 1/200 working condition, the corresponding vehicle acceleration response is larger than that of other vehicles.

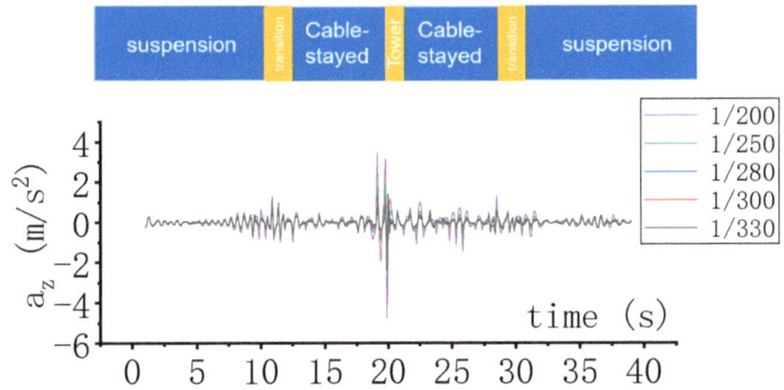

Figure 15. Vertical acceleration time history of vehicles under normal driving conditions.

The vertical acceleration response of vehicles in the most unfavorable situation in the fleet under extreme working conditions is shown in Figure 16. Owing to the large deformation of the bridge beam under static and live load, the driving alignment of vehicles under each working condition is quite different, and its curvature changes faster than that under ordinary driving conditions. As a result, the acceleration response of vehicles under extreme working conditions is obviously greater than that under ordinary driving conditions. When the bridge stiffness decreases, because the suspension cable segment itself is more flexible, its ability to resist deformation is weaker than that of the cable-stayed segment when the overall stiffness decreases uniformly. Therefore, the greater the overall stiffness decreases, the greater the deflection ratio corresponding to vehicle load, the more obvious the deformation of the suspension cable segment, and the greater the vertical acceleration response of the corresponding vehicle. When the vehicle travels to the cable-stayed section, the vertical acceleration response also increases with the decrease of stiffness, but its increase is weaker than that of the cable-stayed section.

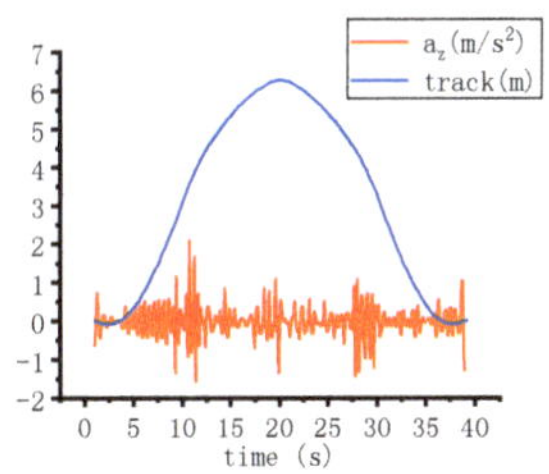

(**a**) Deflection–span ratio 1/330

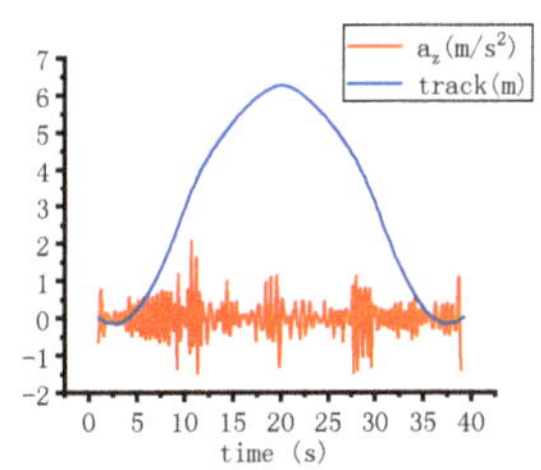

(**b**) Deflection–span ratio 1/3030

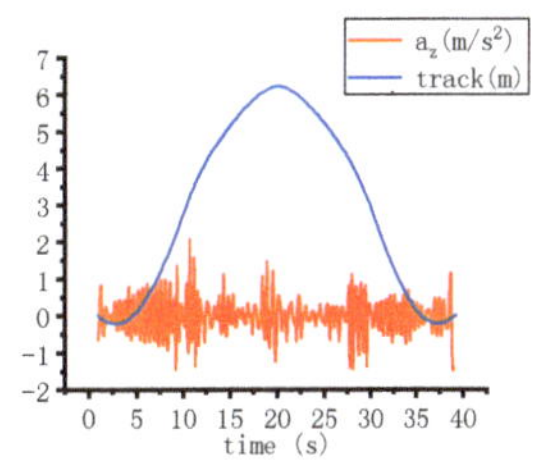

(**c**) Deflection–span ratio 1/280

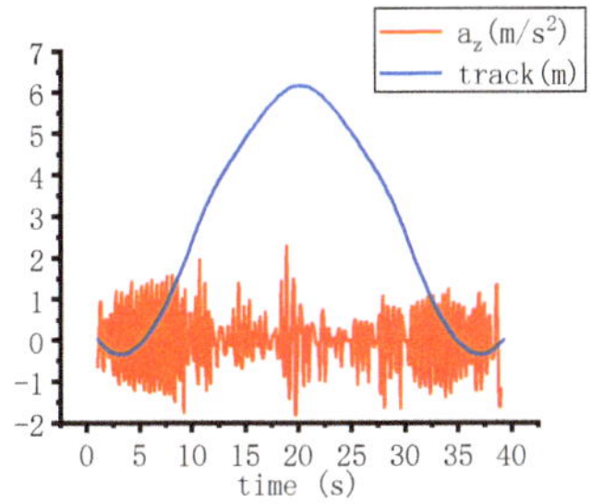

(**d**) Deflection–span ratio 1/250

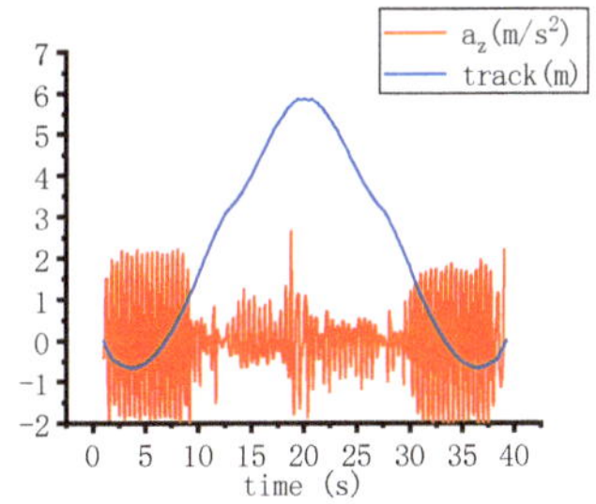

(**e**) Deflection–span ratio 1/200

Figure 16. Vertical acceleration time history of vehicles under extreme driving conditions.

Compared with ordinary driving conditions, the peak value of vertical acceleration of the vehicle is smaller under extreme working conditions. At this time, because there is only one lane under extreme working conditions, the load value is small and the corresponding dynamic displacement is smaller. The main reason the dynamic response of vehicles under extreme driving conditions is generally greater than that under ordinary driving conditions in other sections is that the initial alignment of the bridge is deteriorated owing to the full load on the driveway under extreme driving conditions. In addition, the main girder alignment in the sling area and stay cable area is quite different, which increases the deck irregularity. As the self-weight of random traffic increases the irregularity of the bridge deck, to reduce the interference of self-weight, the single-lane load in extreme working conditions is replaced by a V1 vehicle with a self-weight of 2.4 t, so as to explore the influence of bridge alignment change on the vertical weighted acceleration of the vehicle. The vertical acceleration time history of vehicles under various working conditions is obtained by analysis. The weighted acceleration root mean square value is compared with the extreme working conditions, as shown in Table 6:

Table 6. Root mean square value of vertical acceleration under extreme driving conditions a_z (m/s^2).

Deflection–Span Ratio	Single Car	Random Traffic Flow
1/330	0.108	0.242
1/300	0.110	0.239
1/280	0.125	0.263
1/250	0.180	0.320
1/200	0.516	0.514

The change of static alignment has an obvious influence on the root mean square value of vertical acceleration of vehicles. When the deflection is less than 1/250, the bridge alignment is relatively smooth, and the comfort decreases slightly with the increase of deflection. When the stiffness decreases to 1/200 corresponding to vehicle load (10.5 kN/m per lane), the linear deformation of the bridge is serious, and the driving comfort is seriously reduced.

4.3. Evaluation of Driving Safety and Comfort and Limit Value of Bridge Deflection Ratio

According to Formulas (13) to (16), the comfort and safety indexes of the vehicle are obtained after the vehicle response processing, as shown in Tables 7 and 8. Under normal driving conditions, the driving safety of vehicles meets the requirements, and the comfort is more sensitive to the change of vertical stiffness than safety. When the vertical deflection ratio is controlled below 1/250, the driving comfort still meets the requirements of the specification. When the vertical deflection ratio reaches 1/200, the vehicle still has a high safety reserve, but its driving comfort exceeds the limit.

Table 7. Comfort and safety index of vehicles under ordinary driving conditions.

Torsion Span Ratio	a_w (m/s^2)	RSF	SSF
1/330	0.188	2.688	2.666
1/300	0.192	2.681	2.663
1/280	0.236	2.677	2.652
1/250	0.291	2.584	2.637
1/200	0.406	2.336	2.544

Compared with ordinary driving conditions, the driving safety and comfort of vehicles will decline under extreme driving conditions, but the safety index is still in a reasonable range. When the vertical torsion span ratio is greater than 1/280, the root mean square value a_w of vehicle weighted acceleration reaches 0.315, which no longer meets the comfort requirements.

Table 8. Vehicle comfort and safety index under extreme driving conditions.

Torsion Span Ratio	a_w (m/s^2)	RSF	SSF
1/330	0.278	2.295	2.477
1/300	0.293	2.245	2.455
1/280	0.315	2.231	2.446
1/250	0.41	2.184	2.434
1/200	0.639	2.158	2.247

As can be seen from the above, for the cable-stayed cooperative system bridge studied here, the vehicle driving safety index is easier to meet under each deflection–span ratio, and the driving comfort becomes the limiting factor of the deflection–span ratio. The comparison of comfort indexes under various working conditions is shown in Figure 17. From the point of view of driving comfort, the bridge deflection ratio must be controlled to not be greater than 1/280.

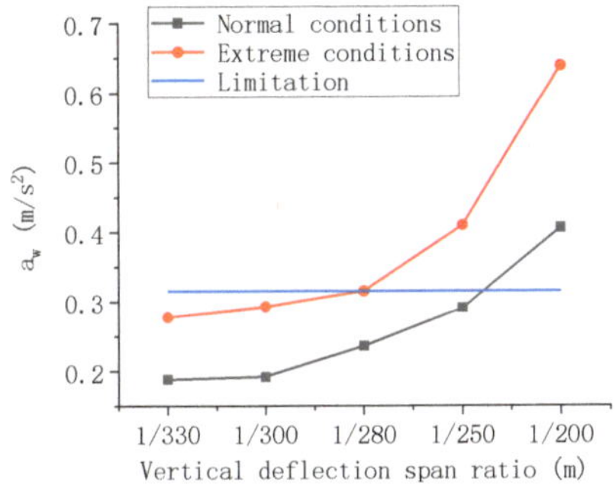

Figure 17. Correspondence between ride comfort and vertical deflection–span ratio.

The distribution of vertical stiffness index of the bridge under construction and completed cable-stayed cooperative system is shown in Figure 18, the blue signs represent the values. The figure shows that the vertical deflection span ratio of the cable-stayed cooperative system in the existing projects is between 1/619 and 1/402 [1,26,27], mostly for highway-railway dual-purpose bridges.

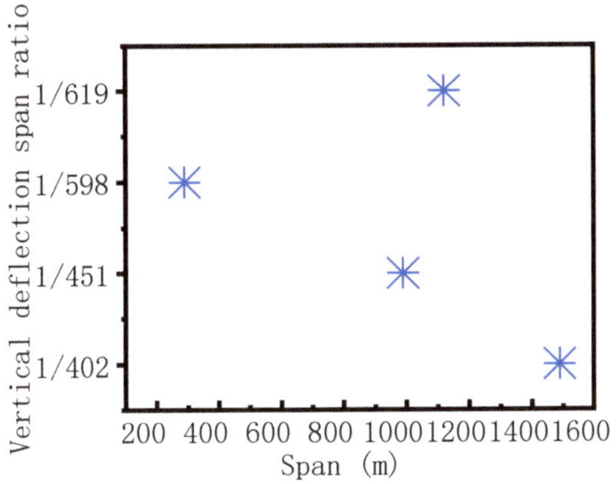

Figure 18. Investigation on engineering experience of vertical stiffness index.

Combining the requirements of driving safety and comfort and the existing engineering practice experience and considering a certain safety reserve, the limit value of the deflection–span ratio of the cable-stayed-suspension cooperative system bridge is suggested to be 1/300. The designed vertical deflection–span ratio of the bridge should be 1/330 to meet the requirements of driving safety and comfort.

5. Conclusions

1. The evaluation of driving comfort and safety of bridges based on vehicle–bridge coupling vibration theory can be used as a reference method for determining bridge stiffness of long-span flexible bridges. Using random traffic simulation and a more accurate tire model is beneficial to improve the accuracy of road vehicle–bridge coupling vibration analysis.

2. The vertical stiffness difference between the suspension cable area and the stay cable area of the cable-stayed cooperative system bridge is obvious. The vertical acceleration response increases obviously when the vehicle travels to the transition section of the stay cable. The suspension cable section is more flexible than the stay cable section and more sensitive to the change of overall stiffness.

3. The bridge deflection ratio has a significant impact on vehicle ride comfort. The greater the deflection ratio, the worse the ride comfort index. From the point of view of ride comfort, the vertical deflection ratio of the single-tower space cable-stayed suspension cable cooperative system studied here is suggested to not be greater than 1/300.

4. On the basis of this study, the numerical analysis method proposed in this paper can be used to study the mechanical behavior of bridges under different loads and its influence on the driving performance in the future. A scaled model can be established for testing, providing a theoretical basis for the design of bridges with a cable-stayed-suspension cooperative system.

Author Contributions: Conceptualization, L.X. and Y.H.; methodology, X.W.; software, L.X.; validation, X.W.; formal analysis, Y.H.; investigation, Y.H.; resources, L.X.; data curation, Y.H.; writing—original draft preparation, L.X.; writing—review and editing, Y.H.; visualization, Y.H.; supervision, X.W.; project administration, L.X. All authors have read and agreed to the published version of the manuscript.

Funding: The support from the National Natural Science Foundation of China (Grant No. 52078424, 52178170), Natural Science Foundation of Sichuan Province (Grant No. 2022NSFSC0426), National Key R&D Program of China (Grant No. 2022YFB3706703) is gratefully acknowledged.

Data Availability Statement: No new data were created or analyzed in this study. Data sharing is not applicable to this article.

Acknowledgments: Thanks to all authors for their cooperation and innovation.

Conflicts of Interest: The authors declare no conflict of interest.

References

1. Liu, F.; Huang, Z.; Guo, S.; Lei, J. Research on rail-road cable-stayed suspension bridge. *Steel Constr. Chin. Engl.* **2019**, *34*, 49–55. (In Chinese)
2. Bryja, D.; Knawa, M. Computational model of an inclined aerial ropeway and numerical method for analyzing nonlinear cable-car interaction. *Comput. Struct.* **2011**, *89*, 1895–1905. [CrossRef]
3. Song, Y.; Liu, Z.G.; Rxnnquist, A.; Navik, P.; Liu, Z.D. Contact Wire Irregularity Stochastics and Effect on High-speed Railway Pantograph-Catenary Interactions. *IEEE Trans. Instrum. Meas.* **2020**, *69*, 8196–8206. [CrossRef]
4. Duan, F.; Song, Y.; Gao, S.; Liu, Y.; Chu, W.; Lu, X.; Liu, Z. Study on Aerodynamic Instability and Galloping Response of Rail Overhead Contact Line Based on Wind Tunnel Tests. *IEEE Trans. Veh. Technol.* **2023**, 1–11. [CrossRef]
5. Larsen, A.; Larose, G.L. Dynamic wind effects on suspension and cable-stayed bridges. *J. Sound Vib.* **2015**, *334*, 2–28. [CrossRef]
6. Huajun, B.; Wangqing, W.; Xu, S. Research on high-speed railway bridge design based on stiffness deformation requirements. *World Bridges* **2020**, *48*, 27–33. (In Chinese)
7. Chen, J.; Lei, J.; Golden, O.; Hu, C. Study on structural characteristics and stiffness indexes of thousand-meter-scale road-rail suspension bridge. *Railw. Stand. Des.* **2022**, *66*, 54–61. (In Chinese)
8. Lu, D.; Wei, H.; Yang, Y.; Wei, W. Research Progress on Theory and Application of Coupled Vibration of Highway Bridges and Vehicles. *China J. Highw. Transp.* **2018**, *31*, 38–54.
9. Li, Y.; Long, J.; Xiang, H.; Ma, H.; He, M.; Xie, J. Transverse deflection-span ratio suggested value of an urban rail transit bridge based on a wind-vehicle-bridge system. *J. Vib. Shock* **2020**, *39*, 211–217. (In Chinese)

10. Li, Y.; Xiang, H.; Liao, H. Study on wind shielding effect of wind screens based on coupling vibration theory of wind-vehicle-bridge(lines) systems. *China Civ. Eng. J.* **2014**, *47*, 97–102. (In Chinese)
11. Li, X.; Gong, Z. Analysis of vehicle-included structure-borne noise characteristics of large-span steel bridge with different structure forms. *J. Vib. Shock* **2021**, *40*, 91–99. (In Chinese)
12. Chen, H.; Zhang, W.; Li, J. Coupled dynamic response analysis of high-speed railway vehicle and track system under random track irregularities. *J. Vib. Control* **2020**, *26*, 1855–1869.
13. Deng, L.; Duan, L.L.; He, W.; Ji, W. Study on Vehicle Model for vehicle-bridge coupling vibration of highway bridges in China. *China J. Highw. Transp.* **2018**, *31*, 92–100. (In Chinese)
14. Kwasniewski, L.; Li, H.; Wekezer, J.; Malachowski, J. Finite element analysis of vehicle–bridge interaction. *Finite Elem. Anal. Des.* **2006**, *42*, 950–959. [CrossRef]
15. Yuan, S. Development and Programing of Typical Vehicles Equation of Motion for Coupling Vibration between Highway Bridge and Vehicles. Master's Thesis, Changan University, Xi'an, China, 2012. (In Chinese).
16. Yu, M.; Yao, T.; Kong, L.; Shilin, L. A Review of Tire-asphalt Pavement Friction Measurement and Skid Resistance Model Study. *J. Highw. Transp. Res. Dev.* **2020**, *37*, 12–24. (In Chinese)
17. Xiaohui, S.; Zeng, J.; Xin, J.; Zhao, J.; Zhao, P. Dynamic Dharacteristics and Parameter Analysis of Tire Model. *J. Chongqing Univ. Technol. Nat. Sci.* **2020**, *34*, 1–8. (In Chinese)
18. Li, Y.; Xiang, H.; Qiang, S. Review on Coupling Vibration of Wind-vehicle-bridge System. *China J. Highw. Transp.* **2018**, *31*, 24–37. (In Chinese)
19. Zhang, W.; Zhang, Z.; Zhang, X. Numerical investigation on the wind-field-dependence property of bridge section aero dynamic admittance. *Acta Aerodyn. Sin.* **2018**, *36*, 677–686. (In Chinese)
20. Han, W.; Chen, A. Ride comfort assesment of vehicle running on long-span cable-stayed bridge under cross wind. *China J. Highw. Transp.* **2008**, *21*, 54–60.
21. Lee, S.; Lee, S.; Lee, J.; Choi, S. A study on the evaluation of vehicle ride comfort using RMSV method. *Int. J. Automot. Technol.* **2018**, *19*, 69–76.
22. Chen, N. The Safety of Road Vehicles Driven on Bridge Deck under Crosswind and Its Windproof Precautions. Ph.D. Thesis, Southwest Jiaotong University, Cheng'du, China, 2015. (In Chinese).
23. Wang, J.; Huang, P.; Han, W.; Yuan, Y.; Zhou, G.; Xu, X. Fatigue Life Prediction of Concrete Bridges Based on Improved Damage Algorithm and MCMC Traffic Flow Simulation. *J. Hunan Univ. Nat. Sci.* **2021**, *48*, 31–43. (In Chinese)
24. Qu, Z.; Liu, M.; Liu, J.; Cao, H.; Yang, G.; Liu, T. Latin hypercube sampling monte carlo simulation for the information flow time-delay of railway power supply system. *China Railw. Sci.* **2012**, *33*, 82–90. (In Chinese)
25. Wang, D. Analysis of Vehicle-Bridge Coupled Vibration for Long-Span Suspension Bridge with Finite Element Model Updating under Random Traffic Flowing. Master's Thesis, Chang'an University, Xi'an, China, 2008.
26. Liu, F.; Luo, X. Influence of central buckle of main cable on mechanical properties of three-tower cable-stayed suspension cooperative system bridge. *J. China Foreign Highw.* **2016**, *36*, 143–146. (In Chinese)
27. He, D.; Xiao, H. Influence of central buckle on mechanical property of combined cable-stayed and suspension bridge. *Bridge Constr.* **2020**, *50*, 23–28. (In Chinese)

infrastructures

MDPI

Article

Micro-Scale Experimental Approach for the Seismic Performance Evaluation of RC Frames with Improper Lap Splices

Ali Javed [1,*], Chaitanya Krishna [2], Khawaja Ali [3], Muhammad Faheem Ud Din Afzal [1], Armin Mehrabi [1,*] and Kimiro Meguro [4]

1 Department of Civil and Environmental Engineering, Florida International University, Miami, FL 33174, USA
2 School of Engineering and Technology, Asian Institute of Technology, Pathum Thani 12120, Thailand
3 Transport Development Division, CTI Engineering International, Tokyo 136-0071, Japan; khawaja.ali@ctii.co.jp
4 Institute of Industrial Science, The University of Tokyo, Tokyo 153-8505, Japan
* Correspondence: ajave010@fiu.edu (A.J.); amehrabi@fiu.edu (A.M.)

Abstract: Reinforced concrete (RC) frames are an integral part of modern construction as they resist both gravity and lateral loads in beams and columns. However, the construction methodologies of RC frames are vulnerable to non-engineering defects, particularly in developing countries. The most common non-engineering defect occurs due to improper lap splice, which can compromise the structural integrity. This research demonstrates an easy, low-cost, and verifiable experimental technique incorporating micro-concrete to evaluate the seismic performance of a completely engineered RC frame with the defect of improper lap splice. The micro-concrete was prepared by using the locally available material for a target compressive strength and then two scaled-down RC frames (1/16 scale) were prepared, including one proper frame and another with improper lap splice. Finally, these frames were tested on a shake table to study their behavior under various seismic loading conditions. This study quantifies the severity of high-risk structural systems due to non-engineering defects. The experimental results demonstrate that improper lap splice can alter the frame's damage points, triggering the failure of the whole structure.

Keywords: reinforced concrete frame; non-engineering defects; improper lap splice; Applied Element Method (AEM); micro-concrete; shake table test

Citation: Javed, A.; Krishna, C.; Ali, K.; Afzal, M.F.U.D.; Mehrabi, A.; Meguro, K. Micro-Scale Experimental Approach for the Seismic Performance Evaluation of RC Frames with Improper Lap Splices. *Infrastructures* **2023**, *8*, 56. https://doi.org/10.3390/infrastructures8030056

Academic Editor: Chris Goodier

Received: 16 February 2023
Revised: 6 March 2023
Accepted: 13 March 2023
Published: 15 March 2023

1. Introduction

The reinforced concrete (RC) moment resisting frame (MRF) is among the most common building construction practices in the modern world [1]. In recent earthquakes, these buildings have suffered damages mainly due to their inadequate design and flawed engineering practices in developing countries [2,3]. Many countries follow international building codes which have stringent building code requirements for high seismic zones. However, the construction practices in developing countries are slack, which has caused a severe threat to these buildings. This vulnerability could be attributed to various kinds of non-engineering defects in the buildings and has been widely accepted by experts in the past [4,5]. In this research, the authors have investigated improper lap splicing as an existing defect in RC frame construction which can occur due to incorrect or inadequate design.

It is imperative to note that through the experiences of experts [6], it is found that the main causes of these various non-engineering defects are due to a lack of (1) concern for safety, (2) awareness to build earthquake-resistant buildings, (3) skills in seismic design and workmanship, and (4) scarcity in construction materials. While some non-governmental organizations (NGOs) and academicians have tried to study and understand the cause of these problems [7–9], especially when built by local contractors, many others have tried to provide various retrofitting measures for these buildings to make them earthquake-resistant [10–13]. However, only a few researchers have examined these defects experimentally or analytically for dynamic testing to achieve a tailor-made solution in the form of retrofitting [14]. The main reason for this is the cost and complications involved in manufacturing RC frames of different cases and conducting a shake table test.

Various researchers have attempted to study different inadequacies in the construction of RC frames concerning their design and code provisions [15,16]. They have primarily focused on the health monitoring of such structures including the effects of such inadequacies on the structural integrity [17–19]. Filiatrault et. al. [20] compared a ductile RC frame with a non-ductile frame using a 1:4 scaled two-story two-bay model and provided some recommendations to improve the seismic protection of ductile and non-ductile RC frames. Caccese and Harris [21] conducted a shaking table test using 1:5 scaled RC models by employing non-linear ultimate strength models. Similarly, Zamic et al. [22] and Kim et al. [23] also conducted static and dynamic tests of RC models with 1:4 and 1:6 scales, respectively, and evaluated various aspects including cyclic degradation of stiffness and energy absorption of the RC frame. Noor et al. [24] explicated the procedures, factors, merits, limitations, and drawbacks of scaled concrete testing which are later used in this article to derive the scaled parameters.

The experimental approaches enlightened in the above-mentioned research works were not qualified for the current requirements on the cost of the setup, which would eventually lead to a series of multiple experiments to investigate other non-engineering defects. To reduce the cost, one of the most lucrative ideas is to have an RC model with a scale higher than 1:10. This requires expertise in the design and construction of frames involving micro-concrete and small-diameter rebars. The possibility of utilizing micro-concrete for small-scale testing is evaluated by several researchers including Dejian et al. [25], Boutemeur et al. [26], and Tobar et al. [27]. This research explains a simple and demonstrable experimental technique using micro-concrete to design and build completely engineered and non-engineered RC frames and test them on a shake table.

Among different non-engineering defects, improper lap splicing has been studied in this research which is one of the most common problems occurring in RC structures, especially in developing countries. Some existing research has focused on the experimental evaluation of this defect concerning columns. For instance, the seismic fragility of RC structures with splice-deficient columns was studied experimentally [28,29]. The results exhibited that the vulnerability of such RC structures is dependent on the failure mode of splice-deficient columns and the limit state under consideration. Similarly, Chronopoulos et al. [30] studied the behavior of inadequate splices in the tension reinforcement of RC elements and recommended their upgradation through welding. Elrakib and Arafa [31] studied these common defects in RC beams under flexural loading. They concluded that the distress produced due to these defects causes an abrupt failure of the structure without the intermediate stage reducing the ductile behavior of the elements. Goksu et al. [32] evaluated the effect of different lap splice lengths in RC columns and suggested the provision of hooks to improve the bond–slip behavior. Cho and Pincheira [33] carried out nonlinear numerical modeling to determine the required splice length in reinforced concrete columns. They presented a model based on the local bond stress–slip relationship which could calculate lateral load resistance and deformation corresponding to splice failure. Chowdhury and Orakcal [34] also performed analytical modeling to investigate the inadequacy of lap splices. This inadequacy of lap splices can lead to an inadequate bond strength causing the reduction of concrete durability and the load-carrying capacity of structural members under various kinds of loadings [35]. Thereby, there should be a sufficient lap splice length with compatible bond strength between concrete and reinforcement rebars to maintain the proper distribution of stresses in the RC structural frames.

2. Code Requirements for Rebar Lap Splices

It is well established that the bond formation between steel reinforcing bars and concrete results from the adhesion, mechanical interlock, and friction produced by the ribs of the reinforcing bar. The mechanical interlock, as compared to adhesion and friction, does not occur with plain reinforcing bars owing to their plain surface [36]. The bond strength of a plain bar is comparatively low and for reaching the yield strength, the lap splices are of paramount importance. Additionally, the reinforcing bars are generally supplied by the manufacturers in shorter lengths and because of it, it is often essential to splice the bars in the field. Several structural design guidelines including ACI 318-19 [37] and ASCE 41-13 [38] stipulate the empirical formulas for the required lap splice length of reinforcing bars.

By definition, the lap splice length is a function of reinforcing steel yield strength (f_y) and concrete strength (f_c'). There are two types of lap splices: (1) tension splices and (2) compression splices. In columns subjected to the axial force and moment, the tensile stresses may generate on one face of the column for eccentric loading. If such stresses occur, ACI 318-19 [37] code provisions require tension splices to be used. The splice requirements are prepared based on that a compression lap splice has a tensile strength of at least 0.25 f_y. Therefore, even if column bars are designed for compression, some tensile strength is inherently provided. The minimum lap splice length in tension (l_{st}) not less than 300 mm is selected based on the class type considering the development length (l_d) of the bar as shown in Table 1 in which Class A and Class B correspond to l_{st} = 1.0 l_d and l_{st} = 1.3 l_d, respectively. For lap splice lengths in tension for column members, Class A and Class B are decided based on the maximum tensile bar stress due to the factored loads. For instance, for tensile stress $\leq 0.5\ f_y$ and maximum percent of A_s spliced within the required lap length $\leq$50%, the splice type is designated as Class A. Otherwise, the lap splice length is taken to be greater than 1.3 l_d and 300 mm for Class B.

Table 1. Classification of tension splices of rebars recommended by ACI 318-19.

Maximum Tensile Bar Stress	The Maximum Percent of A_s Spliced within the Required Lap Length	
	$\leq$50%	>50%
$\leq 0.5\ f_y$	Class A	Class B
$> 0.5\ f_y$	Class B	Class B

On the other hand, compression splices do not need provisions as stringent as those specified for tension splices. Reinforcing bars in compression are spliced mainly in columns, where bars are terminated normally just above each floor or every other floor. This is conducted partly for construction convenience, to avoid handling and supporting long column bars; however, it is also conducted to allow the column steel area to be reduced in steps, as the loads reduce at higher floors. The minimum lap splice length in compression (l_{sc}) not less than 300 mm is selected based on the yield strength f_y of the reinforcement bar and the concrete strength f_c' as shown in Table 2. The required lap splice length increases when f_c' is lower and f_y is higher. As per ACI 318-19 [37], the required lap splice length is to be increased by a multiplying factor of 1.333 for $f_c' < 21$ MPa.

Table 2. Classification of compression splices of rebars recommended by ACI 318-19.

f_c' (MPa)	$f_y \leq 414$ MPa	414 MPa $< f_y \leq 550$ MPa	$f_y > 550$ MPa
$\geq$21	l_{sc} = Larger of (a) 0.071 $f_y d_b$ (b) 300 mm	l_{sc} = Larger of (a) (0.13 f_y − 24)d_b (b) 300 mm	l_{sc} = Larger of (a) (0.13 f_y − 24)d_b (b) l_{st}
<21	$l_{sc} = 1.333 \times 0.071\ f_y d_b$	$l_{sc} = 1.333 \times (0.13\ f_y - 24)d_b$	$l_{sc} = 1.333 \times (0.13\ f_y - 24)d_b$

Note: d_b denotes the diameter of the rebar and l_{st} denotes the lap splice length in tension.

3. Design and Development of Model Structure

3.1. Design of Structure

A simple two-story RC frame was selected and designed based on ACI 318-19 [37] code provisions. A finite-element-based building design software was utilized to find the member forces and section sizes along with the demand for reinforcement bars. The input parameters for the design of the structure were selected from the Building Code of Pakistan because of its frequent issues related to non-engineering defects [4,39].

These seismic provisions stipulated in the Building Code of Pakistan are compatible with the American Concrete Institute ACI 318-08. Figure 1a shows the dimensions and reinforcement bars selected for the actual full-scale frame structure. The connection of the beam-column is designed to be hinged allowing the rotation without any translational displacement, whereas the base of the columns is assumed to be fixed ignoring the soil–structure interaction. The slenderness ratio of the columns (kL_u/r) is determined to be $0.85 \times 3.6/0.139 = 22$, where $k = 0.85$ is the effective length factor for a column with a fixed base and hinged top determined from the Jackson and Moreland Alignment Charts, which provide a graphical determination of k for a column of a constant cross-section in a multi-bay frame; $L_u = 3.6$ m represents the column height of a single story, and $r = 0.139$ m shows the value of the radius of gyration estimated from the basic formula of $r = \left(\frac{I_g}{A_g}\right)^{1/2}$ (where I_g and A_g are the second moment of area (m^4) and cross-sectional area (m^2) of the column, respectively).

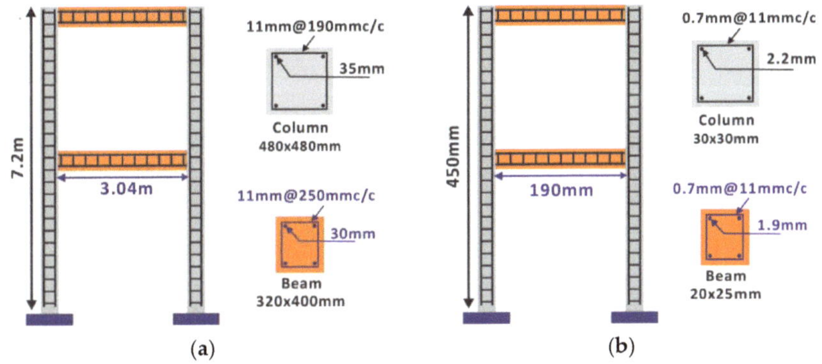

Figure 1. Reinforcement detail. (**a**) Actual size structure and (**b**) 1/16-scaled model.

According to ACI 318-19 [37], slenderness (or second-order) effects can be neglected in unbraced columns (against sideway) if $kL_u/r \leq 22$. The section sizes of structural components, i.e., column and beam, and reinforcement detailing are also shown in Figure 1a. As for the columns, the longitudinal bars of D35 (dia: 35 mm) and stirrups of D11@190 mm c/c are embedded in 480 × 480 square-shaped short columns with a reinforcement ratio of 1.67% while preserving a concrete cover of 35 mm, whereas the singly reinforced beams with a 320 × 400 rectangular-shaped cross-section are designed by providing D30 bars on the top and bottom along with ties of D11@250 mm c/c while preserving a concrete cover of 30 mm. These are the design parameters considered for the full-scale frame structure.

3.2. Scaling of the Structure

The similitude laws are essential for testing the scaled model. Both the dimensional analysis and direct use of the governing equations can be employed to derive the scaling laws [40]. It is a fact that a wholly scaled replica of RC structures is almost impossible because the post-yielding behavior of concrete is governed by the following three significant parameters: (1) the descending part of the stress–strain curve of concrete in compression, (2) the cyclic response degradation, and (3) the steel and concrete bond characteristics. Due to these factors, the use of micro-concrete becomes essential which is also advantageous in terms of saving material and equipment, and helps in better handling [41]. However, by taking the stress and stiffness scale factors equal to unity, the similitude requirement for density cannot be satisfied. Thereby, by selecting micro-concrete as the model material, the similitude requirement for density should be equal to unity. As small-scale models of RC structures in seismic analysis cannot be accurate replica models, one of the following strategies is usually adopted to reproduce similar inertial effects [24] as (1) reducing the material strength, (2) the provision of additional masses at appropriate points on the model and loosely attached so that the stiffness and strength of members are not altered, and (3) changing the density of the materials by keeping other properties constant. The laws of similarity by taking the density as the unity are shown in Table 3, which are applied to develop a small-scale replica of the actual RC frame structure. In other words, the dimensions of the cross-sections of structural components and reinforcement bars are reduced to 1/16-scale to obtain the reinforcement of the small-scale model as shown in Figure 1b.

Table 3. Laws of similarity.

Parameter	1/16 Scaled Value	Target Value
Length	1/16	1/16
Area	$1/16^2$	$1/16^2$
Mass	$1/16^3$	$1/16^3$
Time	$1/\sqrt{16}$	$1/\sqrt{16}$
Density	1	1
Stress	1/16	1
Strain	1/16	1

Furthermore, the same ground motion cannot be applied to an actual model and a scaled model to verify the laws of similarity. In this case, to verify the similarity laws for a scaled model structure, additional weights were calculated based on the weights of actual and scaled models as shown in Table 4. The additional mass calculations are based on Equation (1).

$$DG = \frac{1}{S_L^2}\left\{G_P\left(1 - \frac{1}{S_L}\right) + Q_E\right\} \quad (1)$$

where DG is the additional weight; S_L is the scale of the model (=16); G_p is the weight of the actual size model (=9830 kg); and Q_E is the vertical live load (=0). After putting these values in Equation (1), DG turns to 36 kg. Besides this, the time of the input ground motion was also contracted [24]. The time contraction is estimated to be $S_T = \frac{1}{\sqrt{S_L}} = \frac{1}{4}$.

Table 4. Estimation of weights of actual and scaled models.

Model	Volume (m^3)	Total Weight (kg)
Weight of actual size model	4.096	9830
Weight of scaled model	0.001	2.4

3.3. *Verification of the Similitude Laws*

In order to verify the application of the laws of similarity for a 1/16 scale-down model structure, two analyses, one for the actual size structure and the other for the scaled-down model structure, were carried out and the results were compared to see the effectiveness of AEM in the case of small modeling against earthquake loadings.

3.3.1. Selection of Ground Motion

The selection of ground motions was made on the basis of ASCE 7-16 16.1 and the peak values of the acceleration response spectrum of the ground motions. According to

ASCE 16.1.1 [42], the response history analysis shall consist of not less than three appropriate ground motions. Each ground motion shall consist of a horizontal acceleration history obtained from the records of events having a magnitude, fault distance, and source mechanism that are consistent with those factors which control the maximum considered earthquake. The ground motions shall be scaled such that the average value of the 5% damped response spectra for the suite of motions is not less than the designed response spectrum for the site for periods ranging from 0.2 T to 1.5 T where "T" is the natural period of the structure in the fundamental mode for the direction of the response being analyzed.

For the verification of the laws of similarity, the ground motions were selected based on the time period of the actual size structure. The natural time period of the actual size frame was calculated as 0.275 s by using the ASCE 7-16 12.8.2 analytical formula [42]. On this basis, three earthquakes including the 1999 Chi-Chi earthquake, Taiwan, 1940 El Centro earthquake, California, and 1994 Northridge earthquake, California were selected for the numerical simulation of the actual and scaled structure. The reason for their selection is that the peaks of the response spectrum of these earthquakes are within the range of the time period of the actual size structure as shown in Figure 2.

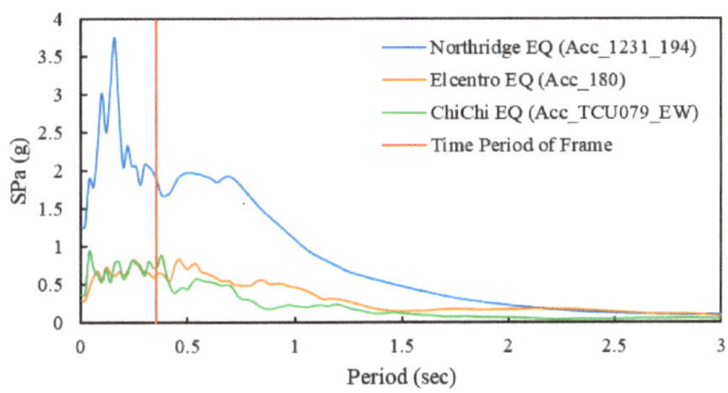

Figure 2. Response spectrum of selected earthquake motions.

3.3.2. Applied Element Method (AEM) Results

The actual size structure and the 1/16 scaled structure were analyzed against the above-mentioned ground motions by using the Applied Element Method (AEM) with similar geometric properties. The properties of the scaled model were reduced according to the 1/16 scale including the size of the elements in the analysis. The additional weight of 36 kg was also applied to the scaled model as per the laws of similarity. Furthermore, the ground motion duration was also contracted with the factor of $\frac{1}{\sqrt{16}}$ as per Table 3. The deformed shape and cracks in the springs for the actual structure and scaled model were compared at three different stages of loading as shown in Figure 3. The analysis results show that both the actual frame and scaled frame exhibited similar crack propagation at different stages of loading. The maximum displacement was obtained for both cases and the time history of both frames is also shown in Figure 3.

For actual structure:	Maximum displacement = 8.55 mm
	Max. Displacement/height of structure = 0.00119
For scaled model:	Maximum displacement = 0.453 mm
	Max. Displacement/height of structure = 0.00101

Difference in results = 2.52%

The drift ratio values of both the actual size frame and the scaled frame are significantly close to each other, which verifies the applicability of the similitude laws and demonstrate that the numerical simulation of the scaled model can lead to fruitful results.

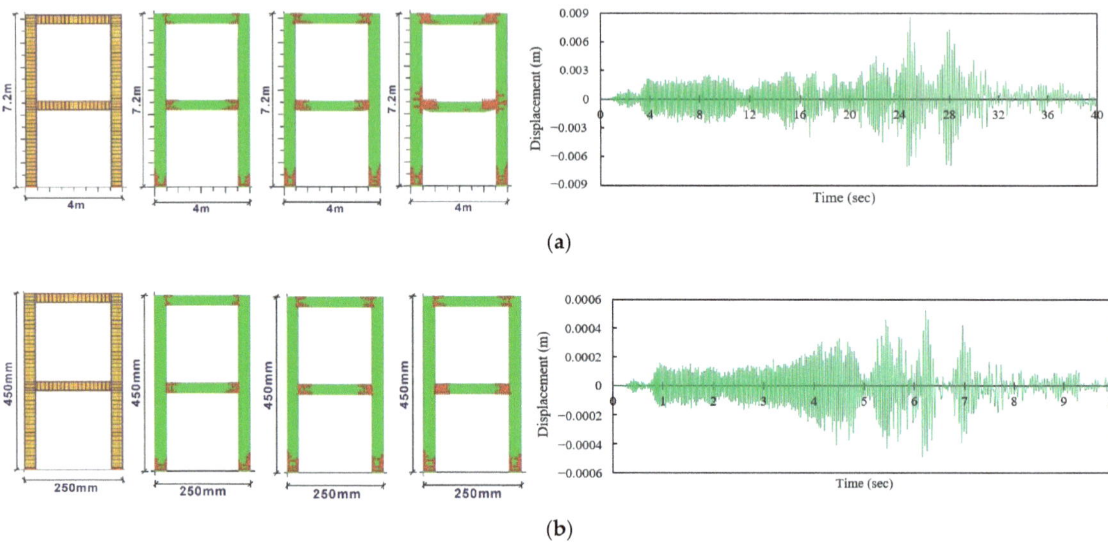

Figure 3. Analysis of actual frame and 1/16 scaled frame using Applied Element Method and time−history of displacement. (**a**) Cracked springs of actual structure at loading stages of 50, 100, and 200, (**b**) cracked springs of 1/16 scaled structure at loading stages of 50, 100, and 200.

3.4. Preparation of Model Structures

The geometric properties of the model are shown in Table 5. Based on these properties, it was decided to fabricate the formwork by aluminum angles with simple nuts and bolts as aluminum angles are lightweight and can give an excellent finish to the surface, which is very important for small-scale models. Due to the small diameter of the reinforcement bars, it was a challenging task to prepare the same size stirrups with a margin of 2 mm clear cover for micro-concrete. Figure 4a,b show the stirrups preparation and spacing between the stirrups. To tie the stirrups with beam and column reinforcement bars, a steel wire of 0.2 mm diameter was used to obtain the same effect as field conditions. To maintain a proper clear cover, small pieces of a 1.9 mm bar were attached to the main reinforcement bars so that the wires did not bend in the formwork. Figure 4c shows the prepared reinforcement cages for the beams and columns.

Table 5. Geometric properties of the model.

Sr. No.	Parameter	Value
1	Height of model	450 mm
2	Width of model	250 mm
3	Dimensions of columns	30 mm × 30 mm
4	Dimensions of beams	20 mm × 25 mm

3.5. Location of Improper Lap Splices

The main focus of this research is to study the effect of improper lap splicing on reinforced concrete frames. It was intended to select that location for improper rebar splicing, where it produces the largest displacement in the structure so that the effect can be easily studied during the experimental program. Different locations of improper lap splices were considered, such as (1) at the center of top bars, (2) at the center of bottom bars, (3) at 1/6th of the beam length (L) from the face of the column for the top bar, and (4) at 1/6th of the beam length (L) from the face of the column for the bottom bar. The maximum displacement and ratio of displacement to the height of the structure were evaluated for each scenario using the Applied Element Method (AEM). Figure 5 shows the plots of maximum displacement achieved in each case which shows that maximum displacement was found to be caused for the case when the improper splice was provided at 1/6th of the beam length (L) from the face of column for the top bar. Thereby, an improper lap splice effect was selected and introduced at 31 mm (1/6 × beam length) from the beam–column interface, as shown in Figure 6. According to the ACI 318, the minimum splice should be 48 d_b where d_b is the diameter of the bar; thereby, in a given case, it should be 90 mm, whereas the provided splice is 9 mm.

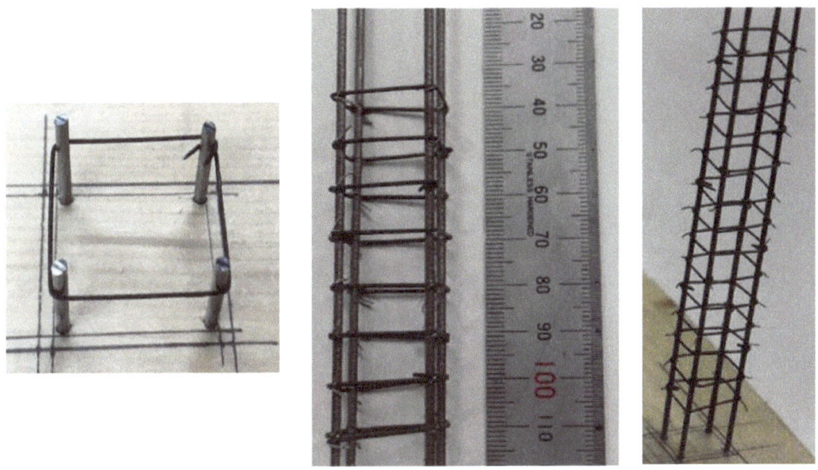

(a) (b) (c)

Figure 4. Reinforcement fabrication (**a**) stirrups preparation, (**b**) spacing between stirrups, and (**c**) column reinforcement caging.

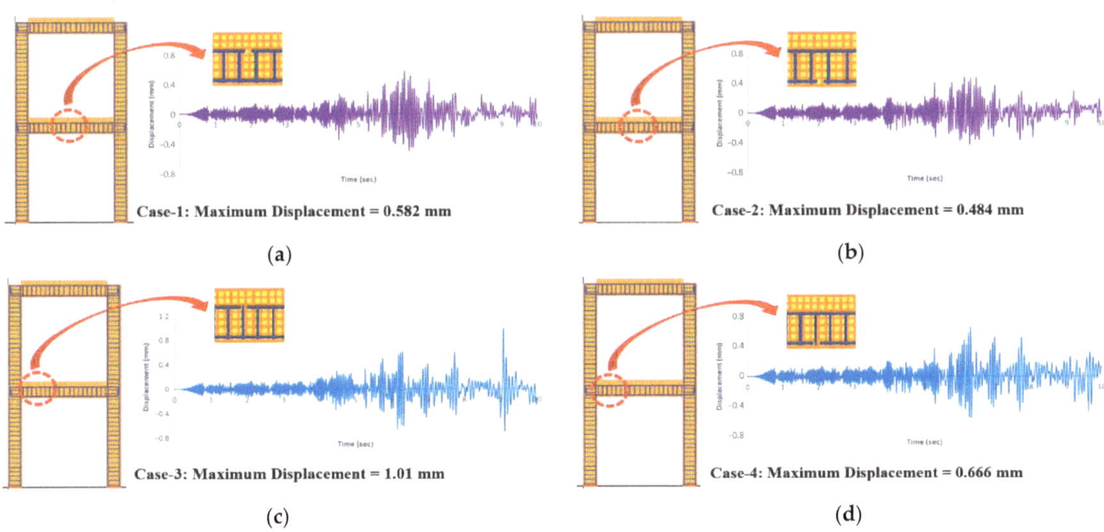

Figure 5. Maximum displacement for different locations of improper lap splice. (**a**) At the center of the top bar, (**b**) at the center of the bottom bar, (**c**) at L/6 of the top bar, and (**d**) at L/6 of the bottom bar.

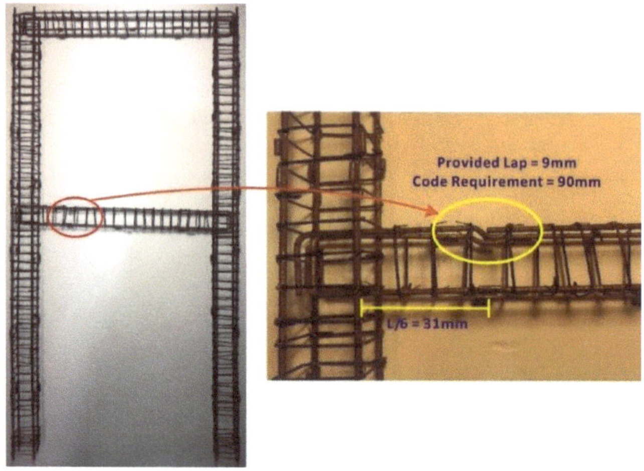

Figure 6. Improper lap splicing location.

4. Materials for Small-Scale Modelling

Micro-concrete is usually used in small-scale RC models, which is a mixture of suitably graded sand, Portland cement, and water [43]. The selection of model materials in this study was made on a practical basis rather than on a theoretical basis considering the inherent complexity of many factors that influence the properties of hardened (cured) concrete. Theoretically, the scaling of concrete properties not only needs the suitable scaling of aggregate size but also the scaling of pore size in the gel and of void size in the hardened mixture. Although the aggregate in the micro-concrete fulfills some of these requirements partially, the cement particle size cannot be scaled. In other words, a precise simulation of small-scale model concrete is not ideally possible [23]. There is no criterion to obtain the values of a mix design against a particular compressive strength for the micro-concrete. The determination of a suitable micro-concrete mix has been the subject of many studies in the past [24] as it is dependent on many factors including the type of locally available sand, varying aggregate-to-cement and water-to-cement ratios, workability required, and most importantly the similitude requirements and scale of the model concrete. The adopted procedure for the 1/16 scale frame with a target strength of 21 MPa is shown in the subsequent subsections.

4.1. Sieve Analysis

The values of the aggregate gradation curve were selected based on the maximum aggregate size given in the Naval Facilities Engineering Command [43]. The aggregate gradation curve used for the model concrete is shown in Figure 7. This gradation curve was achieved by performing the sieve analysis of locally available sand (#2 to #8 sandbags, available in Tokyo, Japan) and then by mixing the required proportions of each size. The curve in Figure 7 shows a monotonic increase of passing aggregate with respect to the aperture size.

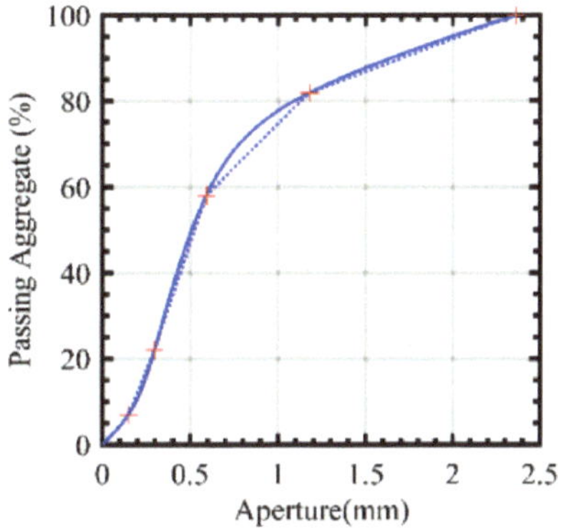

Figure 7. Aggregate gradation curve for micro-concrete.

4.2. Selection of Aggregate-to-Cement and Water-to-Cement Ratios

Following the steps given in the Naval Facilities Engineering Command [43], a value of water-to-cement ratio against the required compressive strength was obtained from the graph shown in Figure 8a. By using this water-to-cement ratio, the aggregate-to-cement ratio was calculated by using the graph shown in Figure 8b. As these values do not confirm whether the required strength will be achieved or not, for this purpose, four different mixes were selected as shown in Table 6, and a compression test was carried out to find the strength achieved for selected mixes. This step-by-step procedure gives an idea about the selections of the aggregate-to-cement ratio and water-to-cement ratio against the targeted strength of micro-concrete with excellent workability. On this basis, four different mixes were selected for the model as shown in Table 6.

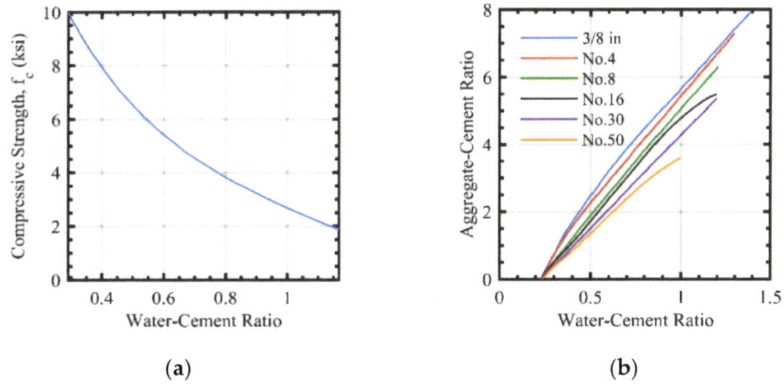

Figure 8. (**a**) Compressive strength vs. w/c ratio and (**b**) workability curves for different aggregate sizes [43].

Table 6. Selected mixes.

Mix No.	Aggregate-to-Cement Ratio	Water-to-Cement Ratio
M-1	3.6	0.76
M-2	3.6	0.72
M-3	4.25	0.8
M-4	4	0.78

4.3. Casting and Capping of Cylinders

A total of 56 cylinders were cast and tested at 7, 14, and 28 days to see the difference in compressive strengths of different mixes. The cylinder size was selected to be 50 × 100 mm so that the size effect does not occur in model concrete and prototype concrete in the controlled compaction and curing conditions of model concrete. Additionally, the variation in mean strength becomes more pronounced for the specimen size below 50 × 100 mm [44]. The casting procedure was adopted according to ASTM C-192 [45]. The mixing of micro-concrete was performed manually by first combining the weighted aggregate and cement and then gradually mixing with water. The micro-concrete was poured in three layers; each layer was tamped by 25 strokes with a steel rod of 10 mm diameter. After smoothing the upper surface of the concrete molds, they were covered to prevent moisture loss and left undisturbed for 24 h. The specimens were put into the curing tank after stripping off the molds. The specimen was capped one day before the test. The whole process of casting, curing, and capping of micro-concrete cylinders is shown in Figure 9.

(**a**) (**b**) (**c**)

Figure 9. Casting of micro-concrete. (**a**) Freshly casted cylinders, (**b**) curing conditions, and (**c**) capping process.

4.4. Compressive Strength of Micro-Concrete

The compression test was conducted under the specifications mentioned in ASTM C-39 [46]. The testing was performed on the AG-100 kN Autograph Shimadzu testing machine. The applied strain rate was 0.3 mm/min. The compression tests of mix numbers M1 to M4 were conducted at the age of 7, 14, and 28 days. The results of the compression test are shown in Table 7. The relationships between the compressive strength and age of micro-concrete are shown in Figure 10a. The results show that mix M1 and M2 had similar strengths which are very close to the target strength, but due to the higher workability, M1 was selected for the model frame. The stress–strain relationship of the cylinders from M2 has been shown in Figure 10b which shows that the selected mix of micro-concrete has a 3% strain.

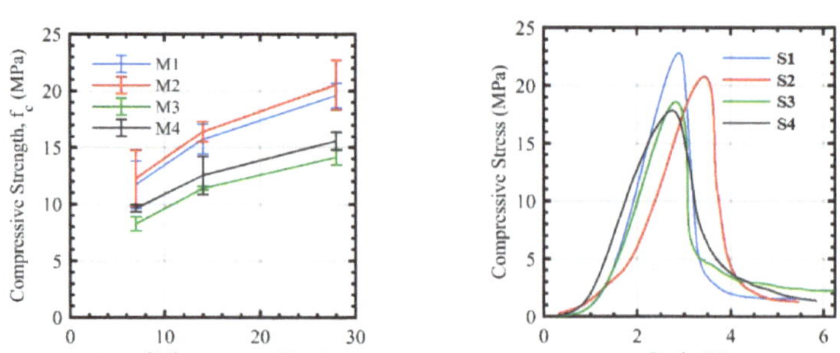

(**a**) (**b**)

Figure 10. (**a**) Compressive strength vs. age of micro-concrete and (**b**) stress vs. strain curve of M2 samples at 28 days.

Table 7. Compressive strength (MPa) of all mixes *.

Mix No.	7 Days	14 Days	28 Days
M1	12.32	15.82	19.27
M2	12.59	16.28	19.89
M3	8.18	11.22	14.39
M4	9.63	12.24	15.39

* Average of 3 samples.

4.5. Size Effect in Compression

To know the influence of age, cement-to-aggregate ratio, relative consistency, specimen shapes, and aggregate fineness on the size effects in concrete, Yi et al. [47] and Sabnis et al. [48] conducted a comprehensive experimental study for cylindrical, cubical, and prismatic shapes of concrete specimens with a wide range of sizes and different height-to-diameter ratios and concluded that, with the increase in cylinder diameter, the strength ratio decreases accordingly. Elfahal et al. [49,50] investigated the size effect in normal-strength concrete cylinders with sizes ranging from 75 × 150 mm to 600 × 1200 mm and validated the existence of the size effect in concrete cylinders. The smaller samples had a higher compressive strength and reached almost full strength in a shorter time than the larger ones.

In this study, cylinders of 38 mm, 50 mm, and 70 mm diameters were cast and tested at the age of 14 days to see the size effect. The average compressive strength of three samples each are shown in Table 8. The trend of the decrease in strength by increasing the specimen size is shown in Figure 11. The decrease in strength for larger specimen sizes is significant. Almost 10% of strength reduces in 50 mm diameter cylinders when compared with 38 mm diameter cylinders. Similarly, a 23% decrease in strength was observed for 70 mm diameter specimens when compared with 38 mm diameter cylinders. This trend is almost similar to the trend of normal concrete.

Table 8. Compressive strength (MPa) of different size cylinders at 14 days *.

	Cylinder Diameter 38 mm	50 mm	75 mm
Average f'_c (MPa)	17.96	16.27	13.76

* Average of 3 samples.

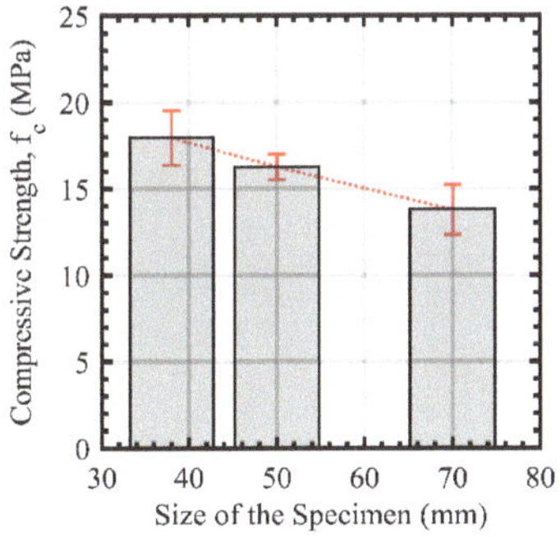

Figure 11. Aggregate gradation curve for micro-concrete.

4.6. Tensile Strength of Scaled Rebars

As the scale of the model structure was selected to be 1/16, the diameter of the reinforcing bars was very small. In this study, three different diameter bars were used for the reinforcement of the model structure. The sizes include 0.7 mm diameter bars for stirrups, 1.9 mm diameter bars for the main reinforcement of beams, and 2.2 mm diameter bars for the column reinforcement. These deformed bars, as shown in Figure 12, were manufactured by Kingo Seisakusho Co. Ltd., Tokyo. The tensile test on these bars was conducted by using the AG-50 kN Autograph Shimadzu testing machine with a loading rate of 1 mm/min. The average tensile strength of these bars is shown in Table 9. The design yield strength for the model structure was 345 MPa. The stress–strain relationships and tested samples are shown in Figure 13 which shows a satisfactory value of the yield strength of the scaled rebars.

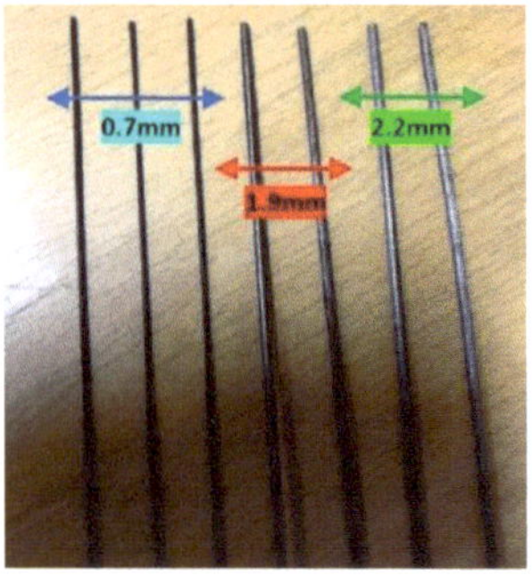

Figure 12. Small-scale reinforcement bars for model structure.

Table 9. Tensile strength of reinforcement bars *.

Diameter (mm)	Stress (MPa)	Strain (%)
0.7	414.7	1.51
1.9	328.7	4.32
2.2	393.28	5.42

* Average of 4 samples.

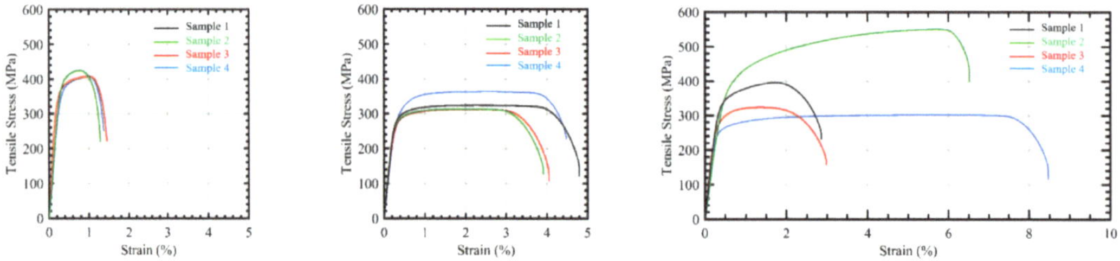

(a) (b) (c)

Figure 13. Stress–strain relationship of scaled reinforcement bars. (**a**) 0.7 mm, (**b**) 1.9 mm, and (**c**) 2.2 mm.

5. Experimental Study

5.1. Instrumentation, Test Setup, and Testing Procedure

The model frames were instrumented with accelerometers and laser displacement sensors to assess their dynamic behavior. A total of 10 AS-2GB cable-integrated accelerometers with a ± 19.61 m/s^2 (± 2 G) rated capacity were installed on both models. In addition to the accelerometers, five LB-300 ultra-long-range type laser displacement sensors were also installed on both models. The experimental setup for the shake table test is shown in Figure 14 which shows the location of all sensors. The model frames were built in 2D which implies that they cannot resist out-of-plane shaking, so to avoid out-of-plane failure of the model frames, frictionless rollers were mounted against the out-of-plane support to avoid out-of-plane failure.

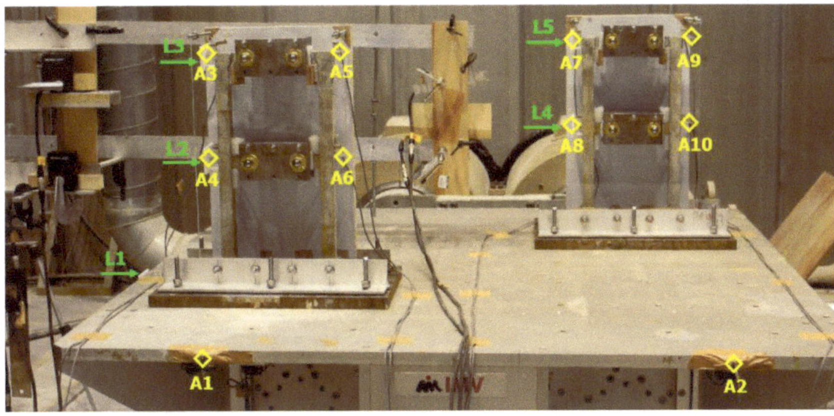

Figure 14. Shake table test setup, where A1–A10 shows locations of accelerometers and L1–L5 shows locations of laser displacement sensors.

5.2. Shake Table Properties

For the shake table experiment, two model frames named a proper frame and an improper frame were cast and tested at the same time to compare the results. The shake table system at the Institute of Industrial Science, The University of Tokyo, can control six degrees of freedom. Its properties are shown in Table 10. In this experiment, the shaking was only limited to one direction.

Table 10. Properties of shake table (Source: ERS, IIS, The University of Tokyo).

Property	Description
Frequency range	DC~100 Hz
Maximum load capacity	2000 kg
Shake table dimensions	1.5 m × 1.5 m
Maximum velocity (XYZ axis)	1.1 m/s
Exciting force (XY axis)	Sine wave: 2000 kgf Pulse wave: 3000 kgf
Maximum displacement (XY axis)	±100 mm
Maximum acceleration (XY axis)	Sine wave: 2.5 G and Pulse wave: 3.7 G (without load) Sine wave: 1.7 G and Pulse wave: 2.6 G (1000 kg) Sine wave: 1.2 G and Pulse wave: 1.0 G (2000 kg)
Maximum rotation angle	±15°

5.3. Sinosoidal Motion Selection

Sinusoidal motions of frequencies ranging from 1.5 Hz to 60 Hz with a maximum of 2.0 g amplitude were applied to obtain the dynamic response of the structure. The rate of sampling frequency was kept at 200 Hz. After each sinusoidal wave, white noise of a 3 min duration was applied to obtain the frequency of the model frames to study the change in frequency with structural damage. The loading sequence of the input motion is shown in Figure 15. In the loading sequence 45–49, a 10 Hz frequency for 40 **s** was applied, and in the loading sequence 55–57, a 10 Hz frequency for 2 min was applied. This loading sequence was obtained by keeping in view the following:

- By obtaining the maximum displacements, for all frequencies, in the proper frame through time history analysis in Staad.pro, which is a structural analysis and design software. The obtained values of displacements against all frequencies are shown in Table 11.
- By judging and following the structure behavior against given inputs on the day of the test.

Table 11. Maximum displacements by time history analysis.

Sr. No.	Frequency (Hz)	Maximum Displacement (mm)	
		(1st Story)	(2nd Story)
1	60	0.065	0.182
2	55	0.095	0.253
3	50	0.203	0.522
4	45	0.559	1.32
5	40	0.264	0.645
6	35	1.962	4.504
7	30	0.778	1.749
8	25	0.455	1.003
9	20	0.336	0.730
10	15	0.278	0.596
11	12	0.257	0.550
12	10	0.254	0.541
13	8	0.240	0.511
14	5	0.237	0.505
15	2	0.144	0.306
16	1.5	0.087	0.184

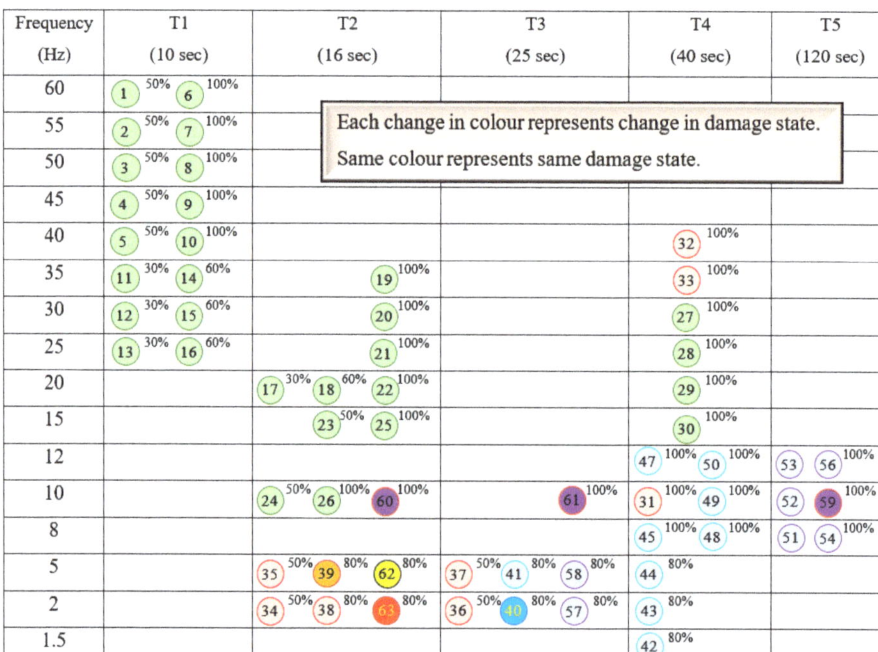

Figure 15. Loading sequence.

5.4. Calibration of Shake Table and Preparation of Data Files

To prepare and calibrate the input data files for the shake table, two wooden dummy frames were fabricated and mounted on the shake table as shown in Figure 16. The additional weight of 63 kg was placed at the shake table to comply with the weights of actual model frames. The other properties were also kept constant as the actual model frames. While keeping in view the limitations of the shake table, sinusoidal waves of frequency ranging from 60 Hz to 1.5 Hz were prepared. These sinusoidal waves were prepared for the maximum achievable amplitude, velocity, and displacement from the shake table. The values of these properties are shown in Table 12.

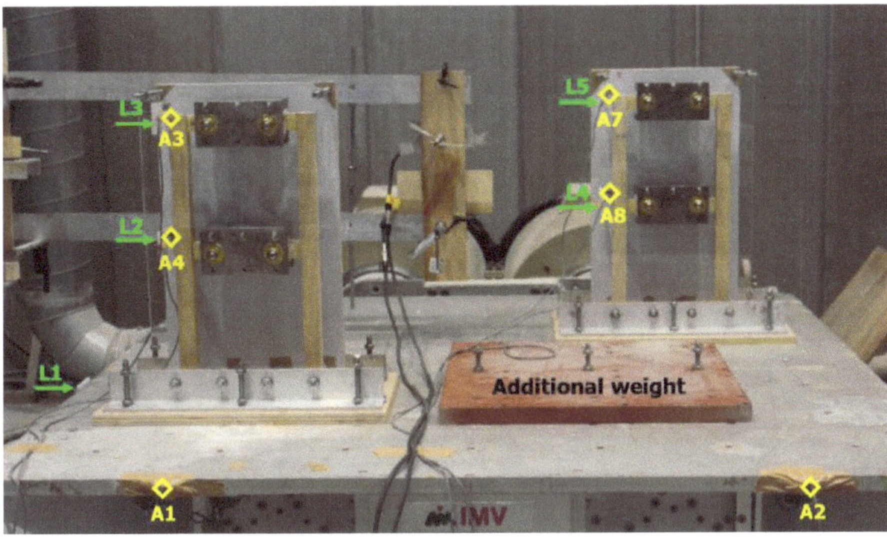

Figure 16. Setup of dummy frames for shake table calibration (L1–L5 = laser displacement sensors and A1–A4, A7–A8 = accelerometers).

Table 12. Properties of sinusoidal waves.

Frequency	Max. Amplitude	Max. Velocity (m/s)	Max. Displacement (mm)
60	2.0 g	0.060	0.11
55	2.0 g	0.065	0.14
50	2.0 g	0.10	0.25
45	2.0 g	0.080	0.24
40	2.0 g	0.086	0.31
35	2.0 g	0.10	0.40
30	2.0 g	0.11	0.52
25	2.0 g	0.14	0.85
20	2.0 g	0.16	1.31
15	2.0 g	0.22	2.30
12	2.0 g	0.27	3.54
10	2.0 g	0.33	5.21
8	2.0 g	0.40	7.91
5	2.0 g	0.65	20.6
2	1.28 g	1.0	79.55
1.5	0.77 g	0.80	88.50

6. Results and Discussions

6.1. Crack Pattern and Energy Dissipation Mechanism

To envisage the crack pattern and energy dissipation mechanism, a total of 63 shake table tests were performed for both proper and improper frame structures. Hereinafter, each test is called a Run as shown in Table 13. For both proper and improper frames, no cracks were observed until Run 30. Some cracks were observed in the improper frame from Run 31, and the complete failure of the improper frame was observed after Run 39, whereas the proper frame did not show any cracks even after Run 55. After Run 57, the proper frame exhibits some cracks in the beam–column interface, which propagate in Run 61 and lead to a complete collapse in Run 62. The detail of each frame after each Run is explained in Table 13.

Table 13. Crack pattern and description of proper and improper frames.

Run	Description	
	Proper Frame	Improper Frame
1–30	No cracks	No significant cracks appear in the frame
31	No cracks	Significant cracks appeared in the beam–column joints when a 10 Hz frequency was applied for 40 s. The cracks in the frame are shown in Figure 17. Partial cracks appear in the base connection.
32–38	No cracks	No significant changes in the cracks. The frame was able to resist higher frequencies (i.e., 35 Hz and 40 Hz) and also low frequencies (i.e., 5 Hz and 2 Hz) with lesser (50%) amplitude.
39	No cracks	The cracks in the joints widen up severely, and additional cracks appear in the base connection. The reinforcement bars in the top right corner of the improper frame had a shear failure. The accelerometer A9 was detached, and reinforcement was visible. Its close-up view is shown in Figure 18.
40	No cracks	Complete failure of the improper frame. Both beams were damaged entirely as shown in Figure 19b. Although there were some cracks at the beam–column interface at point 40-A, but the improper lap splicing dominated the failure at this portion. At all other joints, the failure occurred at the beam–column joints.
41–44	No cracks were observed in the proper frame. After this stage, the 10 Hz frequency loading was repeated several times.	Collapsed
45–58	The 8 Hz, 10 Hz, and 12 Hz frequency waves were repeated several times, but no cracks were observed.	Collapsed
59	Significant cracks appear after repeating the 10 Hz frequency several times for 2 min. There were no significant cracks in the base connection.	Collapsed
60,61	No significant change in cracks.	Collapsed
62	Severe cracks appear in all joints, as shown in Figure 19a. The base connections also had severe cracks. The accelerometers were detached from points 62-a and 62-c.	Collapsed
63	Complete failure of the proper frame. None of the beams were completely detached, but all beam–column joints had severe damage.	Collapsed

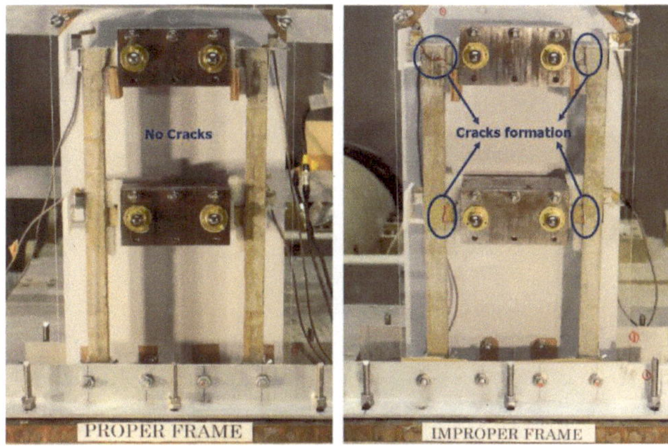

Figure 17. Cracks in the improper frame after Run 31.

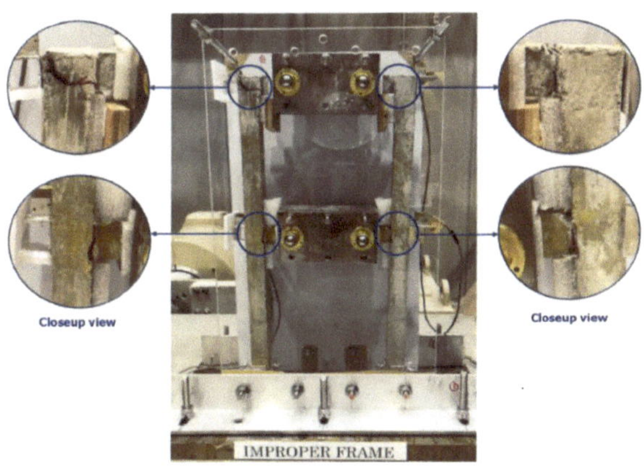

Figure 18. Close-up view of severely widened cracks in the improper frame after Run 39.

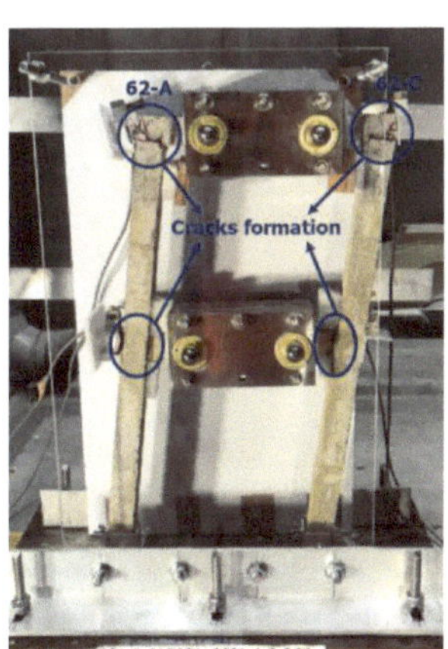

(**a**)

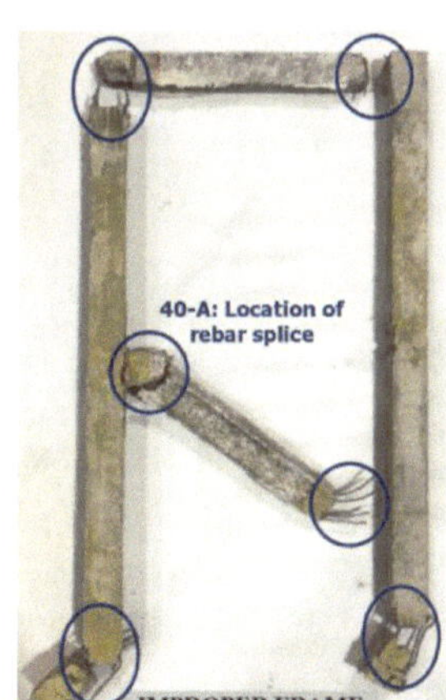

(**b**)

Figure 19. (**a**) Cracks in the proper frame—Run 62 and (**b**) complete failure of the improper frame—Run 40.

6.2. Hysteresis

The hysteresis load–displacement behaviors of both the proper and improper frames are shown in Figure 20. From the hysteresis after Run 31, it is evident that the improper frame has huge roof displacement as compared to the proper frame, indicating the initiation of bond–slip behavior. From this, there is the initiation of gradual strength degradation, due to the deterioration in bond stresses between the concrete and reinforcing steel bars. Ultimately, the hysteresis diagram after Run 40 shows the failure of the improper frame. On the other hand, the proper frame was still behaving structurally fine at Run 40, demonstrating the effectiveness of the lap splices of reinforcing bars. After Run 60, the roof displacement of the proper frame is increased. The difference in displacement at the roof of both frames at Run 20 and Run 30 is shown in Figure 21 which exhibits that the improper frame has a larger displacement since the beginning.

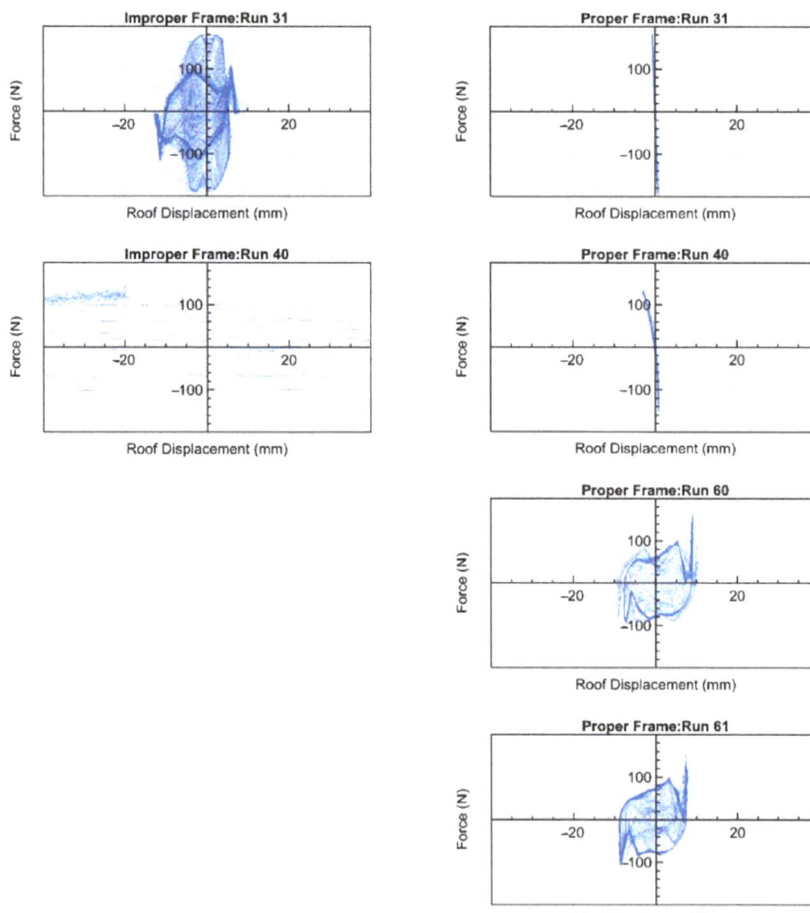

Figure 20. Hysteresis of the proper and improper frame at important runs.

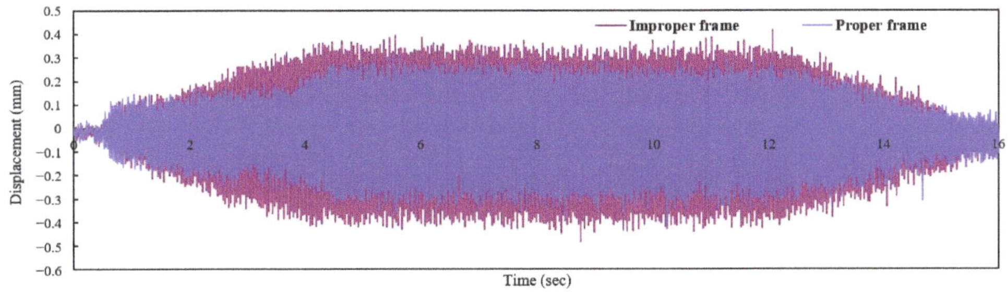

(**a**)

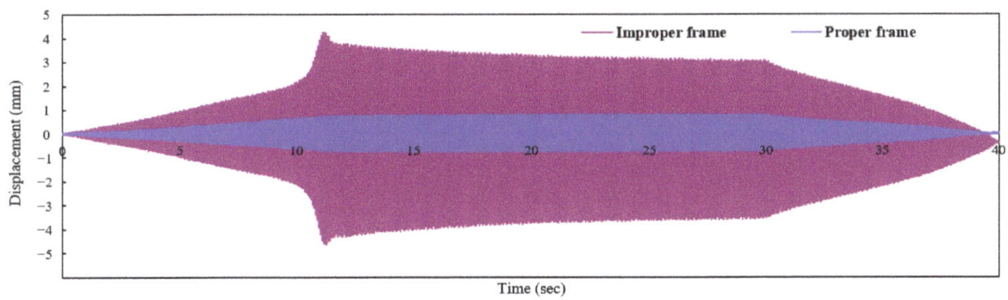

(**b**)

Figure 21. Response time history of displacement. (**a**) Run 20 and (**b**) Run 30.

7. Conclusions

The main goal of this research is to highlight the consequences of improper lap splicing in RC frame structures. To this aim, the authors demonstrated a simple, inexpensive, and evident experimental technique by using micro-concrete with locally available material to assess the seismic performance of a completely engineered RC frame with the defect of an improper lap splice. Based on the experimental investigation, the following conclusions can be drawn:

(1) The experimental results have proved that improper splicing details can reduce the capacity of reinforced concrete structures sufficiently, and the structure is highly vulnerable in case of an earthquake. The shake table test results demonstrate that the plastic hinges and damage points for the proper frame are the beam–column joints and the base connections as shown in Figure 19a, but if there is a provision of improper lap splice, then this can alter the damage point of the structure, triggering complete detachment of the beam with the improper lap splice as observed in Figure 19b at point 40-A.

(2) The experimental results of this research also demonstrate the reduction in strength, stiffness, and energy dissipation of RC structures, which have been through multiple earthquakes, which implies the reduction in the age of the structure. Furthermore, the results from this experiment can be used for awareness among local communities about the vulnerability of structures against earthquakes due to these small, neglected defects. This could reduce poor construction practices in developing countries in future projects.

(3) In addition to the scientific importance of this research, the authors would also like to emphasize the cost-effectiveness of this research. While increasing the burden on the financial component of experimentations, studying problems such as non-engineering defects could not be lucrative with the conventional methods of testing. Therefore, the testing mentioned in this research should pave the way for other research in

micro-modeling to reduce the cost and study non-engineering problems extensively to find scientific and social solutions.

(4) In this research, the shake table test was conducted by limiting the shaking to only one direction to avoid the out-of-plane failure of the 2D frames. Furthermore, it is also important to mention here that this research focuses mainly on the qualitative presentation of the problems caused by non-engineering defects. The idea of a reduction in the capacity of a structure with non-engineering defects can be achieved qualitatively through the numerical simulation presented in the article but the extent of the reduction in strength as compared to the proper structure could not be acquired at this stage. So, there is a need to further improve the numerical analysis techniques by using the experimental results.

This paper will be an additional contribution to the structural engineering community, which focuses on the consequences of commonly encountered non-engineering defects, i.e., improper lap splices in RC frames. Therefore, the results of this research can be useful to prepare the proper guidelines on the non-engineering defects encountered during the construction of RC frame structures for the sake of improving their structural integrity and load-carrying capacity.

Author Contributions: Conceptualization, A.J., C.K. and K.M.; methodology, A.J., C.K., K.A. and K.M.; AEM analysis, A.J. and M.F.U.D.A.; writing—original draft preparation, A.J. and K.A.; writing—review and editing, M.F.U.D.A. and A.M.; supervision, K.M. All authors have read and agreed to the published version of the manuscript.

Funding: This research received no external funding.

Data Availability Statement: Some or all data, models, or analysis results that support the findings of this study are available from the corresponding author upon reasonable request. The item list includes (1) the models of proper and improper frames in the Applied Element Method and (2) the analysis results of the shake table test.

Acknowledgments: We would like to extend our sincere gratitude to the Asian Development Bank (ADB), Japan and the Institute of Industrial Sciences (IIS), The University of Tokyo for providing facilities and support to conduct this research.

Conflicts of Interest: The authors declare no conflict of interest.

References

1. Zafar, A. *Response Modification Factor of Reinforced Concrete Moment Resisting Frames in Developing Countries*; University of Illinois at Urbana-Champaign: Urbana, IL, USA, 2009.
2. Isler, O. Seismic Performances and Typical Damages of Beam-Column Joints in the RC Buildings. In Proceedings of the 14th World Conference on Earthquake Engineering, Beijing, China, 12–17 October 2008.
3. Ates, S.; Kahya, V.; Yurdakul, M.; Adanur, S. Damages on reinforced concrete buildings due to consecutive earthquakes in Van. *Soil Dyn. Earthq. Eng.* **2013**, *53*, 109–118. [CrossRef]
4. Arya, A.S. Non-engineered construction in developing countries—An approach toward earthquake risk preduction. In Proceedings of the 12th World Conference on Earthquake Engineering, Auckland, New Zeland, 30 January–4 February 2000; pp. 1–22.
5. Castillo, A.; López-Almansa, F.; Pujades, L.G. Seismic risk analysis of urban non-engineered buildings: Application to an informal settlement in Mérida, Venezuela. *Nat. Hazards* **2011**, *59*, 891–916. [CrossRef]
6. Arya, A.S.; Boen, T.; Ishiyama, Y. *Guidelines for Earthquake Resistant Non-Engineered Construction*; The United Nations Educational, Scientific and Cultural Organization (UNESCO): Paris, France, 2014.
7. Alam, M.J.; Ahsan, R.; Akhter, F.; Paul, A. Earthquake Resistant Non-Engineered Building Construction for Rural Area in Bangladesh. In Proceedings of the 14th World Conference on Earthquake Engineering, Beijing, China, 12–17 October 2008.
8. Ali, K.; Javed, A.; Mustafa, A.E.; Saleem, A. Blast-Loading Effects on Structural Redundancy of Long-Span Suspension Bridge Using a Simplified Approach. *Pract. Period. Struct. Des. Constr.* **2022**, *27*, 04022024. [CrossRef]
9. Mustafa, A.E.; Javed, A.; Ali, K. Safety Assessment of Cables of Suspension Bridge under Blast Load. *Struct. Congr.* **2022**, *2022*, 79–93. [CrossRef]
10. Chaudhary, M.T.A. Seismic Vulnerability Mitigation of Non-Engineered Buildings. *Int. J. Civ. Struct. Constr. Archit. Eng.* **2014**, *8*, 37–41.

11. Boen, T.; Pribadi, K.S. Engineering of Non-Engineered Masonry Houses for Better Earthquake Resistance in Indonesia. *Asian J. Environ. Disaster Manag.* **2011**, *3*, 65–70. [CrossRef]
12. Blondet, M.; Vargas-Neumann, J.; Tarque, N.; Soto, J.; Sosa, C.; Sarmiento, J. Seismic Reinforcement of Earthen Constructions. In Proceedings of the 16th World Conference on Earthquake Engineering, Santiago, Chile, 9–13 January 2017.
13. Blondet, M.; Vargas, J.; Sosa, C.; Soto, J. Seismic Simulation Tests to Validate a Dual Technique for Repairing Adobe Historical Buildings Damaged by Earthquakes. In Proceedings of the Kerpic—New Generations Earthern Architecture, Istanbul, Turkey, 11–14 September 2013.
14. Fahim, M.; Elkoly, S.; Morad, S. Efficient New Numerical Technique for RC Structures with Construction Defects under Monotonic and Cyclic Loading Conditions. *Asian Trans. Eng.* **2013**, *2*, 7–16.
15. Elishakoff, I.; Ankitha, A.P.; Marzani, A. Rigorous versus naïve implementation of the Galerkin method for stepped beams. *Acta Mech.* **2019**, *230*, 3861–3873. [CrossRef]
16. Elishakoff, I.; Amato, M.; Ankitha, A.P.; Marzani, A. Rigorous implementation of the Galerkin method for stepped structures needs generalized functions. *J. Sound Vib.* **2021**, *490*, 115708. [CrossRef]
17. Javed, A.; Sadeghnejad, A.; Rehmat, S.; Yakel, A.; Azizinamini, A. *Magnetic Flux Leakage (MFL) Method for Damage Detection in Internal Post-Tensioning Tendons*; Florida International University: Miami, FL, USA, 2021. [CrossRef]
18. Afzal, M.F.; Matsumoto, Y.; Nohmi, H.; Sakai, S.; Su, D.; Nagayama, T. Comparison of Radar Based Displacement Measurement Systems with Conventional Systems in Vibration Measurements at a Cable Stayed Bridge. 2016. Available online: https://www.researchgate.net/publication/307932143 (accessed on 31 January 2023).
19. Hameed, A.; Rasool, A.M.; Ibrahim, Y.E.; Afzal, M.F.U.D.; Qazi, A.U.; Hameed, I. Utilization of Fly Ash as a Viscosity-Modifying Agent to Produce Cost-Effective, Self-Compacting Concrete: A Sustainable Solution. *Sustainability* **2022**, *14*, 11559. [CrossRef]
20. Filiatrault, A.; Lachapelle, É.; Lamontagne, P. Seismic performance of ductile and nominally ductile reinforced concrete moment resisting frames. I. Experimental study. *Can. J. Civ. Eng.* **1998**, *25*, 331–341. [CrossRef]
21. Caccese, V.; Harris, H.G. Earthquake simulation testing of small-scale reinforced concrete structures. *Struct. J.* **1990**, *87*, 72–80. [CrossRef]
22. Žarnić, R.; Gostič, S.; Crewe, A.J.; Taylor, C.A. Shaking table tests of 1:4 reduced-scale models of masonry infilled reinforced concrete frame buildings. *Earthq. Eng. Struct. Dyn.* **2001**, *30*, 819–834. [CrossRef]
23. Kim, W.; El-Attar, A.; White, R.N. *Small-Scale Modeling Techniques for Reinforced Concrete Structures Subjected to Seismic Loads*; Technical Report NCEER-88-0041; National Center for Earthquake Engineering Research: New York, NY, USA, 1988; p. 92.
24. Noor, F.A.; Boswell, L.F. *Small Scale Modelling of Concrete Structures*; Elsevier Science Publishers Ltd.: New York, NY, USA, 2005. [CrossRef]
25. Dejian, S.; Lu, X. Experimental study on dynamic compressive properties of microconcrete under different strain rate. In Proceedings of the 14th World Conference on Earthquake Engineering, Beijing, China, 12–17 October 2008.
26. Boutemeur, R.; Demidem, M.; Bali, A.; Benyoussef, E.H. Modelling of Micro-Concrete Behaviour under Static and Dynamic Loading. *Appl. Mech. Mater.* **2014**, *584–586*, 1089–1096. [CrossRef]
27. Tobar, R.; Bonelli, P. Test of micro-concrete model of a building damaged during the 1985 chile earthquake. In Proceedings of the Eleventh World Conference on Earthquake Engineering, Acapulco, Mexico, 23–28 June 1996.
28. Opabola, E.A.; Mangalathu, S. Seismic fragility assessment of bridges with as-built and retrofitted splice-deficient columns. *Bull. Earthq. Eng.* **2022**, *21*, 583–603. [CrossRef]
29. Opabola, E.A.; Elwood, K.J. Seismic assessment of reinforced concrete columns with short lap splices. *Earthq. Spectra* **2021**, *37*, 1726–1757. [CrossRef]
30. Chronopoulos, P.; Trezos, C.; Chronopoulos, M. Behaviour of RC elements with inadequate lap splices, before and after upgrading by welding of reinforcement. In Proceedings of the 4th International Symposium "Bond in Concrete 2012", Brescia, Italy, 17–20 June 2012.
31. Elrakib, T.M.; Arafa, A.I. Experimental evaluation of the common defects in the execution of reinforced concrete beams under flexural loading. *HBRC J.* **2012**, *8*, 47–57. [CrossRef]
32. Goksu, C.; Yilmaz, H.; Chowdhury, S.R.; Orakcal, K.; Ilki, A. The effect of lap splice length on the cyclic lateral load behavior of RC members with low-strength concrete and plain bars. *Adv. Struct. Eng.* **2014**, *17*, 639–658. [CrossRef]
33. Cho, J.-Y.; Pincheira, J.A. Nonlinear modeling of RC columns with short lap splices. In Proceedings of the 13th World Conference on Earthquake Engineering, Vancouver, BC, Canada, 1–6 August 2004.
34. Chowdhury, S.R.; Orakcal, K. Analytical modeling of columns with inadequate lap splices. *ACI Struct. J.* **2014**, *110*, 735–744. [CrossRef]
35. Nawaz, W.; Yehia, S.; Elchalakani, M. Lap splices in confined self-compacting lightweight concrete. *Constr. Build. Mater.* **2020**, *263*, 120619. [CrossRef]
36. *Code CF*; International Federation for Structural Concrete: Lausanne, Switzerland, 2010.
37. *ACI Committee 318, 318–319*; Building Code Requirements for Structural Concrete and Commentary. ACI Committee: Farmington Hills, MI, USA, 2019. [CrossRef]
38. ASCE. *Seismic Evaluation and Retrofit of Existing Buildings ASCE/SEI 41-13*; ASCE: Reston, VA, USA, 2014. [CrossRef]

39. Okazaki, K.; Kusumastuti, D.; Probadi, K.; Saito, T. Comparison of Current Construction Practices of Non-Engineered Buildings in Developing Countries. In Proceedings of the 15th World Conference on Earthquake Engineering (15WCEE), Lisbon, Portugal, 24–28 September 2012; pp. 9962–9971.
40. Rezaeepazhand, J.; Simitses, G.J.; Starnes, J.H. Scale models for laminated cylindrical shells subjected to axial compression. *Compos. Struct.* **1996**, *34*, 371–379. [CrossRef]
41. Thomas, C.; Tamayo, P.; Setién, J.; Ferreño, D.; Polanco, J.A.; Rico, J. Effect of high temperature and accelerated aging in high density micro-concrete. *Constr. Build. Mater.* **2021**, *272*, 121920. [CrossRef]
42. American Society of Civil Engineers. *Minimum Design Loads and Associated Criteria for Buildings and Other Structures*; American Society of Civil Engineers: Reston, VA, USA, 2017; pp. 1–889. [CrossRef]
43. NFEC. *Mix Design for Small-Scale Models of Concrete Structures: Technical Report R564*; NFEC: Port Hueneme, CA, USA, 1968.
44. Talaat, A.; Emad, A.; Tarek, A.; Masbouba, M.; Essam, A.; Kohail, M. Factors affecting the results of concrete compression testing: A review. *Ain Shams Eng. J.* **2021**, *12*, 205–221. [CrossRef]
45. *ASTM C 192/C 192M-02*; Standard Practice for Making and Curing Concrete Test Specimens in the Laboratory. ASTM International, ASTM Committee: Farmington Hills, MI, USA, 2002; Volume 4, pp. 1–8.
46. *ASTM C 39/C 39M-01*; Standard Test Method for Compressive Strength of Cylindrical Concrete Specimens. ASTM International, ASTM Committee: Farmington Hills, MI, USA, 2001; Volume 4, pp. 1–5.
47. Yi, S.T.; Yang, E.I.; Choi, J.C. Effect of specimen sizes, specimen shapes, and placement directions on compressive strength of concrete. *Nucl. Eng. Des.* **2006**, *236*, 115–127. [CrossRef]
48. Sabnis, G.M.; Mirza, S.M. Size Effect in Model Concretes? *ASCE J. Struct. Div.* **1979**, *105*, 1007–1020. [CrossRef]
49. Elfahal, M.M.; Krauthammer, T.; Ohno, T.; Beppu, M.; Mindess, S. Size effect for normal strength concrete cylinders subjected to axial impact. *Int. J. Impact Eng.* **2005**, *31*, 461–481. [CrossRef]
50. Elfahal, M.M.; Theodor, K. Dynamic Size Effect in Normal- and High-Strength Concrete Cylinders. *Mater. J.* **2005**, *102*, 77–85.

infrastructures

Article

Research on the Protection Scheme of a High-Speed Railway Crossing 1000 KV Ultra-High Voltage Transmission Line

Yi Song [1,†] and Wei Xiao [2,*,†]

1 Shaanxi Railway and Underground Traffic Engineering Key Laboratory (FSDI), Xi'an 710043, China; songyi.yy@crcc.com
2 School of Civil Engineering, Chang'an University, Xi'an 710064, China
* Correspondence: wei.xiao@chd.edu.cn; Tel.: +86-15877659599
† These authors contributed equally to this work.

Abstract: The high-speed railway project and the ultra-high-voltage transmission project represent two crucial components of China's "new infrastructure". As the construction of these two projects progresses rapidly, it is inevitable that instances of intersections will occur. Extreme conditions may cause damage to ultra-high voltage transmission cables. When a high-speed train passes by an ultra-high voltage transmission line, it poses a serious safety hazard. To address this issue, engineering examples were utilized to examine the protection structure scheme, protection distance, protection load, and construction procedures when a high-speed railway intersects a 1000 KV ultra-high voltage transmission line. A shed structure form and construction method for the electric power protection were proposed to ensure the safe operation of the high-speed railway while also achieving the safe and rapid construction of the high-speed railway protection structure in the safety zone of the approaching 1000 kV ultra-high voltage transmission line. The results indicated that the protection of high-speed railway crossings and 1000 kV ultra-high voltage transmission lines primarily focuses on line-break protection. The concrete shed structure with a straight wall and a flat roof was designed to meet the requirements of high-speed railway crossings. The line-break protection method enables the construction of an automatic warning protection corridor and a complete movable trolley quickly and safely within the safety zone near the transmission line. The implementation effect is, therefore, positive. It can be used as a reference point for other projects of a similar nature.

Keywords: ultra-high-voltage transmission line; high-speed railway; protective shed tunnel; structure design; safety construction

Citation: Song, Y.; Xiao, W. Research on the Protection Scheme of a High-Speed Railway Crossing 1000 KV Ultra-High Voltage Transmission Line. *Infrastructures* **2024**, *9*, 110. https://doi.org/10.3390/infrastructures9070110

Academic Editors: Armin Mehrabi and Seyed Saman Khedmatgozar Dolati

Received: 6 June 2024
Revised: 30 June 2024
Accepted: 10 July 2024
Published: 15 July 2024

1. Introduction

In 2020, China proposed to take stronger policies and measures to participate in global environmental governance, strive to peak CO_2 emissions by 2030, and strive to achieve carbon neutrality by 2060, and in 2021, the government work report for the National People's Congress and the National People's Congress for the first time included "carbon peaking and carbon neutrality" in the government work report. As an infrastructure strategy in the new stage of China's economic development, the "new infrastructure" can vigorously promote the recovery of the national economy in the wake of the new coronavirus epidemic and provide a green and intelligent energy Internet solution for realizing "carbon peak and carbon neutrality". 2020 is the year when the central government's "new infrastructure" will be launched. In 2020, the central government's "new infrastructure" theme put forward seven major "new infrastructure" areas, including 5G, ultra-high voltage, intercity high-speed railroads, intercity rail transit, charging piles for new energy vehicles, big data centers, artificial intelligence, and industrial Internet [1,2].

As important infrastructure construction projects in the "new infrastructure", ultra-high-voltage (UHV) transmission projects and intercity high-speed railroads will inevitably cross each other and traverse each other in the process of project construction. Since the UHV power transmission project and high-speed railroad project are both projects with high investment, high standards, difficult construction, difficult relocation, and long coordination periods, most of the high-speed railroad crosses the UHV power transmission line by setting up protective structures. For example, the Beijing-Tangshan Intercity Railway and the 800 kV Ximeng-Jiangsu Taizhou special high-voltage line, the 800 kV Zarut-Shandong Qingzhou special high-voltage line, the 1000 kV Ximeng-Shandong special high-voltage line (referred to as the corridor of the three lines), the railroad in the crossing section of the range of 285.4 m, set up a concrete frame structure for protection. Where the Shanghehang Railway crosses three power transmission lines, namely the 1000 kV transmission line from Huainan to Shanghai (Hu'an I/II line), the ±800 kV transmission line from Jinping to Sunan (Jinsu line), and the ±800 kV transmission line from Xiangjiaba to Shanghai (Fufeng line), the railroad is protected by a concrete frame structure in the crossing section within a range of 460 m. The structure of each protection program and the machinery and equipment in the construction process are kept at a certain distance from the safety protection zone of the UHV transmission line, and there have been relatively few studies on the design and construction measures of the high-speed railway protection structure for the close crossing of the 1000 KV UHV transmission line, which is close to the safety protection zone of the UHV [3,4].

Relying on the project of the Jibin high-speed railway crossing 1000 kV Ximeng-Shandong ultra-high-voltage transmission line Hequan I/II line, this paper studies the protection scope, structure design, and construction program of the high-speed railway protection structure close to the safety zone of the 1000 KV UHV transmission line and puts forward a new structure of the electric power protection shed hole close to the safety zone of the UHV transmission line and a safe and fast construction method. This can provide a reference for the design and construction of similar projects in the future.

Project Overview

The Jibin high-speed railway, located in the northern part of the Shandong Peninsula City Cluster, is an important part of the Shandong Peninsula City Cluster Intercity Railway Network, with a total length of 145.299 km and a design speed of 350 km/h. The line crosses the 1000 kV Ximeng-Shandong UHV Transmission Line Hequan I/II line at DK90 + 100.

As one of the 12 key transmission corridors in the national "Action Plan for Prevention and Control of Air Pollution", the Ximeng-Shandong 1000 KV UHV AC Transmission Project is the first UHV project to be put into operation among the "Four Crossing and Four Direct" UHV projects in the national action plan for the prevention and control of air pollution, and the first UHV AC transmission project in North China. As one of the 12 key power transmission corridors in the National Air Pollution Prevention and Control Action Plan, it is the first project to be put into operation among the "four crosses and four directs" of the National Air Pollution Prevention and Control Action Plan, as well as the first ultra-high voltage AC transmission project in North China. The project is responsible for the important mission of transferring electricity from Ximeng to Beijing and Shandong, providing an important energy guarantee for the "Beijing-Tianjin-Hebei Integration" strategy.

For the 1000 kV Hequan I/II line, the power line is a double 1000 kV, the crossing point of the power line from the top of the track is 25.75 m (according to the maximum arc sag of the conductor operating temperature 40 °C case), the north side of the tower is 113.2 m high, the outer edge of the tower to the center of the track is about 374 m, the south side of the tower is 98.8 m high, the outer edge of the tower to the center of the track is about 122 m, the angle of intersection of the power line and the railroad is about 75 degrees. Across the stall does not meet with the independent tension-resistant section, and the wire joints from the north side of the tower have a horizontal distance of 213 m across the section of the insulator string using a double "I" string type; a section of the plan schematic diagram is shown in Figure 1.

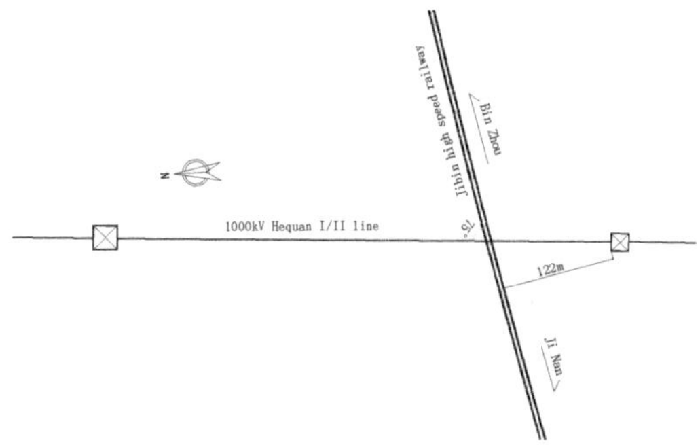

Figure 1. Plan of the intersection of the 1000 KV UHV transmission line and railway.

2. Study of the Scope of High-Speed Rail Protection and Line-Break Load

2.1. Study of the Scope of Protection of High-Speed Rail

According to potential accidents involving UHV transmission lines, two main working conditions are to be considered for the protection of high-speed railways: the downed tower working condition and the broken line working condition.

The relevant provisions of transmission tower load design in the national standards are the "Design Code for 1000 kV Overhead Transmission Lines" GB 50665-2011, the "Design Code for ±800 kV DC Overhead Transmission Lines" GB 50790-2013 (2019 version), and the "Technical Regulations for Design of Ultra-high Voltage Overhead Transmission Line Tower Structures" DL/T5486-2013 [5–7]. In the design of transmission lines, the safety factor of the pylon is extremely high, especially for a 1000 kV UHV pylon. Therefore, the probability of a tower collapse accident is extremely low, and only a wire break accident needs to be considered in the design of the underground railway.

The calculation of the break line (including longitudinal unbalanced tension) of the overhanging type of tower is carried out under meteorological conditions of −5 °C, with ice and no wind. According to the industry standard "Overhead Transmission Line without Spanning Frame Construction Process Guidelines" DL/T5301-2013 [8] and the national standard "1000 kV Overhead Transmission Line Design Specification" GB50665-2011, the protection length for a 1000 kV transmission line in case of conductor breakage impact is calculated using the following formula:

$$
\begin{gathered}
B \geq 2Z_X + b \\
Z_x = w_{4(10)}\left[\frac{m(l-m)}{2H} + \frac{\lambda}{W}\right] \\
w_{4(10)} = 0.0613kd \\
l_{wa} \geq \frac{n}{\sin\alpha} + \frac{B}{\sin\alpha} + 2l_{wb} \\
l_{wb} = h_w \tan 30^\circ \\
L = Kl_{wa}
\end{gathered}
$$

where: B—Seal net width; Z_x—10 m/s-Wind deflection distance at intersections under meteorological conditions; b—Maximum horizontal spacing of bundled conductors; m—Horizontal distance of crossings from neighboring pylons; l—Horizontal straight line distance between two adjacent towers; H—horizontal tension of conductor operation; λ—the length of the pendant insulator string; W—Weight per unit length of wire; d—Outer diameter of wire; k—Wind carrier type coefficient; n—Horizontal width of track; α—Crossing angle; h_w—Vertical height between wire and crossed object; K—Safety coefficient, take 1.5; L—Protection length.

Using the relevant values, it can be calculated that the length of the protective section required for the HSR is not less than 138.5 m.

2.2. Study of Impact Loads on Transmission Line Breaks

To study the loads caused by transmission line breaks, the transmission line breaks are first dynamically analyzed. The power conductor is considered a cable unit, and the deformation property of the cable unit consists of two parts: tension and bending of the axis.

$$U = \frac{1}{2}\int_0^l EA(\varepsilon_{xx})^2 dx + \frac{1}{2}\int_0^l EIk^2 dx$$

where: E denotes the cable material tensile modulus of elasticity; A represents the tensile section area; I denotes the moment of inertia; k represents the bending deformation length; l represents the length of the deformed part.

The tensile section variable, the bending deformation length, is expressed in r vector coordinates as follows:

$$\varepsilon_{xx} = \frac{1}{2}\left(r_x^T r_x - 1\right)$$

$$k = \frac{|r_x * r_{xx}|}{|r_x|^3} = \frac{\left[(r_x * r_{xx})^T (r_x * r_{xx})\right]^{\frac{1}{2}}}{\left(r_x^T * r_x\right)^{\frac{3}{2}}}$$

The kinetic equation of the unit is:

$$M\ddot{e} = Q_{ext}^e - Q_{ela}^e$$

where Q_{ela}^e represents the generalized elastic force of the unit, and Q_{ela}^e represents the external force applied to the unit, including generalized gravity, contact force, and other forms of external loads. When the wire is in sliding contact with the protective structure, the constraint relationship needs to be described using nonlinear constraints. The set of kinetic equations is obtained as a set of second-order differential-algebraic equations.

$$\begin{cases} M\ddot{e} = Q_e - Q_k - C_e^T \lambda \\ C(e,t) = 0 \end{cases}$$

where: C denotes the system constraint equation; Ce represents the Jacobi matrix obtained by partially differentiating the constraint equation with respect to the node coordinates; λ is the Lagrange multiplier. The commonly used methods for solving this equation include the Newmark method, the HHT method, the generalized α method, etc. [2].

The model is based on dynamic equations and numerical analysis. The analysis is conducted under two operating conditions: conductor fracture and in-span conductor joint fracture. The unbroken end of the conductor is controlled by a hinged support, while the broken end is entirely freed, and the bottom of the high-speed rail (HSR) protection construction is completely secure. Figure 2 shows the calculation model 2.

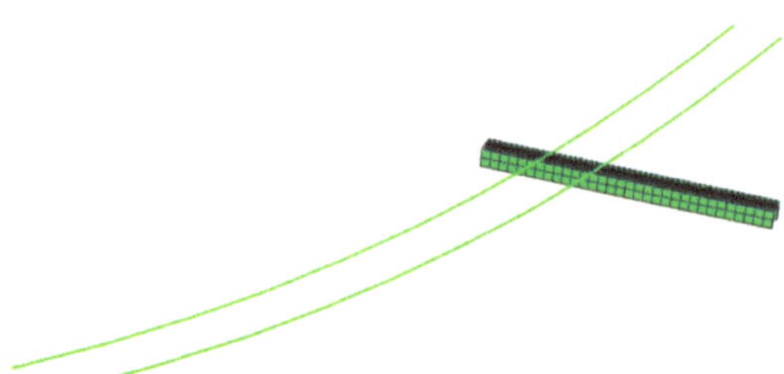

Figure 2. Computation model.

Working Condition 1: Wire breakage after calculation, the changes in wire drop, the protective structure impact force, and structural stress are shown in Figures 3–5:

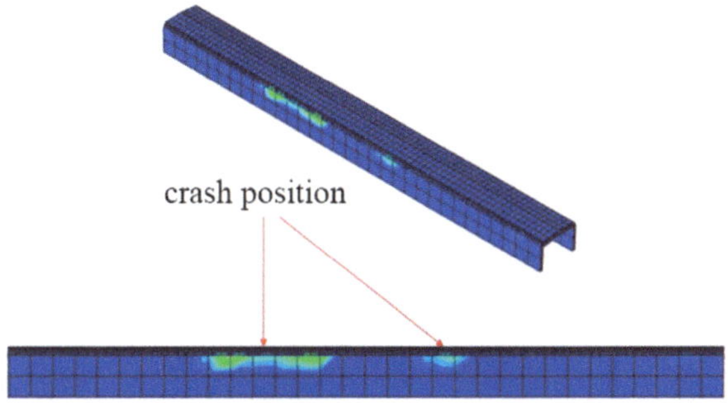

Figure 3. Calculation results of impact structures on transmission lines.

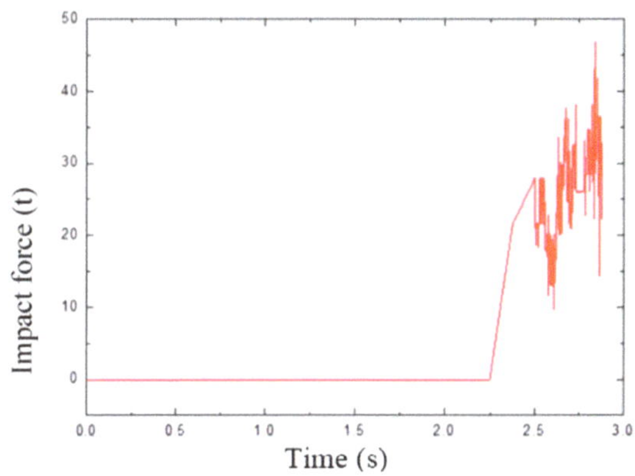

Figure 4. Time evolution curve of transmission line impact force.

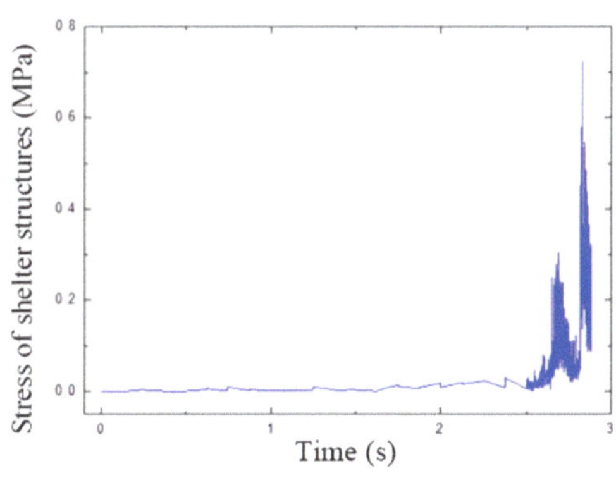

Figure 5. Time evolution curve of structural stress.

The mathematical results show that before the wire makes contact with the structure, the impact force is zero. The impact force is 21.48 tons when the object first makes contact with the structure at 2.37 s. As a result, the wire and the structure clash several times until reaching a maximum impact force of 46.92 tons at 2.84 s. The impact force reduces with time. Before the wire makes contact with the structure, the stress is around zero MPa. The structural stress is 0.03 MPa when the wire makes contact with the structure at 2.37 s. As the wire continually collides with the framework, the structural stress grows until it reaches a high of 0.73 MPa at 2.84 s.

Working Condition 2: Fracture of wire joints within the span.

In this scenario, the in-span conductor connector fails and comes into contact with the structure. The impact will be greater than that in Case I.

After calculation, the impact force and structural stress changes of the falling conductor and protective structure are shown in Figures 6–8:

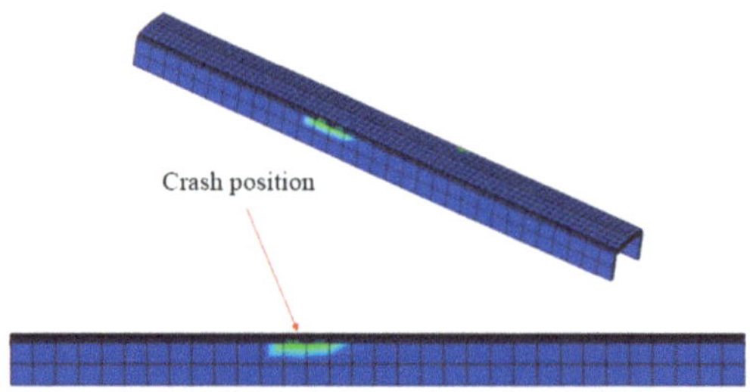

Figure 6. Calculation results of impact structures on transmission lines.

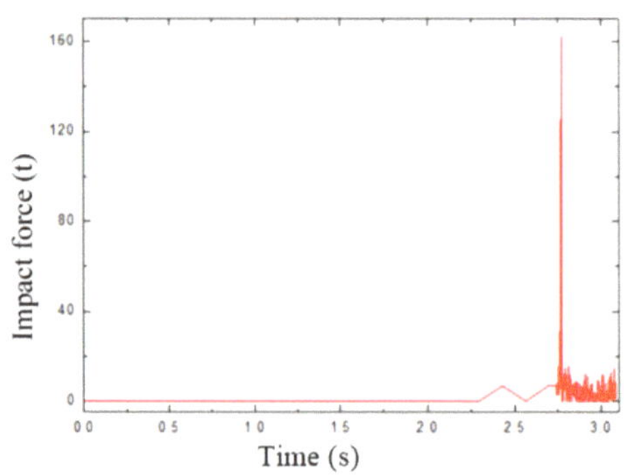

Figure 7. Time evolution curve of transmission line impact force.

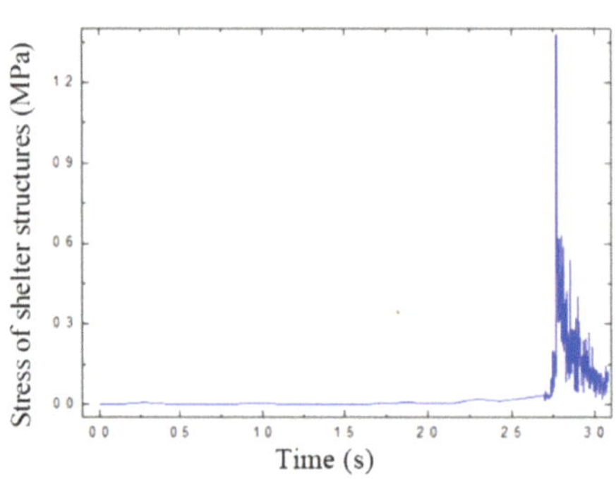

Figure 8. Time evolution curve of structural stress.

From the calculation results, it can be observed that prior to the wire contacting the structure, the impact force is 0. Upon initial contact with the structure at 2.73 s, the impact force is 16.5 tons. Subsequently, as the wire and the structure collide repeatedly, at 2.77 s, the maximum impact force recorded is 94.86 tons. As time passes, the impact force gradually diminishes. Before the wire makes contact with the structure, the stress in the structure is almost 0 MPa. When the wire first makes contact with the structure at 2.73 s, the structural stress is 0.23 MPa. Subsequently, as the wire repeatedly collides with the structure, the structural stress increases to a maximum value of 1.37 MPa at 2.77 s. The comparison of the two conditions reveals that the maximum values of the impact force and structural stress occur when the wire joint fractures within the span. Referring to the "Code for the Design of Concrete Structures" (GB 50010-2010), the standard compressive strength of C35 concrete is 23.4 MPa, and the standard tensile strength is 2.20 MPa. The design compressive strength of C35 concrete is 16.7 MPa, and the design tensile strength is 1.57 MPa. The maximum stress in the structure is smaller than the design value of tensile strength, indicating that the reinforced concrete structure using C35 can meet the safety requirements under all working conditions [9,10].

3. Study of the Form of Protective Structures and Construction Measures

3.1. Study of the Design of Protective Structures for High-Speed Rail

Currently, the protection shed hole serves as the primary protective construction for high-speed railways. Rigid and flexible shelters are the two types of protective shelter structures.

A rigid shelter's major component is its reinforced concrete frame, which is occasionally topped with a layer of cushioning material above the frame's top plate. Examples of this type of material include gravel, clay, or the more recent sand-expandable polyethylene (EPE) composite cushion. By successfully preventing structural damage or deformation brought on by the impact of external foreign objects, these cushions help safeguard the security of high-speed railway operations. This shape is frequently referred to as a classic cushion. The structure of a flexible shelter consists of replacing the rigid shelter's ceiling plate with metal flexible mesh. Additional features include anchor ropes, steel columns, and energy-dissipating sections. The components are assembled using a specific method. The flexible shed hole is increasingly in use because it uses a lot less energy than the rigid shed hole, is less expensive to build and maintain, and is better for the environment [11,12].

Considering the conductivity of the metal material and the interactions between the special environment and the ultra-high voltage transmission line, this design takes into account the use of the reinforced concrete shed hole structural form. The project crosses the 1000 KV ultra-high voltage transmission line, and it is relatively close to the transmission line discharge safety zone.

With reference to the high-speed railway tunnel's shed hole design for protecting against dangerous rockfalls and the overall reinforced concrete structure, there are three types of shed holes: (1) Using a beam-slab-column structural design, creating a concrete frame, and placing an independent foundation beneath the column. (2) To lessen the high-speed rail's aerodynamic effect, a piece of the wall on both sides should have windows that open periodically. The structural form of this section should resemble that of an elongated open hole. The straight wall's arch structure should also be used. (3) Under limited NAC circumstances, a straight-wall, flat-roofed building with windows on both sides of the wall should be used [13–17]. The three schemes' engineering schematic diagrams are displayed in Figures 9–11.

Figure 9. Beam-slab-column structural form.

Figure 10. Straight wall arch structural form.

Figure 11. Straight wall flat top structural form.

Through a comprehensive analysis of the three options based on structural form, dimensions, transmission, construction conditions, etc., the straight-wall flat-roof structure is recommended as the optimal structural option, as illustrated in Table 1.

Table 1. Programme Comparison Table.

	Programme I (Beam-Slab-Column Structure)	**Programme II (Straight-Wall Vaulted Structure)**	**Programme III (Straight-Wall, Flat-Roofed Structure)**
Structural Form	To meet the requirements of traffic clearance, the beam-slab-column rectangular structure does not intrude into the transmission line safety protection zone.	The straight wall arch structure is designed to meet headroom requirements. However, the upper arch occupies more space, causing the structure to intrude into the safety protection zone of the transmission line.	A straight-wall, flat-roofed structure that meets headroom requirements and does not intrude into the transmission line safety protection zone.
Structural Dimensions	Beams and columns are the main load-bearing members designed to support the upper loads, and the structure has small structural dimensions.	Smaller dimensions are required for the arched structure at the top, while larger dimensions are needed for the side wall structure.	The structure features a flat roof at the top, with greater forces and larger dimensions in the center, and larger dimensions in the side wall structure.

Table 1. *Cont.*

	Programme I (Beam-Slab-Column Structure)	Programme II (Straight-Wall Vaulted Structure)	Programme III (Straight-Wall, Flat-Roofed Structure)
Structural Force Transmission	The structural force transfer occurs through a clear path from the slab to the beam, column, and foundation.	The arch loads are transferred to the side walls, which are subjected to significant vertical and horizontal forces.	The load of the roof slab is transferred to the side walls, which experience significant vertical forces and minimal horizontal forces.
Construction Condition	Construction can be carried out using cast-in-place or the pre-supported erection of supports, both of which require large lifting machinery and pose the risk of construction machinery encroaching into the protected area.	It can be constructed using cast-in-place or integral formwork with bracing, which helps prevent the intrusion of construction machinery into the protected area.	It can be constructed by cast-in-place or integral formwork with bracing, which can avoid the intrusion of construction machinery into the protected area.
Analysis	Not recommended	Not recommended	Recommended

C35 concrete is used in the straight-wall flat-roof structure, and the thickness of the wall and roof is 1 m. The base plate adopts a pile foundation with a diameter of 1 m, and the length of the pile is 18 m. The steel bar is a HRB400, φ25@100. The calculation results under fundamental combination, nominal combination, and accidental combination are shown in Table 2.

Table 2. Structural calculation results.

Position	Thickness (m)	Fundamental Combination		Nominal Combination		Accidental Combination		Calculate Reinforcement Area (mm^2)	Actual Reinforcement Area (mm^2)
		Bending Moment (KN*m)	Axial Force (KN)	Bending Moment (KN*m)	Axial Force (KN)	Bending Moment (KN*m)	Axial Force (KN)		
roof	1.0	828	108	513	66	407	335	2417	4909
left side wall	1.0	384	809	244	547	1832	1112	4224	6158
right side wall	1.0	399	809	244	547	2043	691	5417	6158

3.2. *Study of Construction Measures for Protective Structures*

The primary risk factor for the construction of 1000 KV ultra-high-voltage power protection sheds, caves, and related projects is when the distance between construction machinery and power lines is less than the safety distance. This situation can result in induced voltage being generated by the construction equipment, posing a threat to the safety of construction personnel.

According to the safety notice issued by the State Grid Shandong Power Company regarding the construction of protective shelter holes for the Jinwei High-speed Railway Jinan Liaison Line (Jibin High-speed Railway) under the 1000 kV Hequan I/II Line #303–#304, special vehicles such as cranes, transporters, and large-scale machinery involved in construction work under the 1000 kV power line and within the protection zone must maintain a safety distance of more than 16 m from the electrically charged equipment (conductors, fittings, etc.). The protection zone is defined as the area under the conductor and the conductor sideline extending outward, within the two planes formed by the conductor line and the sideline of the conductor line extending horizontally for 30 m and perpendicular to the ground.

Combined with on-site measurements and calculations, the top of the shed structure is only 2 m away from the safety control line, and the nearest point is only 1.5 m. Therefore, it is not feasible to use large cranes for lifting materials and pouring concrete. Due to the proximity to the safety control line, high-altitude protection measures must be implemented to prevent construction personnel and materials from mistakenly entering the protection zone, which could result in safety accidents. In response to the aforementioned points, construction measures such as establishing automatic warning protection corridors and implementing overall movable cart molding were adopted following the study. These measures effectively addressed construction near the safety zone and helped prevent safety accidents. The automatic warning protection corridor involves installing height limiters on the right and left sides before and after the construction area of the shed structure. The height limiters are made of φ180 steel pipes, and their bottoms are connected to the original ground. Grounding measures are implemented to prevent induced electric currents. The infrared alarm device is installed at the top of the height limiters, where the infrared rays create a dense network, reducing inductive currents and enabling automatic warnings for objects exceeding the limit. As shown in Figure 12.

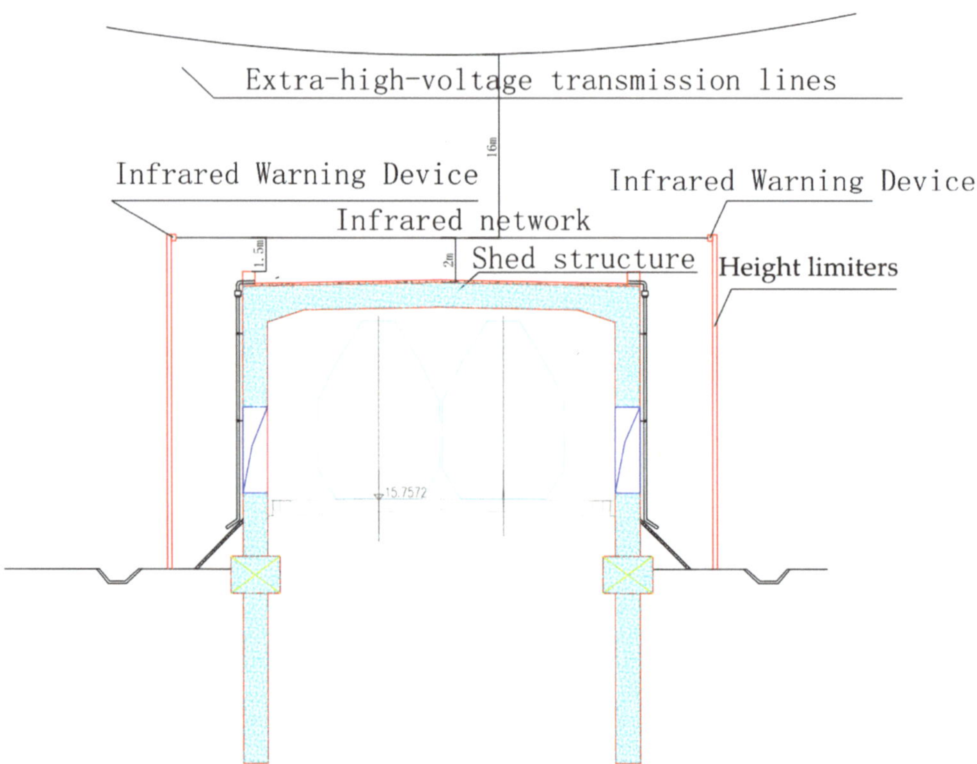

Figure 12. Schematic diagram of the protection corridor.

Due to the proximity of the top of the shed cavern structure to the safety control line, the pouring of concrete in the shed cavern cannot be accomplished using a traditional stent pump truck. To solve the problem, a movable integral cart is adopted, with steel mold plates placed inside and outside, and local windows on the mold plates for concrete pouring and vibration [18–20]. The construction process is as follows: construction preparation → measurement and sampling → steel formwork cart installation in place → wall and roof reinforcement installation → outside truss formwork installation in place → concrete pouring → dismantling and maintenance → automatic traveling of formwork cart to the next section → construction of the next section. The structure of the steel formwork cart is shown in Figure 13.

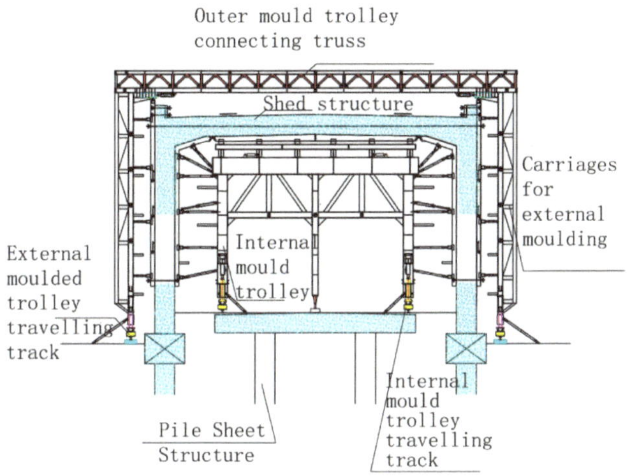

Figure 13. Moving steel mold trolley.

From the field implementation perspective, the adoption of a structural design featuring a straight wall, a flat roof with windows on both sides, and the use of an infrared height-limiting frame and a movable cart for casting concrete during construction have ensured the safe and efficient construction of high-speed rail protection sheds under extra-high-voltage transmission lines. Field implementation has yielded positive results. Figure 14 shows a photo of the protective shed tunnel site.

Figure 14. Site of protective shed tunnel.

4. Conclusions

Through research into the protection scheme for high-speed railways near 1000 KV ultra-high-voltage transmission lines, an equation of motion for the disconnected line, a calculation model, and other parameters were established to determine the load and protection range of the disconnected line. By integrating common high-speed railway protection structures such as sheds and openings, a protection structure design was proposed involving a straight wall with a flat roof and a side wall with open windows near the safety zone of the ultra-high-voltage transmission line. Additionally, the construction method for an automatic early-warning protection corridor and a fully movable cart molding was developed and successfully applied. The study draws the following conclusions:

(1) During the design of the UHV transmission line, the safety factor of the tower is set to be extremely high. Therefore, the protection of the high-speed rail under the 1000 KV UHV line primarily emphasizes safeguarding against line-break accidents in extreme conditions.
(2) The structural form of the C35-reinforced concrete shed hole can not only meet the force requirements under disconnected conditions but also reduce the mutual influence with the UHV transmission line.
(3) The structural form of a straight wall, flat roof, and side wall window can also prevent the structure from intruding into the safety control area of the transmission line while meeting the headroom requirement.
(4) The construction measures, such as establishing an automatic warning protection corridor and utilizing an overall movable cart molding, can guarantee the safe and efficient construction of the high-speed rail protection shelter structure within the safety zone of the adjacent ultra-high-voltage transmission line. The implementation has shown positive results.

Author Contributions: Conceptualization, W.X. and Y.S.; methodology, software, validation, Y.S. and W.X.; formal analysis, Y.S.; data curation, writing—original draft preparation, Y.S.; writing—review and editing, visualization, W.X.; funding acquisition, W.X. All authors have read and agreed to the published version of the manuscript.

Funding: This research was funded by the Natural Science Basic Research Program of Shaanxi Province (Grant No. 2022JQ-277) and The Central Special Fund for scientific research in colleges and universities (No. 300102282110).

Data Availability Statement: Data is provided within the manuscript.

Conflicts of Interest: The authors declare no conflict of interest.

References

1. Zhou, Y.; Chen, J.; Zhang, L.; Zhang, Y.; Teng, C.; Huang, X. Opportunity for developing ultra-high voltage transmission technology under the emission peak, carbon neutrality, and new infrastructure. *High Volt. Eng.* **2021**, *47*, 2396–2408.
2. Li, X. New infrastructure construction and policy orientation for a smart society. *Reform* **2020**, *5*, 34–48.
3. Sun, J.; Jiang, Z.; Yang, J.; Zheng, L.; Zhao, P.F.; Mu, Q.; Cong, X. Research on the line-break influence range of power line crossing electrified railway. *China Railw.* **2023**, *9*, 117–124. [CrossRef]
4. Liu, B. Discussion on the scheme of adding protective sheds to the corridor of the new railway crossing the UHV transmission line. *Electr. Railw.* **2019**, *30*, 72–76.
5. *GB 50665-2011*; Code for Design of 1000kV Overhead Transmission Line. China Planning Press: Beijing, China, 2011.
6. *GB 50790-2013*; Code for Designing of ±800kV DC Overhead Transmission Line. China Planning Press: Beijing, China, 2019.
7. *DL/T5486-2020*; Technical Specification for the Design of Steel Supporting Structures of Overhead Transmission Line. China Planning Press: Beijing, China, 2020.
8. *DL/T5301-2013*; Construction Technology Guide for Tension Stringing of Overhead Transmission Line When Over-Crossing Live Lines without Cross Frame. China Electric Power Press: Beijing, China, 2014.
9. *GB 50010-2010*; Code for Design of Concrete Structures. China Architecture Publishing & Media Co., Ltd.: Beijing, China, 2015.
10. Mo, Z.-L.; Zhang, Z.-F.; Hu, G.-Y. Experimental Research for Broken Conductor Wire of ±800 kV DC EHV Suspension Tower Model. *Electr. Power Constr.* **2010**, *31*, 30–33.
11. He, S.-M.; Shen, J.; Wu, Y. Rock shed dynamic response to impact of rock-fall. *Rock Soil Mech.* **2011**, *32*, 781–787.
12. Jiang, J.; Yang, X.; Luo, Y.; Zhou, Y. Summarize about traditional shed hole compared with flexible shed hole in application research. *Build. Struct.* **2016**, *46* (Suppl. S1), 249–252.
13. Jiang, D.; Zhou, X.; Tang, J.; Liu, K. Mechanical response of large span arched open cut tunnel structure under rockfall impaction. *Railw. Stand. Des.* **2021**, *65*, 122–127.
14. Wang, Y.; Li, S.; Li, W.; Tan, J.; Guo, Z. Study on the Characteristics and Protection Design of Tuyu Tunnel Rockfall on Zhangjiajie-Jishou-Huaihua Railway. *Railw. Stand. Des.* **2023**, *67*, 126–133.
15. Wen, X.; Pei, X.; Pei, Z.; Yang, H.; Chen, Z.; Zhang, Y. Study on structural design and safety of shed tunnels in the scenic area of Changbai mountains. *Mod. Tunn. Technol.* **2021**, *58*, 151–157.
16. Yang, C. Study on protection scheme of dangerous rock falling on high steep slope of chizhou-huangshan high speed railway tunnel. *Railw. Eng.* **2023**, *63*, 124–126.

17. Wang, K.; Gao, C.; Zhou, Y. Research on Design of Railway Bridge-shed Integrated Structure for Preventing Falling Rocks. *Railw. Stand. Des.* **2023**, *67*, 106–112+154.
18. Li, X.; Lin, C.; Li, J. New Construction Method for Secondary Lining of Railway Tunnel and Design of Lining Trolley Scheme. *Tunn. Constr.* **2021**, *41*, 293.
19. Zhang, M.; Jia, D.; Wang, S.; Gao, C. New Automatic Pouring and Vibrating Techniques for the Construction of Railway Tunnel Lining. *Mod. Tunn. Technol.* **2022**, *59*, 29–34+41.
20. Zhang, H. Research on rapid construction technology for arch wall lining inopen-cut hole of large-section loess tunnel. *Railw. Constr. Technol.* **2020**, *1*, 93–96. [CrossRef]

MDPI AG
Grosspeteranlage 5
4052 Basel
Switzerland
Tel.: +41 61 683 77 34

Infrastructures Editorial Office
E-mail: infrastructures@mdpi.com
www.mdpi.com/journal/infrastructures

www.ingramcontent.com/pod-product-compliance
Lightning Source LLC
Chambersburg PA
CBHW042013260926
51850CB00020B/3318
* 9 7 8 3 7 2 5 8 2 2 6 6 9 *